10th International Conference on Multidisciplinary Research (ICMR 2021)

Held in conjunction with the 2nd International and National Symposium on Aquatic Environment and Fisheries

E3S Web of Conferences Volume 339 (2022)

Banda Aceh, Indonesia
27 September 2021

ISBN: 978-1-7138-4281-1

Printed from e-media with permission by:

Curran Associates, Inc.
57 Morehouse Lane
Red Hook, NY 12571

Some format issues inherent in the e-media version may also appear in this print version.

For additional information, please contact EDP Sciences – Web of Conferences
at the address below.

EDP Sciences – Web of Conferences
17, Avenue du Hoggar
Parc d'Activité de Courtabœuf
BP 112
F-91944 Les Ulis Cedex A
France

Phone: +33 (0) 1 69 18 75 75
Fax: +33 (0) 1 69 28 84 91

contact-edps@webofconferences.org

Additional copies of this publication are available from:

Curran Associates, Inc.
57 Morehouse Lane
Red Hook, NY 12571 USA
Phone: 845-758-0400
Fax: 845-758-2633
Email: curran@proceedings.com
Web: www.proceedings.com

TABLE OF CONTENTS

The 10th International Conference on Multidisciplinary Research (ICMR)
in conjunction with
The 2nd International and National Symposium on Aquatic Environment and Fisheries (INSAEF) 2021

More info: https://www.insaef.unsyiah.ac.id/

ICMR x

Third Announcement:

CALL FOR PAPER

Theme:
"A Multidisciplinary Approach Towards a Sustainable Eco Blue Industries 4.0"

Scope*

1. Sciences and Engineering.

2. Environmental and Life Sciences.

3. Social Sciences.

*The detailed scope could be seen on the web site.

Publications:

1. Special issues published in Regional Studies in Marine Science (Scopus indexed)*To be confirmed

2. Environment, Energy and Earth Sciences Web of Conferences (E3S Web of Conferences) (Scopus and WoS indexed).
https://www.e3s-conferences.org/

3. Depik Journal (Accredited SINTA 2).
http://jurnal.unsyiah.ac.id/depik

Keynote Speakers:

Prof. Shu-Kun Hsu
National Central University, Taiwan

Prof. Dr. Kasi Marimuthu
AIMST University, Malaysia

Dr. Ir. Irwan Meilano, M.Sc
Bandung Institute of Technology, Indonesia

Dr. Morelia Urlaub
GEOMAR, Germany

Invited Speakers:

Dr. Do Hai Yen
Tan Trao University, Vietnam

Dr. Zarina Md Nor
USM, Malaysia

Dr. Frauke Klingelhoefer
IFREMER, France

Prof. Marc-André Gutscher
Université de Bretagne Occidentale, France

Prof.Hasrita Lubis. M.Pd
UISU, Indonesia

M. Nasir Badu, Ph.D
Hasanuddin University, Indonesia

Dr. Nur Fadli
Syiah Kuala University, Indonesia

Important Dates:

Abstract submission	= 26 July 2021
Abstract acceptance	= 15 August 2021
Full paper submission	= 31 August 2021
Early-bird registration	= 7 September 2021
Full paper acceptance	= 7-14 September 2021
Conference	= 27 September 2021

More:

- Discounted registration fee for the ICMR member*
- Discounted registration fee for student*
- Competition for the best oral and poster presenters award*
- Outstanding Student Presentation Award (OSPA)*

*Term and condition applied (for more information visit the website)

Organized by: PRKP USK UNIVERSITAS SYIAH KUALA

Supported by: USM

Indexed in: SCOPUS sinta S2

Statement of Peer review

In submitting conference proceedings to E3S Web of Conferences, I certify to the Publisher that I adhere to the Policy on Publishing Integrity of the journal in order to safeguard good scientific practice in publishing.

1. All articles have been subjected to peer review administered by the proceedings editors.
2. Reviews have been conducted by expert referees, who have been requested to provide unbiased and constructive comments aimed, whenever possible, at improving the work.
3. Proceedings editors have taken all reasonable steps to ensure the quality of the materials they publish and their decision to accept or reject a paper for publication has been based only on the merits of the work and the relevance to the journal.

Title, date and place of the conference

Title of conference: The 10th International Conference on Multidisciplinary Research (ICMR) in conjunction with The 2nd International and National Symposium on Aquatic Environment and Fisheries (INSAEF) 2021
Date: 27 September 2021
Place: Banda Aceh (Virtually)

Proceedings editor(s):

1. Dr. Muhammadar, ST, MP
2. Drs. Muhammad, M.Si
3. Dr. Zulkarnain, S.Si., M.Si
4. Dr. Zarina Md Nor
5. Dr. Do Hai Yen
6. Muhammad Nasir Badu, Ph.D
7. Dr. Muhammad Irham, S.Si., M.Si
8. Dr. Nur Fadli
9. Dr. Syafruddin, S.E., MBA
10. Haekal Azief Haridhi, S.Kel., M.Sc., Ph.D

Date and editor's signature

17, avenue du Hoggar - PA de Courtabœuf – BP 112 - 91944 Les Ulis Cedex A (France)
Tél. : 33 (0)1 69 18 75 75 - Fax : 33(0)1 69 07 45 17 – www.edpsciences.org

Alteration in gills, skin, and fins of snakehead fish (*Channa striata*) due to ectoparasite infection: a macropathological and histopathological approach

Nurliza Zaiyana[1], *Arif* Sardi[2], *Muhammad* Ridwan Harahap[3], *Rindhira* Humairani[4], *Epa* Paujiah[5], *Rumondang* Rumondang[6], *Ilham* Zulfahmi[7*]

[1]Center for Aquatic Research and Conservation (CARC), Universitas Islam Negeri Ar-Raniry, Banda Aceh, Indonesia
[2]Department of Biology, Faculty of Science and Technology, Universitas Islam Negeri Ar-Raniry, Banda Aceh, Indonesia
[3]Departement of Chemistry, Faculty of Science and Technology, Universitas Islam Negeri Ar-Raniry Banda Aceh, Indonesia
[4]Department of Aquaculture, Faculty of Agriculture, Almuslim University, Bireuen, Indonesia
[5]Department of Biology Education, Faculty of Education and Teacher Training, Universitas Islam Negeri Sunan Gunung Djati, Bandung, Indonesia
[6]Department of Aquaculture, Faculty of Agriculture, Universitas Asahan, Kisaran, Sumatera Utara, Indonesia
[7]Department of Fisheries Resources Utilization, Faculty of Marine and Fisheries, Universitas Syiah Kuala, Banda Aceh, Indonesia

Abstract. Snakehead fish (*Channa striata*) is a freshwater fish with high economic value both as ornamental and consumed fish. However, the ability to adapt to fluctuating environmental states prompts snakehead fish to be susceptible to parasites, ectoparasites, and endoparasites. Gills, skin, and fins are the major body parts that are often infected with ectoparasites. This study aims to discover the macro pathology in gills, skin, and fins of ectoparasites-infected snakehead fish and its histopathology. This study was conducted from April to June 2021. Observational data on macro pathology and histopathology were analyzed qualitatively. The results indicated that the macro pathology of the skin, gills, and fins of snakehead fish infected with ectoparasites indicates discoloration, excess mucus production, the dull coloration of the gills, and irritation of the gills body surfaces, and skin hemorrhages. In conclusion, infection of ectoparasites affects several alterations in the skin, gills, and fins, both macro pathological and histopathological.

1 Introduction

Snakehead fish (*Channa striata*) is a freshwater fish with high economic value both as ornamental and consumed fish [1]. In addition, this fish also contains higher nutrients and albumin than other fish [2]. Snakehead fish are commonly distributed in freshwater habitats, including rivers, rice fields, ponds, swamps, and even community ditches. In Indonesia's freshwater ecosystem, snakehead fish has a high abundance, including in Aceh water [3]. The ability to adapt to fluctuating environmental conditions makes this fish vurnerable to parasites infection, both ectoparasites, and endoparasites.

Ectoparasites often infect the skin, fins, and gills of fish. Fish infected with ectoparasites usually show specific changes, for instance small and large skin rashes as well as changes in skin color [4]. Gills, skin, and fins are the main body parts of fish that are often infected with ectoparasites. Several types of ectoparasites that infect gills, skin, and fins of snakehead fish are *Trichodina* sp., *Dactylogyrus* sp., *Oodnium* sp., and *Epistylis* sp [5]. Ectoparasite infection causes alteration to the gill filaments, resulting in disrupting the respiratory process [6].

Trichodina sp. infection in fish gill has reported caused excessive mucus secretion, provoking stress and inflammation or hemorrhaging. Meanwhile, *Trichodina* sp. infection on the skin and fins induces inflammation, necrosis, and excessive mucus secretion [7]. Histologically, gills infected with ectoparasites were reported to suffered tissue damage. Previous studies reported alteration in the secondary lamellae of goldfish (*Carassius auratus*) gills, including hyperplasia and necrosis due to infection of ectoparasites of *Argulus* sp. [8]. To date, the study related the macro pathology and histopathology of snakehead fish's gills, skin, and fins due to ectoparasite infection have not been well investigated. Thus, this study aims to investigate the macro pathology and histopathology alteration in gills, skin, and fins of snakehead fish due to ectoparasites infection.

2 Materials and Methods

2.1 Place and Duration of the Research

This study was conducted from April to June 2021. A total of 90 snakehead fish were collecting from rice fields, swamps, and ditches in the Aceh Besar District

*Corresponding author: Ilham.Zulfahmi@unsyiah.ac.id

 https://doi.org/10.1051/e3sconf/202233901001

covering the Kajhu area in Baitussalam Subdistrict (N 5°35'51.11" E 95°22'32.30"), Limpok in Darussalam Subdistrict (N 05°33'43.88" E 095°22'27.56"), Cot Keung area in Kuta Baro Subdistrict (N 5°33'19.54" E 95°24'10.49"), Samahani in Kuta Malaka Subdistrict (N 5°26'39.73" E 95°24'11.16"), and Mata Ie in Montasik Subdistrict (N 5°28'23.51" E 95°23'52.40"). Aceh Besar District was chosen due to it's diversity of habitat of snakehead. Observing parameters was carried out at the Multi-Function Laboratory of the Ar-Raniry State Islamic University, Banda Aceh. Histological preparation was conducted at the Center of Brackish Aquaculture Fisheries (CBAF) Ujong Batee.

2.2 Macro pathology and Histopathology Observation

Ectoparasite infection was observed in the gill, skin, and fins. The mucus was collected and placed on a slide, dripped with 0.85% physiological NaCl solution, covered with a coverslip, and observed under a microscope (Olympus CX 23; Japan). Gills and fins were cut and placed on a slide and dripped with 0.85% physiological NaCl solution then covered with a slide and observed under a microscope. The ectoparasites found were identified by comparing the morphology of ectoparasite with identification references [9]. Macropathological symptoms were analyzed by observing the alteration in snakehead fish's gills, skin, and fins. The ectoparasites found were identified by comparing the morphology of ectoparasite with identification references.

Histological preparations was based on histotechnique method [10]. Briefly, The fish gills were preserved with Davidson's solution. Briefly, histological preparation consists of five stages: fixation, dehydration, clearing, wax infiltration, embedding, cutting, and staining. In the fixation stage, the organs were immersed in Davidson's solution for 24 hours. The organs were then dipped in an alcohol solution with 70%, 80%, 90%, and 100% graded concentrations in the dehydration stage. In the clearing stage, the dehydrated sample was transferred to Xylene/Xylol solution for 2 hours. Paraffin or samples were cut using microtome at 5μm thick. The tissue was stained using hematoxylin and eosin for ten minutes. The histology preparations were then observed with a microscope equipped with a camera (Olympus CX 23; Japan). Observational data on macro pathological and histopathological changes were analyzed qualitatively.

3 Results and Discussion

Snakehead fish infected by ectoparasites showed several macro pathological symptoms, including irritation and hemorrhage on the skin, pale color, white nodules on the gills, and irritation of the tail fin (Table 1 and Figure 1). Normal snakehead fish skin color is greenish-gray. The abdomen is white, the body surface and head have thick scales, and there are no sores or irritations on the entire body surface. The gills of normal snakehead fish is fresh red and without white nodules. In contrast, infected fish was indicated with dull or pale in color and with white nodules. The skin showed irritation, has white nodules, and produces excessive mucus

Table 1. Normal appearance and macro pathological symptoms observed in ectoparasites-infected snakehead fish.

No	Organs	Normal	Infected with Parasites	Ectoparasite
1.	Gill	Reddish Normal mucus secretion	White nodules on gills The gills are pale red. Excessive mucus secretion	*Tricodina sp. Tetrahymena sp. Epistylis sp. Gyrodactylus sp. Dactylogyrus sp.*
2.	Skin	The skin is greenish-grey, and the abdomen is white with thick, rough scales covering the surface of the body and head. No excess mucus secretion	The skin is pale gray, and the fins are peeling off excessive mucus secretion in the whole body irritations on the body surface.	*Tetrahymena sp. Dactylogyrus sp.*
3.	Fin	No excess mucus secretion The fins are blackish gray.	The tail is pale white, and there are sores on the fins and edges of the fins. The fins are reddish.	*Tricodina sp. Epistylis sp.*

High-intensity infection of parasites alters the gill structure, which eventually causes death. It also can attack the others fish organs, including the skin and fins [11]. Apart from *Trichodina* sp., several other parasites can infect fish, such as *Dactylogyrus* sp., which often attacks freshwater fish's gills. Clinical symptoms showed that infected fish tend to be defenseless, mucus production is excessive, and the gills is pale and swollen [11].

The macro pathology observations showed that the skin of infected snakehead fish endures a paler color change accompanied with irritation on the body surface. Clinical symptoms that appear are changes in body color to become dull and infection of the gills, which makes it difficult for fish to breathe. Besides *Tetrahymena* sp., another parasite that attacks the skin of snakehead fish is Dactylogyrus sp. The clinical symptoms it generates are mucus and pale color on the skin's surface due to infection, which prompts excessive mucus production [12]. Macro pathology observation of snakehead fins infected with ectoparasites indicated sores on the caudal fin and pale color on the edges of the fins.

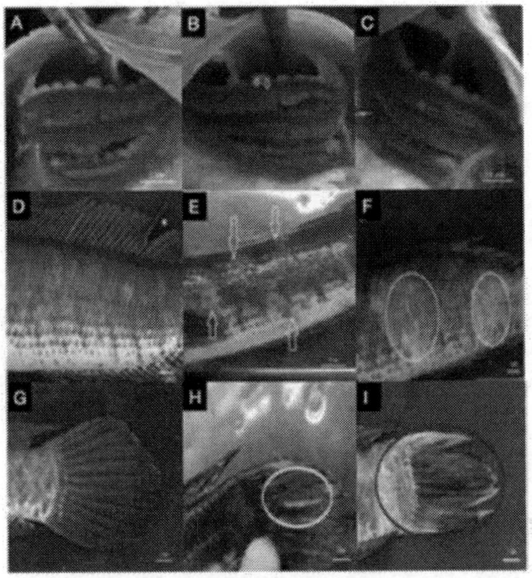

Fig. 1. Macro view of gills, skin, and fins of normal and ectoparasite-infected snakehead fish. (a) Normal snakehead fish gills from rice field habitat (b) snakehead fish gills from ditches habitat are pale in color and have white nodules due to ectoparasite infection. (c) Snakehead fish gills from ditches habitat are pale in color and have white nodules due to ectoparasite infection. (d) Snakehead fish gills from ditches habitat that are dull red. (e) Normal snakehead fish skin from rice field habitat. (f) Irritation and hemorrhage due to ectoparasite infection in snakehead fish from ditches habitat. (g) Exfoliated skin scales due to ectoparasite infection in fish from ditches habitat. (h) Caudal fin of snakehead fish from swamp habitat. (i) Caudal fin irritation of snakehead fish from ditch habitat. (j) Pale white fin edges on snakehead fish from ditches habitat. Scale bar 1 cm.

The shape of the primary lamella (LP) of normal fish is clearly visible, the shape of each secondary lamella (LS) looks neatly arranged, and there is a central sinus venous (SVC) on it. In contrast, in the gills of ectoparasite-infected snakehead fish, the primary lamella (LP) is swollen, the shape of each secondary lamella (SL) is not clearly visible. Ectoparasites Dactylogyrus sp. was observed between the primary and secondary lamellae. The gill structure of normal and ectoparasites-infected snakehead fish is presented in Figure 2.

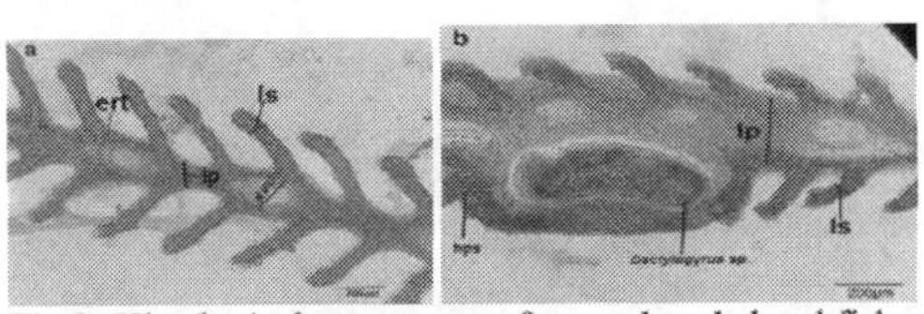

Fig.2. Histological apperrance of normal snakehead fish gills and ectoparasite-infected snakehead fish gills. (a) Primary lamellae (LP), secondary lamellae (LS), erythrocytes (ERT), and central sinus venous (SNC) on the gills of normal snakehead fish. (b) swollen at primary lamellae (LP), secondary lamellae (LS) that are not clearly shaped, and hyperplasia (HPS). Ectoparasites Dactylogyrus sp. in the gills of ectoparasite-infected snakehead fish. Scale bar: 200 μm, Manification 100 X.

The gills are one of the fundamental organs of fish that act as the primary respiratory apparatus that works by diffusion of oxygen dissolved in water [13]. The gills are susceptible to water quality degradation due to their significant contact with water [14]. Gills infected with ectoparasites induce hyperplasia, a disorder that triggers cells between the secondary lamellae. These cells can prompt the secondary lamella to thicken and become irregular in shape. Cells epithelium discovered in the secondary lamellae of the gills might be an adaptive strategy to the presence of parasites.

Histopathologically the process of gill defense from the attachment of parasites using anchors can cause bleeding. Defensive reactions in the lamellae stimulate the rapid growth of epithelial cells of the gill lamellae (hyperplasia) and an increase in mucus secretion by mucus cells [15]. Lamella hyperplasia is not only caused by the growth of epithelial cells but can also synergize with mucus cell proliferation and fusion of secondary lamellae (fused gill lamellae). Proliferation is a condition in which secondary lamellae epithelial cells grow excessively due to ectoparasite injuries, causing increased mucus production and resulting in excess secondary epithelial cells [16]. The proliferation of lamellae cells that occurs is a response to a long or rapid infection. *Dactylogyrus* sp. was known as the parasites that attaches to the surface of the gill lamella by using an opistaptor, a suction device that circulates with blood circulation [17]. Several researchers inform that gills infected with the parasite, including *Dactylogyrus* sp., are characterized by hyperplasia and swelling [18].

4 Conclusion

Macro pathology of gills, skin, and fins of ectoparasites-infected snakehead fish indicates changes in skin color and excessive mucus production. Histopathology of the gills of ectoparasites-infected snakehead fish pointed to the presence of hyperplasia, a disorder that triggers the appearance of cells between the secondary lamellae. These cells can prompt the secondary lamella to thicken and become irregular in shape.

References

1. M. Puspaningdiah, A. Solichin, A. Ghofar. Jurnal of Maquares., **3** (2014)
2. E. Chasanah, M. Nurilmala, A.R. Purnamasari, D. Fithriani. JPBKP., **10** (2015).
3. Z.A. Muchlisin, T. Zairin, N. Fadli, M.A. Sarong, M.N. Siti-Azizah. Acta Ichthyol Piscat., **43** (2013)
4. K.A. Yildiz, Kumantas. J. Vet. Med., 57 (2002);
5. B. Salam, D. Hidayat. J. Sains dan Seni ITS., **6** (2017)
6. A. Chaudhary, C. Harenram, S.S. Hridaya SS. Bioinvasions Rec., **6** (2017)
7. A. Irianto. Gadjah Mada University Press. Yogyakarta, (2005).
8. N. Rifqiyati, M.J. Luthfi, A.N. Miftah. J. Bionature. **18** (2017)

 https://doi.org/10.1051/e3sconf/202233901001

9. C.H. Fernando. University of Waterloo. Canada, (1972)
10. G.L. Humason, Freeman Company, San Francisco, USA. 661 pp. (1979)
11. W. Prince EGC. Jakarta, (2007)
12. M.M.P Camargo, C.B.R Martinez. Neotrop. Ichthyol., **5** (2007)
13. E. Ernita, R. Faumi, Y. Akmal, M. Muliari, I. Zulfahmi. J. Veteriner, **21** (2020).
14. M. Muliari. Y. Akmal, I. Zulfahmi, R. Juanda, N.W.K. Karja, C. Nisa. In IOP Conference Series: Earth and Environmental Science. **216** (2018).
15. R.J. Robert, Edinburgh, Philadelphia, St. Loius, Sydney, Toronto, **427** pp (2001)
16. I.A.N Utami, Wiadyana. J Media Aquakultur., **12** (2017)
17. F.N Fautama, I. Zulfahmi, M. Muliari M, A.A. Anas, Jurnal Biodjati. **4** (2019)
18. F. Mohammadi, S.M. Mousavi, A. Rezaie. Aquaculture, Aquarium, Conservation & Legislation. **5** (2012)

Study of prevalence and intensity of ectoparasites in cultivated green mussel (*Perna viridis* L.) in Alue Naga Waters, Banda Aceh

Ichsan Rusydi[1,4*], *Nadiatul* Salsabila[2], *Cut* Nanda Defira[3], *Sayyid* Afdhal ElRahimi[2,4], *Kavinta* Melanie[2,4], *Dedi* Fazriansyah Putra[2,4], *Afriliana* Afriliana[5], and *Fahmi* Fahmi [6]

[1] Department of Marine Science, Faculty of Marine and Fisheries, Syiah Kuala University, Indonesia
[2] Department of Aquaculture, Faculty of Marine and Fisheries, Syiah Kuala University, Indonesia
[3] Biology Department, Faculty of Mathematics and Natural Sciences Universitas Syiah Kuala Banda Aceh 23111, Indonesia
[4] Research Center for Marine Sciences and Fisheries, Universitas Syiah Kuala, Banda Aceh 23111, Indonesia
[5] Fishery instructor, Ministry Of Marine Affairs And Fisheries Republic Of Indonesia
[6] School of Forestry Science Tgk. Chik Pante Kulu

Abstract. Parasites are a major problem on biota survival or green mussels because parasites are organisms that live in other organisms. Parasites can harm the host they live in because these organisms will absorb nutrients from the host they live on, damage tissues, and cause infection. This study aims to examine the types of ectoparasites and to determine the prevalence and intensity values of cultivated green mussels (Perna Viridis L.) in the Alue Naga reservoir, Banda Aceh. The study was conducted in July 2020, identification of parasites was carried out at the Fish Quarantine Station, Quality Control and Safety of Class 1 Fishery Products in Aceh. The research method was a survey method with a sampling technique using random sampling. Observations were made visually and using a microscope on the target organs in the shell, gills, and mantle. Furthermore, the parasites found in green mussels were identified using the Kabata book (1984). Based on the research results found 4 types of ectoparasites, namely cysts perkinsus sp. (Protozoa) on the coat, Balanus sp. (Crustacea) on the shell, Monogenea (Platyhelminthes) on the gills, Copepoda (Crustacea) on the gills, and it is known that the highest intensity value is 1.16 and the highest prevalence value is 74%. parasite barnacles (Balanus sp.), while the lowest intensity value is 1 and the prevalence value is 1% cysts perkinsus sp. The water quality values were pH 8.7, temperature 29 C, dissolved oxygen 4.8 Ppm, salinity 31 Ppt, depth 1.53 m, brightness 70 cm, and current 17 m / s. The value of water quality parameters in Alue Naga waters is still in the optimum range for the growth of green mussels.

1 Introduction

The presence of green mussels (*Perna Viridis* L.) (Bilvavia Phylum Mollusca Class) is widespread in Indonesian waters and it is abundantly found in coastal waters, reservoirs, mangroves, and river estuaries. Therefore, many coastal communities grow green mussels as a side income in their free time [1]. This statement is also supported by [2], that green mussels have potential value to be cultivated, considering the availability of green mussels in nature will having a decrease due to the increased capture.

Aceh waters are one of the waters that have high diversity. One of the organisms that have high diversity is in the bilvavia phylum mollusks class. The organisms categorized to the bivalve's class are green mussels (*Perna viridis* L.). Alue Naga waters, Banda Aceh is one of the waters in Aceh that green mussels are available. Green mussel (*Perna viridis* L.) is a type of aquatic organism that lives on the bottom substrate of the waters and attaches to a hard substrate. Green mussels (*Perna viridis* L.) are shellfish that have high economic value and it is favorite food by the public because of their high nutrients content.

Corresponding author : ichsanrusydi@unsyiah.ac.id

Parasites are a major problem in the biota survival or green mussels because parasites are organisms that live on the bodies of other organisms. Parasites can harm the host they live in because these organisms will absorb nutrients from the host they are attaching to, damage tissues, and cause infection. Parasites can be protozoa, worms, bacteria, viruses, and fungi [3].

The environment of shellfish resources is a place for shellfish to live includes biota and natural factors surrounding it. One of the often diseases attack green mussels (*Perna viridis* L.) is a parasitic disease. The parasites are divided into two groups, there are groups of ectoparasites and endoparasites group. Ectoparasites are parasites that live attaching to the outside of the host's body, endoparasites are parasites that live inside of the host's body. The types of ectoparasites attacking green mussels are *Balanus sp., Perkinsus sp.* and Nematodes [4].

The cultivated green mussels in Alue Naga need to be examined for parasites to determine the type of parasite that infects green mussels. Because the appearance of disease attacks is caused by many things, one of them is the inappropriate result of the interactions to the environmental conditions that cause disease. In the appropriate interactions, it is strongly influenced by the survival of shellfish. Because the mechanisms of defense of the shell's body do not work optimally so that the infectious disease is easy attacking the body shells.

Management of shrimp ponds that are not environmentally friendly can cause poor quality of marine waters, this can be seen from the existence of ponds that do not have water treatment ponds, where the waste products are directly discharged into the sea. In research [5], in terms of sea water quality, the coastal waters of Probolinggo Regency are categorized as moderate pollution. The results of the study of COD in the sea obtained an average of 17.9 mg/l presented in table 1. This COD value is still in the range of environmental quality standards, namely < 20 mg/l [6].

High levels of COD reflect high concentrations of organic compounds in the waters or a decrease in the health of aquatic ecosystems. The COD ($KMnO_4$) concentration should be below 3 mg/l on the coast and for aquaculture. The COD level in this study, found to be almost the same as that conducted by [7] (14.7-58.9 mg/l) in different areas, because the field of research results is strongly influenced by shrimp pond waste as well as urban pollution and household waste disposal.

Therefore, it is important that this study was conducted with the aim of assessing the types of ectoparasites and to determine the prevalence and intensity values of green mussels (Perna viridis L.) from aquaculture in the Alue Naga reservoir, Banda Aceh.

2 Research Method

2.1 Time and Place

This research is conducted in Alue Naga Reservoir, Banda Aceh, Aceh Province. The research Time performed in July 2020 and the identification will be conducted at stations Quarantine Fish, Control of Quality and Safety Results Fishing Class 1 Aceh. The Research map at Alue Naga Reservoir, Banda Aceh can be seen in Figure 3.1 as follows:

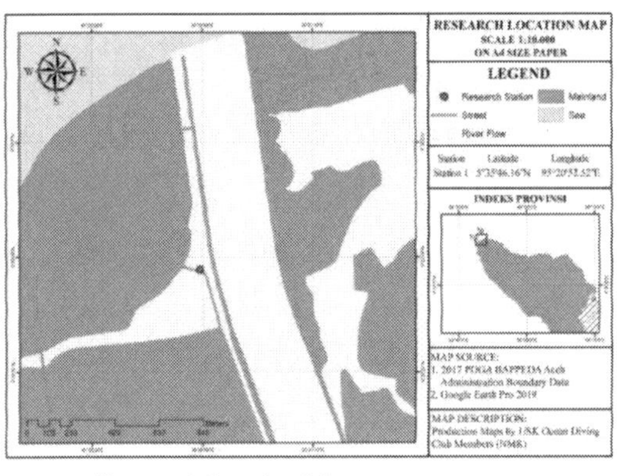

Fig 1. Research location Map

2.2 Experimental Design

2.2.1 Determination Observation Station

The method in the research used the survey method and samples were taken by random sampling technique, the sampling techniques are random, which each population has a similar chance to be selected as the sample. The sampling was conducted on cultivated green mussel containers and the samples used about 100 mussels, with an average long 4 - 7 cm and an average weight of 5– 6 g. The mussels taken are filled into a plastic containing a bit of water and oxygen on it and then it is brought to a laboratory for ectoparasite examination and microscopically examined

2.2.2 Capturing and Sample Treatment

Green mussels sampling was catching by hand then recording the, observing clinical symptoms (the presence of organisms attached to the shell and pale gill color), measuring the length and weight of green mussels, and making range preparations by cutting the target organs, such as the shell and gills. The results of targets organ pieces are placed on the top of slide glass, then the observation would be performed using a microscope. Furthermore, the parasites found in green mussels were identified using the kabata book (1984). The data collected in this study include the value of the prevalence and intensity values, besides, it also retrieved data from the measurement quality of water such as, oxygen dissolved (DO), pH, and temperature of the water.

In this study, efforts to engineer containers and green mussel cultivation technology in the waters of the Alue Naga reservoir were carried out, this is reinforced in References [8], by taking into account the construction of the container properly and correctly, a fish culture container that has a long service life will be obtained. The technical requirements that must be considered in choosing the location of a fish farming business in floating net cages include: Currents, tides, water fertility levels, free from pollution. After obtaining a location that meets technical and socio-economic requirements, planning must be carried out.

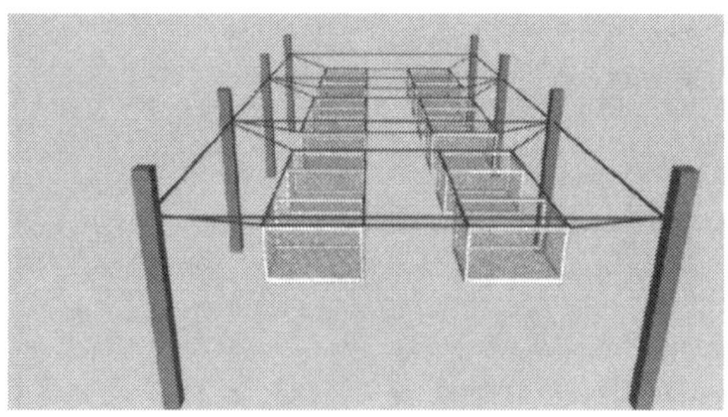

Fig 2. Cultivation Container Illustration

2.3 Data Analysis

2.3.1 Parameters were observed in the study

The parameter observed in the study about prevalence of the ectoparasites in green mussels (*Perna viridis* L.) in Alue Naga, Banda Aceh is the prevalence of ectoparasites. The prevalence of parasites was calculated by using the formula as follows:

$$\text{prevalence (\%)} = \frac{\sum \text{Samples of infected shellfish}}{\sum \text{examined shellfish samples}} \times 100\,\%$$

$$\text{intensity (ind/seed)} = \frac{\sum \text{parasites found}}{\sum \text{infected shellfish}}$$

Prevalence and Intensity category table

Table 1. Prevalence Category

No	Prevalence	Category	description
1	100-99 %	Always	Very severe infection
2	98-90 %	Almost always	Severe infection
3	89-70 %	Usually	Moderate infection
4	69-50%	Very often	Very frequent infection
5	49-30 %	Generally	Common infection
6	29-10 %	Often	Frequent infections
7	9-1 %	Sometimes	Infection sometimes
8	1-0.1 %	Seldom	Rare infection
9	< 0.1-0.1%	Very rare	Very rare infection
10	<0.01	Almost no never	Infections do not ever

Table 2. Intensity category table

No.	Intensity (eng/tail)	Category
1	<1	Very low
2	01-Mei	Low
3	Jun-55	Currently
4	51-100	Critical
5	>100	very bad
6	>1000	super bad

3 Results and Discussion

3.1 Type of parasite that infects green mussels (*Perna viridis* L.)

Based on the results of the research on ectoparasites in cultivated green mussels (*Perna viridis*) in the Alue Naga Reservoir, Banda Aceh, there were four types of ectoparasites were found, there are *Perkinsus sp cysts.* (Protozoa), *Balanus sp.* (Crustacea), Monogenea (Platyhelminthes) and Copepoda (Crustacea). The complete parasite on green mussels can be seen in Figure 4.1.

3.2 Green Mussel Ectoparasites (*Perna viridis*) Based on Microhabitat

Based on the results of the research that has been done, the type and number of parasites that are found by microhabitat is presented in Table 4.1.1. as follows:

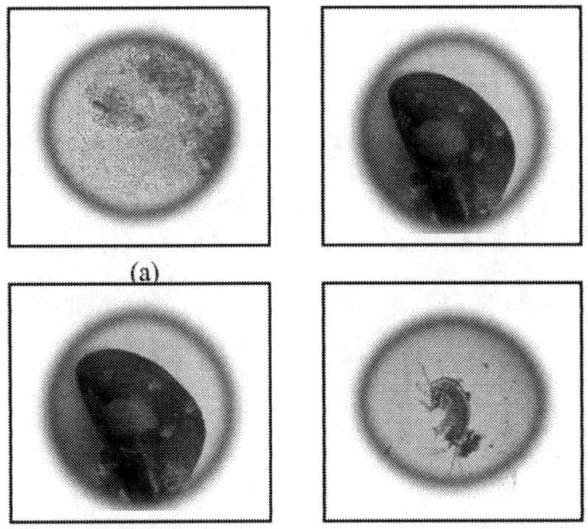

(a)

(b)

(d)

Fig 3. Parasites infect green mussels (*P. viridis*) (a) *Perkinsus sp.* (Protozoa); (b) *Balanus sp* (Crustacea) (c) Monogenea (Platyhelminthes) (d) Copepods (Crustacea).

Table 3. Types and Numbers of Green Mussel Ectoparasites *(P. viridis)*

No.	Name parasite	The number parasite			Total
		Gill	Coat	Shell	
1	*Perkinsuscyst* sp.	-	1	-	1
2	*Balanus* sp.	86	-	-	86
3	Monogenea	21	-	5	26
4	Copepods	12	-	2	14
Total					127

Table 3 shows that the most common parasite found was *Balanus sp.* (figure 4.1b) while the least parasite infecting the green mussels was *Perkinsus sp. cyst,* Balanus sp. that found in the shells of green mussels.

3.3 Intensity and prevalence Value of ectoparasites in green mussels (*Perna viridis*)

Based on table 4, it showed that the highest intensity value is 1.16 and the highest prevalence value is barnacle parasite (*Balanus sp.*) about 74, while the lowest intensity value is 1 and the prevalence value is *Perkinsus sp cyst* about 1%.

Tabel 4. Intensity and prevalence Value of ectoparasites in green mussels (*Perna viridis*)

No.	Parasite	Σ Infected Parasit	Σ samples	Σ samples	I (Ind)	P (%)
1	*Cyst perkinsus sp.*	1	1	100	1	1%
2	*Balanus sp.*	86	74	100	1.16	74%
3	Monogenea	26	24	100	1.08	24%
4	Copepods	14	10	100	1.4	10%

3.4 Water quality parameter results

The results of water quality measurements (pH, temperature, dissolved oxygen, salinity, depth, brightness and current) were carried out at the time of the research and these are fully presented in table 5.

Table 5. Water Quality Parameter Results

No	Parameter Quality Water	Unit	Measurement Results
1	Ph		8.7
2	Temperature	^{0}C	29
3	Dissolved Oxygen	Ppm	4.8
4	Salinity	Ppt	31
5	Depth	Meter	1.53
6	Brightness	Cm	70
7	Current	m/s	17

4 Discussion

4.1 The Type of Parasite Infects Green Mussels (*P. viridis* L.)

Perkinsus sp. cyst parasite. (Figure 4.1a) is a parasite of the phylum Protozoa. The results were obtained during the study's *cyst perkinsus sp. That* attacks the mantle which shows clinical symptoms, namely the coat shrinks and pale. It refers to the research conducted by [9] who found the same parasites in the green mussel. However, Elston found it in Manila clams (*V. philippinarum*) and stated that Perkinsus sp. including protozoa that are usually found on the gills and mantel. This parasite is often found or attacks oysters and shellfish.

The number of the *cysts perkinsus sp.* which is found in the coat scallops green is only one. The number of parasitic *cysts perkinsus sp.* which is found in the coat is influenced by environmental factors such as temperature and salinity. This statement is also supported by [10] which states that environmental factors affect the development of the parasite *Perkinsus sp. cyst.* Due to the high temperature and salinity, this infection will increase in the dry season in the estuary area. Chemical factors such as salinity and temperature exceeding about 35.9 ppt and 32° C affect the life of mussels. In this case, the salinity will be affected in the setting of internal ions, which directly require energy to transport active ions in order to maintain the internal environment. This is a very effect on physiological processes [11]. The onset of disease in shellfish has a close relationship with environmental conditions so that when the water changes just a bit from the environment then it can lead to shellfish stress because the stress mussels will be easier attacked by the disease.

Balanus sp. parasites engaging parasitic organisms has a body that is made from a substance limestone and has a body size of about 1-1.5 cm. According to Jensen [12] said that the *Balanus sp.* has a body size less than 1 cm and a maximum of 2 cm. According to [13], the body of *Balanus sp.* has a calcareous structure. *Balanus sp.* were found attached on the mussel of the examined green mussel samples. This is supported and in line with [13] statement. They said the body of *Balanus sp.* is attached to a host (shell or carapace). In previous research, *Balanus sp.* has been reported to infect blood clams [14]. Barnacles (*Balanus* sp.) live in the tidal zone, ranging from tidal highs up to the average tide receded. The economic impact of barnacles tends to lead to negative to all types of the substrate on which it is attached.

During the identification, the *Balanus sp.* is found a lot in the shell of green mussels because barnacles like a rough substrate and are easy to stick to, even *Balanus sp.* not only attached to the shell but also to the hinge of the shell, so that it is difficult for the shell to open and close its shell valve. How the barnacles (*Balanus sp.*) works is by drilling into the inside of the clamshell so that it can cause death to the clam. These sticking organisms' barnacles (*Balanus sp.*) attacked the shells indirectly due to its damaging and food competition so that the mussels are slowly impaired growth or broken shells and at end die.

Monogenea parasite (figure 4.1c.) is one of the parasites that often infect fish and invertebrates. It is usually found in external organs by plugging haptors. This parasite easily infects aquatic biota because of its direct life cycle, high reproduction, and high pathogenicity, so it often causes death in fish and other biotas. Monogenean parasites are generally ectoparasites and rarely endoparasites. Monogenea is one of the most common parasites that attack the part of the outside body (ectoparasites), seldom attack a part in the body (endoparasites), and usually affects the skin and gills [15]. Monogenea is a flatworm with a length of 0.15 to 20 mm and the body is fusiform, haptor (organs part like hooks) in its posterior and the cycle of central hooks pair and a number of hooks marginal.

Parasitic copepods (figure 4.1d.) is a group of shrimp organism that often invades the body of fish and invertebrates as ectoparasites in outer part. In this research, copepods are found in organ gill that is conducted microscopic. This parasite lives in fresh water and seawater and it is quite difficult to control. Most copepod parasites eat by clawing the surface of the host's body using the mandible [16]. They eat mucus, blood tissue of their host, and their attachment and its feeding activity can lead to the primary disease [17]. Several species are major pests in commercial Bilvavian aquaculture such as Ostrincola koe (Myicolidae) lives in the mussel's mantle and causes disease in mussels in China (Boxshall, 2005).

4.2 Intensity and Prevalence Value of Ectoparasites in Green Mussels (*Perna viridis*)

The intensity value is a value that states or categorizes the level of infected biota from low category (<1) to superinfection (>1000). Meanwhile, the prevalence is the value that states into a category that is often infected ranging from category always (99-100%) and the category is almost not never (<0.01). Based on the results of the study, it was known that from 100 samples of green mussels examined, there were 88 individual green mussels that were attacked by the disease so that the green mussels were free from parasites and were not deformed were counted 12 individuals of green mussels.

Based on table 4.1.2, it is known that the highest intensity value is 1.16 and the highest prevalence value is barnacle parasite type about 74%, (*Balanus sp.*), while the lowest intensity value is 1 and the prevalence value of *Perkinsus sp cyst is 1%.* Prevalence value of 1% with the description "occasional infection" which means the spread of the parasite *Perkinsus sp cyst is* not frequent on green mussels which are still in the "sometimes" category. Prevalence value about 74% with a description "moderate infection" with the category "usually" are parasites *Balanus sp.* (barnacles) that attacked green mussels in the Alue Naga Reservoir, Banda Aceh. Prevalence Value about 10% and 24% showed that there were often occur infections in mussels green with category "frequent". The intensity value of 1- of 1.16 is still in the low category, which means the infection of parasites on each individual is still low or not including infection is very severe.

4.3 Water quality parameter results

Table 5 showed that the temperature was around 29°C and several previous studies stated that the optimal temperature to support the growth of green mussels was 27°C-30°C [19]. Salinity at the location of green mussel cultivation in Alue Naga waters, Banda Aceh is around 31 ppt. According to [20], green mussels can still live in the salinity range between 15-45 ppt. [21] also explained that the mussel green indicates that good growth in the estuary is with a salinity range of 18-33 ppt. The concentration of oxygen dissolved range at the location of the cultivation of green mussels is from 4.8 to 6.5 ppm. According to [22], the concentration of Dissolved oxygen range that can still be tolerated by green mussels is 2-12 mg/L. The level of acidity or pH range in the waters of green mussel cultivation is from 8.7. According to [22], the pH range which still can be tolerated by green mussels is from 7-9. The speed ranges of water currents during the green shell cultivation site are from 17 m/s. The suitable current speed ranges for green shellfish cultivation activities are from 0.1 to 0.6 m/s [15; 21].

5 Conclusion

Based on the results of the research on ectoparasites in cultivated green mussels (*Perna Viridis*) in the Alue Naga Reservoir, Banda Aceh, there were 4 types of ectoparasites, there are *Perkinsus sp cysts.* (Protozoa), *Balanus sp.* (Crustacea), Monogenea (Platyhelminthes), and Copepoda (Crustacea). The barnacle parasite (*Balanus sp.*) has the highest intensity value 1.16 and the highest prevalence value 74%, while the *Perkinsus sp cyst* has the lowest intensity value 1 and the prevalence value 1%.

References

1. Hendrik AWC. *Oceana.* **33**, 33-40. (2008).
2. Prasetyo, A. D. *Determination of Metal Content (Hg, Pb, Cd) with the Addition of Preservatives and Different Soaking Times in Green Mussels (Perna viridis* L.) *in Muara Kamal Waters, Jakarta Bay.* Thesis. Syarif Hidayatullah State Islamic University. (2009).
3. *Kordi, K. Management of Fish Pests and Diseases. Rineka copyright and Bina Adiaksara. Jakarta. (2004).*
4. Bhakti, S. *Prevalence and Identification of Ectoparasites in Koi Fish (Cyprinus carpio) in Several Locations of Ornamental Fish Cultivation in East Java.* Thesis. Faculty of Veterinary Medicine. Airlangga University. Surabaya. (2011).
5. Ichsan, Mahmudi M, Nuddin H. The carrying capacity of the Blue Economy-Based Aquatic Environment on the Production of Vanname Shrimp Aquaculture in the Probolinggo Coastal Area. Postgraduate Program, Faculty of Fisheries and Marine Sciences, Universitas Brawijaya, Proceedings of the National Seminar of the University of Muhammadiyah Gresik pages 79-92. (2013).
6. Decree of the State Minister of the Environment, Number 51 of 2004 Regarding Standard Quality of Seawater, State Minister of the Environment, Republic of Indonesia
7. Nguyen, V.T. The influences of shrimp farming and fishing practices on natural fish conservation in Can Gio, Ho Chi Minh City, Vietnam. Ph.D. Thesis, University of Newcastle. (2008)
8. Ichsan. Receptacle Engineering and Aquaculture Technology, UB Press, Malang. (2012)
9. Elston, R.A, Dungan, C.F, Meyers, T.R, and Reece, K.S. J. of Shellfish Research. **22**, 661665. (2003).
10. Soniat, T.M. *Levels of the parasite Perkinsus marinus in populations of oysters from the Louisiana Public Seed Ground*: Summer 2012, Louisiana: Louisiana Department of Wildlife and Fisheries. (2012)
11. Susiana, A. Niartiningsih and M. A Amran. The *Relationship Between Water Quality Suitability and Kima (Tridacnidae) Abundance in the*

Spermonde Islands. Faculty of Marine Sciences and Fisheries Hasanuddin University Makassar. (2016).

12. Jensen, *KR 2010. NOBANIS – Invasive Alien Species Fact Sheet. Denmark.*
13. Alsaqabi, *S., Albelali, A. S., Eshky, A. A. J. of Scientific Research,* ***30****: 141-149.* (2012)
14. Rahmawan, F. 2013. *Prevalence of Ectoparasites in Blood Shellfish (Anadara granosa) Captured by Fishermen in the Sidoarjo Region, East Java.* Thesis of the Faculty of Marine Affairs and Fisheries, Airlangga University.
15. *Kabata Z. 1985. Parasites and Diseases of Fish Cultured in the Tropics. Taylor & Francis, London & Philadelphia.*
16. Boxshall GA. 2005. Crustacean parasites (Chapter 4). In: K Rohde (Ed). *Marine Parasitology,* 123-169. CABI Publishing, Wallingford.
17. Johnson SC, JW Treasurer, S Bravo, K Nagasawa and Z Kabata. *Zoological Studies.* **43**, 229-243 (2004)
18. Boxshall G.A. *Marine Parasitology,* 123-169. CABI Publishing, (2005).
19. Aypa, *SM 1990. Mussel Culture : Regional sea Farming Development and Demonstration project (p. 210). Bangkok. National Inland Fisheries Institute. Kasetsart University.*
20. McGuire, M. & Stevely, J. (2009). *Invasive species of Florida's coastal waters : the asian green mussel (Perna viridis).* The Florida Sea Grant College Program with support from the National Oceanic and Atmospheric Administration. US Department of Commerce.
21. [FIGIS] Fisheries Global Information System. *Species Fact Sheet: Perna viridis (Linnaeus, 1758) Mytilidae* . Rome : FAO. (2005)
22. Sreedevi, P.R., Uthayakumar, V., Jayakumar, R., Joseph, Kumar, D. S. & Ramasubramanian, V. *World Journal of Fish and Marine Science*, **6**, 487 – 493 (2014)
23. Alfaro, *A.C. Aquaculture.* ***246****, 285 – 294, (2005)*

The effect of *Cassia alata* L. leaf extract on egg hatching and survival of *Pangasianodon hypophthalmus* larvae

Iko Imelda Arisa[1,3,4*], *Sri* Agustina[2,3,5], and *Lisa* Handayani[1]

[1] Departement of Aquaculture, Marine and Fisheries Faculty, Universitas Syiah Kuala, Banda Aceh, Aceh, 23111Indonesia
[2] Departement of Marine Science, Marine and Fisheries Faculty, Universitas Syiah Kuala, Banda Aceh, Aceh, 23111Indonesia
[3] Research Center for Marine and Fisheries, Universitas Syiah Kuala, Banda Aceh, Aceh, 23111Indonesia
[4]Animal Histology and Fish Nutrition Laboratory, Marine and Fisheries Faculty, Universitas Syiah Kuala, Banda Aceh, Aceh, 23111 Indonesia
[5]Laboratory of Marine Chemistry and Biotechnology of Fisheries, Marine and Fisheries Faculty, Universitas Syiah Kuala, Banda Aceh, Aceh, 23111 Indonesia

Abstract. This study aims to determine the effect of using *Cassia alata* L. leaf extract on the prevention of fungus on fish eggs *Pangasianodon hypophthalmus* This research was conducted at the Blangpidie Fish Seed Center. Aceh Barat Daya from July to August 2020. Statistical analysis used a completely randomized design (CRD) method with 7 treatment levels and 3 replications. The treatments were conducted by soaking the fish eggs in the leaf extract for 20 minutes. Those treatments were A (control), B (2.5 ppm), C (5 ppm), D (7.5 ppm) E (10 ppm), F (12.5 ppm), and G (15 ppm). ANOVA test results showed that the addition of leaf extract *C. alata* had a significant ($P < 0.05$) effect on hatchability and survival of *P. hypophthalmus* larvae. A concentration of 15 ppm is the best with a hatchability value of 85.33% and a concentration of 10 ppm is the most optimum for the survival value of *P. hypophthalmus* larvae, namely 92.99%.

1 Introduction

Pangasianodon hypophthalmus is one of the potential aquaculture commodities to be developed. However, according to Muhlis *et al.* [1], *P. hypophthalmus* hatchery activities encountered several obstacles, including disease attacks in the egg phase caused by fungi. Fungal infection at the time of hatching eggs affects the damage to the eggs and causes a decrease in hatchability. Furthermore Larasati *et al.* [2] also mentions that the problem that is often faced in catfish farming is the low degree of egg hatching. Low egg hatchability will have an impact on the supply of catfish seeds which are low, so that aquaculture production activities are also low.

Efforts to control disease in cultured fish eggs due to pathogenic infections that use chemicals or antibiotics in the long term can have negative impacts, among others, can cause bacterial and fungal resistance and can pollute the environment [3,4]. So the disease requires alternative treatments such as the use of natural ingredients. One of the natural ingredients that has the potential to inhibit the growth of pathogens in fish is the use of *Cassia alata* leaves. This plant is known by different names in different countries, namely as Gelenggang (Malaysia), Ketepeng Badak or Ketepeng Cina (Indonesia) and bush ringworm (Australia) and has been used as as traditional medicine to treat various diseases [5,6].

2 Material and methods

The research was carried out in July to August 2020 at the Blang Pidie Fish Seeds Center, Aceh Barat Daya District, Aceh Indonesia. The research design used was a completely randomized design (CRD) consisting 7 treatments and 3 replications. As for the types of treatments tested were as follows: A=immersion without *C. alata* leaves extract (0 ppm); B= immersion with *C.alata* leaves extract at a dose of 2.5 ppm; C= immersion with *C.alata* leaves extract at a dose of 5 ppm; D= immersion with C.*alata* leaves extract dose of 7.5 ppm; E= immersion with *C.alata* leaves extract at a dose of 10 ppm; F= immersion with *C.alata* leaves extract at a dose of 12.5 ppm; G= immersion with *C. alata* leaves extract at a dose of 15 ppm.

2.1 *Cassia alata* L. leaves extract

Cassia alata leaves were washed with clean water and dried under the sun light until the leaves were easily crushed. The dried leaves are ground into flour. A total of ±500 g of *C. alata* leaves powder was soaked in 3 L 96% PA absolute ethanol for 24 hours. The results of the immersion are filtered (Whattman No.41), the results are extracted using a rotary evaporator, then the extract is stored at -5 °C until it is used for testing.

*Corresponding author: ikoimeldaarisa@unsyiah.ac.id

2.2 Fish eggs soak

Fish eggs *Pangasianodon hypophthalmus* 50 healthy eggs that have been fertilized are put into each treatment container. Futhermore, as many as 25 fish eggs that have been infected with fungi and put into each of these containers and then left for 6 hours. After the healthy eggs were infected with the fungus, they were transferred to each container which was given a solution of *C. alata* leaf extract in accordance with the provisions of the treatment concentration. The eggs were soaked in a solution of *C. alata* leaf extract for 20 minutes.

2.3 Observed parameters

2.3.1 Eggs condition

Observation of eggs started after soaking using *C. alata* leaf extract until the eggs hatched. Eggs infected with fungi will show clinical symptoms in the form of threads like cotton covering the eggs. Observations of eggs that recovered due to fungal infection after soaking were observed every 1 hour interval using a microscope.

2.3.2 Hatching rate (HR)

Calculation of egg hatching rate parameter in research based on Arfiah *et al.* [9]:

$$HR = \frac{\text{Number of eggs that hatch}}{\text{Number of fertilized eggs}} \; x \; 100\% \qquad (1)$$

Where: HR = Hatching Rate (%)

2.3.3 Survival rate

Calculation of survival rate parameter in research based on Effendie [10]:

$$SR = \frac{No - Nt}{No} \; x \; 100 \qquad (2)$$

Where: SR = Survival Rate (%); Nt = number of test dead fish that died during rearing (ind); No =the total number of test fish observed (ind).

2.3.4 Water quality parameters

Several water quality parameters measured during the study were pH, temperature, and dissolved oxygen. pH parameter measurement using a pH meter (PCSTestr™ 35), temperature parameter measurement was carried out using a thermometer (PCSTestr™ 35) and dissolved oxygen parameter measurement using a DO meter (Thermo Scientific Eutech DO 6+). Water quality parameters measurements are carried out every day.

2.4 Data analysis

Research parameter data such as egg hatchability and survival rate of fish were analyzed using Analysis of Variance (ANOVA), if the results were influential then Duncan's test was performed with a 95% confidence level. While the water quality parameter measurement data is presented descriptively.

3. Results and Discussion

Cassia alata extract has enormous pharmacological value in supporting the use of traditional medicine in curing several important diseases, including antioxidant, anti-fungal and anti-bacterial properties [11]. Based on the results of the study, it was found that the use of *C. alata* leaf extract could increase egg hatchability and survival of *P. hypophthalmus* (Table 1), where the higher the dose of extract used in the treatment will also increase the hatchability of fish eggs. Fatmawati *et al.* [6] stated that *C. alata* leaf extract contains chemical compounds that can be used as antimicrobials. Furthermore, Makinde *et al* [12] said that the antibacterial and antifungal activity of the aqueous and methanolic extracts of *C. alata* leaves found that they had more antifungal properties than antibacterial. Phytochemical analysis of *C. alata* extract including flavonoids, tannins, steroids, phenols and alkaloids [13, 14, 15]. Then Roshida *et al.* [16] stated that tannins are antibacterial by presipitating protein. The antimicrobial effect of tannins through reactions with cell membrane inctivation of enzymes, destruction or inactivation of functions. Meanwhile, flavonoids have antimicrobial properties can form complex compounds with dissolved extracellular proteins and microbial cell walls. Falvonoids also have antibacterial and antioxidant properties that work by increasing the immune system, where leukocytes as antigen-eaters are produced faster and activate the lymphoid system more quickly [17-19].

The data from this study showed that the highest hatchability of fish eggs was found in the 15 ppm treatment, which was 85.33%. This indicates that the antifungal content of *Cassia alata* leaf extract with a concentration of 15 ppm is optimal in inhibiting the fungus *Saprolegnia* sp. to infect *P. Hypophthalmus* fish eggs. *C. alata* is a traditional medicine that has excellent antibacterial and antifungal properties [20-21]. Previous research Bowo *et al.* [22] stated that a concentration of 60 ppm *C. alata* leaf extract showed the best results in inhibiting the growth of the fungus *Saprolegnia* sp, while for egg hatchability and survival of gouramy larvae the highest was obtained at a concentration of 10 ppm.

Tabel 1. Hatching rate and survival rate of *P. hypophthalmus*

Treatments	Parameters	
	Hatching Rate (%)	Survival Rate (%)
A (0 ppm)	22.00 ± 4.0^{a}	55.32 ± 16.4^{a}
B (2.5 ppm)	37.33 ± 6.1^{b}	74.45 ± 12.3^{ab}
C (5 ppm)	45.33 ± 5.0^{b}	81.75 ± 10.7^{ab}
D (7.5 ppm)	61.33 ± 7.6^{c}	90.56 ± 9.4^{b}
E (10 ppm)	67.33 ± 8.1^{cd}	92.99 ± 8.7^{b}
F (12.5 ppm)	75.33 ± 3.0^{de}	90.29 ± 3.9^{b}
G (15 ppm)	85.33 ± 5.0^{e}	92.73 ± 1.3^{b}

Note:writing different superscripts in the column indicates a significant difference between treatments (P<0,05).

Tabel 2 Water quality parameters measurement data

Treatments	Temperature (°C)	pH	DO (mgL^{-1})
A (0 ppm)	26.4 – 26.8	6,7 – 6.5	5.3 – 5.5
B (2.5 ppm)	26.5 – 26.9	6.6	5.4 – 5.6
C (5 ppm)	26.5 – 26.9	6.5 – 6.7	5.4 – 5.5
D (7.5 ppm)	26.6 – 26.8	6.5 – 6.7	5.4 – 5.5
E (10 ppm)	26.6 – 26.9	6.6 – 6.7	5.4 – 5.5
F (12.5 ppm)	26.6 – 26.9	6.5 – 6.7	5.4 – 5.6
G (15 ppm)	26.5 – 26.9	6.5 – 6.7	5.4 – 5.5

The results of the analysis of variance in use of *C. alata* leaf extract had an effect on survival of catfish larvae ($P<0.05$). During the study, the observation of the survival rate of *P. hypophthalmus* larvae was categorized as good. It is suspected that the chemical content in leaf extractof C. alata was able to inhibit the growth of fungi in *P. hypophthalmus* eggs.

Good water quality management can inhibit the spread of fungi in hatchery activities both at time of hatching eggs and rearing post larvae in aquaculture activities. Good management of aquaculture media water quality will suppress the spread of fungal diseases through water management is carried out for all water quality parameters, both chemical, physical and biological. During the research, it was found that the water quality parameters measurement value data still within the tolerance range of *P. hypophthalmus* (Table 2). Based on the Indonesian National Standard [23], the optimal pH value for catfish rearing ranges from 6.5 - 8.5. The value of dissolved oxygen in the media for hatching eggs and rearing *P. hypophthalmus* larvae is > 5 mgL-1. Furthermore, the research of Putra *et al.* [24] states that the water quality for hatching and rearing catfish larvae is a temperature of 26.2 – 27.90 and dissolved oxygen 4.7 – 5.5 mgL^{-1}.

4. Conclusion

The use of *Cassia alata* L. leaf extract make an impact to hatchability of fish eggs and the survival of *Pangasianodon hypophthalmus* fish larvae. The highest fish egg hatchability value of 85.33% was obtained by soaking *C. alata* leaf extract at a dose of 15 ppm and the survival value the highest fish larvae of 92.99% was obtained at a dose of 10 ppm.

Reference

1. Muhlis, M. Ghofur, M. Sugihartono, *J Akuakultur Sungai dan Danau* **4,** 1 (2019)
2. S. Larasati, F. Basuki, T. Yuniarti, *J of Aquaculture Management and Technology* **6,** 4 (2017)
3. Noga, *Pestisida Nabati untuk Penyakit Ikan* (Penebar Swadaya, Jakarta, 2010)
4. D.Q. Pham, H.T. Pham, J.W. Han, T.H. Nguyen, H.T. Nguyen, T.D. Nguyen, T.T.T. Nguyen, C.T. Ho, H.M. Pham, H.D. Vu, G.J. Choi, Q.L. Dang, *Industrial Crops & Products* **166** (2021)
5. Y.Y. Ling, P.S. Fun, A. Yeop, M.M. Yusoff, J. Gimbun, *Material Todays: Proceedings* **19** (2019)
6. S. Fatmawati, Yuliana, A.S. Purnomo, M.F.A. Bakar, *Helion* **6** (2020)
7. M. Bahi., R. Mutia., E. Mustanir, Lukitaningsih, *J Natural* **14,** 1 (2014)
8. S. Saena, S. yanto, A. Amirah, *J Pendidikan Teknologi Pertanian* **6,** 1 (2020)
9. H. Arfiah, Muftucha, O. Caman, *J Akuakultuur Indonesia* **5** (2006)
10. M.I. Effendie, Metode Biologi Perikanan (Yayasan Dewi Sri, Bogor, 1979)
11. O.T. Tcheghebe, V.R.S. Tala, F.N. Tatong, G.K. Tchuente, *Medicine and Medical Sciences (LRJMSS)* **4,** 2 (2017)
12. A.A. Makinde, J.O. Igoli, L. TA'Ama, S.J. Shaibu, A. Garba, *African Journal of Biotechnology* **6,** 13 (2007)
13. R.P.S. Kumar, V. Malayaman, S. Shinduja, *J of Chemical and Pharmaceutical Research* **5,** 11 (2013)
14. R. Dewi, Y. Firza, M. Nashiry, F.S. Al-Suede, A.S.A. Majid, Australian Herbal Insight **2** (2019)
15. J.M.N. Abille, A Review on The Medicinal Plant *Cassia alata* (L.) ROXB. (FABACEAE) (Biological Sciences Department, College of Science & Computer Studies De La Salle University – Dasmarinas, Cavite, Philippines pp 1-12, 2019)
16. Rosidah, W.M. Afizia *J Akuatika* **3,** 1 (2012)
17. R.S. Hardhini, Amir, D.N. Setyowati *J Perikanan* **8,** 1 (2018)
18. F.M. Nur, R.A. Nugroho, S. Fachmy. *AIP Conference Proceedings* **1813,** (2017)
19. R.A. Nugroho, R. Aryani, W.D. Anggraini, E.H. Hardi, R. Rudianto, E. Kusumawati, S. Sudrajat, F.M. Nur, H. Manurung. IOP Conference Series: Earth and Environmental Science **348,** 1(2019)
20. A. Happy, M.Soumya, S.V. Kumar, S. Rajeshkumar, R.D. Sheba, T. Lakshmi, V.D. Nallaswamy, *Biochemistry and Biophysics Reports* **17** (2019)
21. S.A. Gaddam, V.S. Kotakadi, D.V.R.S. Gopal, Y.S. Rao, A.V. Reddy, *J of Nanostructure Chemistry* **4** (2014)
22. A.T. Bowo, Sunarto, Rachimi, *J Ruaya* **1,** 1 (2014)
23. Standar Nasional Indonesia,*Produksi Benih Ikan Patin Siam (Pangasius hyphthalamus)* Kelas Benih Sebar (SNI : 01- 6483.4 – 2000)
24. P.L. Putra, D. Jubaedah, M. Syaifudin, *J Akuakultur Rawa Indonesia* **8,** 1 (2020)

https://doi.org/10.1051/e3sconf/202233901004

Minapadi productivity with a combination of legowo technology and local fish

Farah Diana[1*], *Mahendra* Mahendra[1] and *Giovanni* Oktavinanda[2]

[1]Aquaculture Study Program, Faculty of Fisheries and Marine Science, Teuku Umar University, West Aceh, Indonesia
[2]Marine science Study Program, Faculty of Fisheries and Marine Science, Teuku Umar University, West Aceh, Indonesia

Abstract. Minapadi is a fish rearing system in rice fields that is carried out with rice plants. This study aims to increase the income of farmers with high productivity of rice yields and increase the production of endemic fish. The method used was a Completely Randomized Design (CRD) experiment with four treatments and 3 replications. PThe treatments are: A1B1 (Legowo 2:1 and giant prawns), A1B2 (legowo 2:1 and fish serukan), A2B1 (legowo 4:1 and giant prawns), and A2B2 (legowo 4:1 and fish serukan). This research was conducted for 90 days in the rice fields of Lango Village, Pante Ceureumen District, West Aceh Regency. Parameters observed were rice productivity, fish growth and business analysis.The results showed that the best treatment was A1B1 (4:1 legowo and giant prawns) with generate value rice productivity average as big as 5 kg/plot, and the average value of the growth of giant prawns is 3.62 g/head and fish serukan is 37.17 g/head. From the results of this study, the application of the Minapadi system with giant prawns and fish with legowo 4:1 gave a positive impact on the observed parameters ($P<0.05$), annd provide benefits to farmers.

1 Introduction

Diversity the life of freshwater fish in Indonesia, especially Sumatra today facethreats from various factors anthropogenic which can cause a decrease in the diversity of native fish in the waters [1]. Fish farming businesses in Aceh Province have begun to develop, howeversadly The types that are kept are foreign fish introduced from outside Aceh and evenoutside the country. This condition puts pressure on the local native fish population. Meanwhile, Aceh also has the potential for native fish that are not inferior both in terms of quantity and quality, for example giant prawns (*Macrobrachium rosenbergii*) [2] and serukan (*Osteochilus Vittatus*) [3]. The research team has also succeeded in developing and producing local fish including producing Tawes fish (*Barbonymus gonionotus*) [4], fish serukan.

The rapid increase in population can cause more limited large areaproduction that can be used for various fields of fishery business, it is necessary to utilize the existing land through intensification. One way that can be done is by implementing the minapadi system.

The jajar legowo planting system in minapadi provides a loose growing space as well as a higher population. With this planting system, it is able to provide air circulation and better utilization of sunlight for planting. Besides therefore, weed control and fertilization efforts can be done more easily. The best type to get minapadi performance is 2:1 and 4:1 [5]

The potential for developing minapadi cultivation is still very wide. Potential data which released by the directorate general of aquaculture, nationally land use for cultivation minapadi only amounted to 127,944 hectares of the potential land area of 1,538,379 hectares. So, the level of land use for rice cultivation is only about 8.3 percent. Therefore, the Ministry of Marine Affairs and Fisheries through the Directorate General of Aquaculture has again promoted the minapadi cultivation program to support the target of aquaculture production [6]

Based on the above, the author is interested in conducting a research entitled "minapadi productivity with a combination of legowo technology and local fish". The purpose of this study was to determine whether the use of different combinations of the legowo system and Aceh endemic fish had an effect on minapadi productivity and which combination was the best for minapadi productivity.

2 Research Methods

2.1 Research sites

This research was conducted in the rice fields of Pante Ceureumen District, West Aceh Regency from preparation to data collection.

2.2 Experimental design

The method used in this study is an experimental method and the design used is a Completely Randomized Design (CRD) consisting of four treatments with 3 replications

* Corresponding author: farahdiana@utu.ac.id

each. The treatment of each material used in this study includes:

The method used in this study is an experimental method and the design used is Completely Randomized Design (CRD). The two factorial was repeated three times.

Factor first tested are:
A1 = legowo system 2: 1
A2 = legowo system 4: 1
The second factor is:
B1 = Giant prawns
B2 = Fish serukan

Combination There were 4 treatments in total, namely: A1B1, A1B2, A2B1, A2B2.

2.3 Research procedure

This research was conducted in rice fields with a total of 9 plots with an area per plot of 5 x 10 meters. Make a rice field bund with a height of 60 cm, a base width of 60 cm and an upper width of 30 cm. While the trench (kemalir) according to the legowo system treatment with the size: 30-60 cm wide, 40 cm deep. Planting is done by transplanting (Tapin). Rice is planted after 30 days. The variety used is resistant to waterlogging and pest attack, namely MAPAN. Spacing 25 x 12.5 cm. The number of seedlings planted 2 - 3 stems per clump. The water level is 10 cm. Shrimp stocking was done when the plants were 30 days old with a water level of 10 cm.

Stocking density was carried out according to the treatment. Stocking is done in the afternoon to avoid stress fish. The length of shrimp rearing in the Minapadi system is 90 days with feeding 3 times a day, namely in the morning, afternoon and evening. The maximum dose given is 3% of the total biomass with the frequency of administration. Shrimp harvest is done 90 days after planting rice. Observation of rice seen from its productivity. While shrimp and fish are seen from their growth.

2.4 Research Parameters

2.4.1 Rice productivity

Productivity rice according to [7] is calculated using the formula:

$$Rice\ Productivity = \frac{Production\ quantity\ (Kg)}{land\ area\ (ha)} \quad (1)$$

2.4.2 Absolute Weight Growth

The calculation of absolute weight growth uses the following [8] formula:

$$W = W_t - W_0 \quad (2)$$

where, w is growth absolute weight (g); wt is final weight of maintenance (g); w_0 is initial weight of maintenance (g).

2.5 Analysis Data

The data obtained will be grouped, tabulated, performed statistical tests ANOVA and SPSS to see the effect of the experiment. If there is a significant difference, Tukey's further test will be carried out.

3 Results and Discussion

3.1 Rice productivity

Based on the results of the ANOVA which showed that the combination of legowo technology and local fish had no significant effect (P > 0.05) on minapadi productivity. The results obtained from this study during 90 days of maintenance can be seen in Figure 1.

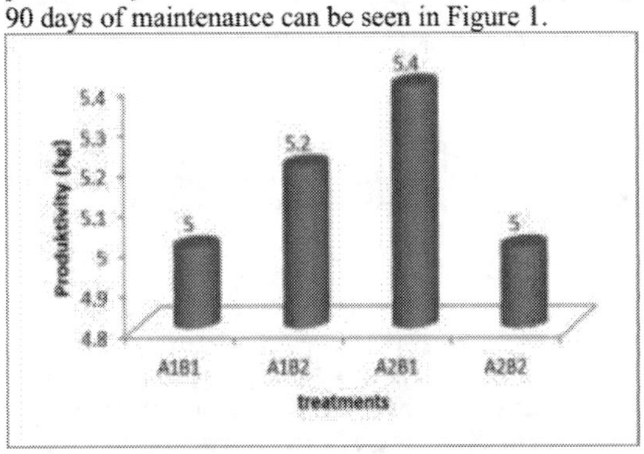

Fig.1. Rice Productivity

Productivity is the process of converting factors of production into a product. Productivity reflects the ratio between total production and land area. The results of the study based on Figure 1 show that the highest rice productivity value in this study was A2B1 treatment (Legowo 4:1 with giant prawns) with an average value of 5 kg/plot. According to research by [9] that through Minapadi there was an increase in land productivity from 5,652 kg/ha to 7,612 kg/ha. The area of rice and Minapadi is the same, but the proportion of Minapadi is 90% for rice cultivation and 10% for fish cultivation.

According to [10], Minapadi cultivation is said to be successful when rice production is at least the same as production before Minapadi. The ability of land in production is getting better when implementing Minapadi. Efforts to increase productivity are achieved through improving soil conditions and intensive handling of pests and weeds.

The high productivity of rice in the A2B1 treatment is suspected to be the pellet feed given is not consumed by shrimp where the pellets used are the trademark FF999 for fish, so that shrimp prefer abundant natural feed. The remaining pellets will accumulate into fertilizer which is absorbed by the rice plant. According to [11] Nitrogen is the main nutrient for plant growth, nitrogen is absorbed by plant roots in the form of NO3 - (nitrate) and NH4 + (ammonium). Nitrogen deficiency can be bad for plants such as stunted plant growth, yellowing of plant leaves, and limited root system, while excess nitrogen causes vegetative growth to be elongated, easy to fall, reduce grain quality and response to pests and diseases.

3.2 Giant Shrimp Growth

The growth results of giant prawns in each treatment for 90 days of rearing period are presented in Figure 2.

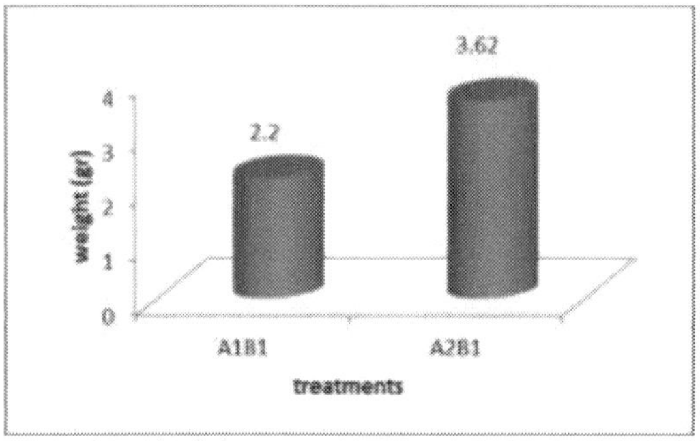

Fig.2. Growth of giant prawns

The results of the study based on Figure 1 showed that the lowest growth value in this study was A1B1 treatment (Legowo 2:1 with giant prawns stocking) with a value of 2.2 gr. Meanwhile, the highest growth value was found in the A2B1 treatment (4:1 legowo with giant prawns stocking) with a value of 3.62 gr. The results of this study indicate that the A2B1 treatment plays a role in supporting the growth of giant prawns.

Some aspects of physiological processes related to individual growth include regeneration, metamorphosis and moulting. Moulting is the process of periodically releasing old shells and forming new, larger shells. In crustaceans, growth occurs periodically after molting. Body weight gain will be hampered if it is not preceded by the molting process (Affandi and Tang, 2002).

High growth is characterized by a faster molting process. The process of energy transfer from hemolymph to shrimp shell requires a large amount of energy. This large energy requirement is obtained from the feed consumed. Feeding is an energy requirement to support faster growth. This is closely related to growth where if feed consumption is high, there is a lot of energy available that can be used for various life needs and by reducing energy expenditure, so that the portion of energy available for growth is greater. If the physiology of shrimp goes well, including metabolism, then the utilization of feed is more efficient which ultimately increases growth [12]. Therefore, the legowo 4:1 system is more suitable for growth, because of the protection of the rice plant 4.

3.3 Fish Growth

The results of the growth of fish in each treatment during the 90-day rearing period are presented in Figure 3.

The results of the study based on Figure 3 it was found that the growth value of the fish serukan was related to the shrimp where the lowest value in this study was the A1B2 treatment (Legowo 2:1 with the distribution of the serukan) with a value of 26.48 gr. Meanwhile, the highest growth value was found in the A2B2 treatment (4:1 legowo with fish serukan) with a value of 37.17 gr. The results of this study indicate that the A2B2 treatment plays a role in supporting the growth of the fish.

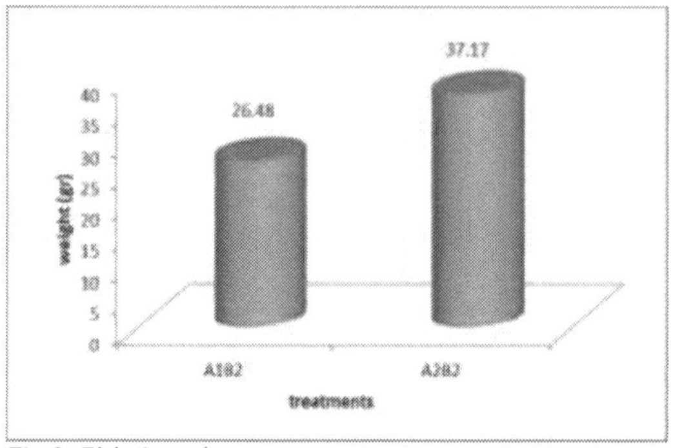

Fig.3. Fish Growth

The growth of fish and shrimp in Minapadi and Legowo systems was higher with Legowo 4:1 compared to Legowo 2:1. It is suspected that the protection of the legowo 4:1 system of rice plants is more maximal compared to the legowo 2:1 and it is also suspected that the number of kemalir 4: is less, so that the activity of fish such as swimming is reduced, therefore less energy is spent. Feeding is an energy requirement to support faster growth. This is closely related to growth where if feed consumption is high, there is a lot of energy available that can be used for various life needs and by reducing energy expenditure, so that the portion of energy available for growth is greater. If the shrimp physiology goes well, including metabolism.

3.4 Analysis Farmers' income

Minapadi system farming analysis for farmers needs to be done to find out how much added value can be obtained from the business, then economic analysis is part of finding out whether the system is feasible or not to be developed and to find out whether the Minapadi system provides benefits or not. So below, a simple analysis of the Minapadi farming system applied can be seen in Tables 1 and 2.

Table 1. Rice production results from the legowo Minapadi system

Model	Rice Production [13]	unit	Rice Production (processed data)	Unit
Legowo 2:1	3.04 a	ton/ha	5.1	Kg/50m2 a
Legowo 4:1	2.94 a	ton/ha	5.2	Kg/50m2 a

The assumption of income from the Minapadi system fish farming business is that farmers need 60 kg of pellets each to raise fish and shrimp. The price of 1 sack of 30 kg pellets is Rp. 200,000. If the farmer uses 100% pellets for feed, the cost required is (60 kg/ 30 kg x Rp. 200,000 = Rp. 400,000). The cost of the farmer's profit if the results obtained are 60%, the cost of income minus the cost of expenses. Where if the assumption of expenditure costs includes the price of 250 giant prawns at a price of @ 500, as much as Rp. 125,000. so that the total assumption issued for giant prawns is (Rp. 400,000 + Rp. 125,000 = Rp. 525,000). While the assumption that the results of the expenditure on the fish of chorus includes 250 fish with a price of @1000, as much as Rp. 250,000 so that the total assumption issued is (Rp. 400,000 + Rp. 250,000 = Rp. 650,000).

Farmers' incomes other than rice yields are assumed to yield 60% of the stocking density of 150 shrimp and fish each with a yield range of 20 per kilogram of shrimp, so the final yield of shrimp is 7.5 kg, with a selling price of Rp. .185.000/kg (Coal and Gustianty. 2016). The total selling price for shrimp is (7.5 kg x 185,000/kg = Rp. 1,387,500. So the total profit income for shrimp is (Rp. 1,387,500 - Rp. 525,000 = Rp. 862,500. While for fish with 60% yields, fish with a yield range of 10 per kilogram of fish, the final yield of fish is 15 kg, with the selling price of fish Rp. 60,000/kg (local market price) The total selling price of fish is (15 kg x Rp. 60,000/kg = Rp. 900,000. So the total profit income for fish is (Rp. 900,000 - Rp. 650,000 = Rp. 250.

The results above are for 1 rice field plot measuring 5x 10 m2, so that the results of this study are 16 plots of rice fields (8 plots for shrimp and 8 plots for fish). Therefore, the results obtained by farmers other than rice ranged, among others, for shrimp (Rp. 862,500 x 8 = Rp. 6,900,000). As for fish it is (Rp. 250,000 x 8 = Rp. 2,000,000)

So, the total that farmers get is (Rp. 6,900,000 + Rp. 2,000,000 = Rp. 8,900,000), these results are in addition to the rice harvest. Farmers' profits can be large if the Minapadi system is applied with a large and wide number of rice fields, so that the stocking density is higher.

Table 2. Rice production results from the legowo Minapadi system

Details	Unit	Amount	Price per unit	Total
Variable Cost				
Fish seeds	Tail	250	500	125,000
Rice Seeds	kg	5	100,000	500,000
Feed	Sak	2	200,000	400,000
Fertilizer	kg	15	10,000	150,000
				1,175,000
Shrinkage	%	20		235,000
Production				-
Fish	kg	7.5	185,000	1,387,500
Paddy	kg	5	10,000	50,000
				1,437,500
Profit				27,500

4 Conclusion

Minapadi's 4:1 legowo system has an effect on the growth of giant prawns and fish serukan.with generate value rice productivity average as big as 5 kg/plot, and the average value of the growth of giant prawns was 3.62 g/head while the growth of fish was 37.17 g/head.

Acknowledgments. The author's research was funded by internal research grants from Teuku Umar University, especially LPPM and Quality Assurance as well as the Aquaculture Study Program, Faculty of Fisheries and Marine Sciences.

References

1. HasriI, M.M. Kamal and Zairion, J Ichthyol, 11(1), 2010)
2. Erlangga, J Terubuk 44(1), (2016)
3. DKP Aceh, Identification of Endemic Fish Species in Aceh Waters South West Region (West Aceh) (2015)
4. F. Diana, and E Safutra Journal of Aquaculture 2(1), (2018)
5. J. Bobihoe, The Legowo Jajar Rice Planting System (2013)
6. KKP, Minapadi System Fish Cultivation. Directorate General of Aquaculture (2018)
7. E. Silamat, Yuwana, and M.Z. Yuliarso, Analysis of Paddy Rice Farming Productivity Using Hand Tractors and Conventional Methods in Rejang Lebong Regency, (2014)
8. R. Affandi, U.M. Tang, Physiology of Aquatic Animals (2002)
9. D.R. Julistia, Analysis of Minapadi Farming Income in Margoluwih Village, Seyegan District, Sleman Regency, (2017)
10. L.F.I. Tysno, LFI, Analysis of Minapadi Farming Income, (2018)
11. E. Kaya, Agrologia, 2(1), (2013)
12. A.B. Zaidy. Utilization of Aquatic Calcium in the Molting Process and Its Consequences for the Growth of Giant Prawns (Macrobrachium rosenbergii de Man) (2007)
13. I.M Sudiarta, E. Syam'un, and R. Syamsuddin, J. Science & Technology, **16**(1), (2016)

The effect of different stock densities on the performance of giant prawns (*Macrobracium rosenbergii*) and rice production in the minapadi system

Mahendra Mahendra[1*], *Farah* Diana[1] and *Giovanni* Oktavinanda[2]

[1]*Aquaculture* Study Program, Faculty of Fisheries and Marine Science, Teuku Umar University, West Aceh, Indonesia.
[2]Marine science Study Program, Faculty of Fisheries and Marine Science, Teuku Umar University, West Aceh, Indonesia.

Abstract. Minapadi is a prawn rearing system in rice fields that is carried out with rice plants. This study aims to increase the income of farmers with high productivity of rice yields and increase the production of giant prawns. The method used is a Completely Randomized Design (CRD) experiment with 3 treatments and 3 replications. The treatments were : P1 = stocking density of 100 ind/plot, P2 = stocking density of 200 ind/plot, P3 = stocking density of 300 ind/plot. This research was conducted for 90 days. Parameters observed were rice productivity and growth. The results showed that the best treatment was P2 = stocking density of 200 ind/plot. By producing an average rice productivity of 5.5 kg/50 m^2 and the average relative growth rate of 4.72%. From the results of this study, the application of the Minapadi system with giant prawns had a positive impact on the observed parameters ($P<0.05$) and thus will provide benefits for farmers.

1 Introduction

Fish farming businesses in Aceh Province have begun to develop, Howeversadly the types that are kept are foreign fish introduced from outside Aceh and evenoutside the Indonesia. This condition puts pressure on the local native fish population. Meanwhile, Aceh also has the potential for native fish that are not inferior both in terms of quantity and quality, for example giant prawns (*Macrobrachium rosenbergii*) [1].

The very fast population increase can cause more limited area production that can be used for various fields of fishery business, Therefore,it is necessary to utilize the existing land through intensification. One way that can be done is by implementing the Minapadi system.

Minapadi is a system of growing rice with fish cultivation in one rice field simultaneously. The potential for developing Minapadi cultivation is still very huge. Nationally the use of land for Minapadi cultivation only amounts to 127,944 hectares of the potential land area of 1,538,379 hectares. Thus, the level of land use for rice cultivation is only about 8.3 %. Therefore, the Ministry of Marine Affairs and Fisheries through the Directorate General of Aquaculture has promoted the Minapadi cultivation program to support the target of aquaculture production [2].

Giant prawns (*Macrobracium rosenbergii*) is one of the most potential aquatic species to be reared in the Minapadi system. The implementation of the Minapadi system with giant prawns has been implemented by a research team whose growth is very good in the Nagan Raya area [3]. However, the optimum stocking density of a high production minapadi system is still unknown. Therefore, we conducted this study to investigate the ideal stocking density of *Macrobracium resonbergii* that able to generate the highest productivity of paddy and the prawns.

2 Research method

2.1 Research sites

This research was conducted in the rice fields of Pante Ceureumen District, West Aceh Regency. The research location is at the coordinates of 4.430299, 96.299837.

2.2 Experimental design

The experimental method used in this study is a Completely Randomized Design (CRD) consisting of 3 treatments with 3 replications each. The treatment of each material used in this study includes:
P1 = Stocking density of 100 ind/plot
P2 = Stocking density of 200 ind/plot
P3 = Stocking density of 300 ind/plot

2.3 Research procedure

This research was conducted in rice fields with a total of 9 plots with an area of 5 x 10 meters per plot. Rice field bunds were constructed with height of 60 cm, base width of 60 cm and upper width of 30 cm, while the trenches (kemalir) according to the Legowo system treatment

*Corresponding author : mahendra@utu.ac.id

 https://doi.org/10.1051/e3sconf/202233901005

were dug with the size of 30-60 cm wide and 40 cm deep. The rice variety used is resistant to waterlogging and pest attack, namely MAPAN which was planted after 30 days with spaces of 25 x 12.5 cm. The number of seedlings planted was 2 - 3 stems per clump. Shrimp (*Macrobracium rosenbergii*) stocking was done when the plants were 30 days old with a water level of 10 cm. Stocking density treatment according to the experimental design was done in the afternoons to avoid stress to the Giant prawns. The shrimp (*Macrobracium rosenbergii*) were reared in the Minapadi system for 90 days with daily feedings of 3 times, namely in the morning, afternoon and evening. The maximum dose given was 3% of the total biomass with the frequency of administration three times a day. Shrimp (*Macrobracium rosenbergii*) harvest iswas done 90 days after the rice planting rice. Rice was observed for its productivity, while the shrimp (*Macrobracium rosenbergii*) was its growth and survival.

2.4 Research Parameters

2.4.1 Rice productivity

Productivity of the rice was calculated according to [4] using the formula:

$$Rice\ productivity = \frac{Production\ quantity\ (Kg)}{land\ area\ (ha)} \quad (1)$$

2.4.2 Relative Growth Rate (RGR)

The relative growth rate of giant prawns was calculated using the following formula by [3]:

$$RGR = \frac{Wt - W0}{W0\ x\ t}\ x\ 100\ \% \quad (2)$$

Information:
SGR = Relative growth rate
Wt = Prawn's Weight at the end of research
W0 = Prawn's Weight at the beginning research
t = Research time (90 day)

2.4.3 Analysis Data

The data obtained Data obtained were grouped, tabulated and statistically tested using ANOVA and SPSS to see the effect of the experiment. If there was a significant difference, a further Tukey test was be carried out.

3 Results and Discussion

3.1 Rice productivity

The results obtained from the research on rice productivity during maintenance can be seen in Figure 1. The highest rice productivity value in this study was P1 treatment with a value of 5.5 kg/plot. While the lowest in the P3 treatment was 4.6 kg/plot. According to [5] there was an increase in land productivity from 5,652 kg/ha to 7,612 kg/ha. Rice and Minapadi area are the same, but the proportion of Minapadi is 90% for rice cultivation and 10% for fish farming.

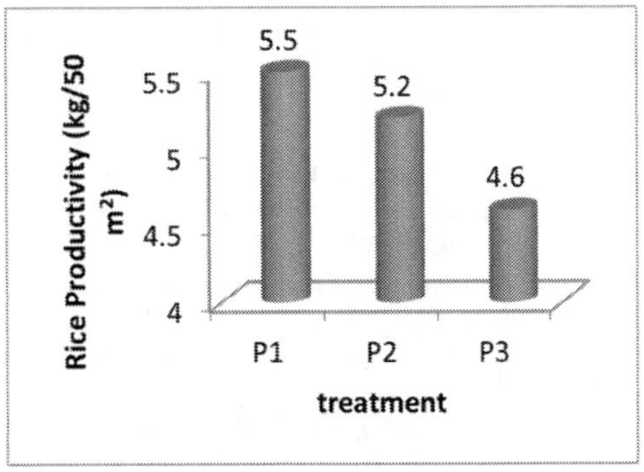

Fig. 1. Rice Productivity of Minapadi system by rearing *Macrobracium rosenbergii*

Minapadi cultivation is said to be successful when rice production is at least the same as production before implementation of Minapadi system. The ability of land in production is better when implementing Minapadi system. Efforts to increase productivity are achieved through improving soil conditions and intensive handling of pests and weeds [6].

The high productivity of rice in the P1 treatment is suspected to be due to the pellet feed when the pellets given were not consumed by the shrimp (*Macrobracium rosenbergii*) because there was an abundance of natural feed that was preferred by shrimp (*Macrobracium rosenbergii*). Thus the finished pellets will accumulate to become fertilizer which was readily absorbed by the rice plants. According to [7], nitrogen which is the main nutrient for plant growth, is absorbed by plant roots in the form of NO3- (nitrate) and NH4+ (ammonium).

3.1.1 Relative Growth Rate

The results of the relative growth rate of giant prawns in each treatment and replication for 90 days of rearing period are presented in Figure 2.

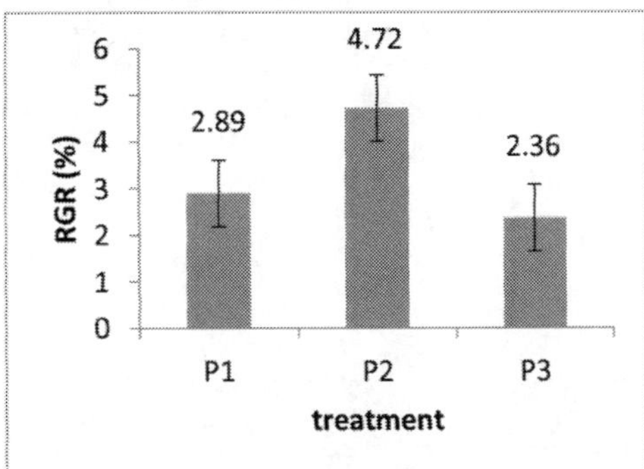

Fig. 2. Relative growth rate of giant prawns in the three treatments

Growth is a combination of behavioral processes and physiological processes. The physiological process is the process of using the consumed feed, while the behavior is the way of consuming the feed. The amount of food consumed by giant prawns is influenced by species, age, size, food availability, nutritional value, and environmental conditions [8].

Figure 2 show that the lowest relative growth rate value in this study was the P3 treatment with a value of 2.36%. Meanwhile, the highest relative growth rate was found in P2 treatment with a value of 4.72%. The results of this study indicate that different stocking densities in the Minapadi system play a role in supporting the growth of giant prawns.

The stocking density of 200 ind/plot was the most effective with a size of 50 m^2, compared to the stocking densities of 100 and 300 ind/plot. The higher stocking density will affect the competition for feed consumption, oxygen and space for amongst the shrimp (*Macrobracium rosenbergii*). Therefore, the stocking density of moderate P2 was more effective. On the other hand, the stocking density of 100 ind/plot is the least with the growth being less effective. It is suspected that the feed in the form of pellets given is was not consumed by the shrimp (*Macrobracium rosenbergii*), because there was an abundance of natural feed which was preferred by the shrimp (*Macrobracium rosenbergii*). Thus the remaining pellets will accumulate into absorbed fertilizer by the rice plants. This can be seen in the productivity of the rice plants in P1 treatment which was higher than other treatments (Figure 1).

Some aspects of physiological processes related to individual growth of the giant prawns include regeneration, metamorphosis and moulting. Moulting is the process of periodically releasing old shells and the formation of new, larger shells. In crustaceans, growth occurs periodically after moulting. Body weight gain will be hampered if it is not preceded by the moulting process [9].

A high growth rate is characterized by a faster molting process. The process of transferring minerals from hemolymph to shrimp shells requires a lot of energy. This large energy requirement is obtained from the feed consumed. Feeding is an energy requirement to support a faster growth rate. This is closely related to growth where if feed consumption is high, there is a lot of energy available that can be used for various life needs and by reducing energy expenditure, so that the portion of energy available for growth is greater. If the physiology of shrimp goes well, including metabolism, then the utilization of feed is more efficient which ultimately increases growth [10].

4 Conclusion

Different stocking densities affect the growth of giant prawns. The best stocking density of 200 ind/plot which resulted in the maximum growth of giant prawns at 4.72 % and the average rice productivity is 5.5 kg/plot.

Acknowledgments. The author's research was funded by internal research grants from Teuku Umar University, especially LPPM and Quality Assurance as well as the Aquaculture Study Program, Faculty of Fisheries and Marine Sciences.

References

1. E. Erlangga. J. BPT. 44, 1 (2016)
2. KKP Indonesia.. Minapadi System Fish Cultivation. Directorate General of Aquaculture. Jakarta, (2018)
3. M. Mahendra, F. Saputra, C.D. Febrina, D. Islama. J. KAM. 3,2 : 286 – 298, (2019)
4. E. Silamat, Yuwana, M.Z. Yuliarso. ISSN: 1412-8837, (2014)
5. D.R. Julistia. Analysis of Minapadi Farming Income in Margoluwih Village, Seyegan District, Sleman Regency. [*Essay*]. Bogor Agricultural University, (2017)
6. L.F.I. Tysno,. Analysis of Minapadi Farming Income. [*Essay*]. Bogor: Bogor Agricultural University, (2018)
7. E. Kaya. J. Agrologia, 2, 1 (2013)
8. T. Kadarini. Effect of Salinity and Calcium on Survival and Seed Growth of Balashark (*Balanthiocheilus melanopterus*). Thesis. Bogor Agricultural University, (2009)
9. R. Affandi, U.M. Tang. Physiology of Aquatic Animals. Riau University, Riau, (2002)
10. A.B. Zaidy. Utilization of Aquatic Calcium in the Molting Process and Its Consequences for the Growth of Giant Prawns (*Macrobrachium rosenbergii* de Man). Dissertation. Bogor Agricultural University, (2007)

Growth of green mussel (*Perna viridis* L.) culture at two different depths in estuary waters of Alue Naga Village, Banda Aceh

Dedi Fazriansyah Putra[1]*, *Akmal* Rizqullah[1], *Adli* Waliul Perdana[1]

[1]Department of Aquaculture, Faculty of Marine and Fisheries, Universitas Syiah Kuala, Aceh, Indonesia.
Jl. Teuku Nyak Arief No.441, Kopelma Darussalam, Kec. Syiah Kuala, Kota Banda Aceh, Aceh. 23111

Abstract. Green mussel (*Perna viridis* L.) is one of economically valuable shellfish that can be potentially developed to support the people economic income. This study aimed to investigate the growth performance of green mussels at two different groups. Two groups of 15-individual green mussels' culture was cultured repeatedly forthree times.. The two groups treatment used were 0.9 m depth and on the surface of the water. TThe growth performances were quantified in terms ofweight gain, length gain, daily growth rate, daily length growth rate, specific growth rate, specific length rate. Result showed that green mussels with a depth of 0.9 m and water surface were weight gain, length gain, daily growth rate, daily length growth rate, specific growth rate, specific length rate by values of 10.33±0.43 g, 38.12±2.02 mm, 0.11±0.004 gr/day, 0.39±0.02 mm/day, 1.60±0.11 %/day, and 0.80±0.02 %/day and 8.09±0.21 g, 33.66±0.92 mm, 0.08±0.002 g/day, 0,34±0.01, 1.27±0.03 %/day and 0.74±0.02 %/day, respectively. The growth performance of 0.9 m depth culture were significantly better that water surface culture. Therefore, it is concluded that the 0.9 m of green mussel culture is suggested to obtain the better growth performance.

1 Introduction

Mussels are belong to one of fisheries product that can be found in tropical regions and are a source of animal protein and have an economic price among the people. According the Ministry data [1], the production of aquaculture since 2014 to 2018 increased by 3,36% a year, one of its product, shellfish that reached a production of 1 million tons a year. Green mussel (*Perna viridis*, Linnaeus. 1758) is known as one of the popular seafood that contribute the significant economical income for Indonesian people that can be found from Sumatera, Jawa, Kalimantan, Bali, and Sulawesi [2].

The estuaria of Alue Naga is one of the biggest coastal area located in the city of Banda Aceh. It is near to the community settlement, so it's very efficient to be managed and has a potential to improve people's economies. Nowadays, the Alue Naga reservoir is used by surrounding villagers as a location for fishing and marine cultivation such as oysters and green mussel.

Green mussels can be found everywhere around the Alue Naga estuaries but they are very few in number. It should be noted that local people take green mussels in all sizes, both small and large sizes. Small green mussels will be cultivated first, and will be harvested within the next 3 or 4 months. Most of the people of Alue Naga Village cultivate green mussels on the surface of the waters, but there are also those who cultivate close to the bottom of the water. Green mussels can be harvested when they are 6-9 cm length [3].

The people of Alue Naga village catch green mussels when the water conditions at low tide, and slow current. The difficulty of getting the right momentum to catch green mussels makes the fishermen take all sizes of green mussles without grade selection. According to research conducted by various authors, the morphometcric characteristics of green mussels contribute as an important reference for monitoring and evaluating the growth and condition of species, especially in natural populations in cultivated or wild environments [4].

Growth is defined as an increase in size, both lenght and weight. Growth is influenced by genetic, hormonal, and environmental factors (nutrients). These three factors work to influence each other, both in the sense of mutually supporting and interfering with each other to control developments that will occur [4]. The growth performance of green mussels in different depth in Banda Aceh coastal area has not been explored yet. Therefore, the aim of this research is to investigate the growth performance of green mussels at different depths.

* Corresponding author: dfputra@unsyiah.ac.id

https://doi.org/10.1051/e3sconf/202233901006

2 Research Methods

2.1 Place and Time

This research was conducted for fourteen weeks starting from October to December 2020 at the estuary waters of Alue Naga Village, Syiah Kuala District, Banda Aceh City, Aceh Province. Measurement of the weight and length of green mussels has been carried out directly at the cultivation site.

2.2 Research Procedure

2.2.1 Catching green mussels

The capture of mussel samples was carried out using hands and a knife as a tool to separate the green mussels from the substrate. The green mussels samples taken were as many as 90 ind.

2.2.2 Experimental design

Two different depth of green mussels' culture were repeated three times with 15 individuals each. The depth treatment used were 0.9 m depth and on the surface of the water as shown in figure 2.

2.2.3 Research container

The green mussels that have been obtained then put into six baskets that have been prepared, each basket used has a square shape with a size of (30 x 50) cm, which has been lined with nets with the intention that shells that are too small will not fall or disappear from the container. Stuffed with 15 green mussels. This study used step on the chart in one column the same waters, a total of 3 containers has been tied to the rafters at the depth of 0.9 meters and 3 containers has been again tied on the surface of the waters.

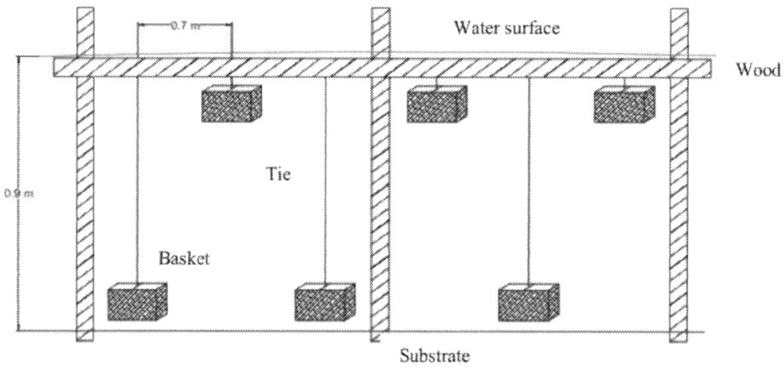

Figure 2. Illustration of a cultivation container

2.2.4 Measurement of length and weight

Measurements of the length and weight of mussels were carried out directly at the location of cultivation. In measuring the length of the shell, the tool used was a digital caliper (1 mm accuracy). Meanwhile, in measuring the total weight of shellfish, the tool used was a digital scale in grams with an accuracy of 0.1 grams.

2.3 Data analysis

2.3.1 Absolute weight gain

The absolute weight gain is the difference between the total weight at the end and the beginning of maintenance. Absolute weight growth can be calculated using the following formula [5]:

$$Wm = Wt - Wo \tag{1}$$

Information :
Wm = Absolute weight gain (gr)
Wt = Average weight at the end of the study (gr)
Wo = Average weight at the beginning of the study (gr)

2.3.2 Absolute Length Gain

The absolute length gain is the difference between the total length at the end and the beginning of the rearing. Absolute length growth can be calculated using the following formula [4]:

$$Lm = Lt - Lo \tag{2}$$

Information :
Lm = Absolute length gain (mm)
Lt = Average length at the end of the study (mm)
Lo = Average length at the beginning of the study (mm)

2.3.3 Daily Weight Rate/DWR

The daily weight rate was calculated using the following formula [6]:

$$DGR\ (W) = \frac{Wm}{t} \tag{3}$$

Information :
DGR (W) = Daily weight growth rate (gr/day)
Wm = Total weight gain (gr)
t = Maintenance time (day)

2.3.4 Daily Length Rate/DLR

The daily length rate was calculated using the following formula [6]:

$$DGR\ (L) = \frac{Lm}{t} \tag{4}$$

Information :
DGR (L) = Daily weight growth rate (mm/day)
Lm = Absolute lenght gain (mm)
t = Maintenance time (day)

2.3.5 Spesific Weight Growth Rate/SGR

Spesific Weight Growth Rate (SGR %/day) of the weight (SGR G) of green mussels, which was calculated according to [7] as follows:

$$SGR\ (W) = \frac{LnWt - LnW0}{t} \times 100\% \tag{5}$$

Information : W_0 = Initial weight (gr)
W_t = Final weight (gr)
t = Length of cultivation of green mussels

2.3.6 Spesific Length Growth Rate/SGR

Spesific Length Growth Rate (SGR %/day) of the length (SGR L) of green mussels, which was calculated as [8,9]:

$$SGR\ (L) = \frac{LnLt - LnL0}{t} \times 100\% \tag{6}$$

Information: L_0 = Initial length (mm)
L_t = Final length (mm)
t = Length of cultivation of green mussels

2.3.7 Survival Rate (SR)

Survival rate (SR) is calculated based on the following formula [7]:

$$SR = \frac{No - Nt}{No} \times 100\% \tag{7}$$

Information:
SR = Survival rate (%)
Nt = Total of shrimp at the end of the study (tails)
No = Total of shrimp at the beginning of the study (tails)

2.4 Data analysis

Data analysis of the effect of treatment on research parameters using the *Independent samples T-test..* Basis for decision making test *Independent samples T-test* with 95% confidence level . If the value is significant (2-tailed) > p-value (0.05), then accept H0 and reject H1. If the value is Significant (2-tailed) < p-value (0.05), then reject H_0 and accept H_1.
Hypothesis:
H_0 = The growth of treatment A and treatment B was not significantly different.
H_1 = The growth of treatment A with treatment B there is a significant difference

3 Results and Discussions

The results showed that the absolute weight gain during the fourteen weeks of the study period in treatment (A) with a depth of 0.9 meters on average was 10.33±0.43 grams, and in treatment (B) or on the surface an average of 8.09±0.21 grams. While in treatment (A) with a depth of 0.9 meters the average absolute length increase was about 38.12 ± 2.02 mm, and in treatment (B) the average was about 33.66 ± 0.92 mm. These results can be seen that the cultivation treatment with different depths has an effect on the speed of growth, the treatment located at a depth of 0.9 meters experienced faster growth (figure 3).

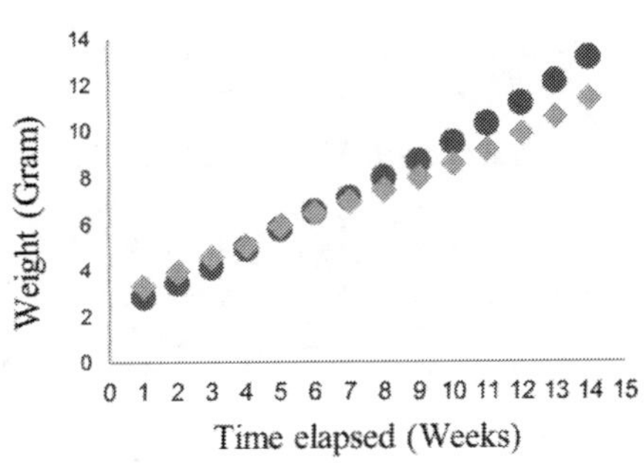

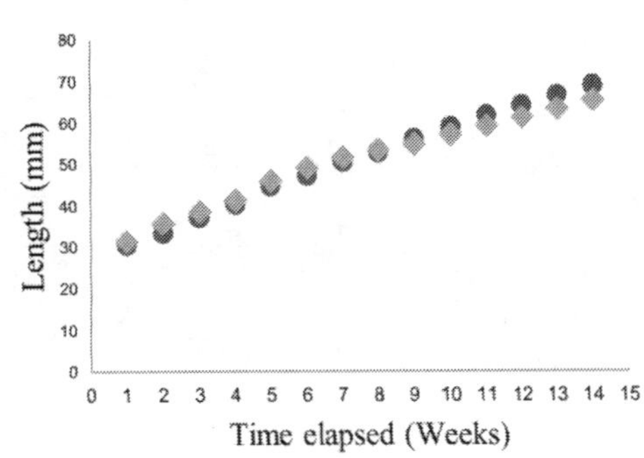

Fig.3. Weight and length growth trend (Treatment A)

The results of research on the growth of green mussels have been presented in table 1 as follows:

Tabel 1. The measurement results of research parameters

Treatment	Wm (gr)	Lm (mm)	DGR (W) (gr/day)	DGR (L) (mm/day)	SGR (W) (%/day)	SGR (L) (%/day)	SR (%)
Treatment (A)	10,33 ± 0,43	38,12 ± 2,02	0,11 ± 0,004	0,39 ± 0,02	1,60 ± 0,11	0,80 ± 0,02	100
Treatment (B)	8,09 ± 0,21	33,66 ± 0,92	0,08 ± 0,002	0,34 ± 0,01	1,27 ± 0,03	0,74 ± 0,02	100

The growth rate of daily weight at a depth of 0,9 meters averaging 0.11 ± 0.004 grams/day, and daily weight growth rate in surface waters by an average of 0.08 ± 0.002 gram/day. The average daily length growth rate at a depth of 0.9 meters was 0.39±0.02 mm/day, and the daily length growth rate at the surface of the waters averaged 0.34±0.01 mm/day. The growth rate of a specific weight at a depth of 0,9 meters averaging of 1.60 ± 0.11%/day, and the rate of postpartum harbor waters specific gravity on the surface an average of 1.27 ± 0.03%/day. The growth rate of a specific length at a depth of 0,9 meters averaging of 0.80 ± 0.02%/day, and the rate of growth of a specific length on surface waters by an average of 0.74 ± 0.02%/day.

The treatment of green mussel cultivation for fourteen weeks which experienced the best weight and length gain was found in treatment (A) with an absolute weight gain of 10.33 ± 0.43 g, and an absolute length increase of 38.12 ± 2.02 mm. The daily weight growth rate was 0.11±0.004 g/day, and the daily length growth rate was 0.39±0.02 mm/day. The specific weight growth rate was 1.60±0.11%/day, the specific length growth rate was 0.80±0.02%/day, and the *survival rate* (SR) was 100%. This is different from the results of the study of [10], green mussel cultivation carried out for 60 days with different stocking densities with a depth of 1.5 meters showed the best weight and length gain with 20 individuals with a specific weight growth rate of 1.18±0.04%/day and the specific length growth rate was 0.86±0.01%/day.

The final weight at the fourteenth week of the results of this study had an average weight of all green mussels ranging from 12.22±1.11 g/head with an average length of 66.87±2.53 mm. This is in accordance with the results of research by [3] which states that the maximum length of green mussels can reach more than 10 cm. The increase in length from this study has an average length of 2.5 mm per week or 1 cm per month. This is reinforced by the results of research [11] namely the effect of different types of rope on the level of attachment of green mussel seeds to obtain optimal results the best average length of green mussels on natural fibers with a length of 0.5-1 cm in a month. This is reinforced by the results of the research by [3] that green mussels can grow between 0.5 to 0.9 cm in a month, but there is an abnormal growth rate with a growth speed of 1.5 cm in a month against the moon. third and fourth from the beginning of cultivation, and the size of green mussels ready for harvest ranged from 6 to 9 cm. This is different from the results of Andriyani [12], the growth rate of green mussels with a depth of 3 m, 5 m, and 10 m cultivated offshore, the best results were obtained at a depth of 5 meters with an increase in length of 1.3 cm in a month. while at a depth of 3 m and 10 m each experienced an increase in length of 1 cm in a month.

The deeper the waters, the less light that enters the waters, so this can also be a different factor in the growth speed of the green mussels. This study shows that on the surface of the waters green mussels grow slower than those at a depth of 0.9 meters. This is also supported by the statement of [13], so that the shells of clams are more open and the food filtering process can maximize the shellfish both in waters where the light intensity is not too high, but if the waters are too deep it can result in waves that are too big.

Tabel 2. Water quality parameters data

No	Water Quality Parameters	Unit	Measurement results
1	Current	m/s	17
2	Brightness	Meter	0,7
3	Salinity	Ppt	31
4	Depth	Meter	Low tide= 0,9 High tide= 1,7
5	pH	-	8,7
6	Temperature	°C	29
7	Dissolved Oxygen	Mg/L	4,8

One of the water quality parameters used was brightness, the brightness is determined visually to measure the transparency in a waters, the tool used is a sechi disk. The unit used to measure brightness is the meter. Brightness was measured in this study is 0,9 meters with sandy mud waters substrate condition. This brightness level is also due to the condition of the waters, the presence of rivers and *run-off* from land which empties into the research location so that it can affect the brightness of the waters. This is reinforced by the results of research by [14], the level of turbidity in waters is influenced by organic and inorganic materials, dissolved particles, and the amount of sediment supply through *run-off* flows from the mainland.

One of the parameters that determine the ecosystem of an aquatic ecosystem is dissolved oxygen (DO). Measurement of dissolved oxygen content in the farming area of green mussels are 4.8 mg/L. According to water quality standards from the [15], marine biota is > 5 mg/l for the survival of biota and for optimal development. According to [16], this is due to the small number of aquatic plants that act as the largest oxygen supplier in growing waters.

Factors that also affect the survival of green mussels are temperature, high temperatures will have an impact on growth and survival. The result of the measurement of the temperature parameter is 29°C. According to [5], a good place for green mussel cultivation is waters that have a natural temperature ranging from 23-32°C.

According to [17], the pH of the waters greatly affects the life of the biota that live in these waters, changes in the pH itself can affect the activity and growth of these biota, because a pH that tends to be acidic will have a low DO. The results of measuring pH using a pH meter in this study were 8.7 and still classified as good for the growth of green mussels. The pH of the waters that are suitable and suitable for use as a place for cultivating green mussels are waters that have a pH ranging from 6.5 to 9 [17].

An important factor that affects the success of marine cultivation, one of which is green mussels is salinity. The measurement results show that the salinity in the waters of the study area is around 31 ppt. The salinity at the location of this study is classified as good for green mussel cultivation, this is supported by the statement of [18] which states that, for the selected location for green mussel cultivation, the salinity is classified as suitable between 27-32 ppt.

4 Conclusions

The test results *Independent samples t-test* to the p rowth green mussel treatment (A) into an 0.9 m by treatment (B) in the surface waters significantly affect weight gain, the length , the growth rate of daily weight, daily length growth rate, the rate of specific weight growth and specific length growth rate. Treatment (A) with a depth of 0.9 meters was the best with a final weight gain of 10.3 3 ± 0.43 g, and a final length increase of 38.12 ± 2.02 mm. The daily weight growth rate was 0.11±0.004 g/day, and the daily length growth rate was 0.39±0.02 mm/day. The growth rate of the specific weight of 1.60 ± 0.1 1 %/day, and the rate of growth of a specific length of 0.80 ± 0.02%/day.

References

1. KKP, Laporan Tahunan Kementerian Kelautan dan Perikanan (2018)
2. H.A.W. Cappenberg, Oseana, 33, 1 (2008)
3. Yonvitner, S. Sukimin, J. Ilm. Pendidik Biol. (Biologi Edukasi), 1, 2 (2009)
4. S.M. Aban, F. Albert, T. Argente, R.S. Raguindin, A.C. Garcia, C.E. Ibarra, PSU J .Nat Allied. Sci, 1, 1 (2017)
5. H. Effendi, Telaah Kualitas Air Bagi Pengelolaan Sumber Daya Dan Lingkungan Perairan (2003)
6. I. Rusydi, I. Dewiyanti, M. Maisuri, D.F. Putra, C. Octavina, N. Nurfadillah, IOP Conf. Ser. Earth Environ. Sci, 1 (2021)
7. D.F. Putra, M. Muhsinah, I.I. Arisa, IOP Conf. Ser. Earth Environ. Sci, 1 (2021)
8. K.D. Hopkins, Reporting Fish Growth: A Review of the Basics, J. World Aquac. Soc, 23, 3 (1992)
9. G. Naik, M.G.G. Aziz. M, V. Ambily, S.B.B. Nandan, J.V. Thomas, B.M. Kurup, India, Indian J. Fish, 5, 2 (2013)
10. A. Sagita, R. Kurnia, S. Sulistiono, J Ris. Akuakultur, 12, 1 (2017)
11. Sulvina, N.M. Noor, H, Wijayanti, S. Hudaidah, e-Jurnal Rekayasa dan Teknol. Budid. Perair. 4, 1 (2015)
12. S. Andriyani, Undergraduate thesis, UIN Sunan Ampel Surabaya (2019)
13. I.G.Y. Wisnawa, JST, Jurnal Sains dan Teknol, 2, 2 (2014)
14. B. Hamuna, R.H.R. Tanjung, S. Suwito, H.K. Maury, A. Alianto, J Ilmu Lingkung, 16, 1 (2018)
15. Menteri Lingkungan Hidup RI, Keputusan Menteri Lingkungan Hidup No 51 Tahun 2004 Tentang Baku Mutu Air Laut, Lembaran Negara Republik Indonesia, 51 (2004)
16. M. Ali, H.W. Maharani, S. Hudaidah, H. Fornando, Mispari J, 7, 2 (2015)
17. J.S.S. Hikmah, H.L. Ika, Res. J. Life Sci, 2, 1 (2015)
18. TPW. Indonesia, Budidaya Kerang Hijau (*Perna viridis*) (2015)

Ontogeny and morphological development of Tilapia (*Oreochromis niloticus*) larvae in Aceh Province, Indonesia

Ismarica Ismarica, *Dedi* Fazriansyah Putra*, and *Ulfa* Khairina

Department of Aquaculture, Faculty of Marine and Fisheries, Universitas Syiah Kuala, Banda Aceh, Indonesia

Abstract. Tilapia (*Oreochromis niloticus*) is one type of fish that has high economic value and is widely consumed by the community. Therefore, the growth and development of tilapia larvae need to be considered. This study aims to determine the growth and development of tilapia larvae. The parameters in this study were observations of total larval length (TL), daily growth rate (LPH), and morphology of tilapia larvae. This research was conducted for 30 days. Measurements of total length growth and morphological development of tilapia larvae were measured on days 1, 5, 10, 15, 20, 25 and 30 DAH (Days After Hatching). Data were analysed descriptively. The results of this study showed that the growth of tilapia larvae from day 1 to day 30 increased. The development of tilapia larvae in the yolk sac phase occurred on day 1-7 and the transition phase (from endogenous feeding to exogenous feeding) occurred on day 6-7 DAH, and on day 8 the yolk sac had been exhausted. The morphological development of tilapia larvae began to form completely on day 10 DAH.

1 Introduction

Tilapia (*Oreochromis niloticus*) is a type of fish with high economic value and is largely consumed by the society. According to [1], Tilapia (*Oreochromis niloticus* L.) is one of the freshwater fish that has a high selling value. *Oreochromis niloticus* is one of the economic commodities in Indonesia [2]. Tilapia cultivation is much done by the community. However, the cultivators have some problems on the larval stage maintainance. In hatchery, larval stage is one of the critical factors, in which there are still many obstacles in increasing the growth [3]. The larval stage is divided into 2 stages, namely the pralarva and postlarva stages. According to [4], at the prelarva stage the fish body is still transparent and then at the postlarva stage the fish body has developed to be complete. [5] revealed that immature larval growth resulted in mortality. The mortality that occurs is due to the immature morphology of fish larvae so that the larvae are susceptible to disease, and the critical period of larvae during the transition from egg yolk to food from outside (*exsogenous feeding*).

Increased growth and development of larvae is very important for larval survival and success in cultivation. The early development of fishes is usually regarded as a steady accumulation of small changes to produce an adult phenotype [6]. According to [7], The growth rate of tilapia larvae was greater in larvae reared at a temperature of 31°C (day): 25°C (night) due to increased synchronization of digestion and feed utilization. Increased growth of tilapia larvae is associated with high feed conversion efficiency [8]. In addition to growth, the rapid development of larval morphology is also important. According to [9] that rearing at 30°C has faster embryo development and hatching time for orange-spotted grouper eggs. Optimal growth and fastest morphogenesis in the interaction treatment of 30°C temperature with 3‰ salinity [10]. Therefore, this study of growth and morphological development of tilapia aims to determine the growth and morphological development that supports the success of rearing tilapia larvae.

2 Materials and Methods

2.1 Tilapia larvae spawning and rearing

Parent male and female tilapia were obtained from the Regional Technical Implementation Unit or Unit Pelaksana Teknis Daerah (UPTD) Fish Seed Center or Balai Benih Ikan (BBI) Lukup Badak, Pegasing District, Central Aceh. In tilapia spawning, the feed used was added with vitamin E with the aim of stimulating the gonad maturity of the tilapia parent. The vitamin E used were as much as 5grams / 1 kg of feed. Vitamin E was first dissolved with alcohol then mixed into the feed, then dried. Feeding was carried out for 14 days with a frequency of 2 times. The spawning container were a 3m x 5m concrete pond, and a 30cm x 20cm x 25cm pond as larval rearing container filled with 10 L of water and equipped with aeration. The natural feed given in the early stages of larvae comes from green water in tilapia rearing ponds with a frequency of feeding 3 times a day. Furthermore, feeding with flour pellets was carried out.

* Corresponding author : dfputra@unsyiah.ac.id

2.2 The Total Length of Larvae

The total length of larvae was measured on day 1, 5, 10, 15, 20, 25 and 30 DAH using a digital calliper (mm) with an accuracy of ± 0.2 mm. 5 larvae were taken randomly every day to measure the total body length, then the average total length was calculated with ± standard deviation (SD). Furthermore, the pattern of growth in the length of tilapia fish larvae was described using Microsoft Excel.
The daily growth rate (LPH) was calculated according to [11]:

$$LPH = Lt\text{-}Lo/t \quad (1)$$

Information:
LPH = Daily growth rate (mm/day),
Lt = Average total length at the end of the study (mm)
Lo = Average total length at the beginning of the study (mm)

2.3 Larvae Morphological Observations

Sampling of larvae for morphological analysis was carried out on day 1, 5, 10, 15, 20, 25 and 30 DAH (Days After Hatching). A total of 15 samples were taken on each observation day from the larval rearing container, then put into a sample bottle and preserved with 10% NBF (Neutral Buffer Formalin). Morphological analysis, 5 larvae samples were taken randomly from the sample bottle, then observed under a stereo microscope and photos were taken.

3 Results and Discussion

3.1 The total Length of Tilapia (*Oreochromis niloticus*) Larvae

Results of the daily growth rate of tilapia larvae was 0.6 mm/day. Newly hatched larvae had an average total length of 7.3±0.7 mm at 1 DAH, and the average total length at 30 DAH was 26±1.1 mm. Growth in larval length for 30 days is shown in (Figure 1).

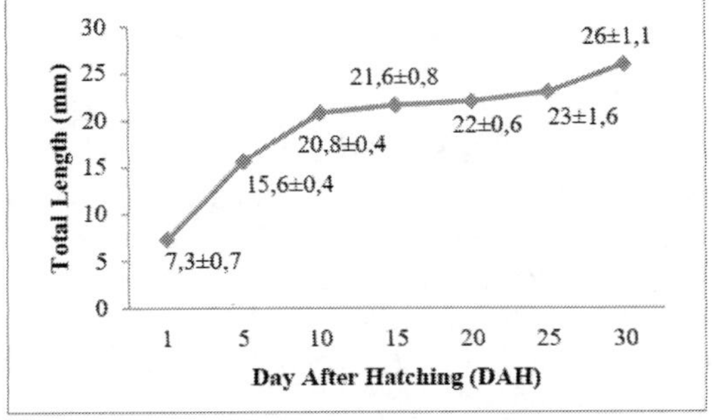

Fig. 1. The growth pattern of tilapia larvae length from 1 DAH to 30 DAH

The results of the growth of tilapia larvae length from the first day of hatching to 30 DAH increased along with the addition of the age of tilapia larvae. In Figure 1, the graph shows that the growth pattern of tilapia larvae length is exponential. This absolute length exponential growth pattern also occurs in white tilapia larvae using different doses of the thyroxine hormone [12]. This absolute length exponential growth pattern also occurs in white tilapia larvae using different doses of the thyroxine hormone The larval length growth of Rainbow Kurumoi (*Melanotaenia parva*) for 7 days increased along with the increasing age of the fish [13]. On gift tilapia, the growth performance of fish larvae increased because the larvae were fed with lipid levels of 4.5%, 8% and 11.5% [14].

The length growth of tilapia larvae at the yolk sac stage (1-5 DAH) was 7.3±0.7 mm – 15.6±0.4 mm. The growth of the average length of tilapia larvae increased at the yolk sac stage. The yolk sac stage ends on day 7, when the yolk reserves are almost depleted. The volume of yolk decreases in the larval phase is closely related to the development and growth of larvae [15]. Before the yolk of silver arowana fish runs out, silver arowana fish must be fed to avoid gaps in the utilization of feed from inside and outside, and to increase the growth and morphological development of silver arowana fish larvae [10]. In the preflexion to postflexion stage, the length growth is increasing. The growth in length of larvae increased due to the good use of natural feed.

3.2 Development of Tilapia Larvae Morphology

The results of observations regarding the morphological development of tilapia larvae were divided into 4 stages based on the structural development of the body of tilapia larvae (Table 1). The stages of larval development are *Yolk sac, preflexion, flexion, and postflexion*. *Yolk sac* stage of Tilapia larvae was on day 1-7 DAH and at this stage the yolk reserves began to run out. The *reflexion* stage of tilapia larvae was on day 8-14 DAH and at this stage the larvae start to eat natural feed. Tilapia larvae *flexion* stage was on day 15-20 DAH. *Postflexion* stage of tilapia larvae was on day 21-30 DAH.

Table 1. Stages of morphological development of tilapia larvae (*Oreochromis niloticus*)

No	Stages	Characteristics of Morphological Development	DAH
1.	*Yolk sac*	The body was still transparent. The fins were not formed, the mouth was closed, the eyes and skin were not pigmented, the tip of the caudal fin was still rounded, and the dorsal fin, anal fin and abdominal fin were undifferentiated.	1-3
		The body was still transparent, the mouth began to open, the eyes began to become pigmented, the tip of the caudal fin as still rounded, the pectoral fins were formed, while the dorsal fins, anal fins, and pelvic fins (abdominal fin) were undifferentiated	4-7

No	Stages	Characteristics of Morphological Development	DAH
2.	*Preflexion*	Yolk reserves were running out, body organs such as fins were almost perfect, the tailbone began to bend.	8-14
3.	*Flexion*	Pigmentation began to appear on the head, abdomen and fins.	15-20
4.	*Postflexion*	The dorsal, pectoral, abdominal and tail fins were perfect so that the larval movement was fast.	21-30

Based on the results obtained from morphological observations of tilapia larvae (Tabel 1), larvae aged 1-3 DAH have a transparent body, fins are not formed, mouth is closed, eyes and skin are not pigmented, the tip of the caudal fin is still rounded, and the dorsal fin, anal fin and abdominal fin are undifferentiated. On the larvae of the Rainbow Kurumoi fish the fins have been formed including the tail fin but the shape is not perfect, the tail and dorsal fins are fused and only the pectoral fins are formed, all fins are perfect and the caudal fin is C-shaped [13]. Nike fish (*Awaous melanocephalus*) larvae on the first day of hatching have a transparent body, smooth and thin pectoral and dorsal fins, black eye spots adjoining the white part, while on the next day all fins are in the developing phase [4]. On *Cyprinus carpio*, mouth and the anus does not open, the pectoral fins are formed, and the membranous fins are formed from cephalus top to bottom stomach.

On larvae aged 4-7 DAH has transparent body, the mouth begins to open, the eyes begin to become pigmented, the tip of the caudal fin is still rounded, the pectoral fins are formed, while the dorsal fins, anal fins, and pelvic fins (abdominal fin) undifferentiated. According to [16], on nilem fish (*Osteochilus hasselti C.V.*) which has a transparent body, unpigmented skin, unilobed vesica metatoria, mouth is open, intestine is straight (not filled), the pectoral fin has been formed, the apical end of the caudal fin is still rounded, while the dorsal fin, anal fin and abdominal fin have not been differentiated.

Tilapia larvae on day 8-14 DAH have almost perfect body organs. On day 15-20, pigmentation begins to appear on the head, abdomen and fins of tilapia larvae. On days 21-30, the dorsal, pectoral, abdominal and tail fins of the tilapia larvae are complete so that the larval movement is fast. According to [4], at the postlarva stage, the morphological development of the fish body is complete.

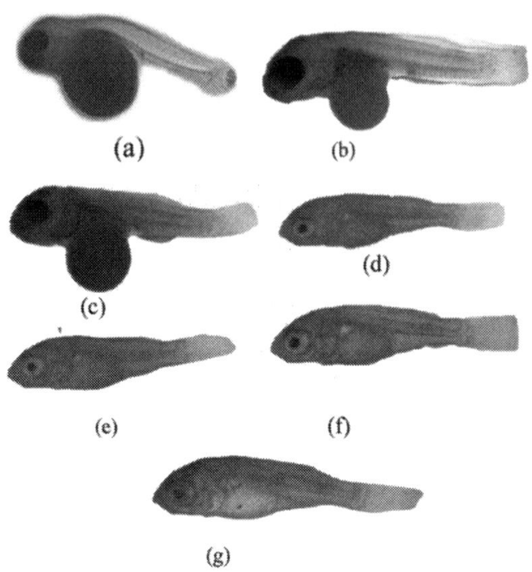

Fig. 2. Morphological development of tilapia larvae (*Oreochromis niloticus*). a) 1 DAH (7,3±0,7 mm) ; (b) 5 DAH (15,6±0,4 mm); (c) 10 DAH (20,8±0,4 mm); (d) 15 DAH (21,6±0,8 mm); (e) 20 DAH (22±0,6 mm);(f) 25 DAH (23±1,6 mm); (h) 30 DAH (26±1,1 mm).

4 Conclusion

The length growth of tilapia (*Oreochromis niloticus*) larvae from 1 DAH to 30 DAH increased along with the increasing age. Tilapia pre-larvae phase on day 1-7 DAH and post-larvae phase on 8-30 DAH. The yolk sac stage occurs at 1-7 DAH, the pre-flexion stage occurs at 8-14 DAH, the flexion stage occurs at 15-20 DAH, and the post-flexion stage occurs at 21-30 DAH.

Acknowledgments. We thank to Mr. Iwan Hasri and all BBI Lukup Badak staff due to their support and assistances during the research. The research was supported by H-Indeks Research grant Number : 169/UN11/SPK/PNBP/2021 tanggal 19 Pebruari 2021.

References

1. S. Paramadina, J. Triastuti, and A. T. Mukti. Eco. Env. & Cons. **26 (1)** (2020)
2. A. Saifulloh. IOP Conference Series (Earth and Environmental Science). **236(1)** (2019)
3. Z. A. Muchlisin. Songklanakarin J. Sci. & Technol. **35(4)** (2013)
4. N. Pasisingi, A. H. Olii, and S. A. Habibie. Tomini J. Aquat. Sci.**1(1)** (2020)
5. R. Melianawati and R. Pratiwi. J Ris Akuakultur, **6 (1)** (2011)
6. D. F. Putra, A. B. Abol-Munafi, Z. A. Muchlisin, and J. C. Chen, Aquaculture, Aquarium, Conservation & Legislation. **5 (1)** (2012)
7. A. H. E. Santo, G. D. Alba, Y. D. S. Reis, L. S. Costa, F. J. S. Vázquez, R. K. Luz, P. A. P. Ribeiro, and J. F. L. Olmeda. Aquaculture **528** (2020)

https://doi.org/10.1051/e3sconf/202233901007

8. J. Maithya, N. M. Mbithi and P. Wanjala. Int. J. Fish. Aquac. **9(2)** (2017)
9. A. Andriyanto and A. Marzuki. J Ris Akuakultur. **8(1)** (2013)
10. Y. W. T. Mulyani, D. D. Solihin and R. Affandi. Jurnal Iktiologi Indonesia. **15(3)** (2015)
11. K. D. Hopkins. JWAS. **23 (3)** (1992)
12. R. Andriawan, F. Basuki, and T. Yuniarti. Jurnal Sains Akuakultur Tropis. **4 (1)** (2020)
13. L. Sholichah. and E Yunita. IOP Conference Series (Journal of Physics. Ser)**1097 012035** (2018)
14. J. Qiang , J. He, H. Yang, Y.-L. Sun, Y.-F. Tao, P. Xu, and Z.-X. Zhu. Aquaculture Research **1–14** (2016)
15. T. Kadarini, M. Zamroni, and E. K. Pambayuningrum. Jurnal Riset Akuakultur, **8(1)** (2013)
16. D. H. Yusuf, Sugiharto, and G. E. Wijayanti. Scripta Biologica, **1(3)** (2014)

https://doi.org/10.1051/e3sconf/202233901008

Polyculture of tilapia (*Oreochromis niloticus*) and *lemeduk* (*Barbonymus schwanenfeldii*) in floating net cages as a strategy for utilizing natural food

Nurfadillah Nurfadillah[1,2,3*], *Iwan* Hasri[4], and *Fahrial* Fahma[1]

[1] Department of Aquaculture, Faculty of Marine and Fisheries Universitas Syiah Kuala, Banda Aceh, Indonesia
[2] Laboratory of Marine Biology, Faculty of Marine and Fisheries, Universitas Syiah Kuala, Banda Aceh
[3] Research Center for Marine and Fishery, Universitas Syiah Kuala, Banda Aceh, Indonesia
[4] Department of Fisheries Aceh Tengah, Aceh Tengah

Abstract. Polyculture is a sustainable aquaculture system because it can reduce environmental pollution and increase feeding efficiency. The purpose of this study was to analyze the performance of growth, biomass and food competition for tilapia as the main commodity and *lemeduk* fish as a secondary commodity with a polyculture system in floating net cages. The research method used a Completely Randomized Design (CRD) consisting of 5 treatments and 3 replications, namely: treatment A (60 tilapia fish), Treatment B (40 tilapia fish + 20 *lemeduk* fish), Treatment C (30 tilapia fish + 30 *lemeduk* fish), Treatment D (20 tilapia fish + 40 *lemeduk* fish), and Treatment E (60 *lemeduk* fish). The research container uses floating net cages with a size of 0.5 × 0.5 × 1m. Parameters observed were absolute length growth, absolute weight growth, specific growth rate, biomass production, competition index, and water quality. The results showed that the polyculture system of tilapia and *lemeduk* at different stocking densities had a significant effect on absolute length growth, specific growth rate, biomass production, and had low competition between tilapia and *lemeduk*. The best treatment was found in treatment B (N 40+ L 20) and treatment C (N 30+ L 30) with absolute length growth value 3.20 cm±0.17, specific growth rate 3.83%±0.36, biomass of tilapia 94.36 g/0.5m^2±16.45 and total biomass 100.69 g/0.5m^2±5.49. Hence, the polyculture system is better than the monoculture system and the absence of competition between tilapia and *lemeduk* in using food.

1 Introduction

The growth of floating net cages in flooded waters is very rapid and often causes environmental problems. The impact of fish farming activities in cage culture can cause the water quality of stagnant waters to decrease. The decrease happened because of the leftover feed waste. So far, the majority of fish farming systems in cage culture use a monoculture system. Just one type of fish that we can keep is Tilapia. Because Tilapia is continuous demand and availability of seeds. In addition, tilapia is one type of fish that has economic value and important commodity in the freshwater fish business because tilapia has beneficial properties, such as: easy to breed, fast growth, and easy to grow in intensive aquaculture systems [1].

Polyculture is a sustainable aquaculture system because it reduces the environmental impact of activities by increasing the efficiency of feeding. A polyculture system in which one contains two or more species are raised together [2]. The benefit of using this system is a high level of land productivity. In principle, there are several things related to products that must be regulated so that there is no competition between products in obtaining their feed. Besides, each product is expected to be able to take advantage of each other so that circulation occurs in one cultivation location [3].

Lemeduk fish (*B. schwanenfeldii*) is a native fish in the inland of Aceh, one of which is Laut Tawar Lake. This fish is called *Tengadak* fish in Kalimantan and *Kapiat* in Riau. This fish is a herbivorous species [4 – 5]. Based on Pulungan [6] said that this fish is classified as an omnivorous fish and does not interfere with the types of small fish in the waters where it lives. Based on the eating habits of these fish, it is expected that they can eat the rest of the feed and natural feed derived from organic matter that is wasted in the waters. In addition, *lemeduk* fish are expected to be able to take advantage of the moss that grows in the net. Lemeduk fish is also a promising aquaculture commodity in the future due to the relatively fast growth of large fish, good meat taste, and low trophic level.

The polyculture of tilapia and *lemeduk* has never been carried out in cage culture, so it is necessary to research the polyculture of tilapia and *lemeduk* with different stocking densities. Research on polyculture has been carried out on several commodities including Windu Shrimp with Milkfish [7], *Larasati* Red Tilapia and Milkfish [8], Goldfish and *Nilem* [9], Tilapia with native fish in Laut Tawar lake (*Lemeduk*, *Depik*, and *Peres* fish) [10]. This study is to analyze monoculture

* Corresponding author: nurfadillah@unsyiah.ac.id

and polyculture systems of Tilapia (*Oreochromis niloticus)* and *Lemeduk* (*Barbonymus schwanenfeldii)* with different stocking densities in cage culture covering the performance of growth, biomass and competition of tilapia as the main commodity and *lemeduk* as a secondary commodity.

2 Methodology

2.1 Sampling and experimental design

The research was carried out at the UPTD Fish Seed xfrom August - September 2020. The research method used is the experimental method. The aim was obtained data and test the feasibility of fish farming with polyculture and monoculture systems using the floating net cages that has been proposed. The study used a one-way Completely Randomized Design (CRD), consisting of five treatments and three replications, namely N60 (60 Tilapia fish /0.25m^2), N40+L20 (40 Tilapia fish /0.25m^2 and 20 *Lemeduk* fish /0.25m^2), N30+L30 (30 Tilapia fish /0.25m^2 and 30 *Lemeduk* fish /0.25m^2), N20+L40 (20 Tilapia fish /0.25m^2 and 40 *Lemeduk* fish /0.25m^2), L60 (60 Lemeduk fish /0.25m^2).

2.2 Research procedure

This study used 15 happa containers. Fish seeds to be stocked are 1-3 cm long, then kept in 2x2 m^2 nets. The fish are acclimatized so that the fish can adapt to the environment. Feed given the fish for tilapia seeds during the study was in the form of floating type commercial pellets. The frequency of feeding was twice a day, in the morning and evening with a total of 8% of the weight of fish biomass per experimental unit. Data collection was carried out 4 times during the study with a duration of once per 10 days.

2.3 Evaluation of growth performance of *Tilapia* and *Lemeduk*

The parameters of growth performance were obtained by according to study Thongprajukaew [11] and Ferosekhan [12] using the following equations:

2.3.1. Absolute length growth

Absolute length growth is used to calculate fish length gain during rearing:

$$L = Lt - Lo \quad (1)$$

Note:
L = absolute length growth (cm)
Lt = Final average length (cm)
Lo = Initial average length (cm)

2.3.2. Absolute weight growth

Absolute weight gain was used to calculate the weight gain of fish biomass during rearing:

$$W = Wt - Wo \quad (2)$$

Note:
W = absolute weight growth (g)
Wt = Average weight of fish at time t (g)
Wo = Average weight of fish at the start (g)

2.3.3. Fish Specific Growth Rate

The specific growth rate is the % of the difference between the final weight and the initial weight, divided by the length of maintenance time:

$$SGR = \frac{LnWt - LnWo}{t} \, x \, 100\% \quad (3)$$

Note:
SGR = Specific growth rate(%/day)
Wt = final body weight (g)
Wo = initial body weight (g)
t = Maintenance time (days).

2.3.4. Biomass

Biomass is the resultant between survival and average final weight:

$$BM = Wt \, x \, Nt \quad (4)$$

Note:
BM = Biomass
Wt = Total Weight of Fish contained in the culture container (g)
Nt = Total number of fish at the end of the study (fish)

2.3.5. Competition Index

Criteria for determining the opportunity for competition between fish species:

$$IK = \frac{a-b}{a} \quad (5)$$

Note:
IK = Competition Index
a = single cultivation biomass (g)
b = Production of mixed cultivation biomass (g)

2.4 Water quality monitoring

Water quality measurements will be carried out at the beginning of the study, in the middle and at the end of the study. Measurements were carried out on day 1, day 20 and day 40. The parameters observed were temperature and degree of acidity (pH).

2.5 Data analysis

The data in this study were subjected to the one-way analysis of variant (one-way ANOVA) test and followed by Least Significant Difference tests (LSD) and Duncans test. The significant difference was determined at $P < 0.05$.

3 Result

The results of polyculture and monoculture studies of tilapia and *lemeduk* using the cage culture system showed the increase in length and weight with increasing time (Figure 1a and 1b). The results showed that the highest length increase was in the N20+L40 treatment and the lowest was in the N60 treatment.

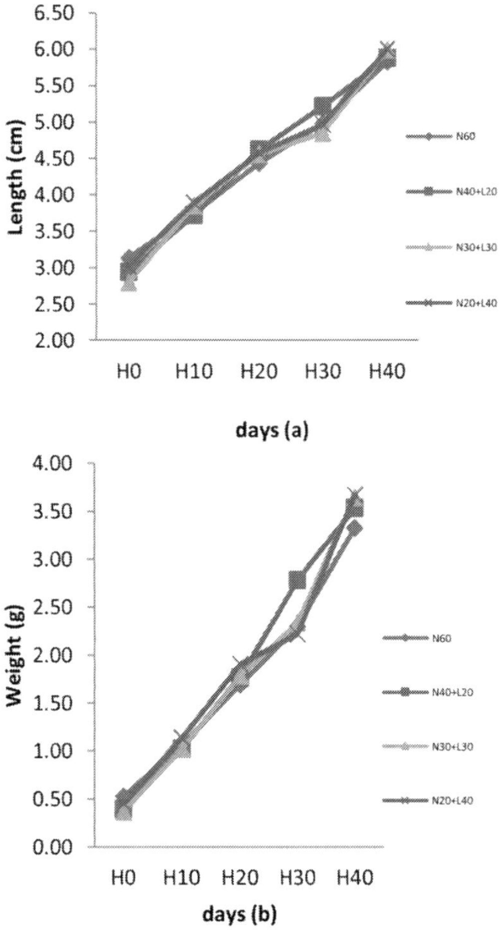

Fig. 1. Growth Performance of monoculture and polyculture of *Oreochromis niloticus* with *Barbonymus schwanenfeldii* in floating net cages (a) Increase in fish length; (b) Fish weight gain.

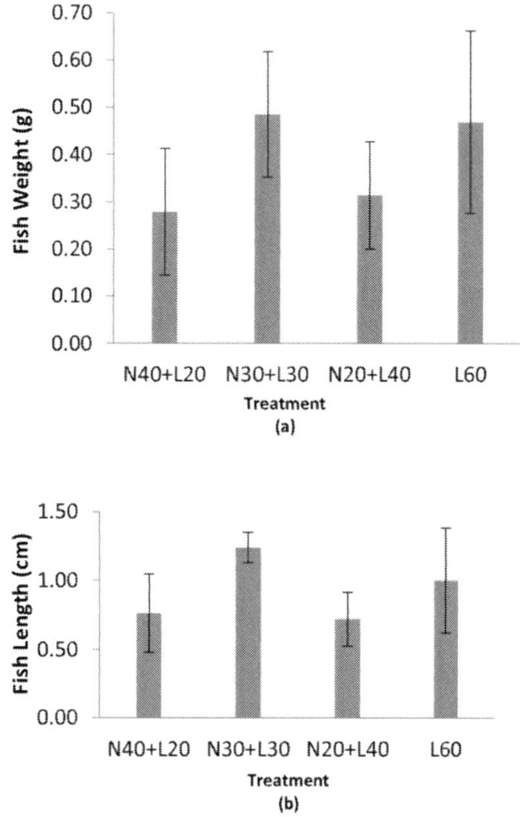

Fig. 2. (a) Absolute Weight Growth; (b) Growth of Absolute Length Monoculture and Polyculture of *Oreochromis niloticus* with *Barbonymus schwanenfeldii* in floating net cages.

The results also showed that the growth of *lemeduk* in both length and weight in polyculture and monoculture systems with tilapia in the cage culture system showed that the growth in length and weight increased with time. The highest absolute weight growth was in the treatment N30+L30 (0.48g±0.13) and the smallest was in the treatment N40+N20 (0.28g±0.13) (Figure 2a). Length growth was also highest in the treatment N30+N30 (1.24cm±0.11), then the lowest in the treatment N20+L40 (0.72cm±0.20).

Based on the results of the ANOVA test, the treatment of tilapia reared in a with *lemeduk* showed a significant effect in increasing the absolute length of tilapia, the specific growth of tilapia, and the biomass of tilapia and the total biomass of the polyculture system for each treatment ($p<0.05$). So it can be concluded that the tilapia polyculture system with *lemeduk* fish succeeded in increasing the biomass and it means that there is one treatment (polyculture) which has a significant effect when compared to the effect of the non-polyculture system and with other treatments (Table 1).

Tilapia reared by a polyculture system with *lemeduk* showed no significant effect in increasing the absolute weight gain of tilapia ($p>0.05$). So it can be concluded that the tilapia polyculture system with *lemeduk* did not succeed in increasing the absolute weight of tilapia.

The highest absolute length growth value was in the N30+L30 treatment (3.20cm±0.17) and the lowest was in the N60 treatment (2.69 cm±0.10). The results of further LSD tests showed that the treatment N30+L30 was not significantly different from the treatment N40+L20 and N20+L40 but significantly different from N60.

The specific growth of tilapia was highest in the treatment N30+L30 (3.83%±0.36) and the lowest was in the treatment N60 (2.83%±0.05). The results of the further LSD test showed that the best treatment was treatment N30+L30 not significantly different from treatment N40+L20 and N20+L40 but significantly different from treatment N60. The highest biomass of tilapia in each treatment was in the N60 treatment (94.36g/0.5m^2 ±16.45) and the lowest was in the treatment N20+L40 (41.23 g/0.5m^2 ±5.31).

The competition index obtained in this study shows that the competition index for tilapia and lemeduk is negative in all treatments (Table 2). This negative value shows that in the applied polyculture the presence of lemeduk fish does not affect tilapia, especially competition in terms of getting food, oxygen and space.

Tabel 1. Absolute length growth, absolute weight growth, specific growth of tilapia, tilapia biomass, and total biomass Monoculture and Polyculture of *Oreochromis niloticus* with *Barbonymus schwanenfeldii* in floating net cages.

Treatment	Research Parameters				
	Tilapia Absolute Length Gain (cm)	Tilapia Absolute Weight Gain (g)	Tilapia Specific Growth Rate (%)	Tilapia Biomass (g/0.25m²)	Total Biomass (g/0.25m²)
N60	2.69±0.10ᵃ	2.82±0.01ᵃ	2.83±0.05ᵃ	94.36±16.45ᵇ	100.69±5.49b
N40+L20	2.94±0.19ᵃᵇ	3.15±0.26ᵃ	3.60±0.39ᵇ	72.61±12.72ᵇ	75.85±6.96b
N30+L30	3.20±0.17ᵇ	3.34±0.42ᵃ	3.83±0.36ᵇ	78.64±18.30ᵇ	80.26±11.24b
N20+L40	2.99±0.14ᵃᵇ	3.15±0.23ᵃ	3.29±0.22ᵃᵇ	41.23±5.31ᵃ	51.66±2.16a

Note: Different superscript letters in the same column show significant differences (P<0.05)

Table 2. Results of the competition index parameters.

Treatment	Competition Index	
	Tilapia	Lemeduk
Tilapia	-	-
Tilapia 30+Lemeduk 30	-0.23	-6.47
Tilapia 20+ Lemeduk 40	-0.43	-2.98
Tilapia 40+ Lemeduk 20	-0.09	-13.36

The water quality observed in the research were temperature parameters and degrees of acidity (pH). The temperature values range from 22.5 - 23.7 °C, pH values range from 6.9 – 7.1, the values of the two parameters are still in optimal conditions for fish growth [27;28].

4 Discussion

The increase in absolute length growth and specific growth of tilapia in polyculture studies of tilapia and lemeduk with different stocking densities showed that the treatments of N30+L30, N20+L40 and N40+L20 were not significantly different. This shows that the difference in stocking density of lemeduk fish does not cause slow growth of tilapia length. It can be said that the lemeduk polyculture fish did not interfere with the growth of tilapia. The increase in the length of the lemeduk, along with the length of the tilapia, showed that the polyculture system applied to fish was able to grow the main fish and by-products simultaneously, resulting in the same growth synergism between fish species. Growth is characterized by changes in length and weight in a certain period [13].

The factors that affect the absolute length growth and specific growth of fish during the study are feed availability. The feed was given a floating type with a feed ration of 8% of the weight of the fish biomass being reared.. Factors that affect growth are feed rations and fish weight while other factors are external and internal factors [14-16]. Tilapia growth is good because the feed provided contains protein following the body's needs for both energy and growth [17-19]. In addition, the water quality in the research container is good to support the growth of pet fish. Herawati and Suantika [20] increased growth rate was related to the increasing average weight of tilapia seeds and water quality.

The absolute weight growth was not significantly different between the treatments N30+L30, N20+L40 and N40+L20. However, the absolute weight growth of fish was highest in the N30+L30 treatment. The high rate of this treatment indicates that the balanced stocking density of tilapia and *lemeduk* causes no competition so that the polyculture system allows fish to take advantage of the food available in their rearing environment. The high density of sinking water plants in the form of *Hydrilla verticilata* in the research location container also causes the availability of natural food for *lemeduk* fish to be well available. According to Mudjiman [21] statement that *Tawes* fish when they are still seeds like to eat plankton. After growing up, *Tawes* fish like to eat moss and young algae shoot in line with Muryanto and Sumarno [22] that *Tawes* fish are herbivores that eat aquatic plants, phytoplankton, molluscs, insects, periphyton, moss and detritus that are in a net container as a maintenance container. The implication is that water recirculation from and outside the net goes well so that the quality of polyculture water for tilapia and *lemeduk* is relatively good, resulting in better growth of tilapia with *lemeduk*.

Tilapia biomass and total biomass in polyculture systems with *lemeduk* were influenced by survival and fish weight at the end of rearing. The biomass of tilapia in monoculture treatment was higher than that of fish in polyculture because the density of tilapia in monoculture was higher than that of tilapia in polyculture. Tilapia biomass value is determined by survival rate, growth rate and stocking density [23-25]. Lante *et al* [23] polyculture vaname shrimp had higher production than monoculture with rostris shrimp and tiger prawns. The high biomass of *lemeduk* fish is caused because the fish feed given to tilapia can be utilized optimally by tilapia.

This study showed that with the addition of *lemeduk* there was no competition between tilapia and *lemeduk* in all different stocking density compositions because *lemeduk* fish were waiting for feed in the water column, in contrast to tilapia which took feed to the surface and then brought it to the water column. The research Nurfadillah *et al.* [10] that there is no competition between tilapia and native fish in Laut Tawar lake. Lante *et al* [23] and Bayu *et al* [26] also stated that polyculture-reared tiger prawns, Vaname and Rostris did not cause competition between shrimps reared. The absence of competition can be seen in negative values (Table 2). During the maintenance period, it can be seen that at the time of feeding there was no competition. Tilapia are more active in taking food and bringing it to the water column than *lemeduk* fish. In addition, fish that are polyculture with tilapia can take advantage of the natural feed which is available in the container.

5 Conclusion

A polyculture system is better than monoculture. The best polycultures with different stocking densities were shown in the treatment of N40+L20 and N30+L30 on the parameters of absolute length, specific growth rate, tilapia biomass and total biomass.

References

1. S. E. Wardoyo, Media Akuakultur, **2** (2007)
2. L. V. Arana, Fundamentos de aquicultura. Florianopolis, UFSC. (2004)
3. U. Kholifah, N. Trisyani, I. Yuniar, Neptunus J, **14** (2008)
4. I. I. Kusmini, R. Gustiano, I. Mulyasari, G.H. Huwoyon, *Ikan lokal tengadak (Barbonymus schwanenfeldii) asal Kali-mantan sebagai andalan untuk ikan budi daya* dalam Prosiding Seminar Nasional Ikan ke-8, 3-4 Juni 2014, Bogor, Indonesia (2015)
5. R. H. Gunawan, Z.A. Muchlisin, S. Mellisa, Jurnal Ilmiah Mahasiswa Kelautan dan Perikanan Unsyiah, **2** (2017)
6. C. P. Pulungan, The potential for kapiek fish cultivation from the Kampar River, Riau (Research Center Universitas Riau. Pekanbaru 73, 1987).
7. M. Mangampa, *Polikultur udang windu (penaeus monodon), bandeng (Chanos chanos), nila srikandi (Oreochromis aureus x o. niloticus), dan rumput laut (Gracilaria verrucosa) di tambak tanah sulfat masam (TSM)* dalam Prosiding Forum Inovasi Teknologi Akuakultur, **1**, 13 (2014)
8. I. Samidjan, D. Rachmawati, A. Indarjo, H. Panggono, *Pengembangan teknologi polikultur ikan nila merah larasati (Oreochromis niloticus) dan ikan bandeng (Chanos chanos forskal) berbasis penggunaan probiotik untuk percepatan pertumbuhan dan kelulushidupan dalam menunjang agromina kota pekalongan,* dalam Prosiding Seminar Nasional Kelautan, 27 Juli 2016, Universitas Trunojoyo Madura, Indonesia (2016)
9. A. K. Putri, Zahidah, S.A. Harahap, Journal of Marine Fisheries, **7**, 1 (2016)
10. N. Nurfadillah, I. Hasri, R. Fahmi, Misran, *The competition index and growth performance between tilapia (Oreochromis niloticus) and native fish spesies Laut Tawar Lake in polyculture system,* in IOP Conference Series: Earth and Environmental Science, 24 September 2020, Banda Aceh, Indonesia (2021)
11. K. Thongprajukaew, S. Pettawee, S. Muangthong, S. Saekhow, W. Phromkunthong, Aquaculture Research. J, **50** (2019)
12. S. Ferosekhan, S. K. Sahoo, K. Radhakrishnan, P. Velmurugan, N. Shamna, S.S. Giri, B.R. Pillai, Aquaculture, **521** (2020)
13. P. Yuliati, T. Kadarini, S. Subandiyah, Jurnal Ikhtiologi, **3** (2017)
14. M. I. Effendi, Metode Biologi Perikanan (Yayasan Dwi Sri, Foundation, Bogor, 1997)
15. B. Hepher, Nutrition of Pond Fishes, Formerly of Fish and Aquaculture Research Station, (Cambridge, University Press, 1988)
16. C. J. Jaspe, C. Caipang, B.J.G. Elle, Animal Biology & Animal Husbandry, **3**, 2 (2011)
17. C. Miroslav, T. Dejana, L. Dragana, D.J. Vesna, Tehnologija Mesa, **52**, 1 (2011)
18. H. Djajasewaka, Fish Feed, (CV Yasaguna, Jakarta, 1995)
19. S. S. De Silva, A. Anderson, Fish nurtrition in Aquaculture (The Fish Edition, Chapman and Hall, London 1995)
20. Herawati, G. Suantika, DisSainTek, **1**, 1 (2007)
21. Mujiman, Makanan Ikan, (Jakarta: CV. Simplex, 2000)
22. T. Muryanto, D. Sumarno, Jurnal Teknis BPP-KSD Jatiluhur, **11**, 1 (2014)
23. S. Lante, H.S. Suwoyo, M. Mangapa, (Balai Besar Riset Perikanan Jakarta: Pengembangan Teknologi Budidaya Perikanan, 2012).
24. R. Iskandar, Elrifadah, Jurnal Ziraa"ah, **40**, 1 (2015)
25. M. K. Anam, F. Basuki, L.L. Widowati, Jurnal sains akuakultur tropis, **1**, 1 (2017)
26. R. Bayu, H. Hamdani, Yuniarti, Jurnal Perikanan dan Kelautan, **3**, 3 (2012)
27. Farida, E.I. Raharjo, M. Sahrio, Jurnal Ruaya, **4**, 2 (2016)
28. G. H. Huwoyon, I.I. Kusmini, Jurnal Ikhtiologi Indonesia, **10**, 1 (2010)

Induction of oocyte developer hormones (oodev) on the maturity of *Poropuntius tawarensis*

Siska Mellisa[1*], *Iwan* Hasri[2], and *Kasturi* Ramdayani[1]

[1]Department of Aquaculture, Faculty of Marine and Fisheries, Universitas Syiah Kuala, Banda Aceh 23111, Indonesia.
[2]Fisheries Department of Central Aceh Regency 24552, Indonesia.

Abstract. *Poropuntius tawarensis* is an endemic fish belonging to the Cyprinidae family found in Laut Tawar Lake, Central Aceh, Indonesia. *P. tawarensis* has a limited distribution and the fertilization only occurs in the rainy season. The purpose of this research was to observe the effect of Oodev hormone with different doses on the maturity of *P. tawarensis*. This study was used completely randomized design (CRD) consisting of 3 treatments and 3 replicates. Treatments used were the hormone doses of 0.25 mL/kg, 0.50 mL/kg and treatment without hormone (control). The parameters used in this study were gonadosomatic index (GSI), egg diameter, coefficients of variation (CV) of egg diameter, fecundity, and hepatosomatic index. The study revealed the highest value found in a dose of 0.50 mL/kg which the value of egg diameter was 0.8791±0.02 mm, coefficients of variation of egg was 6.46±0.49% and fecundity was 869±33.99 egg/kg. While the highest gonadosomatic index was found in a dose of 0.25 ml/kg (10.41±2.03%). Moreover, the induction of Oocyte developer (Oodev) hormone on the maturity of *P. tawarensis* did not significantly affect the hepatosomatic index (HSI), but Oodev hormone was efficient to accelerate the gonad maturity of fish.

1 Introduction

Poropuntius tawarensis is an endemic fish belonging to the Cyprinidae family found in Laut Tawar lake, Central Aceh, Indonesia. *P. tawarensis* are freshwater fish that have characteristics of benthopelagic. *Poropuntius tawarensis* are currently very difficult to find due to environmental damage and the introduction of new species [1, 2].

Domestication is an effort for adapting fish from the wild (original habitat) to the cultivated environment [3, 4]. According to Effendi [5], fish domestication is divided into three stages, firstly adapting to the survival of fish in aquaculture containers, second strive fish growth, and third strive fish reproduction/breeding.

The problem of the reduction or extinction of endemic fish and native fish has a risk to damage the sustainability of aquatic biodiversity. Therefore, it is needed to preserve the natural wealth in order to support a friendly life with nature. The Government of Central Aceh Regency has tried to domesticate and spawn *P. tawarensis*, but they are still constrained by the process of gonad maturation of fish which only occured during the rainy season, therefore *P. tawarensis* fish was unable to spawn throughout the year [1]. Hormone induction is one of the alternative that can be done to ensure the availability of reproduction hormone for fish maturation. Furthermore, one of the hormones that plays a role in maturation is the Oodev hormone (Oocyte developer), a commercial hormone, which has been produced on a large scale.

The use of Oodev hormone is known to accelerate the maturation of various types of fish. According to Farastuti et al [6], Oodev hormone is a combination of pregnant mare's serum gonadotropin (PMSG) and antidopamine (AD). PMSG hormone contains more Follicle Stimulating Hormone (FSH) than Luteinizing Hormone (LH) which is good for gonad maturation of fish. FSH from outside stimulates the gonads and then gonadotropin releasing hormone (GnRH) stimulates the pituitary to secrete gonadotropin hormones for the function of early gonadal maturation or vitellogenesis. Antidopamine is a chemical that works to stop dopamine, while dopamine is an inhibitor of gonadal maturation activity. Inhibited dopamine activity stimulates the synthesis and release of the hormone lutein, which is initiated by gonadotropin releasing hormone (GnRH) in the pituitary gland [7, 8]. The previous studies of Oodev hormone were done in several fish such as *Tor soro* fish [6], snakehead fish [9], *Helostoma teminkii* [10], barramundi [11] and *Tor douronesis* [12]. The objective of this study was to observe the effect of Oodev hormone with different doses on the maturity of *P. tawarensis*.

2 Materials and Methods

2.1 Time and Site

This study was carried out from August to September 2020, located at the Technical Implementation Unit of

* Corresponding author : siska_mellisa@unsyiah.ac.id

the Fish Seed Agency (UPTD BBI) Lukup Badak, Fisheries Service, Pegasing District, Central Aceh Regency.

2.1.1 Research Procedure

This research used completely randomized design method consisting of 3 treatments and 3 replications, namely treatment A (control), B (0.25 mL/kg Oodev hormone), and C (0.50 mL/kg Oodev hormone).
It was conducted at BBI Lukup Badak water reservoir. The container used is a net with size of 1x1x1m. Before used, the net was dried in the sun and then randomly installed in the maintenance size pond. The minimum pond size was 1.5 meters to 2.5 meters depth.

The broodstocks used were 94 females of *P. tawarensis*. The brood fish was obtained from Laut Tawar Lake with an average weight of 3.31g. Before doing the treatment, the broodstock were acclimated for 10 days. Acclimation helps to adjust or adapt fish into a new environment.

The injection of *P. tawarensis* was carried out once in the early week. Injections were performed intramuscularly using a 1 mL syringe. Oodev used in each treatment dose was dissolved in NaCl with a ratio of 1:1. Before being injected, *P. tawarensis* were anesthetized with a stabilizer solution (0.015 mL/L water) for 5-10 minutes. After the fish passed out, the weight were measured, then fish placed on a wet towel for injection. The addition of stabilizer can increase the stability of the emulsion. This stabilizer is a surface active compound that is able to reduce the surface tension between the liquid air mass and the liquid.

Gonadosomatic Index (GSI)

Gonadosomatic index is defined as the percentage of gonadal weight divided by the body weight [13]. The gonadosomatic index was calculated using the following formula [14]:

$$GSI = \frac{total\ weight\ of\ gonad}{total\ weight\ of\ fish} x\ 100\% \quad (1)$$

Egg Diameter

A total of 30 eggs from each sample fish were measured the size of the diameter of the eggs. Egg diameter will be measured by the following formula:

$$Ds = Dt - D0 \quad (2)$$

where: Ds is Actual egg diameter, Dt is Final egg diameter and D0 is Initial egg diameter [6].

Coefficients of variation of egg diameter describes the size of the diameter of fish eggs in a population spread from the average value. Coefficients of variation of egg were calculated according to Steel and Torrie [15] as follow :

$$kk = (S/Y) \times 100\% \quad (3)$$

where : kk is coefficients of variation of egg diameter, S is square root of sample, Y is average of sample.

Fecundity

Fecundity was total amount of eggs of matured female fish and ready to spawn. According to Biswas [16] relative fecundity and Total fecundity was calculated using the following formula:

$$\text{Total fecundity} = n \times (wt/ws) \quad (4)$$

where n is total amount of eggs in sub-sampled, wt is total weight of gonad (g) and ws is total weight of sub-sampled (g).

Hepatosomatic Index (HSI)

Hepatosomatic Index (HSI) is defined as the ratio of liver weight to total body weight, It was calculated based on Effendi [17] as follow:

$$HSI = \frac{liver\ weight}{total\ body\ weight} x100\% \quad (5)$$

Gonadal Maturity Stages

Gonad maturity can be determined by observing the shape, size, colour, smoothness and filling of the ovaries in the body cavity, clarity of shape, colour of the eggs in the ovaries. Analysis of gonad maturity was defined following the stages of Cassie (1965) in Effendie [17].

Table 1. Gonadal maturity stages based on Cassie (1965) in Effendie [17].

Stage of Maturity	Gonad Morphology
I Imature	The ovary is like a thread. The eggs are not yet distinguishable. The length of the gonads varies between 1/3-1/2 the length of the body cavity.
II Maturing	There is a milky white tissue, the eggs are still fused and cannot be separated. The length of the gonads varies between 1/3–2/3 of the length of the body cavity.
III Maturing ripe	Larger size, widened anteriorly and tapered posteriorly, the eggs can be separated, darker in color. The length of the gonads varies between 1/3–2/3 of the length of the body cavity.
IV Ripe	Egg diameter is getting bigger and clearly visible under the microscope. The eggs are yellow. The length of the gonads varies between 2/3–3/4 of the length of the body cavity.
V Spent	Crimped ovaries, leftover eggs in the posterior. Ovaries are reddish.

3 Result and Discussion

3.1 Result

The study of induction of Oodev hormone on the maturity of *Poropuntius tawarensis* with doses of 0 mL/kg, 0.25 mL/kg, and 0.50 mL/kg showed the growth of gonad during 28 rearing days. The highest percentage of gonad maturity was reached by treatment C (0.50 mL/kg) (Table 2). The percentage of gonad maturity achieved by using induction of Oodev hormone was better than treatment without hormone induction.

Table 2. The percentage of gonad maturity (%) of *Poropuntius tawarensis*.

Treatment	Day 14 (%)	Day 28 (%)
A (0 mL/kg)	33.33	44.44
B (0.25 mL/kg)	55.55	66.66
C (0.50 mL/kg)	66.66	88.88

Table 3. Observation of gonad maturity stages on *P. tawarensis* induced Oodev hormone.

Treatment	Stage of maturity	Observation	Figure
A (0 mL/kg)	II Maturing	Egg diameter is about 0.25 mm to 0.54 mm.	
B (0,25 mL/kg)	III,IV Maturing ripe, ripe	Egg diameter is about 0.54 mm to 0.80 mm	
C (0,50 mL/kg)	V Spent	Egg diameter is about 0.80 mm to 0.87 mm	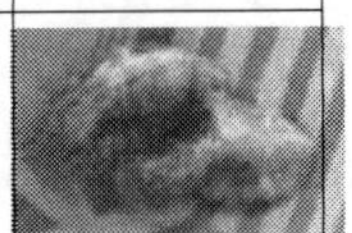

Based on Table 3, treatment A (0 mL/kg) only reached gonad maturity stages II with characterize by small gonad, eggs were not clearly seen. Then, treatment B (0.25mL/kg) reached gonad maturity stages III and IV with characterized can be distinguished. While, treatment C (0.50 mL/kg) achieved gonad maturity stages V with the characteristic shape of round eggs, some of which were clear and ripe.

ANOVA showed that induction of oocyte developer (oodev) on *P. tawarensis* had a significant effect on gonadosomatic index, eggs diameter, coefficients of variation of egg and fecundity. But it did not have a significant effect on hepatosomatic index.

Based on the Duncan's test, the highest gonadosomatic index was found in treatment B (10.41±2.03%), but did not significantly different with treatment C (9.06±0.04%) (Table 4). While, eggs diameter and coefficients of variation of egg had the highest value on treatment C but did not have significantly different with treatment B. Furthermore, the best value of fecundity in this study was found on treatment C (869 ±33.99 egg/kg).

Table 4. Gonadosomatic index (GSI), egg diameter (mm), coefficients of variation of egg (CV), fecundity and hepatosomatic index (HSI).

Treatment	GSI (%)	Egg diameter (mm)	CV (%)	Fecundity (egg/kg)	HSI (%)
A	1.62± 1.34a	0.548 ± 0.22a	4.41± 0.21a	101±92 .25a	0.33± 0.09a
B	10.41± 2.03b	0.8697± 0.01b	5.90± 0.07b	614±13 9.12b	0.27± 0.06a
C	9.06± 0.04b	0.8791± 0.02b	6.46± 0.49b	869±33 .99c	0.22± 0.07a

Note: Different superscripts in the same column are significantly different (p <0.5).

4 Discussion

Induction of Oodev hormone on *Poropuntius tawarensis* for 28 days of rearing had better results than without the use of Oodev hormone. Oodev hormone is a combination of pregnant mare serum gonadotropin (PMSG) and antidopamine (AD). According to Mayasari [18] PMSG contains follicle stimulating hormone (FSH) and luteinizing hormone (LH), follicle stimulating hormone (FSH) plays a role in early gonadal maturation or vitellogenesis, while luteinizing hormone (LH) is a hormone that helps the process of releasing eggs (ovulation). Antidopamine (AD) functions to inhibit the work of dopamine, where dopamine itself acts as an inhibitor of gonadal maturation [8]. Induction of Oodev hormone maturation carried out on *P. tawarensis* was able to provide stimulation so ovulation occured. Moreover, the similar results was found on *Tor douronesis*, Mellisa et al. [12] studied on induction of gonadal maturation of *Tor douronesis* using Oodev hormone with the best dose of 1 mL/kg fish.

The development of gonadal maturity was determined by morphological observation by looking at the size of the egg diameter level which according to Cassie (1965) in Effendie [17]. Nurmahdi [19] mention that egg size is an indicator of egg quality and gonadal maturity level. Egg diameter showed a difference at each level of gonad maturity. The larger the diameter of the egg, the higher the percentage of gonad maturity. The results showed that the best treatment was treatment with the addition of Oodev at a dose of 0.50 mL/kg (treatment C) which could increase the level of gonad maturity. As the opinion of Darliansyah et al. [20] injection of Oodev hormone with different doses resulted in different egg diameter sizes which indicated different stages of gonadal maturity.

The value of the gonadosomatic index (GSI) can vary because it is closely related to the weight of the gonads and the body depending on the individual. Mukti et al. [21] revealed that GSI increases with the increasing gonadal maturity stage. It was showed in this study, the high GSI were found in treatment B and C with the stage III, IV and V. In this study, the GSI value obtained was less than 20%. Therefore, its indicated that *P. tawarensis* are included in the group of fish with low GSI values and can be categorized as fish that can spawn more than once per year, according to Fatah et al. [22], the GSI value ≤ 20% is a group of fish that can spawn more than once a year. The use of Oodev hormone on treatment B and C was able to ripen brood fish of *P. tawarensis*. This is due to the weight of the gonads in this treatment is greater than the other treatments and the body weight of the fish also increased. In accordance with the opinion of Effendie [23], gonadosomatic index (GSI) is the value in % (percent) as a result of the comparison of gonad weight with fish body weight. In line with the growth of the gonads produced will increase in size and weight to the maximum level when spawning occurs. In a study of Sihaloho [24], it was shown that the injection of a broodstock of 3 kg Siamese catfish (*P. hypopthalamus*) using Oodev hormone at a dose of 0.25 mL/kg at intervals of 2 weeks was effective in accelerating gonad maturity.

Furthermore, Egg diameter increased in each treatment from the beginning to the end of the study. The results of this study indicated that the high value of egg diameter were treatment C. Yulianto et al. [25] concluded that Oodev hormone contains PMSG hormone, where the FSH content is greater and the LH is less. FSH functions for egg maturation and LH functions for egg production. So it can be concluded that the dose of Oodev hormone of 0.25 mL/kg and 0.50 mL/kg resulted in an increase in egg diameter. Giving Oodev hormone to *P. tawarensis* affected the diameter of the eggs and caused the process of vitellogenesis. According to Libzens et al. [26] Vittelogenesis is the incorporation of vitellogenin proteins by oocytes and processes them into egg yolks, causing an increase in the size of the fish gonads. In comparison with other research, Mellisa et al. [12] stated that injection of Oodev hormone into Tor douronensis at a dose of 1 mL/kg had the highest value of egg diameter and Darliansyah et al. [20] showed the highest value of egg diameter of *Osteochilus kappeni* was at a dose of 0.4 mL/kg.

Fecundity of *P. tawarensis* was determined based on the results of counting eggs in fish gonads. Observation of fecundity induced by Oodev hormone is related to egg diameter. In this study, the best fecundity results were treatment C (0.50mL/kg) where the higher the dose of induced of Oodev hormone, the higher the egg diameter and fecundity. This is similar to the opinion of Djuhanda [27], which states that the size of fecundity is influenced by food, fish size and environmental conditions, and can also be influenced by egg diameter. This is also supported by the opinion of Suwarso et al. [28] which stated that the number of eggs produced by fish will increase in line with the size of the gonads. So that the selection of the dose 0.50 mL/kg in *P. tawarensis* was appropriate to calculate the fecundity. Tinus [29] revealed that Oodev induction in vitellogenesis in rematured catfish (*Pangasius hypophthalmus*) with the best fecundity was found at a dose of 0.3 mL/kg.

Hepatosomatic index is the percentage comparison between liver weight and body weight which describes the metabolic processes in the liver [30]. The HSI value increased at the beginning of treatment A (0 mL/kg) and decreased in treatment C (0.50 mL/kg) at the end of the study, this was because in treatment A (0 mL/kg) the vitellogenesis process was not occurred so that the liver stores fat obtained from feed which then caused an increase in liver weight. Then in treatment C (0.50 mL/kg) HSI decreased, it was because the fat stored in the liver was passed on to the gonads to be used during the vitellogenesis process. In accordance with the statement of Sulistyo et al. [31] stated that the HSI value will begin to increase when the fish experience the beginning of the vitellogenesis process and will begin to decrease at the time of gonadal maturation.

5 Conclusion

Induction of Oodev hormone on maturity of *Poropuntius tawarensis* had the better results than treatment without hormone induction. The dose of 0.50 mL/kg was the best treatment but it was not significantly different from the 0.25mL/kg and a dose of 0.25 mL/kg efficiently used to accelerate the gonadal maturity of *P. tawarensis*.

References

1. Z.A. Muchlisin, I. Hasri, *Karakteristik biologi ikan dominan Danau Laut Tawar. In: Pengelolaan sumberdaya perikanan Danau Laut Tawar Aceh Tengah,* 252-281 (2015)
2. Z.A. Muchlisin, N. Fadli, E. Rudi, T. Mendo, M. N. Siti-Azizah, AACL Bioflux, **4**, 5, (2015)
3. F. Teletchea, Natural Resources, **7**, 399-404 (2016)
4. S. Anggoro, S. Rudiyanti, I.Y. Rahmawati, Water Resour. Manag., **2**, 3 (2013)
5. I. Effendi, Pengantar Akuakultur, **198** (2004)
6. E.R. Farastuti, O.S. Agus, S. Rudhy, Limnotek, **21**, 1 (2014)
7. S. Dufour, M.E. Sebert, F.A. Weltzien, K. Rousseau, C. Pasqualini, J. Fish. Biol. **76**, (2010)
8. C. Chen, D. Fernald, J. Fish. Biol. **73**, 5 (2008)
9. R.A. Hutagalung, M.S. Widodo, A.R. Faqih, JAI, **14**, 1 (2013)
10. Farida, T. Puji Lestari, H. Hasan, J. Arismunanda, Jurnal Ruaya, **7**,1 (2019)
11. S. Saputra, W. K. A. Putra, H. Irawan, Intek Akuakultur, **3**, 1 (2019)
12. S. Mellisa, I. Hasri, W. Agung, N. Nurfadillah, D. F. Putra, *IOP Conference Series: Earth and Evironmental Science*, **348**, 012101 (2019)
13. M. Hasan, M.H.A Hosen, M.I. Miah, Z.F. Ahmed, M. S. Chhanda, S. I. M. Shahriar Egypt. J. Aquat. Res. **46**, (2020)

14. M.I. Effendie. Metode Biologi Perikanan, (1979)
15. R.G.D. Steel, J.H. Torrie, *Prinsip dan Prosedur Statistik. Soemantri (Penerjemah),* (1993)
16. S.P. Biswas, *Manual of Methods in Fish Biology, 2nd* 157p (1993)
17. M.I. Effendie. *Biologi perikanan*, 163p (2002)
18. N. Mayasari, A.O. Sudrajat, T. Chrismadha, Jurnal Limnotek **19,** (2012)
19. T. Nurmahdi, *Pengaruh Penggunaan Hormon HCG Dengan Dosis Yang Berbeda Terhadap Perkembangan Gonad Ikan Baung (Hemibagrus nemurus). [Tesis]. Bogor: Pascasarjana Institut Pertanian Bogor* (2005)
20. R. Darliansyah, S. A. E. Rahimi, I. Hasri, JIM **2**, 2, (2017)
21. A.T. Mukti, Y.G.P. Sari, G.S.R. Agusdinata, W.H. Satyantini, A.S. Mubarak, E.M. Luqman, Widjiati. Theriogenology **147,** (2020)
22. K. Fatah, S. Adjie, Bawal, **5**, 2 (2013)
23. M.I. Effendi, *Biologi Perikanan*, (1997)
24. Sihaloho, I.S. Ovie, Induksi Pematangan Gonad *Calon Induk Ikan Patin Siam (Pangasianodon hypophthalmus) ukuran 3 Kg Menggunakan Oodev Melalui Penyuntikan. [Tesis]. Departemen Budidaya Perairan. Fakultas Perikanan dan Ilmu Kelautan, Institut Pertanian Bogor,* (2014)
25. T. Yulianto, W.K.A. Putra, S. Miranti, T. Hardiyanti, S. Fitriana, Fauzanadi JPK **9**,1 (2019)
26. E. Libzens, G. Young, J. Bobe, J. Cerda, General and Comparative Endocrinology **165** (2010)
27. T. Djuhanda, *Dunia Ikan,* (1981)
28. D. Suwarso, W. Pralampita, M. Wahyono. Biologi reproduksi malalugis biru, Decapterus macarellus di Sulawesi Utara in *Prosiding seminar Hasil Penelitian Perikanan 1999/2000. Pusat Penelitian dan Pengembangan Esplorasi Laut dan Perikanan* (2000)
29. A. Tinus, Scientiae Journal, **3**, 1 (2013)
30. S. Sadekarpawar, P. Parikh. J. of Zoology **8**, 1 (2013)
31. I. Sulistyo, P. Fontaine, J. Rincarh, J.N. Gardeur, H. Migaud, B. Capdeville, P. Kestemont, J. Aqu. Res. **13**, (2000)

The effect of concentration of ab mix and zpt solutions on the growth and production of mustard plants (*Brassica juncea* L.) in hydroponic wick systems

Muhammad Reza Subakti[1], *Nurhayati*[2*], and *Murni* Sari Rahayu[3]

[1]Student Department of Agrotechnology, Faculty of Agriculture, Universitas Islam Sumatera Utara, Address Karya Wisata Gedung Johor, Medan 20144, Indonesia
[2]Department of Agrotechnology, Faculty of Agriculture, Universitas Islam Sumatera Utara, Address Karya Wisata Gedung Johor, Medan 20144, Indonesia
[3]Department of Agrotechnology, Faculty of Agriculture, Universitas Islam Sumatera Utara, Address Karya Wisata Gedung Johor, Medan 20144, Indonesia

Abstract. This research was implemented in the greenhouse at the Experimental Garden of the Faculty of Agriculture Universitas Islam Sumatera Utara, Sub-distric Gedung Johor, Medan Johor District, Medan City, North Sumatra Province. This research was implemented from November to December 2020. This study aims to determine the effect of concentration of AB Mix and ZPT solutions on the growth and production of mustard greens. This study used a completely randomized design (CRD) with two treatment factors the concentration of AB Mix solution and ZPT application. The first factors, the concentration of AB Mix solution consisting of 3 levels: K_0 = control, K_1 = 550 ppm, K_2 = 1100 ppm. The second factor was giving ZPT the earliest in 3 levels: Z_0 = control, Z_1 = 0.5 ml, Z_2 = 1 ml. The parameters observed were plant height, number of leaves, leaf length, leaf width, wet weight and dry weight. Its is know from the results concentrations of AB Mix solution significantly affected plant height, number of leaves, leaf length, leaf width, wet weight and dry weight. Giving of ZPT had a significant effect on leaf width but had no significant effect on plant height, number of leaves, leaf length, wet weight and dry weight. The interaction of AB Mix solution concentration and ZPT giving had no significant effect on the growth and production of mustard greens.

1 Preliminary

Vegetables are plant commodities that are able to contribute to national development in order to welfare the community, such as fulfilling community nutrition as a complement to the four healthy five perfect foods, also very potential and prospective to be cultivated because cultivation methods tend to be easy and simple. One of the easiest vegetables to cultivate is mustard [1].

Green mustard (*Brassica juncea* L.) is one of the most important vegetable commodities in the world. Although green mustard is not native to Indonesia, the development of this agribusiness and agro-industry patterned crop commodity can be categorized as a source of income in the agricultural sector in Indonesia [1].

Green mustard plants are known to have high economic value because this vegetable is one of Indonesia's main export commodities. However, until now, mustard plants production has not been able to meet the needs of domestic and foreign markets. This is because the average national mustard plants production is still very low. The potential yield of mustard greens can reach 40 tons/ha, while the average yield of mustard plant in Indonesia is 9 tons/ha [2].

The low production of mustard greens in Indonesia can be caused by several reasons, such as the application of cultivation technology that is still simple, or because the land for growing crops is decreasing. As is known, today's industrial development is progressing rapidly. These development have shifted a lot of agricultural land, especially in urban areas. As a result, agricultural land is getting smaller. To overcome this, various ways can be taken to increase plant productivity. In this way, it is hoped that from a narrow area, large production can be produced [3].

Hydroponics comes from the Greek, namely hydro which means water and ponos which means power. Hydroponics is also known as soilless culture or plant cultivation without soil. So hydroponics is a method of growing crops without using soil media, but using a nutrient solution of minerals or other materials containing nutrients such as coconut fiber, mineral fiber, sand, broken bricks, sawdust and others as a substitute for soil media. Hydroponic plants can be done on a small scale at home as a hobby or on a large scale with commercial purposes [4].

One of the simplest hydroponic systems is the wick system. Hydroponics wicks (wicks) is a simple method by using a wick as a link between nutrients and the roots

* Corresponding author: nurhayatijb@yahoo.co.id

https://doi.org/10.1051/e3sconf/202233901010

of the growing media. This hydroponic system is passive, because there are no moving parts in this medium [5].

In the production process, proper planting media and growth regulators are needed to obtain good growth [6]. The use of growth regulators plays a role in increasing the growth and production of vegetable crops. Plant regulatory substances are organic compounds that are not active plant nutrients in small amounts that are synthesized in certain parts of the plant and are generally transported to other parts of the plant, where these growth regulators cause biochemical, physiological and morphological responses [7].

The purpose of the study was to determine the right concentration of nutrients and the right level of ZPT in increasing the growth and production of mustard greens (*Brassica juncue* L.) in the Hydroponic Wick system.

2 Materials and Methods

2.1 Place and Time Research

This research was conducted at the Experimental Garden Greenhouse, Faculty of Agriculture, Islamic University of North Sumatra, Jalan Karya Wisata, Gedung Johor Village, Medan Johor District, Medan City, North Sumatra Province. This research was carried out in October until Nopember 2020.

2.2 Materials and Tools

The materials used in this study were green mustard seeds as planting material, Styrofoam used as a planting medium and a place to accommodate nutrient water solutions, a netpot for placing mustard plants, rockwool as a planting medium for mustard plants, seed seeding tanks, AB mix nutrient solution. Atonic as material to be used for ZPT solution.

The tools used in this study are PH measuring devices that will be used to measure the PH of water, TDS tools that will be used to measure the concentration of nutrient solutions, digital scales used to weigh the wet and dry weight of mustard plants, oven to dry mustard plants, drill to make planting holes in Styrofoam boxes, markers used to mark the location of planting holes, gauze, tape measure or rollers to measure plant spacing and to measure plant parameters, buckets, gembors or spray guns for watering mustard plants, a 20ml syringe is used as a substitute for a measuring cup that will be used as a measure of the AB MIX and ZPT solutions that will be used.

2.3. Research Methods

The design model used in this study was a factorial completely randomized design (CRD) consisting of factor I: Concentration of AB mix solution consisting of 3 levels, namely: K_0: Control, K_1: 550 ppm, K_2: 1100 ppm and Factor II: Addition ZPT consists of 3 levels, namely: Z_0: Control, Z_1: 0.5 ml, Z_2: 1 ml. Observation parameters were plant height, number of leaves, leaf length, leaf width, wet weight and dry weight of mustard plants.

Application of Nutrition and ZPT. Provision of nutrients and ZPT is carried out before the plants are transferred to the planting plot, before the nutrients and PGR are entered, first enter the water into the Styrofoam box as much as 8 liters/plot, then after that input the ABmix nutrient solution at a rate of 8 ml/liter of water for K2 and 4 ml treatment/liters of water for treatment K1 and ZPT solution with a dose of 1 ml/liters of water for treatment Z2 and 0.5 ml/liters of water for treatment K1 then mix the solution into a Styrofoam box that already contains water then after that stir and measure whether the content the nutrients entered are in accordance with the doses determined using the TDS tool.

3 Research Result

3.1 Plant Height (cm)

The results of the analysis of variance show that the concentration of the AB Mix solution had a significant effect. The treatment of giving ZPT and the interaction of the two treatment factors had no significant effect on the height of the mustard plant. The average height of mustard plants is presented in Table 1.

In Table 1 it can be seen that the concentration of the AB Mix solution had a significant effect on the height of the mustard plant. The highest plants were obtained in treatment K_2 (1100 ppm) which was 7.18 cm, which was significantly different from treatment K_0 (control) which was 5.69 cm, but not significantly different from treatment K_1 (550 ppm) which was 6.54 cm.

The relationship between the height of the mustard plant and the concentration of the AB Mix solution is linear with the equation = 5.725 + 0.0014 K, r = 0.99. The equation describes a positive linear relationship between the concentration of AB Mix solution and plant height. Thus, the nutrients contained in the AB Mix solution have a positive effect on mustard plants, namely being able to trigger vegetative development.

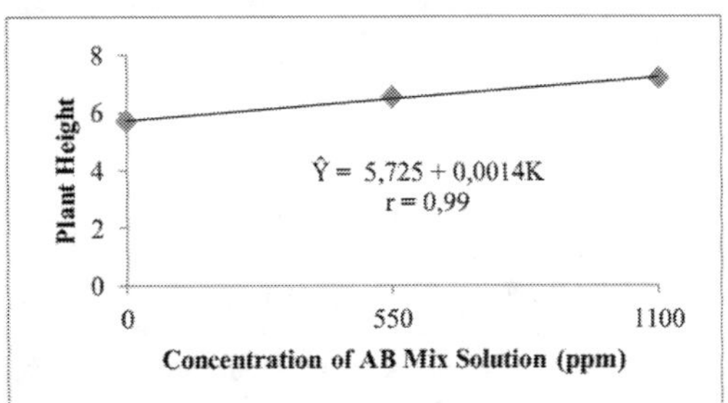

Fig.1. Relationship of Mustard Plant Height (cm) with AB Mix Solution Concentration (ppm)

The provision of ZPT had no significant effect on the height of the mustard plant. The highest plant height was obtained in treatment Z_2 (1 ml) which was 6.59 cm, followed by treatment Z_1 (0.5 ml) which was 6.49 cm and treatment Z_0 (control) was 6.32 cm.

3.2 Number of Leaves (sheet)

The results of the analysis of variance show that the concentration of the AB Mix solution had a significant effect. The treatment of giving ZPT and the interaction of the two treatment factors had no significant effect on the number of mustard leaves. The average number of mustard leaves is presented in Table 1.

In Table 1 it can be seen that the concentration of the AB Mix solution had a significant effect on the number of mustard leaves. The highest number of leaves was obtained in treatment K_2 (1100 ppm) which was 9.39 leaves, which was significantly different from treatment K_0 (control) which was 8.39 leaves, but not significantly different from treatment K_1 (550 ppm) which was 9.00 leaves.

The relationship between the number of mustard leaves and the concentration of the AB Mix solution is linear with the equation = 8.426 + 0.0009 K, r = 0.99. Based on these equations, it describes a positive linear relationship between the concentration of AB Mix solution and the number of leaves. The figure shows that the number of leaves will continue to increase along with the concentration of AB Mix solution.

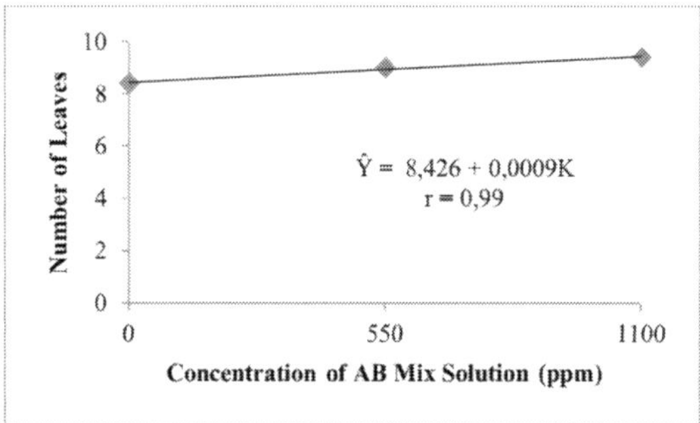

Fig.2. The Relationship between the Number of Mustard Leaves (strands) and the Concentration of AB Mix Solution (ppm)

Giving ZPT had no significant effect on the number of mustard leaves. The highest number of leaves was obtained in treatment Z_2 (1 ml) with 9.08 leaves, followed by treatment Z_1 (0.5 ml) with 8.89 leaves and treatment Z_0 (control) with 8.81 leaves.

3.3 Leaf Length (cm)

The results of the analysis of variance show that the concentration of the AB Mix solution had a significant effect. The treatment of giving ZPT and the interaction of the two treatment factors had no significant effect on the length of the mustard leaf. The average mustard leaf length is presented in Table 1.

In Table 1 it can be seen that the concentration of the AB Mix solution had a significant effect on the length of the mustard leaf. The longest leaf length was obtained in treatment K_2 (1100 ppm) which was 12.61 cm, which was significantly different from treatment K_0 (control) which was 10.68 cm, but not significantly different from treatment K_1 (550 ppm) which was 11.71 cm.

The relationship between the length of the mustard leaf and the concentration of the AB Mix solution is linear with the equation = 10.698 + 0.0018 K, r = 0.99. This equation illustrates that there is a positive linear relationship between the concentration of AB Mix solution and leaf length. The higher the concentration of the AB Mix solution, the longer the leaf length.

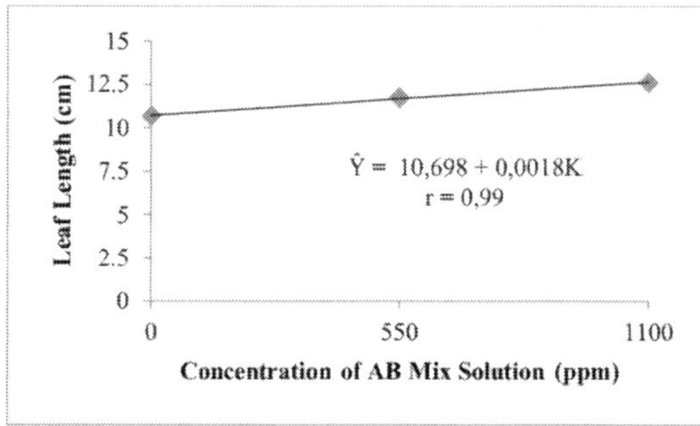

Fig.3. The Relationship of Mustard Leaf Length (cm) with AB Mix Solution Concentration (ppm)

Giving ZPT did not significantly affect the length of mustard leaves. The longest leaf length was obtained in treatment Z_2 (1 ml) which was 12.19 cm, followed by treatment Z_1 (0.5 ml) which was 11.78 cm and treatment Z_0 (control) was 11.02 cm.

3.4 Leaf Width (cm)

The results of the analysis of variance show that the treatment with the concentration of AB Mix solution and the administration of ZPT had a significant effect. The interaction of the two treatment factors had no significant effect on the width of the mustard leaf. The average mustard leaf width is presented in Table 1.

In Table 1 it can be seen that the concentration of the AB Mix solution had a significant effect on the width of the mustard leaf. The widest leaf width was obtained in treatment K_2 (1100 ppm) which was 10.87 cm, which was significantly different from treatment K_0 (control) which was 8.46 cm, but not significantly different from treatment K_1 (550 ppm) which was 10.43 cm.

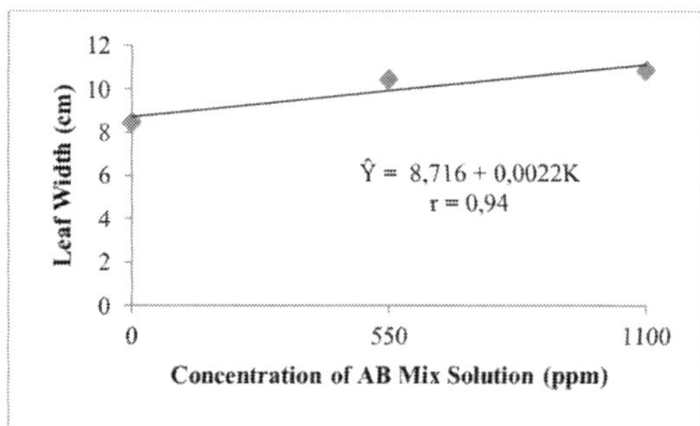

Fig.4. The Relationship of Mustard Leaf Width (cm) with Provision of AB Mix Solution Concentration (ppm)

The relationship between mustard leaf width and the concentration of AB Mix solution is linear with the equation = 8.716 + 0.0022 K, r = 0.94. This equation illustrates that there is a positive linear relationship between the concentration of AB Mix solution and leaf width. The higher the concentration of the AB Mix solution, the wider the leaf width.

Giving ZPT significantly affected the width of mustard leaves. The widest leaf width was obtained in treatment Z_2 (1 ml) which was 10.81 cm, which was significantly different from treatment Z_1 (0.5 ml) which was 9.58 cm and treatment Z_0 (control) was 9.36 cm.

The relationship between mustard leaf width and ZPT concentration was linear with the equation = 9.1917 + 1.45 Z, r = 0.93. This equation illustrates that there is a positive linear relationship between the administration of ZPT concentration and leaf width.

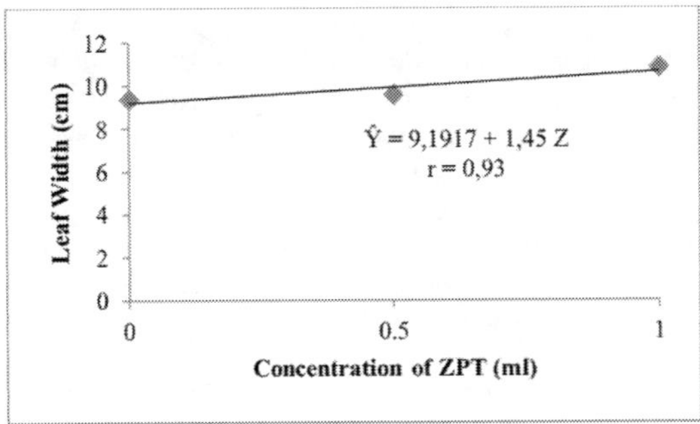

Fig.5. Relationship of Mustard Leaf Width (cm) with ZPT Concentration (ml)

3.5 Wet Weight (g)

The results of the analysis of variance show that the concentration of the AB Mix solution had a significant effect. The treatment of giving ZPT and the interaction of the two treatment factors had no significant effect on the wet weight of mustard greens. The average wet weight of mustard greens is presented in Table 1.

In Table 1, it can be seen that the concentration of the AB Mix solution had a significant effect on the wet weight of mustard greens. The heaviest wet weight was obtained in treatment K_2 (1100 ppm) which was 159.22 g, which was significantly different from treatment K_0 (control) which was 84.89 g and treatment K_1 (550 ppm) which was 125.55 g.

The relationship between the wet weight of mustard greens and the concentration of the AB Mix solution is linear with the equation = 86.056 + 0.0676 K, r = 0.99. The equation describes a positive linear relationship between the concentration of the AB Mix solution and the wet weight.

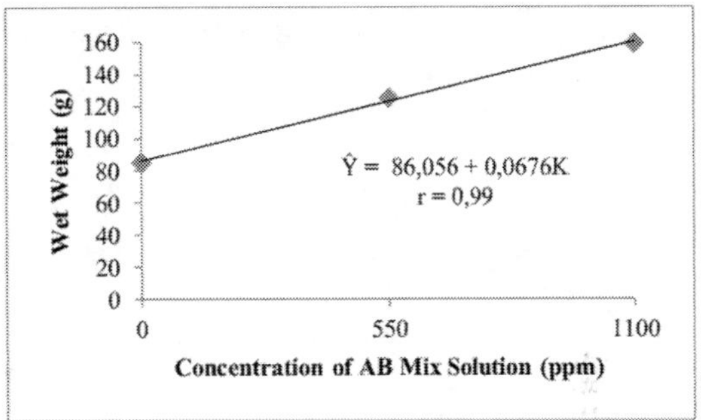

Fig.6. Relationship between Wet Weight (g) and Concentration of AB Mix Solution (ppm)

Giving ZPT had no significant effect on the wet weight of mustard greens. The heaviest wet weight was obtained in treatment Z_2 (1 ml) which was 141.33 g, followed by treatment Z_1 (0.5 ml) which was 120.22 g and treatment Z_0 (control) was 108.11 g.

Table 1. Average Effect of Concentration of AB Mix Solution and ZPT on Growth and Production of Mustard Plants (*Brassica juncea* L) in Hydroponic Wick Systems

Treatment	Plant Height (cm)	Number of Leaves (sheet)	Leaf Length (cm)	Leaf Width (cm)	Wet Weight (g)	Dry Weight (g)
AB Mix Solution						
K_0	5,69 b	8,39 b	10,68 b	8,46 b	84,89 c	6,56 c
K_1	6,54 a	9,00 a	11,71 a	10,43 a	125,55 b	10,56 b
K_2	7,18 a	9,39 a	12,61 a	10,87 a	159,22 a	12,33 a
Giving of ZPT						
Z_0	6,32	8,81	11,02	9,36 b	108,11	8,55
Z_1	6,49	8,89	11,78	9,58 b	120,22	9,67
Z_2	6,59	9,08	12,19	10,81 a	141,33	11,22
Interaction						
K_0Z_0	6,20	8,25	9,84	8,05	60,00	4,33
K_0Z_1	5,40	8,25	10,94	8,31	82,00	6,67
K_0Z_2	5,47	8,67	11,25	9,02	112,67	8,67
K_1Z_0	6,04	8,92	11,20	9,86	120,00	10,00
K_1Z_1	6,73	9,00	11,92	10,11	127,33	10,67
K_1Z_2	6,85	9,08	12,00	11,32	129,33	11,00
K_2Z_0	6,73	9,25	12,03	10,18	144,33	11,33
K_2Z_1	7,35	9,42	12,47	10,33	151,33	11,67
K_2Z_2	7,46	9,50	13,33	12,09	182,00	14,00

Note: the numbers followed by letters that are not the same in the same treatment group are significantly different at 5% with the DMRT test

3.6 Dry Weight (g)

The results of the analysis of variance show that the concentration of the AB Mix solution had a significant effect. The treatment of giving ZPT and the interaction of the two treatment factors had no significant effect on the dry weight of mustard greens. The average dry weight of mustard greens is presented in Table 1.

In Table 1 it can be seen that the concentration of the AB Mix solution had a significant effect on the dry weight of mustard greens. The heaviest dry weight was obtained in treatment K2 (1100 ppm) which was 12.33 g, which was significantly different from treatment K0 (control) which was 10.56 g and treatment K1 (550 ppm) which was 6.56 g.

The relationship between dry weight of mustard greens and the concentration of AB Mix solution is linear with the equation = 6.927 + 0.0053 K, r = 0.97. The equation describes a positive linear relationship between the concentration of the AB Mix solution and the dry weight.

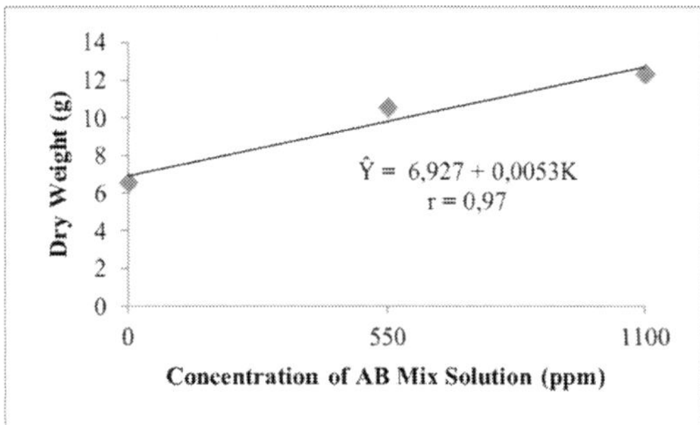

Fig.7. The Relationship of Mustard Dry Weight (g) with Provision of AB Mix Solution Concentration (ppm)

Giving of ZPT had no significant effect on the dry weight of mustard greens. The heaviest dry weight was obtained in treatment Z_2 (1 ml) which was 11.22 g, followed by treatment Z_1 (0.5 ml) which was 9.67 g and treatment Z_0 (control) was 8.55 g.

4 Discussion

4.1 Effect of Concentration of AB Mix Solution on Growth and Production of Mustard Plants (*Brassica juncea* L) in Hydroponic Wick Systems

The concentration of AB Mix solution significantly affected plant height, number of leaves, leaf width, leaf length, wet weight and dry weight.

Giving AB MIX ocean with a concentration of 1100 ppm significantly increased the concentration of the AB Mix solution significantly affected plant height, number of leaves, leaf width, leaf length, wet weight and dry weight, with increasing concentration of the given solution from 0 ppm, 550 ppm and 1100 ppm the increase in both growth and production of mustard greens. This is because the nutrient content in the AB Mix solution plays an important role in increasing growth and production.

The results showed that the treatment with the concentration of AB Mix 1100 ppm (K2) gave the highest plant height, which was 7.18 cm, this was related to giving the right concentration of AB Mix solution for mustard greens. According to Nugraha (2014) states that the composition of macro and micro nutrients is very influential on plants. Therefore, in the application of fertilizers, it must be balanced according to the needs of the plant [8].

The results showed that the concentration treatment of AB Mix 1100 (K_2) solution gave the highest number of leaves, namely 9.39 strands. This is because the AB Mix solution is able to trigger the growth of the number of leaf stalks more than the 1200 ppm treatment and the 1300 ppm treatment used. Based on the difference in concentration which is considered sufficient in plant tissue, the higher the availability of plant nutrients, the higher the possibility of plant production if other factors also support it, if the availability of plant nutrients is large but other factors do not support the plant will not grow perfectly [9].

The results showed that the concentration treatment of AB Mix 1100 (K_2) solution gave the highest leaf length of 12.61 cm and the highest leaf width of 10.87 cm. This is because the AB Mix solution given is a concentration that suits the needs of plants, such as mustard greens grown with a hydroponic system. This opinion is in line with the results of research by Akasika *et al.*, (2014), which states that the provision of nutrients with a concentration of 1000 ppm will Accelerate the rate of leaf formation, because plants need nutrients, especially N, which plays a major role in the vegetative phase have had enough. In plant vegetative growth as indicated by the increase in plant length, the nutrient that plays a role is nitrogen (N) [10].

The results showed that the concentration treatment of AB Mix 1100 (K_2) solution gave the highest wet weight of 159.22 g. The increase in wet weight was related to other growth parameters such as plant height, number of leaves, roots and chlorophyll content. Suarsana *et al.*, (2019) states that the rate of cell division and tissue formation is proportional to the growth of stems, leaves and root systems. This depends on the availability of carbohydrates in plants [11].

The results showed that the concentration treatment of AB Mix 1100 (K2) solution gave the highest dry weight of 12.33 g. Dry weight decreased with increasing nutrient concentration. This dry weight loss showed no signs of toxicity. The most likely thing is that increasing the concentration of nutrients will also increase the availability and absorption of N and K elements, where these two elements play a very important role in regulating cell turgor and succulence of plant organs [12]. Succulent plants usually have high fresh weight but low dry weight due to the high water content in their organs.

4.2 Effect of Concentration of ZPT on Growth and Production of Mustard Plants (*Brassica juncea* L) in Hydroponic Wick Systems

Giving ZPT had a significant effect on leaf width, but had no significant effect on plant height, number of leaves, leaf length, wet weight and dry weight.

The best ZPT treatment was obtained at a concentration of 1 ml. Increasing the concentration of PGR significantly increased leaf width. This increase was caused by the hormone content in the ZPT, resulting in a process of elongation of new cells and eventually increasing the amount of tissue in plants resulting in faster plant growth. This is in line with Djamhari (2010) who explained that the application of exogenous ZPT in plants can function to stimulate the formation of phytohormones, so that they can encourage a biochemical activity. Phytohormones as organic compounds that work actively in small amounts are usually transformed throughout the plant so that they can affect plant growth or physiological processes [13].

Provision of ZPT had no significant effect on plant height, leaf length, wet weight and dry weight. According to Ardiana (2009), in an in vitro experiment, it was stated that giving ZPT with a concentration of ppm gave the best results, however, if the PGR concentration was increased to 1–1½ ppm, plant growth would be stunted [14]. According to Suyanti *et al.*, (2013), cytokinins can inhibit the initiation and growth of lateral roots (lateral root primordium) but the mechanism of inhibitory regulation of these compounds is not known with certainty [15]. This is in accordance with the results obtained where the lowest results were in the ZPT treatment with a concentration of 0.5 ml even though the results were still higher than the control.

Giving ZPT did not significantly affect the number of leaves. This was indicated by the addition of higher ZPT which did not increase the number of leaves. These results are supported by research conducted by Widiastoety (2014), who wrote that excessive cytokinin concentrations resulted in inhibition of shoot differentiation and growth, in addition to in vitro studies showing that often the use of cytokinins can stimulate the biosynthesis of ethylene compounds which causes inhibition of shoot growth, inhibits cell division. and cell elongation which results in the absence of other organs [16].

Cytokinins play a role in slowing plant aging. However, an increase in cytokinin content will inhibit shoot growth, reduce the accumulation of total chlorophyll, and delay the induction of flowering [17]. In a study conducted by Handayani *et al.*, (2020), there was an increase in the number of cytokinins during the aging process, so that the yellowing process in plants was delayed. This is related to chlorophyll-degrading enzymes such as chlorophyllase, Mg-dechlatase, and peroxidase, where these three enzymes will increase their activity by ethylene while cytokinins decrease it [18]. This statement is in accordance with the results obtained where the addition of ZPT with a concentration of 0.5 ml is not much different from the treatment of giving ZPT with a concentration of 1 ml.

4.3 Effect of Concentration of AB Mix Solution and ZPT on Growth and Production of Mustard Plants (Brassica juncea L) in Hydroponic Wick Systems

The interaction of giving the concentration of AB Mix solution and ZPT had no significant effect on the growth and production of mustard greens (*Brassica juncea* L.) in the Hydroponic Wick system.

Although it had no significant effect, the two factors interacted positively to increase the growth and production of mustard greens. Giving AB Mix solution with ZPT will support each other. This is because the availability of nutrients greatly affects the growth and development of plants, especially N for mustard plants. According to Gardner et al., (1991) in Zamriyeti et al., (2019) the essential function of nitrogen nutrients in plant tissue is cell division and enlargement [19].

5 Conclusion

The concentration of AB Mix solution significantly affected plant height, number of leaves, leaf width, leaf length, wet weight and dry weight mustard plants. The best concentration of AB Mix solution was in the K_2 (1100 ppm) treatment with a plant height of 7.18 cm, leaf number 9.39, leaf width 10.87 cm, leaf length 12.61 cm, wet weight 159.22 g and dry weight 2.33 g.

Giving ZPT had a significant effect on leaf width, but had no significant effect on plant height, number of leaves, leaf length, wet weight and dry weight mustard plants. The best ZPT treatment was given to Z_2 (1 ml) with a leaf width of 10.87 cm, plant height 6.59 cm, leaf number 9.08, leaf length 12.19 cm, wet weight 141.33 g and dry weight 11 ,22 g. The interaction of the concentration of the AB Mix solution and the addition of ZPT had no significant effect on the growth and production of mustard plants.

References

1. A. Fuad, Budidaya tanaman sawi (Brassica juncea L.) (Skripsi, Universitas Sebelas Maret Surakarta, 2010)
2. Badan Pusat Statistik. Statistik Tanaman Hortikultura Indonesia.http:www.bps.go.id.
3. A. D. Susila, Sistem Hidroponik (Departemen Agronomi dan Hortikultura, Fakultas Pertanian, Institut Pertanian Bogor, 2013)
4. H. Sunarjono, Bertanam Sawi dan Selada (Penebar Swadaya, Jakarta, 2014)
5. S. Cahyo dan S. Rini, Panduan Lengkap Budidaya Ikan dan Sayuran Dengan Sistem Aquaponik (Yogyakarta, Lily Publisher, 2014)
6. Y. Nurlaeni dan M.I. Surya, Pros Sem Nas Masy Biodiv Indon **1**, 5 (2015)
7. M. Dwiati, Peran Zat Pengatur Tumbuh Auksin Dan Sitokinin Terhadap Pertumbuhan Semai Anggrek Phalaenopsis (Fakultas Biologi Unsoed, Banteran, 2016)

https://doi.org/10.1051/e3sconf/202233901010

8. R. U. Nugraha, Sumber Hara Sebagai Pengganti AB Mix pada Budidaya Sayuran Daun Secara Hidroponik (Departemen Agronomi dan Holtikultura, Institut Pertanian Bogor, 2014)
9. H. Fitriani, Iskandar M. Lapanjang Ramal Yusuf. Respon Pertumbuhan Tanaman Sawi (Brassica juncea L.) Secara Hidroponik Terhadap Komposisi Media Tanam dan Konsentrasi Pupuk Organik Cair (Program Studi Agroteknologi Fakultas Pertanian Universitas Tadulako, Palu, 2015)
10. R. Akasika, Samekto dan Siswadi, J. Inov. Pertan, **13**, 2 (2014)
11. M. Suarsana, I. P. Parmila, K. A. Gunawan, Agro Bali (Agricultural J.) **2**, 2 (2019)
12. M. N. Subandi, Purnama dan B, Frasetya, J. Agroekoteknologi UIN Sunan Gunung Djati Bandung **9**, 2 (2015)
13. S. J. Djamhari S, J, Sains dan Teknologi Indonesia **12**, 1 (2010)
14. D. W. Ardiana, Buletin Teknik Pertanian **14**, 2 (2009)
15. Suyanti, Mukarlina, Rizalinda, Protobiont **2**, 2 (2013)
16. D. Widiastoety, J. Hort **24**, 3 (2014)
17. E. Mutryarny, J. Ilmiah Pertan **14**, 2 (2018)
18. K. P. Handayani, Safruddin, S., Hasibuan, S. Bernas **15**, 1 (2020)
19. M. Zamriyetti, Siregar, Refnizuida. Agrium, J. Ilmu Pertan **22**, 1 (2019)

The currents, sea surface temperature, and salinity patterns in the Bay of Bengal

Nadia Amsalia[1], *Yudi* Haditiar[1], *Reza* Wafdan[1], *Muhammad* Ikhwan[2], *Makwiyah* A. Chaliluddin[3], *Sugianto* Sugianto[4], and *Syamsul* Rizal[1,2,5*]

[1]Department of Marine Sciences, Faculty of Marine and Fisheries, Universitas Syiah Kuala, Banda Aceh, 23111, Indonesia
[2]Graduate School of Mathematics and Applied Sciences, Universitas Syiah Kuala, Banda Aceh, 23111, Indonesia
[3]Department of Fisheries Resource Utilization, Faculty of Marine and Fisheries, Universitas Syiah Kuala, Banda Aceh, 23111, Indonesia
[4]Department of Soil Sciences, Faculty of Agriculture, Universitas Syiah Kuala, Banda Aceh, 23111, Indonesia
[5]Research Center for Marine Sciences and Fisheries, Universitas Syiah Kuala, Banda Aceh 23111, Indonesia

Abstract. The Bay of Bengal is a semi-enclosed basin, with its southern side open to the Indian Ocean. Therefore, the current circulation is very influential in the condition of the ocean parameters. This study aims to examine the effect of current velocity on the distribution pattern of sea surface temperature (SST) and sea surface salinity (SSS) in February and August in 2017 in the Bay of Bengal. This study uses the results of Hybrid Coordinate Ocean Model (HYCOM) data for the main parameters, namely current, temperature and salinity which were analyzed with the calculation results of precipitation and evaporation data. The results of the data obtained in February showed the average current velocity conditions and low SST and relatively high SSS in the south. In August the average current velocity and SST is high and SSS is low. The results of the average data of low rainfall and relatively low evaporation are produced in February, the results of the average high rainfall and relatively high evaporation are produced in August. The resulting value gives the results of the relationship between current circulation and changes in parameter conditions in the waters.

1 Introduction

The Bay of Bengal (BoB) is a bay in the northeastern Indian Ocean that resembles a triangle bordering Bangladesh, eastern India, and the Andaman Sea, and is located at coordinates 6°N-24°N and 78°E-98°E [1]. BoB is one of the world's largest bays, with a semi-enclosed shape. BoB is in the path of a tropical cyclone with a wide and shallow continental shelf. Current circulation in the northern Indian Ocean is influenced by the Indian monsoon current drift system, namely the northeast monsoon current (NMC) and southwest monsoon current (SMC). The northern boundary of the BoB is bounded by mainland Asia, and the ocean dynamics are dominated by monsoon winds, which are seasonal winds that periodically reverse direction. The BoB has a complicated water exchange mechanism because it is affected by wind variations, sea level dynamics, bathymetry, and mass distribution throughout the year. Monsoon currents reverse in April-May (NMC to SMC) and October-November (SMC to NMC) [2,3].

The distribution of salinity in BoB is a distinct feature at a specific time in the northern Indian Ocean. BoB is known for its unique surface waters, which are influenced by mainland rivers that bring fresh water to the bay [4]. The distribution of salinity in BoB water mass is influenced by the waters of the Red Sea and Persian Gulf, which have high salinity and evaporation [3].

Surface temperature in BoB is influenced by seasons, particularly during the monsoon season, which consists of winter (east monsoon) and summer (southwest monsoon) periods with intermonsoon (transitional) periods [5,6]. BoB is distinguished by the presence of cyclones, which frequently occur between April and October [7]. Due to current instability, the West Boundary Current (WBC) in BoB is characterized by eddy activity. The vortex's kinetic energy is greater than that of other oceans [8–10]. Rizal *et al.* [11] carried out monthly simulations in BoB using the HAMSOM model, which revealed the pattern of SST and SSS in 2007. The simulation was completed successfully by interpreting that the highest SST occurred in April and the lowest in October. The SST pattern depends on the wind vector from January to June.

Because of the unique circulation in BoB, current research is important. Salinity estimates in the bay's central region are heavily influenced by freshwater flows, seasonal currents, eddies, and vertical displacements of surface and subsurface waters [5,12]. The influx of abundant fresh water in the waters of BoB is generated by the runoff of rivers such as the Ganges, Brahmaputra, Irrawaddy, Meghna, Godavari, Mahanadi, Krishna and Cauvery [13].

* Corresponding author: srizal@unsyiah.ac.id

The data and descriptions provide information on current velocity, temperature, salinity, and precipitation, as well as surface evaporation, through spatial mapping, which plays an important role in water circulation and climate, as well as the life of organisms in BoB waters. Current conditions affect the distribution of parameters in BoB and its relationship to evaporation and precipitation to temperature and sea surface salinity parameters. The purpose of this study is to investigate the characteristics and influence of current velocity on the salinity and temperature distribution patterns caused by circulation factors in the BoB. This detailed examination of the findings of research on the effect of sea water velocity on the distribution pattern of surface temperature and salinity in BoB is expected to provide research support and a better understanding, and it can be demonstrated through mapping.

2 Materials and Methods

The research was conducted at the Ocean Modeling Laboratory, Faculty of Marine and Fisheries, Universitas Syiah Kuala. The research began in January 2021. The research location is in BoB with a domain of 6.6°-24.60°N and 78.20°-96.70°E. Topographic data obtained from ERDDAP data from the Shuttle Radar Topography Mission (STRM30) which is accessed on the website https://topex.ucsd.edu/WWWhtml/srtm30_plus.html and then plotted to the Matrix Laboratory (MATLAB), so that the topographic map in Fig. 1.

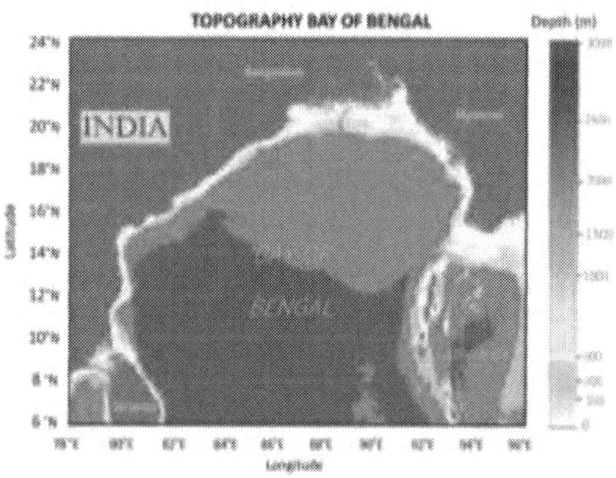

Fig. 1. BoB Topography.

2.1 Hybrid coordinate ocean model

Current, temperature and salinity data in the study area were obtained from the output of the Hybrid Coordinate Ocean Model (HYCOM) + NCODA Global 1/12° Reanalysis (source: https://www.hycom.org/data/glbu0pt08/expt-91pt2). Data was extracted over a period of two months, namely February and August 2017. HYCOM + NCODA global near real time model data. This data can be used as daily data because it can be accessed within 48 hours of initial runtime. This HYCOM data has a resolution of 1/120 x 1/120 (~9.25 km x 9.25 km). Details of HYCOM data specifications in providing data in this research in Table 1 are as follows.

Table 1. HYCOM data specification.

Title	HYCOM GLBu0.08
Resolution	**0.08°**
Institution	**Naval Research Laboratory**
Period	**18 April 2016 to 20 November 2018**
Grid	**80.48 S and 80.48 N**
Temporal Frequency	**Daily and per 3 Hours**

2.2 Precipitation and evaporation data

The precipitation and evaporation model data in this study were obtained from the European Center for Medium-Range Weather Forecast (ECMWF) data which was downloaded at the following link (source: https://www.ecm wf.int). The data used is daily data with units (m of water equivalent) in February and August 2017 (see Table 2).

Table 2. ECMWF data specification.

Title	Eropean Centre for Medium-Range Weather Forecast
Parameter	**Total Precipitation and Evaporation**
Period	January 1979 to December 2019
Grid	0.125° x 0.125° (ocean waves) Mean
Temporal Frequency	**Daily**

2.3 Data Analysis

Sea surface current data has a net common data file (NetCdf) data format obtained from HYCOM+NCODA in the form of daily data for February and August in 2017. Data extraction was carried out using MATLAB software to obtain two main current components, namely u and v. The formula for determining the velocity of the current direction is: $V = \sqrt{u^2 + v^2}$. Global sea surface temperature and salinity data as well as evaporation and precipitation data are processed using MATLAB software for monthly average calculations. Image cropping is carried out according to the study area. The cropping result is exported into image formats so that it can be read and displayed.

3 Results and Discussion

3.1 Current direction and velocity

The pattern of sea surface current circulation varies in the waters of BoB depending on the seasonal circulation. February and August are the months that represent winter (east season) and summer (southwest

https://doi.org/10.1051/e3sconf/202233902001

wind season). Current circulation in BoB is influenced by Northeast Monsoon Current (NMC) and South Monsoon Current (SMC) [13]. The current circulation pattern generated by the plot in the current pattern mapping provides an efficient average of the results. In the east monsoon, the northeast monsoon blows bringing the ocean currents to the southwest (NMC-SMC).

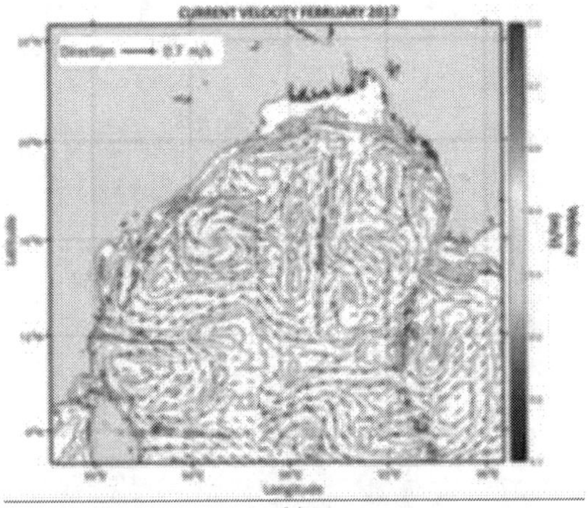

(a)

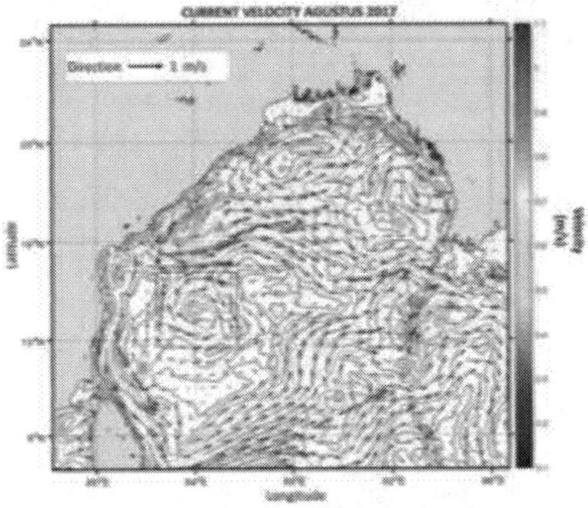

(b)

Fig. 2. (a) The direction and velocity of the current in February and (b) the direction and velocity of the current in August 2017.

Fig. 2(a) pattern of current direction and velocity in February. The current shown in February has an average velocity slightly lower than the current in August. It is due to the difference in wind pressure blowing from the northeastern mainland. An enlarged current vector as a comparison of a current direction with current velocity contours, surface geostrophic currents range from 0.4 to 0.6 m/s. A vortex (anti-cyclone) occurs in the western part of the bay at a velocity of about 0.5 m/s at 16°N,85°E. In the west of the bay (condition A) at latitude 12°N the direction of the current turns right, the pattern is formed due to a collision with the western part of the concave mainland called WBC. The direction of the current along the west coast that interacts with the eddies flows towards the north of the bay (WBC-north) then the current is blocked by currents originating from the east so that it flows towards the middle of the bay (condition B) forming a linear direction pattern, this is influenced by the anti-slip eddies. The cyclone produces a spread towards the continental surf, resulting in a reversal of the current with an irregular flow [14]. The current deflection to the right occurs in the south of the bay (condition C) where the curving current collides with the southwest boundary of the bay, namely the mainland of Sri Lanka.

Fig. 2(b) shows the current circulation, namely the direction and velocity in August. In the southwest monsoon the direction of the current reverses which is influenced by the west monsoon, where the wind blows moving the ocean currents to the northeast through the south of the bay (SMC-NMC). The current shown in August has a slightly higher velocity than the current in February, this is because the wind that blows comes from the southwest sea. Surface geostrophic currents in the west monsoon range from 0.4 to 0.8 m/s. In the west monsoon, the vortex (anti-cyclone) is observed to undergo a slight transition to the point 14°N.84°E west of the bay at a velocity of about 0.3 m/s. The geostrophic current pattern (condition A) has a strong contribution to bring the flow to the east with an average velocity of 0.6 m/s, the maximum current velocity of 0.8 m/s is around the point 15°N,83°E. The maximum current velocity is about 0.7 m/s (condition B) covering the southwestern part of the bay basin, where the current is heading right towards south through east Sri Lanka. This is caused by the Ekman pump and local winds around the coast. The flow of geostrophic currents from the south of about 1 m/s carried will enter the Andaman waters (conditions C and D).

3.2 Distribution of sea surface temperature

Winter identifies the average condition of SST in February with a low range which is influenced by seasonal currents. Fig. 3(a) shows the distribution of the SST in February. The lowest SST covers the northernmost waters of the bay ranging from 23°C - 26°C and extends to the shallow waters of the western continental. High SST conditions cover the south of the bay which is around 27°C -29°C. The distribution of SST seen is influenced by several seasonal current patterns that form in Figure 2.(a) It can be seen in Figure 3.(a) there is a variation in surface temperature in the western boundary area influenced by condition A on the rightward deflection of the current which brings a temperature of 27° C to the western boundary, the presence of an anti-cyclonic eddy produces an upwelling of 26°C. In the northern part of the bay, it is seen that the linear current pattern forms a temperature distribution pattern where the mass of water with low temperature flows towards the center of the bay (condition B).

Fig. 3(b) shows the distribution of SST in August. Summer identifies the average condition of SST in August with a high range that dominates along the continental surf. The lowest SST around 25°C covers the waters of the west coast of the bay. Low SST also dominates the eastern part of the bay at the Andaman Sea land boundary averaging 27°C-28°C.

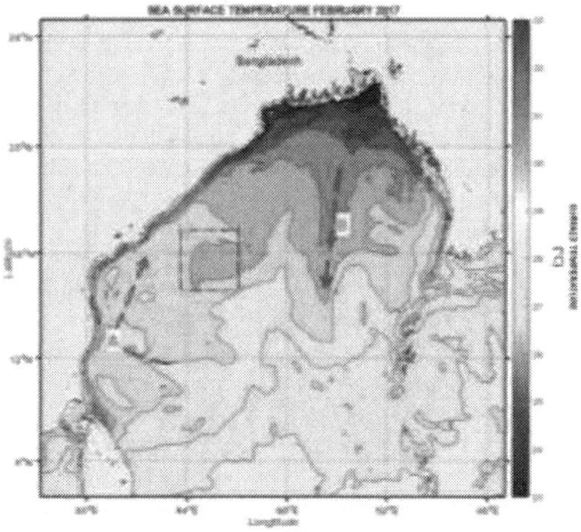

(a)

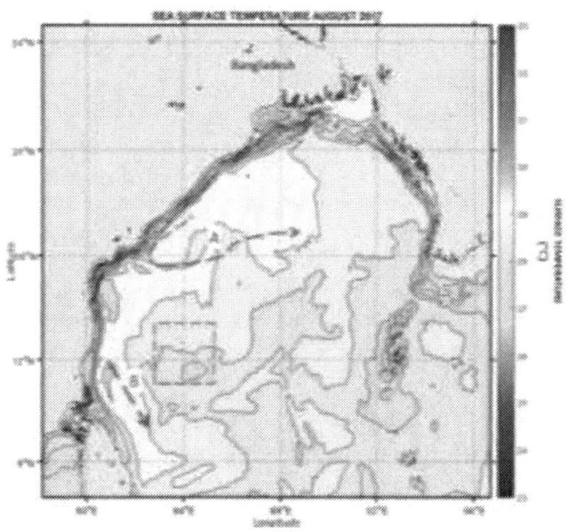

(b)

Fig. 3. (a) SST in February and (b) SST in August 2017.

In the northernmost area of the bay, the SST range is very high at 29.5°C-30°C due to atmospheric convection and turbidity. The high SST dominance covers the western part of the bay at an average of 29.5°C and interacts with anti-cyclonic eddies to produce an upwelling of 27°C. Some of the western seasonal current patterns that form in Fig. 2(b). It provides a pattern of distribution of high surface temperatures along the western boundary. Geostrophic currents in condition A push the hot water mass to spread to the middle of the bay and in condition B the hot water mass goes to the south influenced by currents in the southwest basin.

3.3 Sea surface salinity

Surface salinity in February is influenced by east monsoon currents and evaporation. Fig. 4(a) shows the distribution of Sea Surface Salinity (SSS) in February. The SSS plot shows the lowest salinity distribution covering the northern and northeastern waters of the bay at 29-32.5 psu. High salinity covers the waters southwest and south of the bay (33.5-34 psu) which is maintained by high evaporation and brine mass from the oceans. The pattern of salinity distribution seen in the image plot almost follows the shape of the current pattern in February. The spread of low salinity to the center is hampered by the presence of the western circulation, it is influenced by the magnitude of the backflow in the continental surf as in conditions A and eddies. The distribution of low salinity towards the middle is seen in condition B and the distribution of high salinity in the southern area of the bay is 33.5-34 psu from the open water mass.

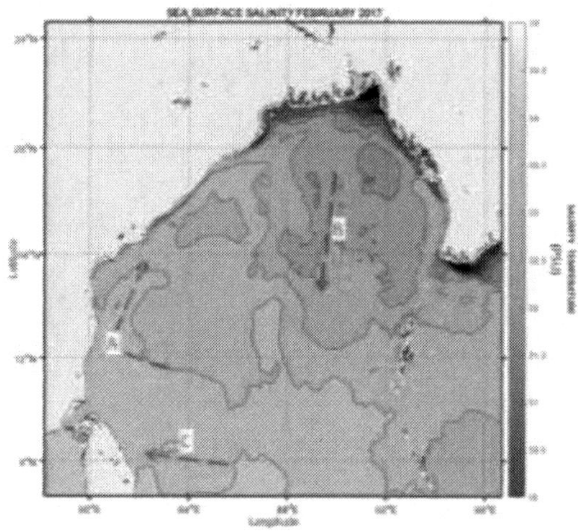

(a)

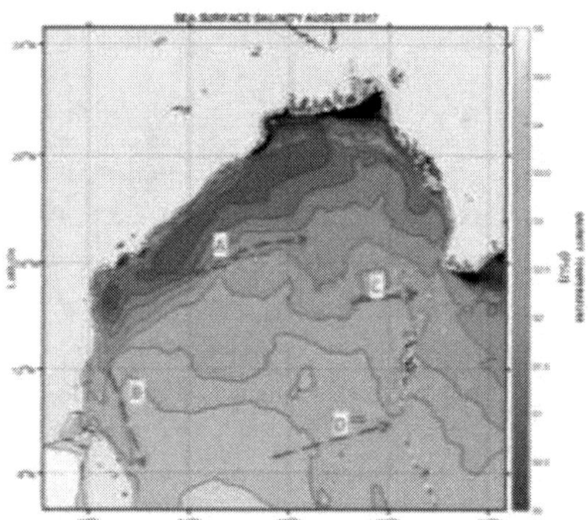

(b)

Fig. 4. (a) SSS in February and (b) SSS in August 2017.

Fig. 4(b) shows the condition of the distribution of SSS in August. In the southwest monsoon, the lowest salinity covers the north to west of the bay, which is 20-31.5 psu while high salinity covers the waters of the southwest and south of the bay (33.5-34.5 psu) maintained by evaporation, where high salinity currents from the ocean enter BoB. The surface salinity distribution pattern that occurs in the plot results shows that the geostrophic current in condition A maintains the mass distribution of low salinity water dominates the

 https://doi.org/10.1051/e3sconf/202233902001

bay. the forcing of low salinity flow to the south occurs in condition B by right-hand currents in the basin but the current from south to east is greater pushing the high salinity water mass of 34.5 psu. Conditions C and D show that each of the relatively high salinity water masses originating from the influence of the gulf geostrophic current flows into the Andaman Sea.

3.4 Relationship of precipitation and evaporation to SST and SSS

The convention that applies to precipitation is in the form of a positive value and evaporation in the form of a negative value (-) indicating that there is an accumulation of water mass in units of mm water of equivalent. The variation of precipitation in each season is related to the distribution of low salinity, precipitation in August (summer) is higher than February (winter). Figure 5.(a) shows that the precipitation produced in February is very small, ranging from 1-3 mm/day, a positive value of 3.5 mm is only covered in the southern part of the bay. Therefore, the distribution of salinity in February is more influenced by fresh water discharge.

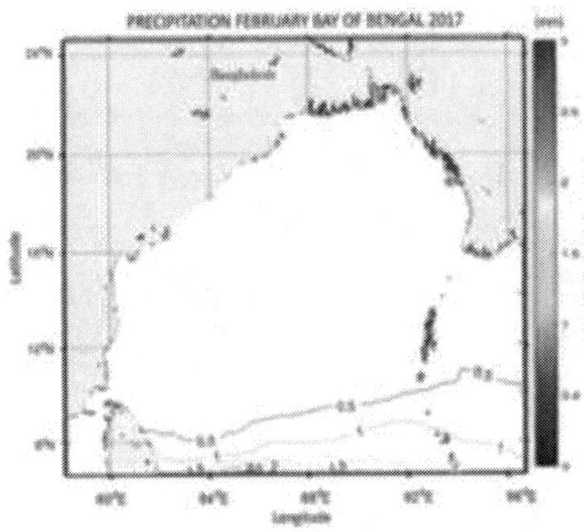

(a)

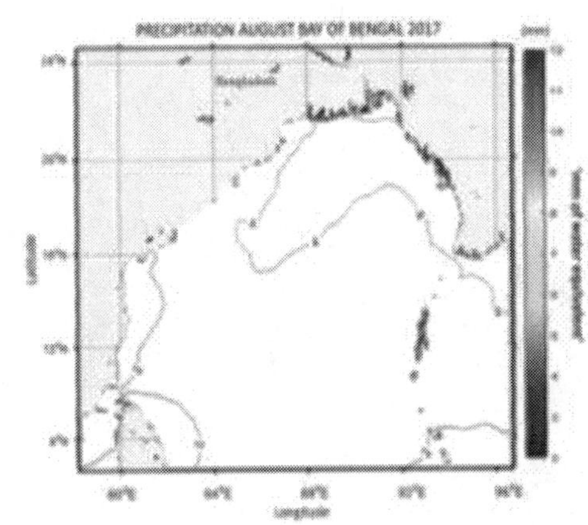

(b)

Fig. 5. (a) Precipitation in February and (b) SSS in August 2017.

In August (Fig. 5(b)), the predominance of rainfall over the bay averaged around 4 mm, strongly supporting the maintenance of a low SSS in the north, a more positive range of precipitation values towards the mainland of the bay.

In Fig. 6(a) shows the evaporation value in February. High evaporation is indicated by increasingly negative values that dominate the center of the bay, namely -1.4 to -1.6 mm. The smallest evaporation is -1.8 covering the south of the bay. The value of evaporation is relatively positive to the north where the temperature is lowest. In August (Fig. 6(b)), the negative value of evaporation covers the south of -1.6 mm, but the negative value extends to the central bay towards the north of -1.4 mm.

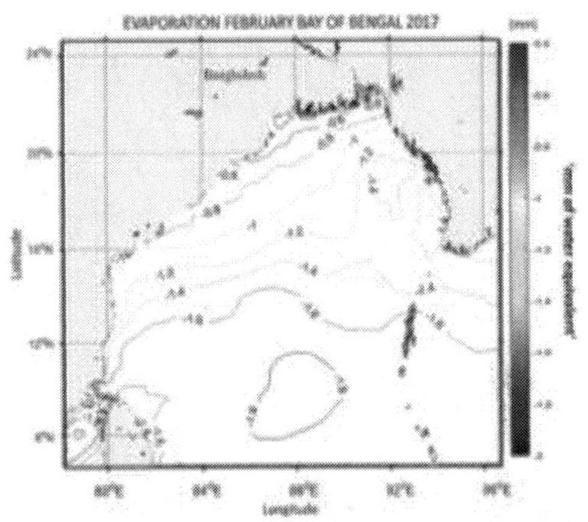

(a)

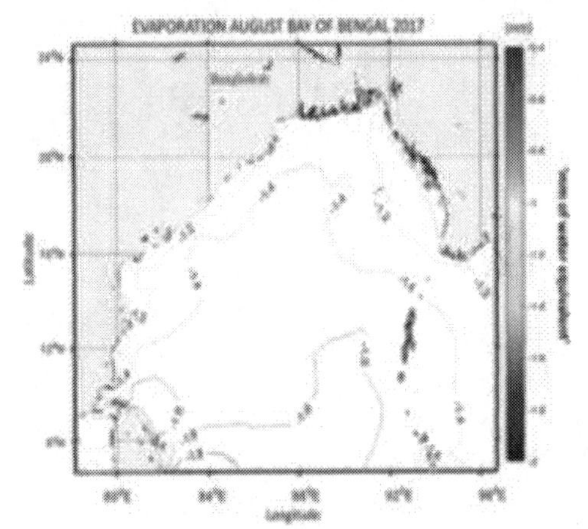

(b)

Fig. 6. (a) Evaporation in February and (b) SSS in August 2017.

In February the absolute positive value reached 4.5 mm towards the waters, while in August the positive value was 2 mm to 4 mm towards the mainland. High evaporation is supported by high SST. High SSS is supported by evaporation and low SSS is supported by high precipitation. The effect of high average rainfall values supports low SSS in the north and high evaporation supports high SSS in the south. The plot of Fig. 7 shows that the average value of the resulting precipitation exceeds the average value of evaporation. Therefore, the results of the combination of SST, SSS, precipitation and evaporation seem to have an impact on each other.

 https://doi.org/10.1051/e3sconf/202233902001

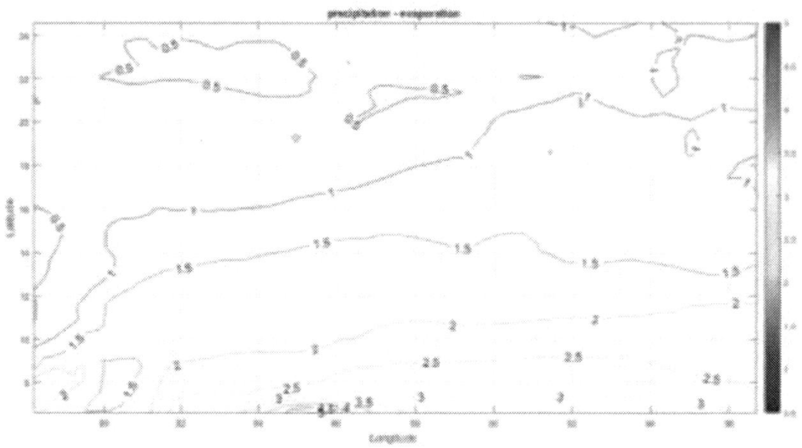

(a)

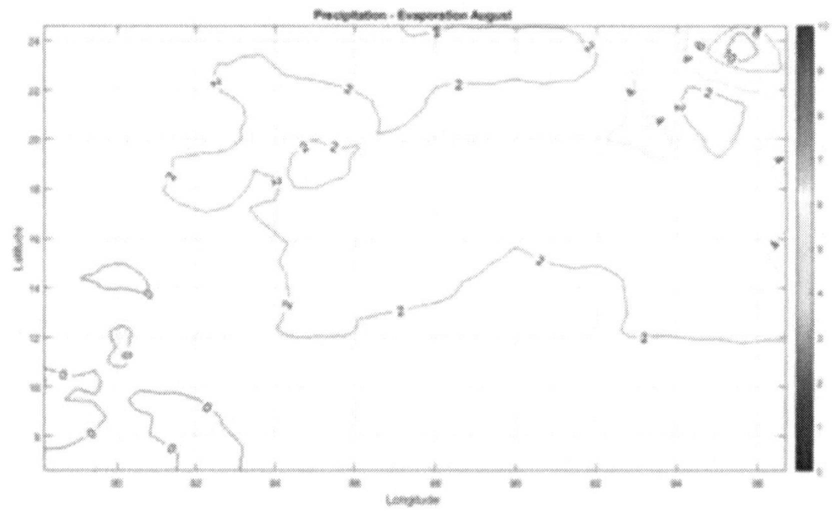

(b)

Fig. 7. Value of absolute difference between precipitation and evaporation (a) February (b) August.

4 Conclusions

Based on the results of the research that has been carried out, it can be concluded that in February the average sea water velocity is 0.4 - 0.6 m/s forming a backflow pattern from the western boundary and in August the average sea water velocity is 0.5 - 0.8 m/s. In February the distribution of SST is low and SSS is relatively high following the east monsoon current pattern. In August, the SST is high and the SSS is low, following the pattern of the southwest monsoon conditions. SST and SSS variations with precipitation and evaporation conditions support each other. The negative value of evaporation is supported by high SST. High SSS is supported by negative values of evaporation and low SSS by positive values of precipitation. Marine parameter data from HYCOM at sea level gives very good calculation results. It is recommended for more further research on monsoon currents, surface temperature and surface salinity stratification in the waters of BoB for detailed information of the impact provided is useful for research in the field of Marine Science.

Acknowledgment. Authors would like to express gratitude to Universitas Syiah Kuala for financial assistance in term of '*Penelitian Terapan Unggulan Perguruan Tinggi*', under contract number: 11/UN11.2/PT.01.03/DPRM /2021. The authors would like also to thank the Faculty of Marine and Fisheries for providing research support facilities. This research is carried out in the Ocean Modelling Laboratory, Department of Marine Sciences, Universitas Syiah Kuala, Indonesia.

References

1. G. S. Bhat and H. J. S. Fernando, Oceanography **29**, 232 (2016)
2. L. M. Beal, V. Hormann, R. Lumpkin, and G. R. Foltz, J. Phys. Oceanogr. **43**, 2008 (2013)
3. A. Chatterjee, D. Shankar, S. S. C. Shenoi, G. V. Reddy, G. S. Michael, M. Ravichandran, V. V. Gopalkrishna, E. P. Rama Rao, T. V. S.

Udaya Bhaskar, and V. N. Sanjeevan, J. Earth Syst. Sci. **121**, 559 (2012)
4. A. Buckley, J. M., Mingels, B., Tandon, Deep. Res. Part II Top. Stud. Oceanogr. **172**, 104653 (2020)
5. J. Sree Lekha, J. M. Buckley, A. Tandon, and D. Sengupta, J. Geophys. Res. Ocean. **123**, 6330 (2018)
6. Sree Lekha, J. Clim. **32**, (2019)
7. Gibilisco, Remote Sens. Environ. **248**, 111964 (2020)
8. K. Wyrtki, Physical Oceanography of the Southeast Asian waters (Naga Report. 1961)
9. K. Wyrtki, L. Magaard, and J. Hager, J. Geophys. Res. **81**, 2641 (1976)
10. D. Ardila, Y. Haditiar, M. Ikhwan, R. Wafdan, Muhammad, S. Sugianto, and S. Rizal, IOP Conf. Ser. Earth Environ. Sci. **348**, 012063 (2019)
11. S. Rizal, Muhammad, T. Iskandar, I. Setiawan, A. Satriadi, and Radinal, Proc. AIC Syiah Kuala Univ. Banda Aceh (2011)
12. V. P. Akhil, F. Durand, M. Lengaigne, J. Vialard, M. G. Keerthi, V. V. Gopalakrishna, C. Deltel, F. Papa, and C. De Boyer Montégut, J. Geophys. Res. Ocean. **119**, (2014)
13. P. N. Vinayachandran, D. Shankar, S. Vernekar, K. K. Sandeep, P. Amol, C. P. Neema, and A. Chatterjee, Geophys. Res. Lett. **40**, 1777 (2013)
14. S. Mandal, S. Pramanik, S. Sil, K. S. Arunraj, and B. K. Jena, J. Coast. Res. **89**, 132 (2020)

Mapping potential fishing zones based on sea surface temperature and Chlorophyll-A in the Waters of Aceh Besar, Indonesia

Muhammad Muhammad[1,2], *Agus* Widi Priana[1,3], *Junaidi* M. Affan[1,2], *Haekal* Azief Haridhi[1,2], *Irwan* Irwan[1,2], *Syarifah* Meurah Yuni[4], *Ichsan* Setiawan[1,2*]

[1]Departement of Marine Sciences, Faculty of Marine and Fisheries, Universitas Syiah Kuala, Banda Aceh 23111, Indonesia
[2]Research Center for Marine and Fisheries, Universitas Syiah Kuala, Banda Aceh 23111, Indonesia
[3]Ocean Diving Club, Universitas Syiah Kuala, Banda Aceh 23111, Indonesia
[4]Departement of Mathematics, Faculty of Mathematics and Natural Sciences, Universitas Syiah Kuala, Banda Aceh 23111, Indonesia

Abstract. Utilization of potential fishery and marine resources is still not maximally used by fishermen. Sea surface temperatures and the spread of chlorophyll-a through remote sensing systems can be used as potential fishing zones. This study aims to determine potential fishing zones using geographic information systems through the distribution of sea surface temperatures and the value of chlorophyll-a concentrations in Aceh Besar waters which is expected to improve the sustainability of fishing efforts. This research was conducted in January-December 2019 using geographic information system by extracting from Aqua-MODIS satellite image with resolution of 4 km and sea surface temperature overlay method with chlorophyll-a concentration. The results found the highest sea surface temperature value was in April at 31.62°C and the lowest temperature was in December at 28.45°C. The highest average chlorophyll-a concentration in Aceh Besar waters was 1.53 mg/m3 in December and the lowest concentration of chlorophyll-a in June and September was 0.20 mg/m3. Suspected potential fishing zones during January to December 2019 were found as many as 62 potential fishing zones, namely in the Western, Northwest and Eastern waters of Aceh Besar. The most zones are in December as many as eight zones, namely in the Western, Northwest and East parts of Aceh Besar waters and the least found in July as many as one zone that is in the Western part of Aceh Besar waters. When compared to fishing operations areas, the potential fishing zone is in the western part of Aceh Besar waters.

1 Introduction

Aceh Besar Regency is the westernmost region after Sabang City. Geographically, Aceh Besar Regency is located at coordinates 5.05° - 5.75° North Latitude and 94.99° - 95.93° East Longitude. To the north it is bordered by the Malacca Strait and Banda Aceh City, to the south by Aceh Jaya Regency, to the east by Pidie Regency, and to the west by the Indian Ocean. Aceh Besar Regency has a coastline of 296.16 km and a sea area of 1,351.01 km2 [1].

Aceh Besar District has a large potential for natural wealth, specially in the field of fisheries with a potential of around 5,158 tons/year [2]. The lack of technology used by fishermen, resulted in the utilization of the potential of fisheries and marine resources less than the maximum. The problem that is commonly faced in the utilization of fish resources is the existence of dynamic fishing areas, always changing and moving following changes in environmental parameters [3].

Fishermen usually determine the catch zone by looking at the foam or ripples on the sea surface and by looking at the birds flying on the sea surface. The abundance of fish in an area can also be predicted based on the oceanographic conditions of the waters [4]. Oceanographic conditions that greatly affect the abundance of fish are the distribution of chlorophyll-a and sea surface temperature. The location of waters with high chlorophyll-a content can be indicated in these waters rich in fish because the location is rich in food [5]. Meanwhile, Sea Surface Temperature (SST) is one of the factors that affect the life of organisms in the ocean, because temperature can affect the metabolism and reproduction of organisms in the sea. Each type of fish has a different tolerance to SST for their survival [6].

One way to make it easier for fishermen to find fishing locations, fishermen must have a map of areas that have the potential to carry out fishing operations. Remote sensing technology can determine water areas that have the potential for profitable fish resources. By knowing the sea surface temperature and the distribution of chlorophyll-a through remote sensing systems, it is possible to predict Potential Fishing Zones (PFZ) [7].

2 Materials and Methods

2.1 Time and location

This research was carried out by monitoring sea surface temperature (SST) and chlorophyll-a from January 2019 to December 2019 in the sea waters of Aceh Besar. Geographically, the research location is located at the coordinates as shown in Figure 1. The tools and materials used in this research are Personal Computer (PC), SeaDAS Software, and ArcGIS Software.

*Corresponding author : ichsansetiawan@unsyiah.ac.id

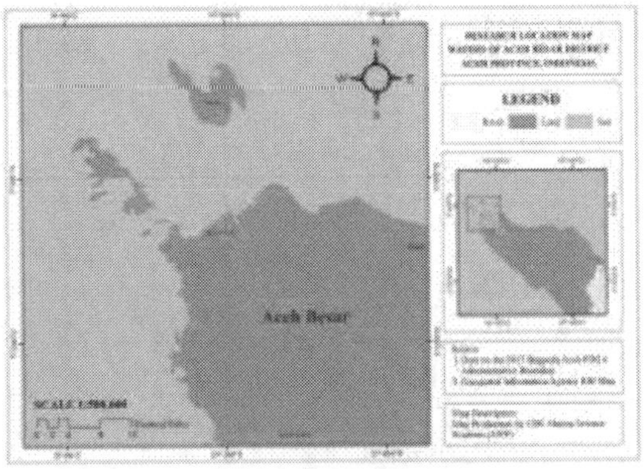

Fig. 1. Research map

2.2 Research data

The research data collected includes fishing grounds obtained by interviewing fishermen and image data from the Aqua-MODIS satellite in the form of chlorophyll-a (OCI Algorithm) and monthly sea surface temperatures from January 2019 to December 2019. The image data used is Browser Level data. 3 with a spatial resolution of 4 km. Satellite image data is accessed from the website www.oceancolor.gsfc.nasa.gov.

2.3 Estimated *fishing area*

If a water with a sea surface temperature range between 24°C - 27°C, is categorized as potential, the sea surface temperature range between 27°C - 30°C is categorized as moderate and the temperature range is less than 24°C and more than 30°C, then the water is categorized as less potential. Meanwhile, if the chlorophyll-a content in the waters is greater than 0.2 mg/m^3, it is categorized as potential, the chlorophyll-a content 0.1 mg/m^3 - 0.2 mg/m^3 is categorized as moderate and the chlorophyll-a content is less than 0.1 mg/m^3 (Table 1).

Table 1. PFZ assessment through SST and Cholorophyll-a indicators

No	PFZ Category	Temperature Criteria (°C)	Chlorophyll-a Criteria (mg/m^3)
1	Potential	24 - 27	> 0.2
2	Medium potential	27 - 30	0.1 – 0.2
3	Less potential	<24 - >30	< 0.1

Source : [8], [9], [10]

2.4 Research procedure

1. Primary data was collected through direct interviews with 10 fishermen.
2. Secondary data was taken in the form of images of sea surface temperature and chlorophyll-a from MODIS (Moderate Resolution Imaging Spectroradiomater) satellite imagery.
3. Cropping is done according to the desired area which includes the waters of Aceh Besar.
4. The process of extracting information from each image using the Seadas 7.5 application, which produces ASCII data on the distribution of chlorophyll-a and SST distribution in Aceh Besar waters as well as tabulated data on the distribution of chlorophyll-a and SST values for each pixel and its coordinates.
5. The interpolation process is carried out to get the value in the empty area (empty pixels).
6. Overlay is performed to combine the contour of the sea surface temperature image and the contour of the chlorophyll-a distribution image.
7. Map of potential fishing zones in Aceh Besar waters is made.

3 Results and Discussion

3.1 Characteristics of sea surface temperature

Sea surface temperature (SST) can be used to determine potential fishing zones in an area. The distribution of sea surface temperatures in the waters of Aceh Besar from January to December 2019 based on the results of MODIS satellite image extraction has varying values. The SST value in the waters of Aceh Besar can be seen in Figure 2.

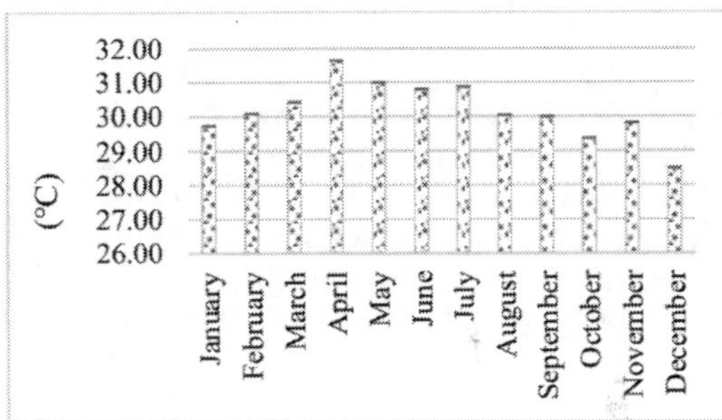

Fig. 2. Characteristics of sea surface temperature

The results of MODIS satellite image data extraction show that the highest SST value is in April, which is 31.62°C. The lowest temperature was detected in December at 28.45°C. The difference in SST values in each month is thought to be caused by differences in weather, season and rainfall. The water temperature becomes low when the rainfall is high [11]. Sea surface temperature in waters can also be influenced by patterns of movement of the sun, wind, and air pressure [12].

3.2 Characteristics of chlorophyll-a

Chlorophyll-a can be used to determine potential fishing zones, where chlorophyll-a can determine the fertility of a waters. The concentration of chlorophyll-a in the waters of Aceh Besar from January to December 2019 based on MODIS satellite image extraction has varied values. The concentration of chlorophyll-a in the waters of Aceh Besar can be seen in Figure 3.

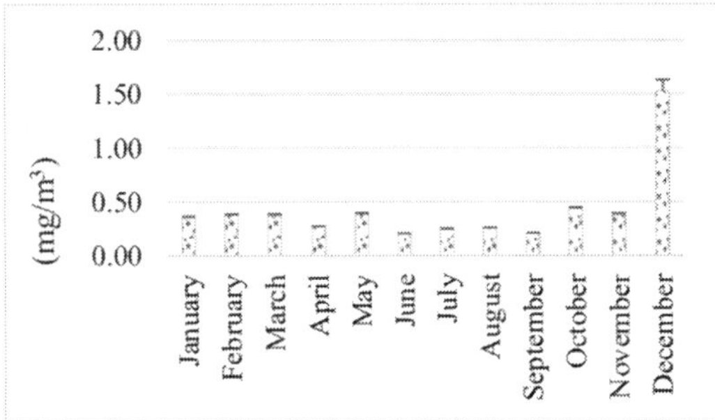

Fig. 3. Chlorophyll-a

The results of MODIS satellite image extraction show that the highest average value of chlorophyll-a concentration in Aceh Besar waters is in December, which is 1.53 mg/m^3. The lowest concentration of chlorophyll-a was in June and September at 0.20 mg/m^3. High and low concentrations of chlorophyll-a are thought to be caused by currents. Chlorophyll-a is a green pigment found in phytoplankton where the movement of phytoplankton is strongly influenced by currents [13]. Fish react directly to the current by orienting themselves directly to the current [14].

3.3 Estimation of potential fishing zones

The results of the estimation of potential fishing zone are shown in Figures 4 to 14. Based on the results of the study, the number of potential fishing zones each month has different values. In January there were 6 potential fishing zones with SST ranging from 28.57°C - 29.68°C classified in the medium category, while the concentration of chlorophyll-a ranged from 0.2 mg/m^3 - 0.8 mg/m^3 belonging to the category potential. In February, there were 6 potential fishing zones with SST ranging from 29.04°C - 29.64°C belonging to the medium category, while the concentration of chlorophyll-a ranged from 0.4 mg/m^3 - 0.7 mg/m^3 classified in the middle category. potential category. In March, there were 4 potential fishing zones with SST) in the potential fishing zones ranging from 29.19°C - 30.11°C belonging to the medium category, while the concentration of chlorophyll-a ranged from 0.5 mg/m^3 - 1, 9 mg/m^3 belongs to the category of potential.

In April, there were 3 potential fishing zones with SST ranging from 29.84°C - 31.30°C classified in the medium category, while the concentration of chlorophyll-a ranged from 0.2 mg/m^3 - 0.8 mg/m^3 classified in the middle category. potential category. In May, there were 5 potential fishing zones with SST ranging from 29.90°C - 30.62°C belonging to the medium category, while the concentration of chlorophyll-a ranged from 0.4 mg/m^3 – 2.3 mg/m^3 classified in the middle category. potential category. In June, there were 3 potential fishing zones with SST ranging from 29.18°C - 30.32°C classified in the medium category, while the concentration of chlorophyll-a ranged from 0.2 mg/m^3 – 0.3 mg/m^3 classified in the middle category. potential category.

In July, 1 fishing potential zone with SST ranging from 29.97°C - 30.34°C was found in the medium category, while the chlorophyll-a concentration of 0.4 mg/m^3 was classified in the potential category. In August, there were 6 potential fishing zones with SST ranging from 28.83°C - 29.65°C classified in the medium category, while the concentration of chlorophyll-a ranged from 0.2 mg/m^3 - 0.8 mg/m^3 classified in the middle category. Potential category. In September, there were 7 potential fishing zones with SST in the fishing potential zone ranging from 28.03°C - 29.98°C classified as medium category, while the concentration of chlorophyll-a ranged from 0.2 mg/m^3 - 0.7 mg/m^3 belongs to the potential category.

In October, there were 7 potential fishing zones with SST ranging from 28.13°C - 28.88°C classified as being in the medium category, while the concentration of chlorophyll-a ranged from 0.3 mg/m^3 – 1.4 mg/m^3 classified in the middle category. Potential category. In November, 6 fishing potential zones were found with SST in the fishing potential zone ranging from 28.71°C - 29.62°C classified as medium category, while the concentration of chlorophyll-a ranged from 0.4 mg/m^3 - 0.7 mg/m^3 belongs to the potential category. In December, there were 8 potential fishing zones with SST ranging from 26.74°C - 29.47°C and classified as being in the medium category, while the concentration of chlorophyll-a ranged from 0.4 mg/m^3 - 8.5 mg/m^3 classified as in the potential category.

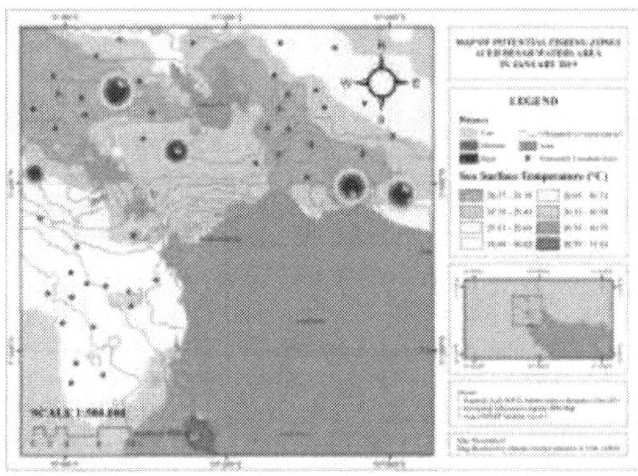

Fig 4. Estimation of potential fishing zones in January

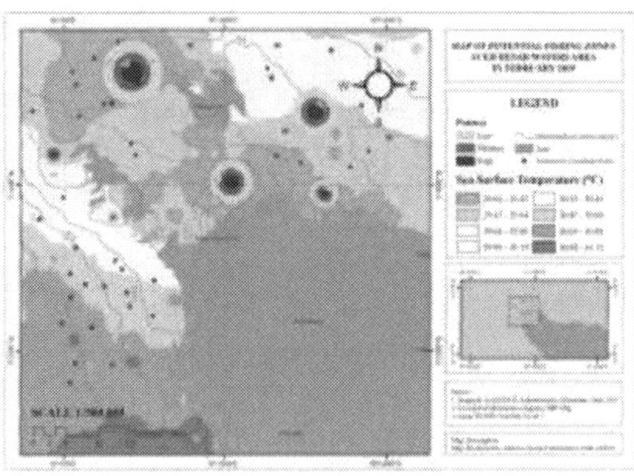

Fig. 5. Estimation of potential fishing zones in February

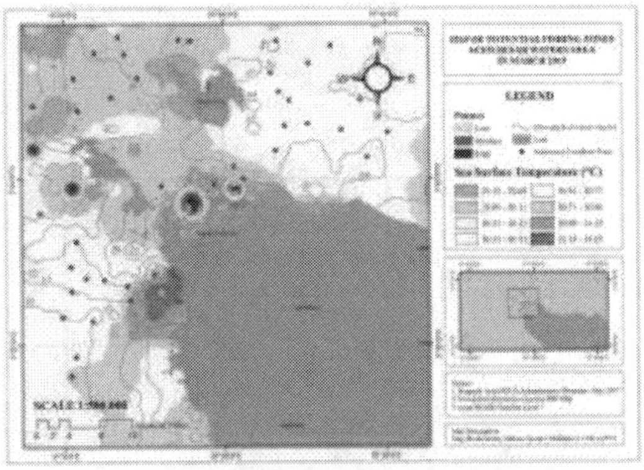

Fig. 6. Estimation of potential fishing zones in March

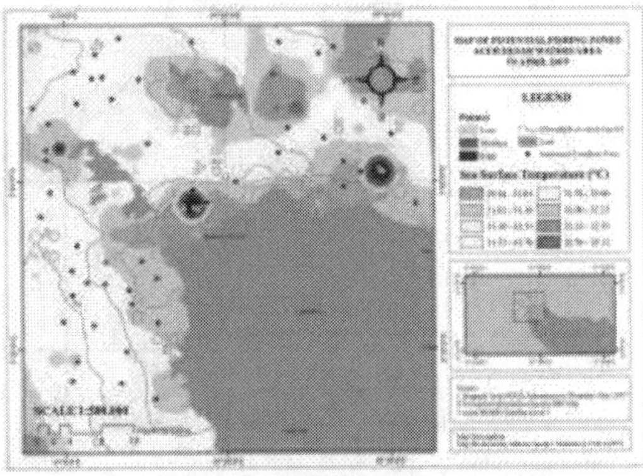

Fig. 7. Estimation of potential fishing zones in April

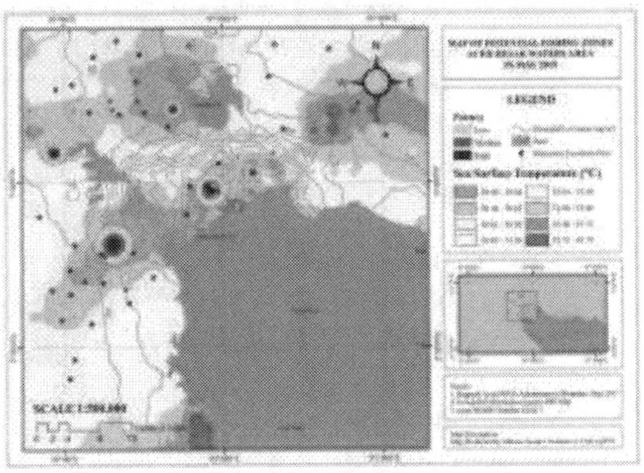

Fig. 8. Estimation of potential fishing zones in May

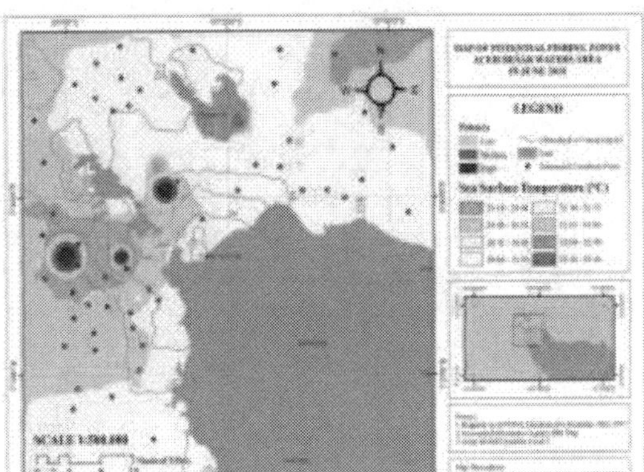

Fig. 9. Estimation of potential fishing zones in June

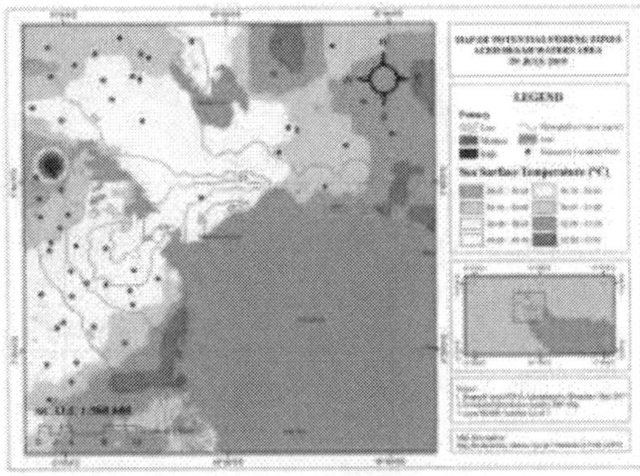

Fig. 10. Estimation of potential fishing zones in July

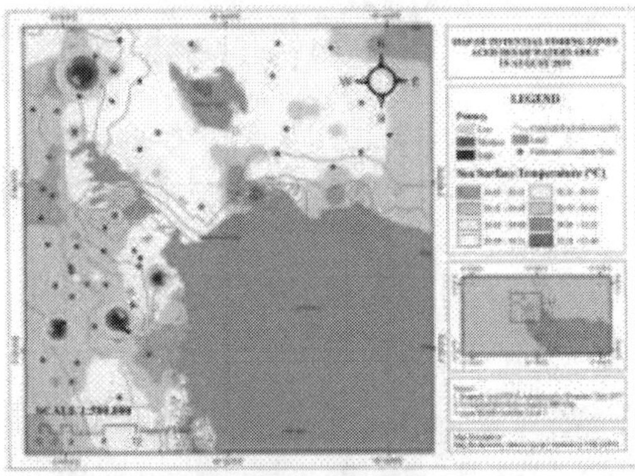

Fig. 11. Estimation of potential fishing zones in August

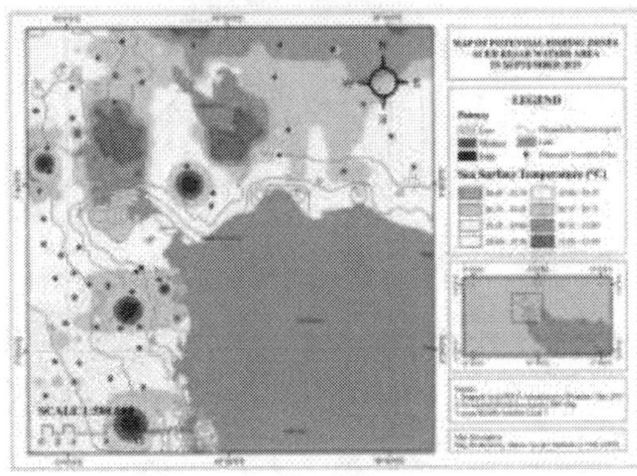

Fig. 12. Estimation of potential fishing zones in September

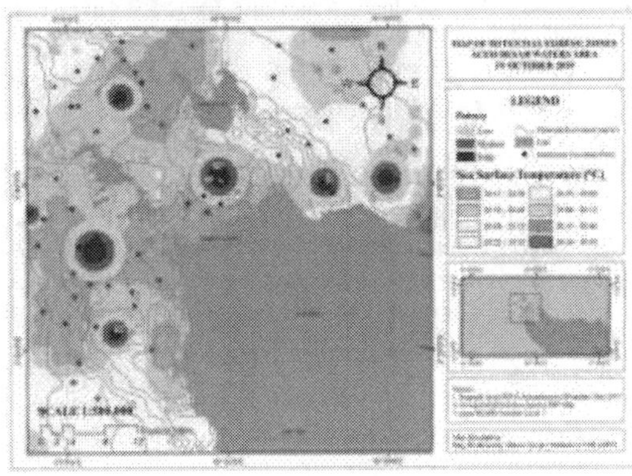

Fig. 13. Estimation of potential fishing zones in October

 https://doi.org/10.1051/e3sconf/202233902002

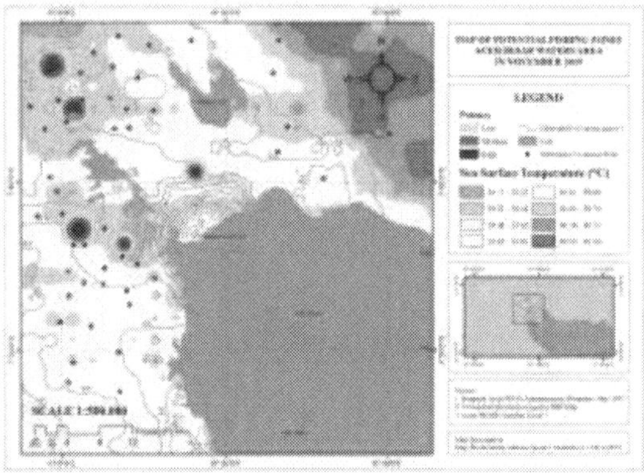

Fig. 14. Estimation of potential fishing zones in November

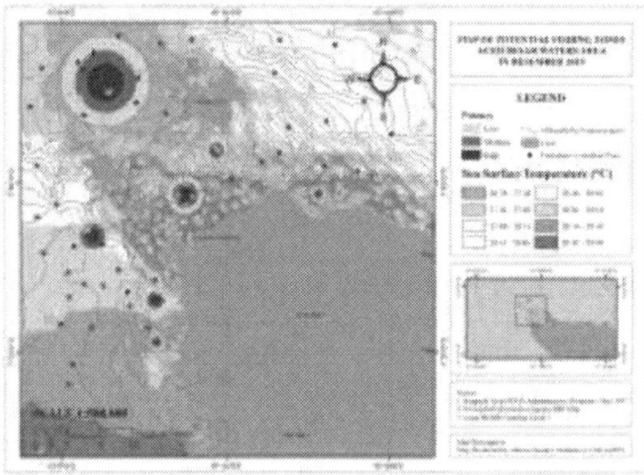

Fig. 15. Estimation of potential fishing zones in December

3.4 Fisherman's fishing operation area

Potential fishing zones based on sea surface temperature and chlorophyll-a that have been analyzed are validated with data from the fishing operation area of fishermen in Aceh Besar. The results of data analysis based on the fishing grounds of fishermen in Aceh Besar differ from the results of data analysis conducted through the Geographic Information System based on sea surface temperature and chlorophyll-a.

The fishing grounds of fishermen are spread over almost all temperature ranges and the amount of chlorophyll-a, which is different from the results of potential fishery zones obtained based on SST and chlorophyll-a from MODIS satellite imagery data. This can be influenced by several factors, one of which is currents that can directly or indirectly affect the marine environment and the biota that live in it, including the distribution of fish [15]. In addition, currents also affect small pelagic fish and the stability of the fishing gear used by fishermen [16]. Another influential factor is other oceanographic factors. The distribution of pelagic fish is strongly influenced by oceanographic conditions, such as temperature, salinity, surface currents, and dissolved oxygen [17].

4 Conclusion

The conclusion of this research is that the highest sea surface temperature is in April, which is 31.62°C and the lowest temperature is in December, which is 28.45°C. MODIS satellite image extraction results show the highest average value of chlorophyll-a concentration in Aceh Besar waters is in December, which is 1.53 mg/m^3 and the lowest chlorophyll-a concentration is in June and September, which is 0.20 mg/m^3. The estimation of potential fishing zones during January to December 2019 found as many as 62 potential fishing zones. The most zones were found in December as many as 8 zones and the least found in July as many as 1 zone.

Acknowledgment. The authors would like to thank the fishermen at TPI Lampulo who have been willing to provide information regarding fishing operation areas and to the Ocean Diving Club of Universitas Syiah Kuala (USK) for providing additional information related to Geographic Information Systems and all parties involved directly or indirectly involved We cannot mention one by one.

References

1 BPS, *Kabupaten Aceh Besar dalam angka,* Aceh Besar (2019)

2 DKP Aceh, *Potensi usaha dan peluang investasi di Provinsi Aceh* (2019)

3 F. Prastianto, Jurnal Ilmu Perikanan Tropis, **21(2)** (2016)

4 Mursyidin, K. Munadi, Z. A. Muchlisin, Jurnal Rekayasa Elektrika, **11(5)** (2015)

5 D. Simbolon, Silvia, P.I. Wahyuningrum, Marine Fisheries, **4(1)** (2013)

6 Kurnia, S. Purnawan, T. Rizwan, Jurnal Ilmiah Mahasiswa Kelautan dan Perikanan Unsyiah, **1(2)** (2016)

7 M. F. Hafiz, T. Imam, A. W. Bambang, Journal of Fisheries Resources Utilization Management and Technology, **6(4)** (2017)

8 Bukhari, W. Adi, Kurniawan, Jurnal Sumberdaya Perairan, **11(1)** (2017)

9 B. Hasyim. Makalah Pribadi Pengantar Kefalsafahan Sains. Sekolah Pasca Sarjana, IPB. Bogor. (2004)

10 Widodo, J. Prosiding seminar validasi data inderaja untuk Bidang Perikanan. Badan Pengkajian dan Penerapan Teknologi. Jakarta. (2019)

11 I. Shadiqin, M. Musman, A. Rahmah, Jurnal Ilmiah Mahasiswa Kelautan dan Perikanan Unsyiah, **1 (3)** (2016)

12 K. Wyrtki, *Physical oceanography of the southeast ecosystem*, California (1961)

13 T. D. Kuswanto, M. L. Syamsuddin, Sunarto, Jurnal Perikanan dan Kelautan, **8(2)** (2017)

14 M. P. M. Reddy, *Influence of the various oceanographic parameters on the abundance of fish catch* in Proceeding of International workshop on Aplication of Satellite Remote Sensing for Identifying and Forecasting Potential Fishing Zones in Developing Countries, India (1993)

15 M. S. Wibisono, *Pengantar ilmu kelautan*, Grasindo, Jakarta (2005)

16 A. R. Jalil, Depik, **2(1)** (2013)

https://doi.org/10.1051/e3sconf/202233902002

17 Edmondri, *Studi penangkapan ikan cakalang dan madidihang di perairan Sumatera Barat pada musim timur*, Skripsi, Institut Pertanian Bogor, Bogor (1999)

Doping nikel (Ni) metal in zeolitic imidazolate framework-8 using green synthesis method

Rahmatul Fajri[1*], *Puji* Wahyuningsih[1], *Jofrishal* Jofrishal[2], and *Ratna* Ediati[3]

[1]Department of Chemistry, Faculty of Engineering, Universitas Samudra, Jl. Meurandeh Langsa Aceh 24416, Indonesia
[2]Department of Chemistry Education, Faculty of Teacher Training and Education, Universitas Samudra, Jl. Meurandeh Langsa Aceh 24416, Indonesia
[3]Department of Chemistry, Faculty of Mathematics and Natural Sciences Institute of Technology of Sepuluh Nopember (ITS), Jl. Arif Rahman Hakim, Surabaya 60111 Indonesia

Abstract. Ni-Doping Ni metal into ZIF-8 material aims to increase the ability of ZIF-8. Ni-ZIF-8 was synthesized using an environmentally friendly method by reusing methanol as a solvent in the previous process and carried out at room temperature. The ratio of the molar ratios used was 1:4:4 for Zinc nitrate:2-methylimidazolate: TEA. As an effort to save the environment from the dangers of waste caused by the filtrate, Ni-ZIF-8 was then synthesized using the same filtrate up to six times. Based on the results of characterization using XRD, FTIR and SEM showed that the Ni-ZIF-8 material synthesized using the filtrate had similarities with Ni-ZIF-8 synthesized with pure methanol.

1 Introduction

Zeolitic Imidazole framework (ZIF-8) is a material synthesized from zinc metal and imidazole ligands with frameworks that are similar to zeolite. [1]. ZIF-8 has many advantages in various fields, such as gas storage materials, adsorbents, and catalysts [2-5]. Synthesis of ZIF-8 can be carried out using several organic solvents such as methanol, DMF, and mixing between the two solvents [6-7]. The addition of other metals such as Ni, Co, and Fe in the synthesis process of ZIF-8 aims to improve the performance of ZIF-8 that's expected it has better performance than before metal addition. For example, the use of Fe metal in the synthesis of ZIF-8 as an adsorbent has been successfully applied to remazol Deep Black RGB pollutants found in water. The results of this study indicate that Fe-ZIF-8 has a better capacity compared to ZIF-8. [8-10].

The excessive use of organic solvents in the synthesis of inorganic materials creates new problems for the environment, that are the waste from the filtrate produced by the synthesis pollutes the environment. One of the efforts made by researchers is to minimize the use of organic solvents in the synthesis, by recycling the remaining solvent from the previous synthesis [11].

2 Method

2.1 Procedures

$Zn(NO_3)_2.4H_2O$ and $NiCl_2.6H_2O$ (5%) dissolved in methanol. In another place, using methanol, triethylamine and 2-methyl imidazole were dissolved. The two solutions were mixed in one place and then stirred for 1 hour at room temperature. Furthermore, the residue obtained was filtered and washed using pure methanol. The sample was dried for 3 hours at 60°C, and again dried for 2 hours at 100°C to remove the TEA present in the sample. The obtained material is labeled P1-Ni-ZIF-8.

Ni-ZIF-8 was then synthesized with solvent from the P1-Ni-ZIF-8 synthesis filtrate. All treatments were carried out the same as before. The obtained material was then labeled P2-Ni-ZIF-8. Synthesis was carried out up to 6 times.

2.2 Characterization of Synthesized Materials

The XRD characterization aims to determine the structure of the resulting solid and compare it to the simulated ZIF-8 diffractogram. The Ni-ZIF-8 material was analyzed using XRD by 1 gram of the crushed sample and placing it in the holder provided. The sample was irradiated with an X-ray with a Cu Kα axis (1 = 1.5405) at 40 kV and 30 mA, at 2θ from 2°-50° with a scan speed of 0.05°/sec. The data obtained are in the form of 2θ, d spacing and intensity of diffraction peaks. Then, the data obtained were compared with references [12]. Characterization with FTIR was used to determine the peak or absorption characteristics of the Ni-ZIF-8 material. FTIR characterization was carried out on solids ZIF-8, P1-Ni-ZIF-8 and P3-Ni-ZIF-8, and recorded at wavenumbers 400-4000 cm-1. The samples were also characterized by scanning electron microscopy (SEM) (20 kV and 9.5 mm).

* Corresponding author: rahmatulfajri@unsam.ac.id

3 Results And Discussion

3.1 Characterization of Zeolitic Imidazole Frameworks-8 Doping Nickel (Ni-ZIF-8)

3.1.1 X-Ray Difraction Analysis

The synthesis of Ni-ZIF-8 begins with the dissolution of Zn $(NO_3)_2.4H_2O$ and $NiCl_2.6H_2O$ in methanol, producing a colorless solution (solution A). then the methylimidazole ligand and TEA were dissolved in methanol solvent, producing a colorless solution (solution B). Next, the two solutions are mixed in a closed vial and stirred for 1 hour. When the two solutions are mixed, the color changes from a colorless solution to a purplish blue. The color change that occurs is a possibility of competition between Zn^{2+} and Ni^{2+} in their interaction with methylimidazole ligands and forming a precipitate. After the mixture was allowed to stand for 24 hours, two layers were formed. The bottom layer is a purplish-blue precipitate, and the top layer is a clear greenish liquid. The separation of the two layers is carried out by filtering. After drying, the solid is obtained in the form of a purplish blue powder given the notation P1-Ni-ZIF-8. The resulting filtrate was used for the subsequent synthesis of Ni-ZIF-8. The purplish blue color seen in the resulting solid is due to the addition of the transition metal Ni which tends to form colored compounds or complex ions. This is related to the excitation of electrons that occurs in the d subshell. An electron jumps from a lower energy d orbital to a higher energy d orbital resulting in an excited electron. The difference in energy of the excited electron is equal to the energy of the absorbed photon, thus showing color at a wavelength that has the same energy as the energy of the excited atom (crystal field theory) [13].

The XRD pattern of Ni-ZIF-8 (Fig. 1). The pattern compare with the simulated pattern of ZIF-8 Gross et. al [14]. P1-Ni-ZIF-8 solid diffractogram synthesized using pure solvent has a higher intensity than the solid synthesized using filtrate. This is occur due to the amount of pure solvent used is more than the solvent from the filtrate. Characteristic peaks of Ni-ZIF-8 were synthesized using filtratte from previous synthesis are the same. Based on Gross et. at [14] diffraction pattern X-Ray has characteristic peaks. There was sliding slightly peaks between ZIF-8 and Ni-ZIF-8. The sliding effect indicate Ni^{2+} has been dopped in ZIF-8, for replacing most between Zn metal with Ni metal. The metal replace between Zn metal with Ni metal doesn`t change of ZIF-8 scructure formed. As well Ni-ZIF-8 has been synthesized using filtrate from frevios synthesis, it capable produce the same sctructure with the first synthesis of Ni-ZIF-8. Repetition of Ni-ZIF-8 synthesis using filtrate from previos synthesis can be done six times.

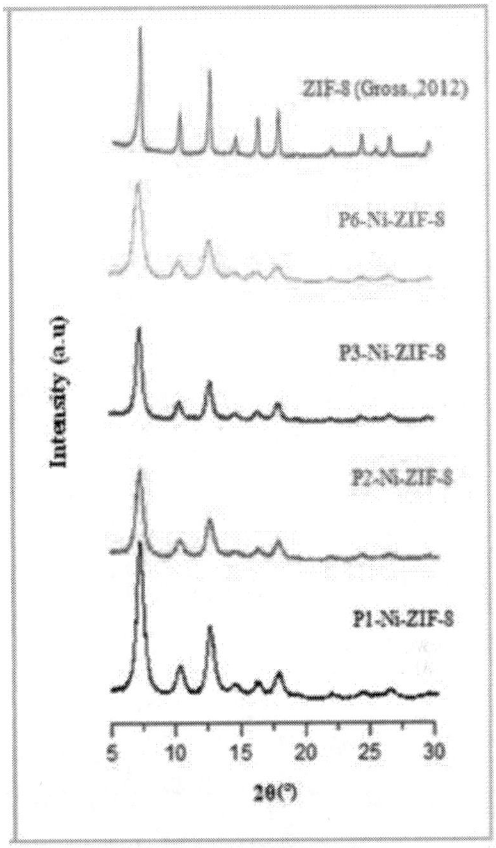

Fig. 1. X-ray diffraction pattern of the sample ZIF-8 and Ni-ZIF-8 synthesized and research Gross et al. (2012).

3.1.2 Spektrofotometer Fourier Transform Infra-Red Analysis

The FTIR spectra of the obtained Ni-ZIF-8 materials (Fig. 2) confirm that the chemical bond structure corresponds to that reported for ZIF-8 [15]. Figure 2 shows 2-methylimidazole has a strong and broadband at wave number 3400 to 2200 cm^{-1}. The confirm vibrations of the Hydrogen bond established between the Nitrogen group which has a hydrogen bond. This indicates that 2-methylimidazole has been deprotonated. FTIR of Ni-ZIF-8 shows the formation of a bond Zink between Nickel-metal the presence of an absorption band at a wavelength of 421 cm^{-1}, and the presence of vibrations from carbon-hydrogen bonds at wavenumber 760 cm^{-1} [16]. Furthermore, there is an absorption band at a wavenumber of 1586 cm^{-1} with a weak intensity which is a bending vibration of Carbon and Nitrogen.

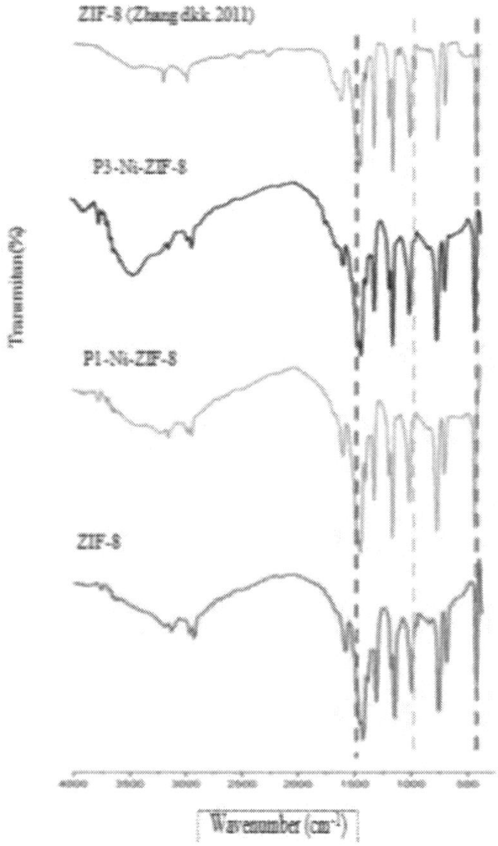

Fig. 2. FTIR spectra of ZIF-8 and Ni-ZIF-8 reported Zhang et al. 2011

Based on FTIR is known that all samples synthesized and reuse have a characteristic absorption band of material ZIF-8 and there was a slight shift as well as a decrease in the peak intensity at several wave numbers due to the addition of Ni^{2+} cations in the synthesis process.

3.1.3 Scanning Electron Microscopy Analysis

SEM image (Fig. 3) showed that they were aggregated as globular particles of Ni-ZIF-8 material. The material of Ni-ZIF-8 has a microsize and an imperfect cube shape and some are round [5]. The SEM results of the P6-Ni-ZIF-8 material show a variety of surface morphology, some are box-shaped, some are slightly rectangular and some are even irregular. P1-Ni-ZIF-8 has imperfect cube morphology and confirm about particle size smaller than ZIF-8. The particle size of P1-Ni-ZIF-8 is ± 22.66 m while that of ZIF-8 is ± 46 m. Fig. 4(b) and 4(c) shows the P3-Ni-ZIF-8 and P6-Ni-ZIF-8 morphology respectively. Both P3-Ni-ZIF-8 and P6-Ni-ZIF-8 have cube morphology and particle size around 46.25 83.3 m. This result shows the possibility of synthesis Ni-ZIF-8 using filtrate or we call green method.

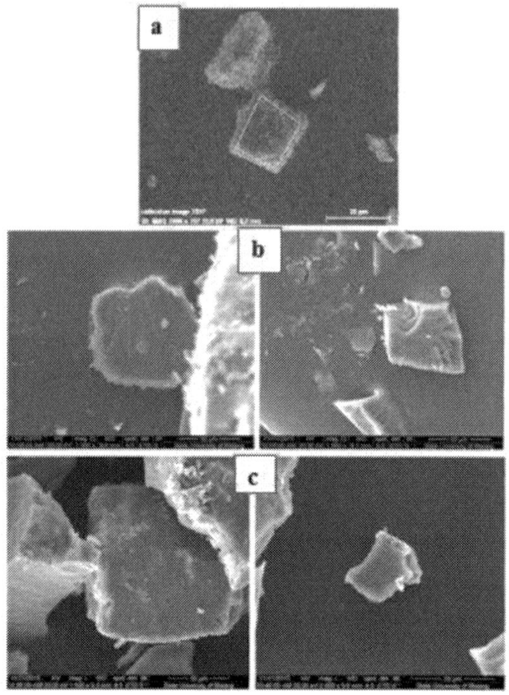

Fig. 3. SEM image of (a) P1-Ni-ZIF-8, (b) P3-Ni-ZIF-8 and (c) P6-Ni-ZIF-8

4 Conclusions

Ni-ZIF-8 material has been successfully synthesized using an environmentally friendly method, with the addition of TEA as a deprotonating agent, and synthesized at room temperature. Based on the results of the analysis using XRD, FTIR and SEM showed that there was no change in the structure of the resulting Ni-ZIF-8. This synthesis method can be applied as an effort to anticipate environmental damage from the waste generated from the synthesis.

Acknowledgments. This research was funded by Universitas Samudra through the Institute for Research and Community Service (LPPM) for a research grant Penelitian Dasar Unggulan (PDU). We would like to express our gratitude to several institutions, including the Basic Laboratory of Universitas Samudra, Lhokseumawe Polytechnic Laboratory, and the Chemistry Laboratory of Universitas Syiah Kuala.

References

1. Q. Bao., Y. lou., T. Xiang., J. Chen, Inorganic Chemistry Communication **37**, 170-173 (2013)
2. K. S. Park., Z. Ni., A.P. Cote., J.Y. Choi., R. Huang., F. J. U. Romo., H.K. Chae., M. O'Keeffe., O. M. Yaghi. PNAS **103**. 10187-10191(2006)
3. B. Chen., F. Bai., Y. Zhu., Y. Xia. Microporous and Mesoporous Material **193**. 7-14 (2014)
4. Y. Pan., Y. Liu., G. Zeng, L. Zhao., Z. Lai. Chem. Commun. **47**. 2017-2073 (2011)
5. L.T.L. Nguyen., K.K.A. Le., N.T.S. Phan. **33**. 688-698 (2012)

6. Y. H. Hu., L. Zhang. Adv. Mater. **22**. E117-E130 (2010)
7. E.I. Bustamante., J.I. Fernandez., J. M. Zamaro. Journal of Colloid and Interface Science. **424**. 37-43 (2014)
8. J. M. Yang., Q. Liu., W. Y. Sun, Microporous and Mesoporous Material **190**, 26-31 (2014)
9. J. A. Botas., G. Calleja., M. Sanchez., M. G. Orcajo. Hydrogen Energy **36**, 10834-10844 (2011)
10. M. T. Thanh., T. V. Thien., V. T. Chau., P. D. Du., N. P. Hung., D. Q. Khieu Hindawi Journal of Chemistry. 1-18 (2017)
11. N.K. Demir., B. Topuz., L. Yilmaz., H. Kalipcilar. Microporous and Mesoporous Material. **198**, 291-300 (2014)
12. Elements of X-ray diffraction, second Ed., B.D. Cullity Ed., USA, 1978.
13. Adlim, Kimia Anorganik. Universitas Syiah Kuala. Banda Aceh. (2009)
14. A.F. Gross., E. Sherman., J.J. Vajo. Dalton Trans. **41**, 5458-5460 (2012)
15. Z. Zhang., S. Xian., H. Xi., H. Wang., Z. Li. Chemical Engineering Science. **66**, 4878-4888.
16. R. M. Silverstein., F. X. Webster., D. J. Kiemle, Spectrometric Identification of Organic Compounds. 7th Ed., John Wiley and Sons, Inc., New York. (2005).

Catching investigation of yellowfin tuna (*Thunnus Albacares*) based on the distribution of chlorophyll-a in the North Waters of Aceh: A November and December analysis

Muhammad Irham[1,2,3]*, *Maulana* Wali Akbar[4], *Mukhlis* Mukhli[5,2], *Afdhal* Fuadi[6], *Muhammad* Authar[7] and *Ichsan* Setiawan[1,3]

[1]Department of Marine Sciences, Faculty of Marine and Fisheries, Universitas Syiah Kuala, Banda Aceh, Indonesia, 23111.
[2]Geographical Information Systems Laboratory, Faculty of Marine and Fisheries, Universitas Syiah Kuala, Banda Aceh, Indonesia, 23111.
[3]Center for Environmental and Natural Resources Research (PPLH-SDA), Universitas Syiah Kuala, Banda Aceh, Indonesia, 23111.
[4]Department of Fisheries Resource Utilization, Faculty of Marine and Fisheries, Universitas Syiah Kuala, Banda Aceh, Indonesia 23111.
[5]Marine and Fishery Resources Monitoring Base (PSDKP) Lampulo, Aceh, Indonesia 23127.
[6]Aquatic Resources Study Program, Faculty of Fisheries and Marine Affairs, Teuku Umar University, Aceh Barat, Indonesia 23615.
[7]Department of Agribusiness, Faculty of Agriculture, Universitas Malikussaleh, Lhokseumawe, Indonesia.

Abstract. The purpose of this study was to analyze the distribution of chlorophyll-a on the catch of yellowfin tuna in the northern waters of Aceh. This research was conducted in the north of Aceh waters starting from November to December 2020. fish catch data was collected by direct survey in the field by following the activities of fishermen and recording the number of catches. Data on chlorophyll-a was obtained from Aqua Modis satellite imagery. The catch data analysis used the CPUE model, while the chlorophyll-a analysis was carried out using the SeaDas 7.5 and ArcGIS. The results showed that the catch of yellowfin tuna was higher in November than in December as well as the distribution of chlorophyll-a content. The coefficient of determination (Adjusted R Square) is 0.09, which means that the relationship between chlorophyll-a and yellowfin tuna catches is only 9%, while 91% is influenced by other factors. These results inform that the content of chlorophyll-a does not show a significant relationship to the catch of yellowfin tuna.

1 Introduction

Aceh is one of the largest producers of tuna commodities, especially Yellowfin Tuna (Thunnus Albacares) [1,2], because it is surrounded by the Indian Ocean and Andaman Sea [3], and has a high selling value [4–6]. In 2012 yellowfin tuna production was 2,719 tons, in 2013 it was 3,656 tons and in 2016 production increased to 7,302 tons [7].

The distribution and abundance of yellowfin tuna is strongly influenced by several oceanographic parameters [8,9], one of which is chlorophyll-a [10,11]. Chlorophyll-a is one of the oceanographic biological parameters that plays an important role in primary productivity in the sea [12]. The distribution of chlorophyll-a contained in phytoplankton can be used as a measure of fertility in a water[13,14]. As a green pigment found in plants, chlorophyll-a use sunlight to carry out photosynthesis [15].

Phytoplankton needs chlorophyll-a, which is available in the photic layer, for the photosynthesis process [16]. Phytoplankton plays an important role as a source of nutrition for yellowfin tuna [17]. Although phytoplankton is not a natural food for yellowfin tuna, but it acts as a basic food chain [18]. Therefore, information about the distribution area of chlorophyll-a plays an important factor to determine the distribution of tuna fishing locations, especially yellowfin tuna [19].

Siregar et al., [20] found that yellowfin tuna were distributed in the southern area of Sumatra based on chlorophyll analysis. Bahri and Sari [21] also conducted research on the catchment area of yellowfin tuna through a morphometric approach from January to August 2017. The research shows that the northern and southern parts of Aceh waters have relatively higher catches compared to the western part of Aceh. However, the determination of fishing areas using the chlorophyll-a distribution approach in the northern waters of Aceh has not yet been carried out. Therefore, a study to determine the range of chlorophyll-a in the northern waters of Aceh which is used as a reference for information on yellowfin tuna fishing areas needs to be carried out. The purpose of this study was to analyze the distribution of chlorophyll-a as a yellowfin tuna fishing area in the northern coast of Aceh.

* Corresponding author: irham@unsyiah.ac.id

https://doi.org/10.1051/e3sconf/202233902004

2 Method

2.1 Location, Time and Data Collection

This research was carried out from November to December 2020 in the northern waters of the Aceh Province. Data collection in this study was carried out by observing primary data (data on fishing grounds by fishermen, time, and number of catches), and secondary data (satellite image data of chlorophyll-a). Data were collected by conducting interviews with 12 samples of ships and following a fishing trip.

The chlorophyll-a distribution data used in this study is Aqua-MODIS satellite image data which has been cleaned from cloud coverage and the data taken was in level 3. The downloaded chlorophyll-a image was a cloud-free image at the same time and location with yellowfin tuna fishing operations.

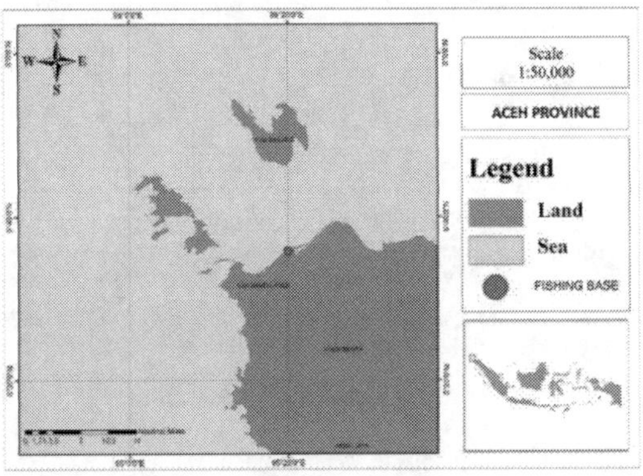

Figure 1. Research Area Map.

2.2 Data Analysis

Yellowfin tuna production data obtained during the study were used to calculate catch per unit effort (CPUE). The formula used to calculate CPUE is [22]:

$$CPUE = \frac{\text{catch}_i}{\text{effort}_i}$$

where:
CPUE = Catch per fishing effort (kg/trip) in i-month
catch_i = Catch (kg) in i-month
effort_i = Attempt to catch (trip) in i-month

The CPUE value is then displayed in the form of a table based on. By looking at the catch, it can be categorized fluctuations in catches based on time and location/fishing area.

The chlorophyll-a data was then analyzed for temporal and spatial distribution. The following are the steps in processing chlorophyll-a which are downloaded from the Aqua Modis satellite image [23].

The relationship of chlorophyll-a to the catch of yellowfin tuna can be done by statistical analysis. The steps taken to determine each variable followed Bahri et al [21]. The relationship between chlorophyll-a and yellowfin tuna catch variables was carried out by simple linear regression analysis. The following are the provisions for the category of correlation values between variables X and Y (Table 1).

Table 1. Correlation value category used for statistical analysis.

Multiple R Value	
Very Low	0.0 – 0.19
Medium	0.40 – 0.59
Strong	0.60 – 0.79
Very Strong	0.80 – 0.99

The yellowfin catch area based on the results of the catch analysis refers to [24] which states that:

1. If the CPUE value in the highest catch class is greater than the average CPUE, then the value for the catch is very good to be used as a very potential fishing area.
2. If the CPUE value is in the medium class, then the value for the catch is included in the potential fishing area.
3. If the CPUE value is in a very low class, then the value for the area can be classified into a fishing area with less potential for fishing grounds.

While criteria for the distribution of chlorophyll-a to yellowfin tuna fishing areas can be seen in the Table 2 below.

Table 2. Criteria for the the distribution of chlorophyll-a to yellowfin tuna fishing areas.

No	chlorophyll-a	Category	DPI criteria
1	0.100 < 1.00 mg^3	Optimum	Potential
2	1.00 < 2.00 mg^3	Minimum	Not potnetial

3 Result and Discussion

3.1 Result

3.1.1 Yellowfin Tuna Catch and CPUE Productivity

The catch of yellowfin tuna in the northern waters of Aceh is fluctuating (Figure 2). Based on data from 12 vessels using handline fishing gear, the catch of yellowfin tuna in November amounted to 5,864 kg. While the catch of yellowfin tuna in December was 3,342 kg.

The decline in the production of catches in December is influenced by the number of fishing trips, weather conditions.

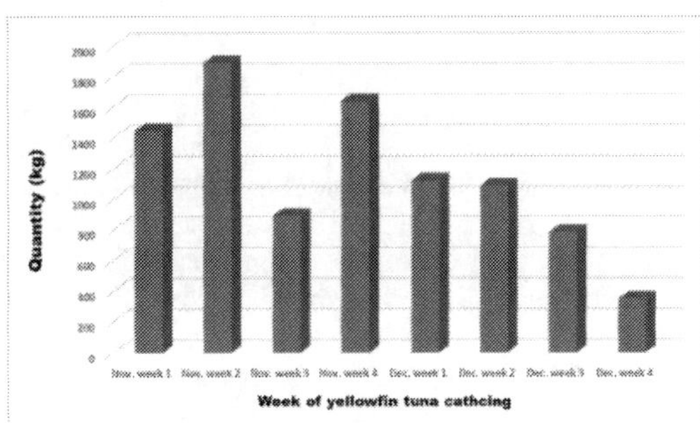

Figure 2. Yellowfin tuna catches in November and December 2020.

The productivity of yellowfin tuna CPUE in November is 177 Kg/trip and in December it is 138 kg/Trip. The total trips of fishing vessels that work in November are 33 trips while in December there are 25 trips.

3.1.2 Distribution of Chlorophyll-a

The distribution of chlorophyll-a content in the northern waters of Aceh can be seen in Figures 3 and 4. The distribution of chlorophyll-a content in the northern waters of Aceh in November 2020 ranged from 0.189 mg/m3 to 1.082 mg/m3, with an average content of 0.635 mg/m3 in November 2020. The content of chlorophyll-a in November was relatively high with various variance values. In December 2020, the content of Chlorophyll-a ranged from 0.101 mg/m3 to 2.568 mg/m3, with an average content of 1.258 mg/m3 in December 2020. The distribution of chlorophyll-a content in December was relatively low because it was dominated by the lowest variance value of 0.101 mg/m3. The data are represented in the Table 3 below.

Table 3. Chlorophyll-a distribution in November and December 2020

No.	Parameter	Distribution of chlorophyll-a	
		November	December
1	Chlorophyll-a	0.189-1.082	0.101-2.568
2	High chlorophyll-a	1.082	2.568
3	Low chlorophyll-a	0.189	0.101
4	Avg. chlorophyll-a	0.635	1,258

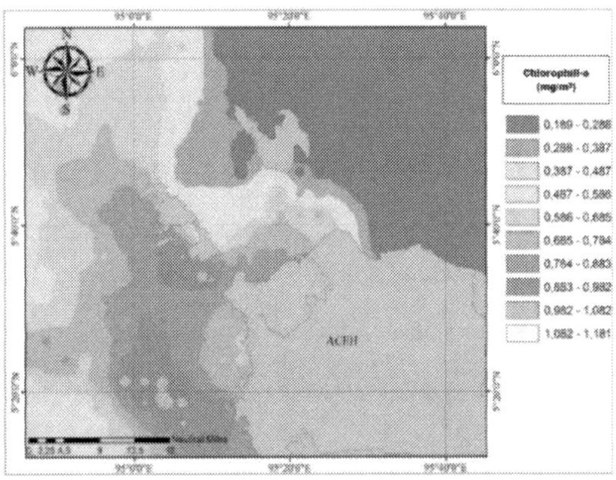

Figure 3. Map of distribution of chlorophyll-a content in the northern waters of Aceh Province in November 2020.

3.1.3 The Relationship of Chlorophyll-A to Yellowfin Tuna Catches

The relationship of chlorophyll-a with the catch of yellowfin tuna was obtained by performing a simple linear regression analysis. The relationship of chlorophyll-a with the catch of yellowfin tuna in November 2020 resulted in a coefficient of determination (R Square) of 0.18 which means the relationship of chlorophyll-a with the catch of yellowfin tuna in November 2020 was 18% and the remaining 82%. influenced by other factors. While the value of the correlation coefficient (Multiple R) between chlorophyll-a and the catch of yellowfin tuna is 0.42, which means that chlorophyll-a has a moderate effect on the catch of yellowfin tuna.

The relationship of chlorophyll-a with the catch of yellowfin tuna in December 2020 resulted in a coefficient of determination (R Square) of 0.02, which means the relationship of chlorophyll-a with the catch of yellowfin tuna in December 2020 was 2% and the remaining 98%. influenced by other factors. While the value of the correlation coefficient (Multiple R) between chlorophyll-a and the catch of yellowfin tuna is 0.14 which means that chlorophyll-a has a very low effect on the catch of yellowfin tuna.

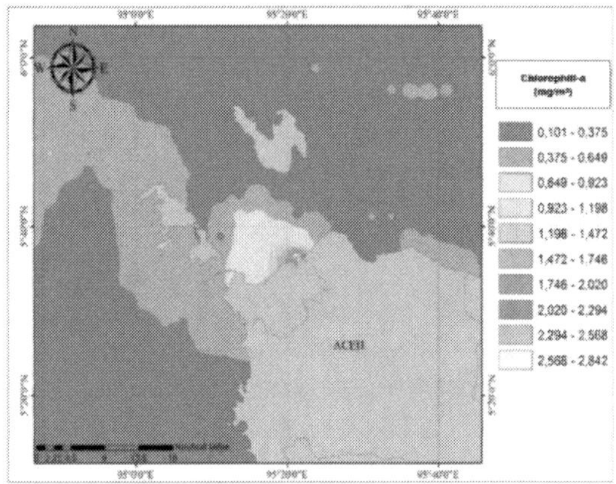

Figure 4. Map of distribution of chlorophyll-a content in the northern waters of Aceh Province in December 2020.

3.1.4 The Potential Yellowfin Tuna Fishing Area

Potential areas for catching yellowfin tuna in November and December 2020 are located in the northern waters of Aceh Island to the south-west waters of Aceh (Figure 5). Based on CPUE productivity, the total catch of the area is classified as a potential fishing area. Judging from the productivity of yellowfin tuna caught in these waters, which is relatively high. Meanwhile, the yellowfin tuna fishing areas that lack potential during November and December 2020 are in the northern waters of Sabang Island to the western waters of Sabang Island because the productivity (CPUE) of the fishing area tends to decrease in November and December 2020.

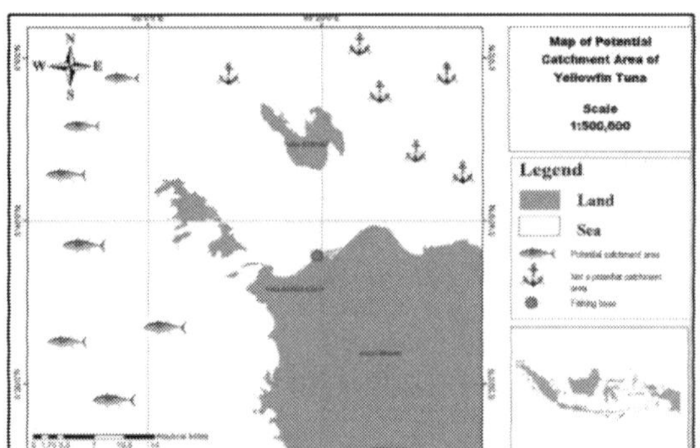

Figure 5. Potential fishing area of yellowfin tuna

3.2 Discussion

The catch of yellowfin tuna in the northern waters of Aceh in November and December 2020 tends to fluctuate. This is due to changing fishing grounds, weather conditions and food availability. However, the main factor in the migratory behavior of large pelagic fish is food availability [25].

The decline in catch productivity in the northern waters of Sabang Island and the western waters of the Sabang Island was caused by weather factors, changing fishing areas, as well as the warning of the year-end Tsunami disaster in Aceh Province which resulted in fewer yellowfin tuna fishing trips by fishermen. However, determination of potential yellowfin tuna fishing areas is not representative enough to describe potential yellowfin tuna fishing areas if only based on productivity (CPUE). Therefore, another parameter is needed, such as the content of chlorophyll-a in these waters. Even though, the high CPUE productivity from yellow tuna fishing occurred in November and in December CPUE productivity was low.

Based on the results of simple linear regression analysis, the distribution of chlorophyll-a in November did not show a significant relationship to the catch of Yellowfin tuna. The results of the analysis show that the coefficient of determination (Adjusted R Square) produces a value of 0.09, which means that the relationship between chlorophyll-a and yellowfin tuna catches in November 2020 is only 9% and the other 91% is influenced by other factors.

The results of linear regression analysis also showed that the distribution of chlorophyll-a in December also did not show a significant relationship to the catch of yellowfin tuna. yellowfin tuna in November 2020 was only 7% and the other 93% was influenced by other factors. The same finding was shown by Adnan [26] who stated that the relationship between chlorophyll-a concentration and the catch, the increase in the concentration of chlorophyll-a was linearly correlated with the increase in catch, and vice versa.

The results of the coordinate point data obtained from GPS fishermen, yellowfin tuna fishing in November 2020 was concentrated in the northern waters of the island of Aceh to the south-west waters of Aceh. In December 2020 yellowfin tuna fishing was concentrated in the northern waters to the western waters of Sabang Island (Figure 3 and 4).

yellowfin tuna catch in November 2020 in the northern waters of Aceh Island to the south-west waters of Aceh can be said to be a potential fishing area. The results of this analysis are supported by satellite image data of the distribution of chlorophyll-a in these areas compared to the data from satellite image in the northern waters of Sabang and the western waters of Sabang (Figure 3). Meanwhile, in December 2020, in the northern waters of the island of Sabang and the western area of the island of Sabang can be said to be a non-potential fishing area, due to reduced catches and and the data of chlorophyll-a distribution from satellite image shows a low value of chlorophyll-a in the area of north of Sabang and west of Sabang waters (Figure 4).

4 Conclusion

The distribution of chlorophyll-a on the catch of yellowfin tuna had no significant effect in November 2020 because the coefficient of determination was only 18% and the other 82% was influenced by other factors. The same thing also happened in December 2020 with the coefficient of determination 2% and the other 98% influenced by other factors. Judging from the distribution of chlorophyll-a, the northern waters of Aceh Island to the south-west waters of Aceh are potential yellowfin tuna fishing areas, while the northern waters of Sabang Island and west of Sabang Island are non-potential yellowfin tuna fishing areas.

References

1. J. Burhanis, Z. Radmi, International Journal of Fisheries and Aquatic Studies, **5**, 264 (2017)
2. Burhanis, Alaudin, Edwarsyah, Jaliadi, A. Rozi, Zulfadhli, R. Fadhillah, and Zulradmi, *Utilization and optimization of the sustainability of yellowfin tuna (Thunnus albacares) in Simeulue waters, Aceh Indonesia*, in IOP Conference Series: Earth and Environmental Science, 28 November 2020, Ambon, Indonesia (2021)
3. L. Campling, Journal of Agrarian Change, **12**, 252 (2012)
4. M. Khaliqi, T.C. Pane, R.B.M.I. Fatoni, *Indonesian tuna position in the international market, in* IOP Conference Series: Earth and Environmental Science, 24-25 Oktober 2018, Sumatera Utara, Indonesia (2019)
5. B. Theophille, S.D. Hutchinson, A. Iton, *Yellowfin Tuna in the Market for High Value Agricultural Products: Competitive Capability in Dominica*, in CAES: 27th West Indies Agricultural Economics Conference, July 2007, Belize City, Belize (2009)
6. [KKP] Kementerian Kelautan dan Perikanan Indonesia, *Statistik Perikanan Indonesia,* (Jakarta, 2020)
7. [KKP] Kementerian Kelautan dan Perikanan Indonesia, *Kelautan Dan Perikanan Dalam Angka 2017* (Jakarta, 2017)
8. M. Cornic, J.R. Rooker, Fisheries Research, **201**, 1 (2018)
9. K. Nimit, N.K. Masuluri, A.M. Berger, R.P. Bright, S. Prakash, U. TVS, S.K.T.P. Rohit, T.A.S. Ghosh, S.P. Varghese, International Journal of Remote Sensing, **41**, 5785 (2020)
10. B. Wiryawan, N. Loneragan, U. Mardhiah, S. Kleinertz, P. Wahyuningrum, J. Pingkan, Wildan, P. Timur, D. Duggan, I. Yulianto, Fishes, **5**, 28 (2020)
11. R. Habibullah, Mubarak, M. Galib, Asian Journal of Aquatic Sciences, **3**, 236 (2020)
12. J. Kovács, P. Tanos, G. Várbíró, A. Anda, S. Molnár, I.G. Hatvani, Ecological Indicators **78**, 311 (2017)

13. F. Aulia, M. Rusdi, A. Deli, A. Fuadi, M. Irham, I. Indra, *The Marxan model for determining no-catch zones based on conservation targets in the north-eastern region of Simeulue District*, in IOP Conference Series: Earth and Environmental Science, 15-16 October 2020, Banda Aceh, Indonesia (2021)
14. M. A. Welliken, E.H. Melmambessy, S.L. Merly, R.D. Pangaribuan, B. Lantang, J. Hutabarat, A. Wirasatriya, *Variability Chlorophyll-A and Sea Surface Temperature As the Fishing Ground Basis of Mackerel Fish in the Arafura Sea*, in E3S Web of Conferences, 14-15 August 2018, Semarang, Indonesia (2018)
15. S. Siregar, A.B. Ternala, A.H. Zulham, Jurnal Aquacoastmarine, **4**, 27 (2016)
16. X. Liu, E.A. Laws, Y. Xie, L. Wang, L. Lin, B. Huang, JGR:Biogeosciences, **124**, 2400 (2019)
17. K. W. Lan, T. Shimada, M.A. Lee, N.J. Su, Y. Chang, Remote Sensing, **9**, 444 (2017)
18. F. Lepori, Fundam. Appl. Limnol, **192**, 305 (2019)
19. M. D. Setiawati, H.A. Rachman, A.R. As-syakur, A. Syahailatua, *Identification of Commercial Tuna Hotspot in the Southern Waters of Java-Bali Through Satellite Remote Sensing Data*, in 2021 IEEE International Geoscience and Remote Sensing Symposium IGARSS, 11-16 July 2021, Brussels, Belgium (2021)
20. Y. Siregar, E. Suri, V.P. Siregar, S.B. Agus, Jurnal Ilmu Dan Teknologi Kelautan Tropis, **10**, 501 (2018)
21. S. Bahri, W.E. Sari, ALBACORE, **2**, 307 (2018)
22. J. M. Petrere, J.P. Giacomini, J.P. De Marco, Braz. J. Biol, **70**, 483 (2010)
23. M. Muklis, J.L. Gaol, D. Simbolon, J. Ilmu Teknologi Kelautan Tropis, **1**, 24 (2009)
24. Y. J. Chang, C.L. Sun, Y. Chen, S.Z. Yeh, G. Dinardo, International Journal of Remote Sensing, **33**, 7523 (2012)
25. S. Minarro, G.N. Ferero, H. Reuter, I.E. van Putten, Marine Policy, **69**, 73 (2016)
26. A. Adnan, J PSP FPIK Unpatti, **1**, 1 (2010)

 https://doi.org/10.1051/e3sconf/202233902005

Proximate composition and fatty acids profiling of Seahorse originated from Simeulue, Aceh-Indonesia

Muhammadar Abdullah Abbas[1*] , *Sofyatuddin* Karina[1], *Ririn* Qofifah[1], *Muhammad* Nasir[2], and *Asmawati* Asmawati[3]

[1] Marine Culture Program, Faculty of Marine and Fisheries, Universitas Syiah Kuala. Aceh. 23111, Indonesia.
[2] Bology department, Faculty of Math and Science, Universitas Syiah Kuala. Aceh. 23111, Indonesia.
[3] Agricultural Product Technolog, Faculty of Agriculture, Universitas Syiah Kuala. Aceh. 23111, Indonesia.

Abstract. The aim of this study was to determine the proximate composition, quantitative of fatty acids analysis as well as the content in the seahorse sample, Hippocampus sp which originated from Sibigo waters, Simeulue, Aceh, Indonesia. In this study, proximate composition and fatty acids of seahorse were determined. Based on the proximate analyses that had been conducted, results showed that protein was the most abundantly found in the seahorse (39.32±0.22%), orderly followed by carbohydrate (28.48±0.25%), ash (25.43±0.16%), moisture (6.29±0.13%) (dry weight basis) and fat (0.47±0.30%). Fatty acids profiling and analysis were also conducted by using the a GC-MS (Gas Chromatography-Mass Spectroscopy). The analysis result showed that sample possessed of saturated fatty acids in big size were palmitic acid, stearic acid, myristic acid, and lauric acid. While unsaturated fatty acids were arachidonic acid, linoleic acid and oleic acid. Traces of saturated fatty acids in small seahorse were found in the sample including palmitic acid, stearic acid, miristic acid and lauric acid, while unsaturated fatty acids were linoleic acid and oleic acid. The highest saturated fatty acid in seahorse was palmitic acid (60.67%), and unsaturated fatty acids were oleic acid (45.5%), arachidonic acid (25.08%) and linoleic acid (0.93%).

1 Introduction

Seahorses (*Hippocampus* sp.) are an outstanding marine species considering their reproductive patterns and other features. Owing to the iconic characteristics of these fishes, aquarium trade, and research efforts have increased in the last years. Types of seahorses found in Indonesia, among others, are the *Hippocampus barbouri*, *Hippocampus bargibanti*, *Hippocampus comes*, *Hippocampus histrix*, *Hippocampus kelloggi*, *Hippocampus kuda*, *Hippocampus spinosissimus*, *Hippocampus trimaculatus* and *Hippocampus* sp. [1]. The types of seahorses found in Sibigo waters, Simeulue, Aceh, Indonesia are *Hippocampus* sp.

Seahorses are superior fishes that has been over-exploited in nature owing to their use in several human activities, habitat loss, and the aquarium and antique trade, at the expense of its conservation [2]. Seahorses have been used in the discovery and development of traditional and allopathic medical treatments for a long time. Seahorses are efficacious as ingredients in traditional Chinese medicine for generations and are believed to strengthen stamina [3]. It is believed that seahorses have the potential to cure infertility, baldness, asthma and arthritis inflammation [4]. Apart from this, seahorses are also believed to have free radical scavenging effects in controlling the aging process [5]. The research result of Sanaye et al. [6], antioxidant activity of yellow seahorse, *Hippocampus kuda* was 24.04%.

The results of other studies stated that seahorses can be used as antifatigue with a cell proliferation activity value of 160% and able to inhibit oxidation activity with an IC_{50} value of 43.8 ppm [7], inhibiting the growth of *E. coli* bacteria with an inhibition zone of 4 mm, and inhibiting the growth of the fungus *Aspergillus flavus* with an inhibitory zone of 2 mm. The many benefits of seahorses are owing to the presence of complex bioactive compounds [8]. Therefore, people often take advantage of seahorse as an alternative in natural medicine.

To being used in pharmaceuticals, other benefits of sea horses are as a food source of animal protein which is full of nutrition and very good for consumption by humans. Protein is a substance that important in the body. Amino acids are the main components that make up proteins has a metabolic function in the body and divided into two groups, namely amino acids essential and non-essential [9]. In addition to amino acids, seahorses also contain fatty acids, both saturated and unsaturated. The difference in the amount of content and the types of fatty acids present in seahorse are very instrumental in provide benefits to humans.

The nutritional content of sea horses other than protein, namely vitamins, minerals, carbohydrates, fiber, and others. The nutritional content of the seahorse has the potential as a dietary supplement for human health [10]. This research is related to proximate analysis and the content of fatty acids and the types contained in seahorses. This is because there is still a

* Corresponding author: muhammadar@unsyiah.ac.id

lack of information about the chemical characteristics of seahorses that are important to the community. The purpose of the study was to examine the proximate composition and fatty acids profiling of seahorse originated from Simeulue, Aceh-Indonesia.

2 Materials and methods

2.1 Materials

Seahorses (*Hippocampus* sp.) were collected from Sibigo waters, Simeulue, Aceh-Indonesia with a size of 3-10 cm to be used for this study. The sizes were divided into 2 group, small (3-5 cm) and large (6-10 cm). The seahorse sample was storage in an ice cooler box. This was to maintain low temperature and preserve the structure of the sample through the journey in bringing the sample to the laboratory. The sample was then thoroughly rinsed with distilled water for several times and oven-dried at temperature of 60 °C for 2 days, until the sample was completely dry. The dried sample was sorted and ground into powder form by using a high power grinder and then tightly-packed and stored at -18 °C cold room until further chemical analyses. All chemical analyses of the seahorses sample were carried out in triplicates to obtain more accurate results.

2.2 Methods

2.2.1 Moisture content analysis.

2 g sample was accurately weighed into the crucible and dried at 105 °C in a universal oven (Memmert) until constant sample weight was obtained [11].

2.2.2 Ash content analysis.

Dried seahorse sample that was obtained from the previous analysis was then burnt and ashed using a muffle furnace at 525 °C overnight. Sample was weighed to its nearest 0.001 g using an analytical balance [11].

2.2.3 Crude protein analysis.

Determination of crude protein content of seahorse sample was using semimicro Kjeldahl method. 0.5 g of dried seahorse sample was accurately weighed into 100mL Kjeldahl tube. 2 g of selenium and 25 mL of concentrated sulphuric acid (H2SO4) was carefully added into the digestion tubes and gently shaken. Destruction procedure was performed at 420 °C for approximately 120 minutes until clear blue/green solution was obtained and cooled for 30 minutes. Distillation procedure was then performed at 100 °C for 10 minutes. The distillate was titrated with 0.01 N HCl with a phenolphthalein as indicator until the color of the solution changes to pink.The percentage of crude protein was calculated by multiplying the percent of nitrogen, N found with a factor of 6.25 [11].

2.2.4 Fat content analysis.

The fat content of seahorse sample was extracted by using the Soxhlet method. 2 g of powder dried seahorse sample was weighed in filter paper and placed in the extraction thimble. The samples were extracted using a Soxhlet extractor by adding the 200 mL of hexane as a solvent in the boiling flask. Extraction process was run for 5 hours and the extracted fat was dried using a rotary evaporator. The boiling flask was dried in an oven at 105 °C for 1 hour and left to cool prior obtaining the final weight [11].

2.2.5 Carbohydrate content.

Carbohydrate content was calculated based on this calculation [% carbohydrate = 100%-%moisture+% ash+% crude protein+ % fat].

2.2.6 The lipids and fatty acids analysis.

The fatty acid content of seahorses identified from the chromatogram (peak) of the fatty acid methyl ester (FAME) fractionation using capillary gas chromatography (50 x 32 mm) SP2330. Methyl ester was obtained from the fat methylation process according to the modified British Standard Method, while the fat for methylation was extracted from the material by the modified Bligh–Dyer Method. Fatty acids were identified by comparing the peaks of the analysis with the peaks of standard fatty acids. Each sample of large and small seahorse was repeated twice with the method equipped with an Rxi-MS column, 0.25 m film thickness, capillary column and with 5973 MSD tissue mass spectrometer, operated in electron impact ionization mode (70 eV). Analysis GC-MS was performed in split mode (split ratio 1:50), using helium as the carrier gas (flow rate 1 mL/min). The injector temperature was set at 250 °C. The sample volume, injected 1 μL. The oven temperature was maintained at 150 °C for 2 minutes and then programmed at 5 °C/min to a final temperature of 280 °C, where it was maintained for 5 minutes, then waited until the separation of all fatty acids was completed [11].

3 Results and discussion

3.1 Proximate compostion

Proximate analyses of the seahorse sample provided useful information on the composition of the seahorse (*Hippocampus* sp.) from Sibigo waters, Simeulue, Aceh-Indonesia. According to [12] the proximate content in the body of the seahorse can characterize the physiological and health conditions of the seahorse. In general, the body composition of the seahorse is obtained by proximate analysis, but measurements of physiological conditions can also be determined using a comparison of the standard weight and length relationship. The determined proximate compositions of seahorse are moisture content, ash content, crude protein, fat content and carbohydrate content which had been done were tabulated as the following. Based on Table 1, shows that seahorse collected in Sibigo waters, Simeulue contained 39.32±0.22% of crude protein, 0.47±0.24% of fat, 25.43±0.16% of ash content, 6.29±0.13% of moisture content (powdered sample), and 28.48±0.25% of carbohydrate content. Seahorse sample contained quite a large amount of crude protein, ash and carbohydrate, but contained low fat.

Table 1. The proximate composition in seahorse from Simeulue.

Composition (dry weight basis)	Persentage (%) ± SD
Crude protein content	39.32±0.22
Fat content	0.47±0.30
Ash content	25.43±0.16
Carbohydrate content	28.48±0.25
Moisture content	6.29±0.13

The proximate composition of the seahorse was calculated based on dry weight. The moisture content of the sampled seahorse was 6.29±0.13. The moisture content contained in each type of seahorse was different which caused by several factors e.g the fishing season and the age of the species. [13] mentioned that the average moisture content in fresh seahorses was 65-76%. In addition, [14] stated that the moisture content in marine biota ranges from 50-85% depend on the type of species and body nutritional conditions. The season of fishing for some species that was suspected to be during the reproductive process, so that the species will loss of moisture content in their bodies.

The proximate composition of seahorses on a dry basis were 70.70% of protein, 1.71% of fat, 20.92% of ash and 38.15 (µg/g) of Zn mineral [12]. The nutritional content of seahorses can also be influenced by their living habitat and food sources. The protein content of the seahorse was 39.32±0.22% which was obtained from mixing several sexes of the seahorse that were sampled. [15] studied the quality of juvenile sea horse (*Hippocampus barbouri*), which fed with a modified diet, the crude protein content was 37.12-61.47%. The value of fat content in seahorses was 0. 0.47±0.30%. According to the results of research from [16-18], it is stated that the average value of fat content in the tuna meat is 0.51%. Fat is one of the excess energy stored by animals so that the amount of fat in the animal's body that is used as food is determined by the energy balance of the animal.

3.2 Fatty acids profiling

The fatty acids content of the seahorse (*Hippocampus sp*) were saturated and saturated fatty acids. The saturated fatty acids were pamitic acid, stearic acid, myristic acid and behenic acid, while the unsaturated fatty acids were arachidonic acid and oleic acid. Table 2 and Tabel 3 shows the fatty acid composition, respectively.

Tabel 2. The fatty acids (FA) composition of big seahorses

No.	Kind of FA	Name of FA	Formula of Molecule	Amount (%)
1	Saturated FA	Tetradecanoic acid (Myristic acid)	$C_{14}H_{28}O_2$	1.6
2		Hexadecanoic acid (Palimitic acid)	$C_{16}H_{32}O_2$	25.2
3		Octadecanoic acid (Stearic acid)	$C_{18}H_{36}O_2$	20.3
4		Docosanoic acid (Behenic acid)	$C_{22}H_{44}O_2$	0.6
Total				47.7
1	Unsaturated FA	Eicosanoic acid (Aracidic acid)	$C_{20}H_{40}O_2$	22.2
2		Oleic Acid	$C_{18}H_{34}O_2$	30.2
Total				52.4

Tabel 3. The fatty acids (FA) composition of small seahorses

No.	Kind of FA	Name of FA	Formula of Molecule	Average
1	Saturated FA	Tetradecanoic acid (Myristic acid)	$C_{14}H_{28}O_2$	0.7
2		Hexadecanoic acid (Palimitic acid)	$C_{16}H_{32}O_2$	49.4
3		Octadecanoic acid (Stearic acid)	$C_{18}H_{36}O_2$	37.3
4		Docosanoic acid (Behenic acid)	$C_{22}H_{44}O_2$	0.7
Total				88.1
1	Unsaturated FA	Oleic Acid	$C_{18}H_{34}O_2$	12.0
Total				12.0

The oil of seahorse was injected twice for each sample. It can be seen from the table that the trans-esterification has been successfully carried out. This is indicated by the presence of a chromatogram that has been identified. All sample of seahorse contain two kinds of fatty acids, namely saturated fatty acids and unsaturated fatty acids. For big size seahorses, 2 unsaturated fatty acids were identified, but for small seahorse only 1 unsaturated fatty acid that is oleic acid. The distribution of fatty acids differs from one species to another, and depends on various factors, such as season, temperature, growing place, fish species, age, sex and feeding habits [16].

Based on Table 2 and Tabel 3, the data obtained from the GC-MS analysis showed that the content of unsaturated fatty acids between large seahorses (52.4%) and small seahorses (12.0%) had quite a significant difference. The higher saturated fatty acid was palmitic acid both in large and small size, then followed by stearic acid. The fatty acid content of seahorses is the same as fish, which are both aquatic animals. The results of fatty acid analysis of oil carp (*Cyprinus carpio*), the higher fatty acid was palmitic acid (C:16-0) of 17,15% [19].

The fatty acids contained in fish consist of saturated fatty acids (15-25%), unsaturated fatty acids (35-60%) and polyunsaturated fatty acids (25-40%) [20]. This happens because according to [21]. the quality of fish oil, especially its fatty acid composition, is influenced by season, fishing area, type of food, sexual maturity and age of fish. According to [22], Arachidonic acid (20:4n-6;AA) is also needed by larvae. Both types of fatty acids, namely AA and EPA, are substrates needed for the formation of eicosanoates which play a role in various physiological functions including ion regulation and egg maturity in female parents. The results of the study on the Arachidonic value of large seahorses of 44.3% which has an important role for the growth of seahorses.

The results of research conducted by [23] in Layur fish, myristic acid levels were only 0.24% of the total fatty acids in the sample. Meanwhile, the fat from mackerel and tuna is physically solid at room temperature because it has more saturated fatty acid composition than unsaturated fatty acids. This is in accordance with the results of research by [24] which shows that the composition of fatty acids is one of them influenced by species differences (species specific).

The results of the comparison test with other seahorses are the first on seahorses from Simeulu Waters while the comparison is on seahorse eggs (*Hippocampus barbouri*). The results of the seahorse eggs when the research was conducted showed that the seahorse eggs against repeated spawning the first treatment on Myristic acid as much as 0.53%, Palmitic acid as much as 26.49%, Stearic acid as much as 0.53% this is different from the analysis of the results of the study This was done because the research used seahorse samples, while in the comparison results using seahorse eggs spawned by seahorses. The content of seahorse eggs also depends on the large diameter of the eggs and the large number of mature eggs that will accommodate higher lipids which of course will have an impact on the high fatty acid content of the female seahorse (*Hippocampus barbouri*). While the results of the study the content of seahorse samples depends on the nutritional value or proximate.

4 Conclusion

This research can be concluded that the seahorse sample contains the nutritional composition and fatty acid composition of the large seahorse and small seahorse consisting of saturated fatty acids and unsaturated fatty acids. Saturated fatty acids contained in seahorses were Palmitic acid, Behenic acid, Myristic acid, and Stearic acid. Unsaturated fatty acids contained in seahorses were Oleic acid and Arachidonic acid.

Acknowledgements. The authors gratefully thanks to the Universitas Syiah Kuala for supporting research costs through the research scheme of Lektor Kepala.

References

1. W. Planas, Sulistyaningrum, Pengaruh Warna Wadah Percobaan Terhadap Sintasan dan Pertumbuhan Kuda Laut (Skripsi Program Studi Ilmu dan Teknologi Kelautan, Fakultas Perikanan dan Ilmu Kelautan, Institut Pertanian Bogor, 2006)
2. M. A. Dwiputra, Pemeliharaan Juwana Kuda Laut (Hippocampus barbouri, Jordan & Richardson, 1908) dengan Sistem Resirkulasi (Skripsi Jurusan Ilmu Kelautan, Fakultas Ilmu Kelautan dan Perikanan, Universitas Hasanuddin, Makassar, 2013)
3. Kumaravel Ghufran, M., K. Kordi. Panduan Lengkap Budidaya Kuda Laut. Ikan Unik yang Berpotensi Obat (Lily Publisher, Yogyakarta, 2010)
4. N. Kang, S.Y. Kim, S. Rho, J.Y. Ko & Y.J. Jeon. Fish Aquat Sci, **20** (2017)
5. S. P. Mandila dan N. Hidajati, UNESA J. of Chemistry **2** 1 (2013)
6. S.V. Sanaye, N.M. Pise, A.P. Pawar, P.P. Parab, R.A. Sreepada, H.B. Pawar & A.D. Revankar. Aquac **434** (2014)
7. Q. Lin, J. Lin, J. Lu, and B. Li. J. World Aquac Soc **39** (2008)
8. E.M. Sari, Identifikasi kuda laut berdasarkan marka molekuler serta potensinya sebagai antioksidan, antiglikasi dan imunomodulator. (Tesis, Bogor, Institut Pertanian Bogor, 2018)
9. F. Ahmad, Syafiuddin, Haryati. Jurnal Ilmu Kelautan, SPERMONDE **5** 2 (2019)
10. Lin, Q., J. Lin, J. Lu, and B. Li. 2008. Biochemical composition of six sea-horse species, Hippocampus sp., from the Chinese coast. J. World Aquac Soc, 39:225–234.
11. AOAC. 1990.
12. E.M. Sari, Identifikasi kuda laut berdasarkan marka molekuler serta potensinya sebagai antioksidan, antiglikasi dan imunomodulator. (Tesis, Bogor, Institut Pertanian Bogor, 2018)
13. F. Ahmad, Syafiuddin, Haryati. Jurnal Ilmu Kelautan, SPERMONDE **5** 2 (2019)
14. M. Harli. 2008. Asam Amino Esensial. http://www.suparmas.com. [20 April 2010].
15. D. Wahyu, S, Dwi T. S, Eddy S. THPI Student J **1** 1 (2013)
16. M. Harli. 2008. Asam Amino Esensial. http://www.suparmas.com. [20 April 2010].
17. A.A. Muhammadar, A.G. Mazlan, A. Samat, and K.D. Simon. AACL Bioflux **4** (2011)
18. A.A. Muhammadar, A.G. Mazlan, A. Samat, M.S. Asmawati, Z.A. Muchlisin, M.A. Rimmer, and K.D. Simon, AACL Bioflux **7** (2014)

https://doi.org/10.1051/e3sconf/202233902005

19. M. Pandiangan, J. Kaban, B. Wirjosentono and J. Silalahi. Analisis Kandungan Asam Lemak Omega 3 dan Omega 6 pada Minyak Ikan Mas (Cyprinus carpio), TALENTA Conference Series (2019) https://talentaconfseries.usu.ac.id
20. J.P. Branathan. Adv. Biochem. Engin/Biotechnol **96** (2005)
21. T. Estiasih, Minyak ikan dan Teknologi dan Penerapan untuk Pangan dan Kesehatan (Yogyakarta, Graha Ilmu, 2009)
22. L.A Copeman, C.C. Parrish, J. A. Brown, and M. Harel, Aquac, **210**, (2002)
23. R. I. Pratama, Yusuf dan Safri, Jurnal Akuatika **2** 2 (2011)
24. S. J. Iverson, Frost, K. J. and S. L. C. Lang, Mar. Ecol. Prog. Ser. **241** (2002)

https://doi.org/10.1051/e3sconf/202233902006

Estimation of the volume of production of turbine vapor of a fuel boiler with stochastic exogenous factors

Seyit Kerimkhulle[1*], *Gulmira* Azieva[1], *Ainur* Saliyeva[2], and *Ayagoz* Mukhanova[3]

[1]L.N. Gumilyov Eurasian National University, 1, Satpaev str., 010000, Nur-Sultan, Kazakhstan
[2]KMG Engineering LLP, 17, Kabanbai batyr str., 010000, Nur-Sultan, Kazakhstan
[3]Al-Farabi Kazakh National University, 71, al-Farabi ave., 050040, Almaty, Kazakhstan

Abstract. The paper is based on the analysis and review of the conceptual principles, methodology and practice of modelling water-steam processes of a fuel boiler of power plants; implementation of algorithms and models of the theory of system dynamics, white and pink noise; using computational technologies of difference-scheme equations and numerical integration methods, the values of the input parameters of the production of turbine vapor of the fuel boiler are determined), which, regardless of stochastic exogenous factors, provide stable states of the necessary and model volumes, norms of water intake and output of turbine vapor of the fuel boiler.

1 Introduction

An urgent problem for the formation of an intelligent system for evaluating the production of turbine vapor by exogenous stochastic factors in the chain of technological processes of heat power plants is the reliable (stable) provision of tools for managing the state of fuel boiler facilities. In particular, it is possible to solve this problem of steam generation of a fuel boiler by creating a clear system of information environment for managing internal resources and external connections of complex objects [1] of water-steam processes.

We conduct a review of the literature on the following studies, for example, the article [2] considers the problem of balancing supply and demand in the Saxony-Anhalt's energy system and issues of ensuring flexibility in energy systems are studied in four categories: flexible generators, demand management, storage, and transmission of electricity at X, and developed a practical method for quantifying and comparing the technical capabilities of flexibility options at the regional level.

In article [3], the concept of application flexibility to energy systems for operational planning of the transmission network in a market environment is proposed. A specific and unique definition of the flexibility of the transmission system in relation to generation changes is proposed. A mathematical model and a method for calculating some indicators of the flexibility of a volumetric system are also presented.

In [4], an assessment of the industrial water consumption of the electronic industry of the People's Republic of China is given and the geolocation of semiconductor manufacturing enterprises by water intakes at various spatial scales is mapped. On the other hand, geolocation of the data also made it possible to calculate the water intake weighted by the deficit to identify a specific catchment area and map hot spots of industrial water use.

Article [5] presents a thermodynamic simulation of the steam compression cycle based on neural network approach. A generalized radial basis function is used for the network, which generates the predicted current state as the output of the network. The trained network is confirmed by untrained data and shows to a satisfactory extent all the characteristics of the steam compression cycle process for an air-water heat pump.

Article [6] analyzes the consequences of probable hydrological fluctuations during climate change in anticipation of water availability and its variability. Conditions for optimal distribution of water resources are determined to maximize the overall social benefit from water use. The results show that the active participation of the regulator in the water market leads to excellent results in terms of both greater overall social benefits and more water available for environmental purposes in the context of climate change.

Article [7] analyzes for a long-term forecast period all available simulated climate scenarios by checking the scenarios using historical data and assessing the impact on water resources over the next 30 years using a hydrological model proposed by the Spanish Agency of Meteorology (AEMET) for the geographical area covered by the river basin, located in eastern Spain.

Article [8] presents the groundwater stability index, which offers a new combination of characteristics in which the signal-to-noise ratio of long-term trends is used, estimated using reliable regression and the

* Corresponding author: kerimkhulle@gmail.com

concentration of the main pollutant causing anxiety of the population.

Article [9] substantiates the methodology for assessing the impact of changes related to groundwater and climate change on the needs for basic and environmental runoff (EFN) and showed that under average meteorological conditions, the basic runoff in the main streams still meets the requirements of EFN. It also determined that anthropogenic changes have a significant impact: drainage caused a 25-40% reduction in base runoff, groundwater abstraction caused 5-28%, and climate change will potentially cause an additional 33-70% reduction by 2050.

2 Materials and Methods

Conceptual principles, methodology and practice of modelling water-steam processes of a fuel boiler of power plants [2-13], algorithms and object-oriented program codes for data processing of problems of system dynamics, white and pink noise [1, 15], as well as models of difference-scheme equations [14, also see (1) and BoilerB in Fig. 1] and methods of numerical integration, such as Euler and Runge-Kutta methods.

3 Results and Discussion

Let $\mathrm{BoilerB_t}$, – an endogenous indicator of the volume of turbine steam production of a fuel boiler for a period of time t = 1, 2,..., then the equilibrium state of the water-steam mass of the fuel boiler depends on the difference of exogenous production indicators between $\mathrm{WaterRateWR_t}$, – the norms of water intake and $\mathrm{SteamRateSR_t}$, – the norms of steam production as a difference-scheme equation of the first order:

$$\frac{\mathrm{BoilerB_t}}{\mathrm{dt}} = \mathrm{WaterRateWR_t} - \mathrm{SteamRateSR_t} \qquad (1)$$

and the necessary initial volume of turbine steam production:

$$\mathrm{BoilerB_t}|_{t=0} = \mathrm{DesiredBoilerB\_}, \qquad (2)$$

Where:

– $\mathrm{WaterRateWR_t}$, – the water intake rate of the input to the water-steam fuel boiler system in the equation of state (1) is defined as the maximum value between "zero" and the required water intake rate:

$$Water\ Rate\ WR_t = \max(0, \mathrm{DesiredWaterRateDWR_t}) \qquad (3)$$

– $\mathrm{DesiredWaterRateDWR_t}$, – the required volume of the water intake rate into the water-steam fuel boiler system in (3) consists of the sum of the expected volume of the steam rate and the adjusted volume of turbine steam production in accordance with the desired level:

$$\begin{aligned}&\mathrm{DesiredWaterRateDWR_t}\\&= \mathrm{ExpectedSteamRateES_t}\\&+\mathrm{AdjustmentForBoilerAB_t},\end{aligned} \qquad (4)$$

– $\mathrm{AdjustmentForBoilerAB_t}$, - the adjustment of the volume of production of turbine vapor of the fuel boiler in (4) is proportional to the difference between the necessary and the actual volume of turbine steam production:

$$\begin{aligned}&\mathrm{AdjustmentForBoilerAB_t}\\&= \frac{\mathrm{DesiredBoilerB\_-BoilerB_t}}{\mathrm{BoilerAdjustmentTimeBAT}},\\&\mathrm{BoilerAdjustmentTimeBAT} = 3,\end{aligned} \qquad (5)$$

– $\mathrm{ExpectedSteamRateES_t}$, – the expected volume of the steam norm is assumed to be equal to the actual level of the developed norm of turbine vapor of a water-steam fuel boiler:

$$\mathrm{ExpectedSteamRateES_t} = \mathrm{SteamRateSR_t}, \qquad (6)$$

– $\mathrm{SteamRateSR_t}$, – the norms of turbine vapor are assumed to be equal to the volume of production of turbine vapor of the fuel boiler in (1) with a constant average production life cycle:

$$\begin{aligned}&\mathrm{SteamRateSR_t} = \frac{\mathrm{BoilerB_t}}{\mathrm{AverageLife}},\\&\mathrm{AverageLife} = 8,\end{aligned} \qquad (7)$$

– DesiredBoilerB_, – the exogenous volume of the expected production of turbine vapor of a fuel boiler is determined as the product of the initial volume of the expected steam and the input exogenous variable:

$$\begin{aligned}&\mathrm{DesiredBoilerB\_}\\&= \mathrm{InitialDesiredBoiler} * \mathrm{Input_t},\\&\mathrm{InitialDesiredBoiler} = 100,\end{aligned} \qquad (8)$$

– $\mathrm{Input_t}$, – a dimensionless variable that provides simulations of input, for example, step, pulse, sinusoidal data and PinkNoise random noise data:

$$\begin{aligned}&\mathrm{Input_t} = 1 + \mathrm{step(StepHeight, StepTime)}\\&+\left(\frac{\mathrm{PulseQuantity}}{\mathrm{TIME\_STEP}}\right)\cdot \mathrm{pulse}\begin{pmatrix}\mathrm{PulseTime,}\\ \mathrm{TIME_{STEP}}\end{pmatrix}\\&+\mathrm{ramp}\begin{pmatrix}\mathrm{RampSlope,}\\ \mathrm{RampStartTime,}\\ \mathrm{RampEndTime}\end{pmatrix}\\&+\mathrm{SineAmplitude}\cdot \sin\left(2\cdot 3.14159\cdot \frac{\mathrm{time(\ )}}{\mathrm{SinePeriod}}\right)\\&+\mathrm{step(1, NoiseStartTime)} * \mathrm{PinkNoise},\end{aligned} \qquad (10)$$

and the value of constants:

$$\begin{aligned}&\mathrm{RampStartTime} = 1, \mathrm{RampEndTime} = 1e\\&+009, \mathrm{StepTime} = 1, \mathrm{SinePeriod} = 50,\\&\mathrm{PulseTime} = 1, \mathrm{NoiseStartTime} = 1.\end{aligned} \qquad (11)$$

The implementation of the above model (1)-(10) hold in the environment of the software product AnyLogic 8.7.6 [15] for this input put the following values:

– PulseQuantity, – number of pulses = 0.38,
– SineAmplitude, – amplitude sine = 0.14,
– StepHeight, – step height = 0.5,
– RampSlope, gradient ramp = 0.00,
– NoiseStandardDeviation, is the standard deviation of the noise = 0.21 and will be used the data obtained in the result of this simulation white and pink noise random variables.

The obtained system management solution-finding the values of the input parameters of development vapor turbine fuel boiler (see Fig. 1) that regardless of stochastic exogenous factors provide a reliable (stable) States of the water-steam process:

– required model and the volume of vapor turbine (see Fig. 2),

– the rules of water flow and the exit vapor turbine fuel boiler (see Fig. 3).

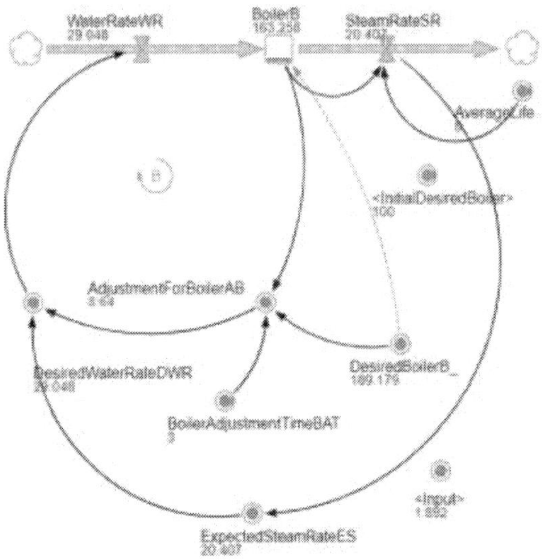

Fig. 1. A system solution for generating turbine vapor of a fuel boiler with stochastic exogenous factors.

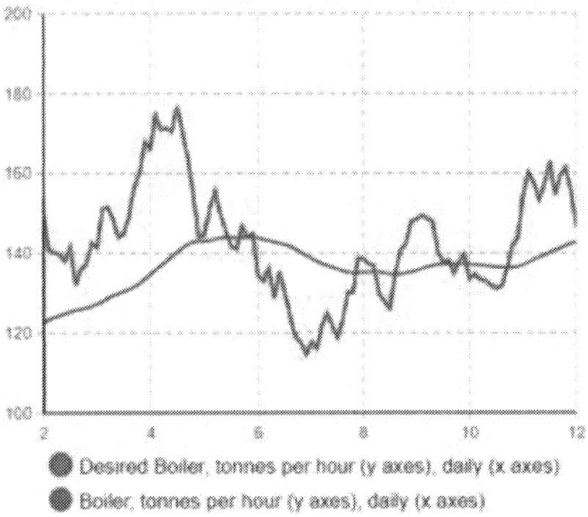

Fig. 2. The necessary and model volumes of steam generation.

Article [10] systematizes and highlights the experience of developing countries in the field of information technologies in marketing, in particular, digital marketing tools classified according to the principle of marketing needs, as well as marketing tools that are suitable and effective in a specific market context are considered.

The article [11] defines the features of "green" bonds and development practices in Europe. The number of climate-conditioned and green bonds in the world in 2018 by region, the distribution of green bond issues in Europe in 2019 by types of regional issuers and the volume of green bond issues in individual European countries in the first half of the year are also determined.

Article [12] substantiates the automation of technological process systems and control of the operation of equipment in the energy sector of the Republic of Kazakhstan, as well as the problem of automation of the entire complex of the industry, including access to networks and efficient transmission of electricity across the country. Article [13] examines a number of problems of the implementation of the state policy in the field of electric power industry, aimed at increasing energy saving So, taking into account all the pros and cons of classical types of electricity production, more and more attention is being paid to new, emerging sources – biofuels and processing of production and consumption waste. Renewable energy sources based on wind-wind energy has become the object of close attention not only of domestic entrepreneurs, but also foreign companies that consider it as a promising profitable industry for long-term investment.

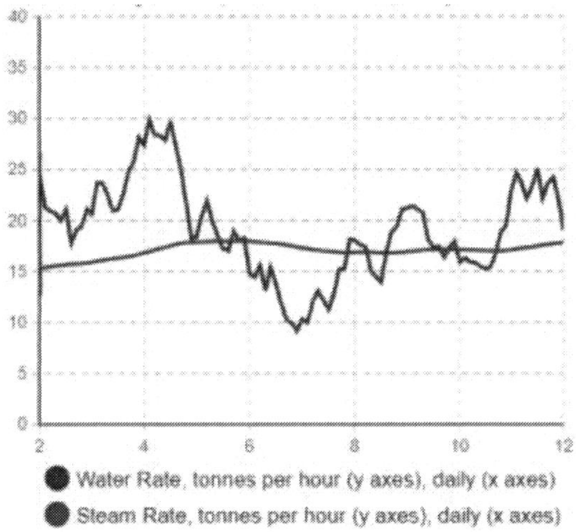

Fig. 3. The norms of water intake and steam output of the fuel boiler.

4 Conclusion

Based on the analysis and review of the conceptual principles, methodology and practice of modeling water-steam processes of the fuel boiler of power plants; implementation of algorithms and models of the theory of system dynamics, white and pink noise; using computational technologies of difference-scheme equations and numerical integration methods, the values of the input parameters of the production of turbine vapor of the fuel boiler are determined), which, regardless of stochastic exogenous factors, provide stable states of the necessary and model volumes, norms of water intake and output of turbine vapor of the fuel boiler.

In conclusion, we note that the scientific results of this study can be recommended in engineering and technical solutions for the implementation of power plants using water steam in post-Soviet countries, including the Republic of Kazakhstan, according to the technical standard Stationary steam boilers: Types and basic parameters [16].

Acknowledgement. This research was funded by the Science Committee of the Ministry of Education and Science of the Republic of Kazakhstan (Grant No. AP09259435).

References

1. J.D. Sterman, Business Dynamics: Systems Thinking and Modeling for a Complex World. McGraw-Hill Inc. (2000)
2. A. Seefried, B. Müller, E. Förster, Journal of Sustainable Development of Energy, Water and Environment Systems, **8(1)**, (2019)
3. A. Capasso, A. Cervone, M.K. Falvo, R. International Journal of Electrical Power & Energy Systems, **56,** 332-339 (2014)
4. Y.-S. Jung, K. Lee, S.-Y. Lee, M.H. Choi, B.H. Lee, International Journal of Control Automation and Systems, **7(4)**, 615-629 (2009)
5. Y.-J. Yoon, M.H. Lee, International Journal of Control Automation and Systems, **8(6)**, 1241-1249 (2011)
6. T. Ancev, Water Economics and Policy, **1(3)**, 1550016 (2015)
7. V.C. Osma, J.E. Capilla, M.Á. Pérez-Martín, Hydrological Sciences Journal, **60(1)**, 1341-1354 (2015)
8. S.W. Fleming, C. Wong, G. Graham, Hydrological Sciences Journal, **59(6)**, 1154-1166 (2014)
9. D.M.D. Hendriks, M.J.M. Kuijper, R. van Ek, Hydrological Sciences Journal, **59(3-4)**, 562-577 (2014)
10. A.I. Leonow, M.N. Koniagina, S.V. Petrova, E. Grunt, S.Y. Kerimkhulle, V.G. Shubaeva, Espacios, **40 (38)** (2019)
11. S. Niyazbekova, L. Moldashbayeva, S. Kerimkhulle, N. Dzholdoshev, T. Dzholdosheva, M. Serikova, E3S Web of Conferences, **244**, 10060 (2021)
12. G. Azieva, S. Kerimkhulle, U. Turusbekova, A. Alimagambetova, S. Niyazbekova, E3S Web of Conferences, **258**, 11003 (2021)
13. S. Niyazbekova, L. Moldashbayeva, S. Kerimkhulle, B. Jazykbayeva, E. Beloussova, B. Suleimenova, E3S Web of Conferences, **258**, 11011 (2021)
14. S. Kerimkhulle, Z. Aitkozha, **1880**, 040016 (2017)
15. The official site of The AnyLogic Company. https://www.anylogic.com (2021)
16. GOST – 3619-89. Stationary steam boilers. Types and basic parameters. USSR State Standard (1989). (in Russian)

A study of sea surface temperature effect on purse seine catch in Aceh North Waters, Indonesia

Ratna Mutia Aprilla[1], *Junaidi* M. Affan[1], *Rahmatul* Fitri[1], *Rianjuanda* Djamani[1], *Afdhal* Fuadi[2], and *Imelda* Agustina[1*]

[1]Fisheries Resource Utilization Study Program, Marine and Fisheries Faculty, Syiah Kuala University, Darussalam, Banda Aceh, 23111, Indonesia.
[2]Program Studi Perikanan, Fakultas Kelautan dan Perikanan, Universitas Teuku Umar, Meulaboh, 23615, Indonesia.

Abstract. Aceh fishermen in Ocean Fishing Port Kutaraja Banda Aceh generally carry out fishing operations around Aceh North waters which have abundant and diverse biological potential. Catches in Aceh North waters are still heavily influenced by several natural factors that often occur in certain seasons, one of which is the factor of Sea Surface Temperature (SST) based on this research about the effect of SST on purse seine catches based in Kutaraja Banda Aceh Ocean Fishery Port. This study aims to determine the catches of purse seine fishermen in Kutaraja Ocean Fishery Port to know the SST distribution, and to determine the effect of SST on purse seine catches based in Kutaraja Banda Aceh Ocean Fishery Port. This research was conducted in November 2019 at Kutaraja Ocean Fishery Port. Data analysis used in this study is purse seine catch data obtained from UPTD from 2015-2019 tabulated in tabular and graphical format, SST distribution obtained from Aqua-MODIS satellites from 2015-2019 using SeaDas software and analysis used to see the effect of SST on catch is using linear regression analysis. The conclusian is based on the results of studies on how ocean SPL affect the fish catches using purse seine with the highest vessel size of 11-20 GT in 2018 with a total of 2,499,634 kg while the lowest catch in 2016 with a total of 1,460,220 kg. SST distribution found in Aceh North waters shows that the highest temperature occurred in 2015 which was 30.24°C while the lowest occurred in 2016 which was 29.04°C. From linear regression analysis results showed that SST in Aceh North waters showed a coefficient of determination (R^2) of 0.2279. This value can give a picture that the SST parameters contribute to fish catches of 22.79% and the rest is influenced by other oceanographic factors. While the correlation coefficient (r) of 0.477 means that this condition can indicate sufficient correlation have a positive

1 Introduction

Aceh Province is located between 01° 58' 37.2" – 06° 04' 33.6" North Latitude and 94° 57' 57.6" – 98° 17' 13.2" East Longitude with an average altitude of 125 meters above sea level. Aceh province boundaries from the north and east are bordered by the Malacca Strait. In the south with the province of North Sumatra and in the west with the Indonesian Ocean [1].

Acehnese fishermen in Ocean Fishery Port Kutaraja generally carry out fishing operations around the Aceh northern waters. There have abundant and diverse potential for biological resources. Therefore, the Aceh northern waters are used as fishing operations areas (Fishing ground). The water was often used for fishing and was a migration line, thus promoting the fishing economy of Aceh [2, 3].

Aceh northern waters have small islands such as Sabang Island, Breueh Island, Tunom Island, Nasi Island and Buntal Island. It has a geographical location of 05°00'00"-06°00'00" North Latitude and 95°00'00"-96°00'00" East Longitude [2]. Oceanographic conditions and economic value of a fishing area are a consideration whether or not to carry out fishing activities in the fishing area. To consider a location in the sea suitable for fishing, several oceanographic parameters are usually considered, i.e the physical, chemical and biological characteristics of the location, including sea surface temperature [4]. The life patterns of a fish are in insparable with environmental conditions. Fluctuations in environmental conditions have had a profoud effeect on fish existence [5]. One such SPL has a biological influence on the metabolism on the fish, judging by its physical effects. Therefore, the purpose of the study is to see and connect the SST studies with catch.

2 Materials and Methods

2.1 Research location and time

This research was conducted at Kutaraja Ocean Fishery Port Banda Aceh in November 2019 with a geographical position located at 5.56"62'110 North Latitude and

* Corresponding author: agustinaimelda1208@yahoo.com

https://doi.org/10.1051/e3sconf/202233902007

95.33”55'210 East Longitude. The research location map is presented in Figure 1.

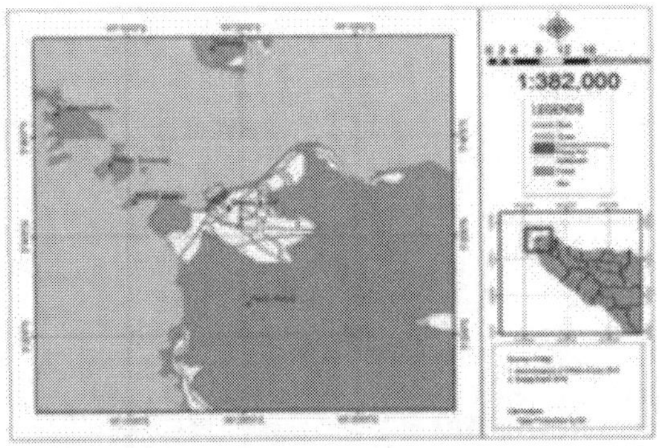

Fig. 1. Research sites

Tools and materials used in this study are: (1) Computer: Media processing data (2) Stationery: recording, (3) Microsoft office excel: Processing data, (4) Capture fisheries statistics book: Supplementary reference, (5) Aqua-MODIS: SST data collection, (6) SeaDas application Type 7.3: Processing SST data (7) Arc Gis Type 10.3: To create SST maps: and (8) Questionnaire: To obtain information from respondents.

Method used in this study is a survey method conducted by conducting interviews based on questionnaires to fishermen in purse seine 11-20 GT. Fishing fleet determination and fishermen as respondents was determined by the purposive sampling method (deliberately) namely purse seine measuring 11 to 20 Gross tonnage (GT) located at the Kutaraja Ocean Fishery Port.

2.2 Data analysis

Based on the purpose above, data analysis is divided into three stages.

1. Fish Catch

Collecting data on the number of catches obtained from the Kutaraja unit per month for the last five years, 2015, 2016, 2017, 2018, and 2019, next is to calculate the total catch of 11-20 GT purse seine that have been classified, tabulated, then reviewed in tabular and graphic formats.

2. Sea Surface Temperature Distribution

Data on sea surface temperatures distribution is known from A-qua-MODIS satellite imagery and processed with SeaDas software (Sea Data Analysis System), then processed again with Arc Gis 10.3 software for making fishing area maps. SST distribution image that has been processed is then issued a concentration value based on the fishing area.

3. SST Effect on Catches

To see SST effect on catch of purse seines, Regression Analyst Method was used with the following formula:

$$Y=a+bx \quad (1)$$

Where:

Y = Fish catch (kg)

a = Constant

b = Regression coefficient

x = Sea surface temperature (SST)

3 Results and Disscussion

Catches number of 11-20 GT purse seine for the last 5 years (2015-2019) at the Kutaraja Ocean Fishery Port (PPS) was the highest in 2018 of 2,499,634 kg and the lowest in 2016 with a total of 1,464,220 kg (Figure 2).

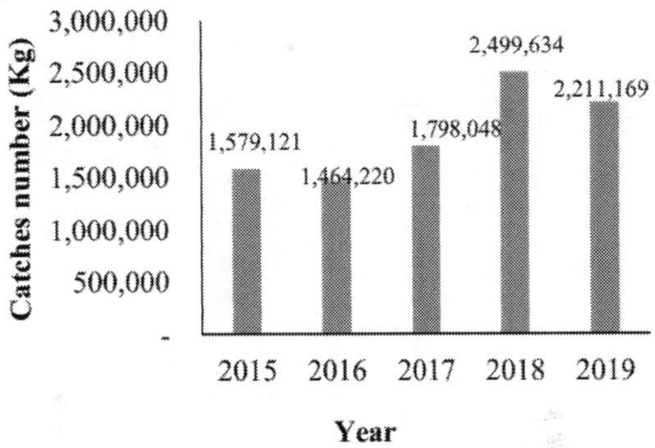

Fig. 2. 11-20 GT purse seine

SST content obtained in Aceh northern waters shows that the sea surface temperature in 2016 to 2019 started from the lowest temperature of 29.04 °C occurred in 2016 and reached the highest temperature of 30.24 °C occurred in 2015 with an average value of the average sea surface temperature of 29.86 °C (Figure 3).

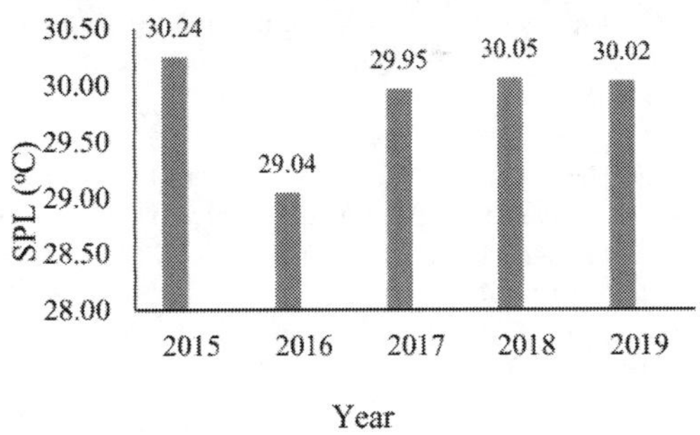

Fig. 3. Average distribution of sea surface temperature in the waters of North Aceh

Table 1. Average distribution of sea surface temperature in the waters of North Aceh

No	Acquisition Time	Average SPL
1	2015	30.24
2	2016	29.04
3	2017	29.95
4	2018	30.05
5	2019	30.02

Source: Satellite imagery (processed)

The average temporal distribution of sea surface temperatures in Aceh North waters is different every year, starting from the highest sea surface temperature to the lowest. It can provide an illustration that there is a dynamic change in dynamics of fishing areas in Aceh

North waters (Table 1). The symptom factors that occur due to changes in condition of these oceanographic parameters are one of the parameters to be able to determine a potential and environmentally friendly fishing area.

Sea surface temperature can be used as an indicator to be able to determine the presence of a type of fish in a water, because each type of fish has a certain temperature tolerance that is preferred for its survival thus it affects their presence and distribution. Therefore, temperature is very important in determining a fishing area.

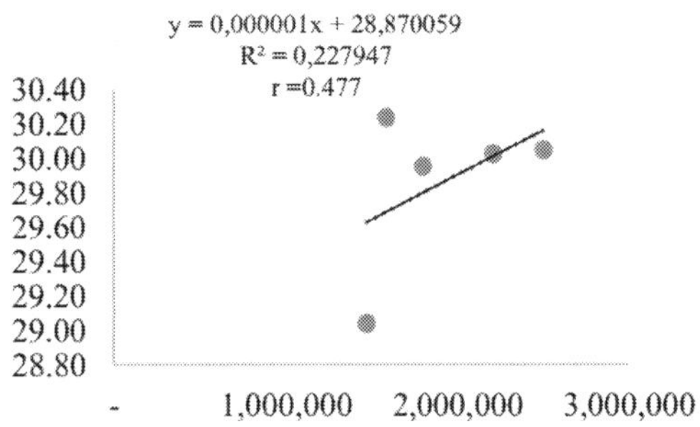

Fig. 4. Effect of sea surface temperature on fish catch

Based on Figure 4. show that the effect of fish catches graph on changes in sea surface temperature at Aceh North waters, the coefficient of determination (R2) is 0.2279. This value can provide an illustration that the sea surface temperature (SST) parameter contributes to fish catches by 22.79% and the rest is influenced by other oceanographic factors and production factors. While the value of the correlation coefficient (r) is 0.477, meaning that this condition can indicate a sufficient correlation have a positive relationship.

The number of fish caught at Kutaraja Ocean Fishing Port (PPS) by purse seine vessel measuring 11 -20 GT for 5 years was the highest in 2018 which was 2,499,634 kg and the lowest occurred in 2016 which was 1,464.220 kg in Aceh north waters. Catch types for the last five years (2015-2019) were dominated by scad, tuna, also mackerel with each total catch of 8,331,785 kg, 1,115,237 kg and 105,170 kg. Either, the most dominant catches of small pelagic fish on land in Kutaraja Ocean Fishing Port are scad and tuna [2, 6]. Changes to the SPLdue to the transition seasoneast - west in the waters of Aceh so that the results catch caused theso the fluctuations are not too much different.

Sea surface temperature that occurs in Indonesia in the East monsoon is lowly, while the sea surface temperature in West monsoon tends to be warmer when compared to temperature in the East monsoon [7]. According to [8], sea surface temperature in the West monsoon (December to February) is on average 4°C hotter than the East monsoon (June to August).

According to [9]), temperature is one of oceanographic parameters that expected to have a relationship with the life of fish or marine biota in particular and marine biological resources in general. Most marine biota like the optimum temperature or according to the body needs so that temperature is one of the most important factors in regulating life processes and organisms spread in the waters [10].

According [5, 11], states that changes in water temperature below the optimum temperature can cause a decrease in fish moving activity and also in feeding activities and can inhibit survival [12], said that temperature can affect the fish spread or marine biota in waters, because temperature plays a big role in regulating metabolic processes in the fish's body, which functions as a regulator of body movement activities, for example swimming speed and as a nerve stimulus. Information on the existence of fish is one of the kays to success in the development of catch fisheries [13].

4 Conclusion

The SST in today's waters is not fluctuating too significantly, thanks to a large surge of fashionable sensors resulting in a SPL range still available in a wide area (low resolution).

Based on the results of studies on how ocean SPL affect the fish catches using 11-20 GT purse seine gear was in 2018 with a total of 2,499,634 kg while the lowest catch was in 2016 with a total of 1,460,220 kg. Sea surface temperatures distribution in Aceh north waters shows that the highest temperature occurred in 2015 which was 30.24°C while the lowest occurred in 2016 which was 29.04°C with an average value of 29.86°C. So, from the simple linear regression analysis results about sea surface temperature effect on fish catch shows that there is a sufficient correlation with positive relationship direction and coefficient of determination value (R^2) is 0.2279 which means that the effect is 22.79% and other is influenced by other oceanographic factors, viz. chlorophyll-a, salinity and current.

Acknowledgment. The author thanked the head and staff units and Syahbandar Kutaraja Ocean Fishing Port also respondents who have helped provide related information and data.

References

1. [BPS Aceh] Central Bureau of Statistics Aceh. Aceh statistics book in Aceh Province regional figures. Banda Aceh (2018)
2. Kurnia., S. Purnawan, T. Rizwan. Mapping of small pelagic fishing areas in northern waters of aceh. Unsyiah Marine and Fisheries Student Scientific Journal, 1, 185-194 (2016)
3. Neliyana. Comparative analysis of purse seine fishing at Lampulau Coastal Fishing Port (PPP) Banda Aceh. Thesis, Spostgraduate school, Bogor Agricultural Institute (IPB) (2014)
4. Hartoko., A. Spatial distribution of thunnus.sp, vertical and horizontal sub-surface multilayer temperature profiles of in-situ agro float data in indian ocean. Journal of Coastal Development. **14**, 61-74 (2010)
5. Jufri A, Amran MA, Zainuddin. Caracteristic of the claw fishing grounds in the wastern season in the

waters of Bone. IPTEKS PSP Journal, **1**, 1-10 (2014).

6. Makwiyah, A., Chaliluddin, M., Aprilla. RM, Affan, JM, Muhammadar, AA, Rahmadani, H., Miswar, E., and Firdus, F. The effectiveness of using FADs as fishing grounds in Pusong Waters, Lhokseumawe City. Depik Journal. 7, 119-126 (2018)
7. Sukresno, B., Kasa IW. Dynamical analysis of Banda Sea concerning with el nino, Indonesian through flow and monsoon by using satellite data and numerical model. Echotrophic Journal. 3, 87-91 (2008)
8. Sulaiman, A. Banda Sea Turbulence (Preliminary Study of ARLINDO Microstructure). Directorate of Natural Resources Invertarization Technology (TISDA) Agency for the Assessment and Application of Technology (BPPT). Jakarta (2008)
9. Nontji, A. Nusantara Sea. Jakarta: Sjambatan Publisher. 372 p (2005) 9
10. Nybakken, J. Marine Biology an Ecological Approach. Jakarta: PT. grammar. 459 pp (1992) 10
11. Laevastu T, Hayes ML. Fisheries Oceanography and Ecology. New York: Fishering News (1981)
12. Baskoro. MS, Revelation. RI, Effendi. A. Migration and Distribution of Fish. Bogor: Fisheries and Marine Sciences Faculty, Bogor Agricultural Institute. 152 pp (2004)
13. Aguslina SE, Tadjuddah M, Mustafa A. Across Sea Surface Temperatures and Tracking Claw Areas in Banda Seawater. Halu Oleo University, **2**, 41-49 (2016)

https://doi.org/10.1051/e3sconf/202233902008

Grain size variation in the swash zone of Lhoong Beach, Aceh Besar

Khairul Akbar[1], *Chitra* Octavina[1,2], and *Syahrul* Purnawan[1*]

[1]Department of Marine Sciences, Faculty of Marine and Fisheries, Universitas Syiah Kuala, Banda Aceh 23111, Indonesia
[2]Laboratory of Marine Biology, Faculty of Marine and Fisheries, Universitas Syiah Kuala, Banda Aceh 23111, Indonesia

Abstract. The spatio-temporal texture characteristics are key parameters for understanding the dynamics of the sedimentary environment. This study was conducted to determine the sediment characteristics in swash zone of Lhoong Beach, Aceh Besar Regency. The analysis emphasizes the seasonal variation of sediment mean grain size (MGS) based on vertical stratification. Sediment samples were taken during November 2018, January 2019 and March 2019, from 3 stations. Samples from station 3 had the coarsest grain characteristics, where the highest MGS value at station 3 was recorded in the mid-layer (5-10 cm) in November. The grain size was found to vary with changing seasons and the depth of the sediment layer. Based on depth stratification, the highest standard deviation was obtained by the mid-layer (5-10 cm); while by period, the highest standard deviation of the MGS value was generated during November.

1 Introduction

Coastal areas are one of the most dynamic areas on the earth's surface [1]. The interaction between land and sea will form a unique morphology in each coastal area [2]–[6]. As a dynamic zone, the coastal zone does not only affect physical factors, but also biology [7], [8] and chemistry [9]. A number of parameters that can affect changes in coastal physical conditions, including: currents, waves, tides, storms, and landscapes [10]–[12].

In general, seasons play a major role in controlling energy in the aquatic environment [13], [14], where in coastal areas seasonal changes result in variations in the strength and intensity of waves on the coast [15]. So it can be further explained that seasonal variations can affect the grain size of sediment through waves on the coast [16]. In particular, several researchers also studied the effect of waves in swash areas on sediment profiles [17], [18].

Although seasons are widely studied for their influence on coastal dynamics in general, specific observations of grain size can be the focus for observing the impact of seasonal changes on coastal conditions [19]. The grain size study of coastal sediments provides information about the intrinsic properties of sediments and their depositional environment [20]. Complex coastal processes operating in the past and in the present have left their traces in sediments [21], [22]. In this regard, coastal sedimentology plays an important role in documenting the depositional history of an aquatic environment [14], [23], [24]. A granulometric study of the Aceh coast has been carried out by several researchers [25], [26]. However, as an area bordering the Indian Ocean, the condition of the west coast of Aceh has a fairly large seasonal variation [27]. It is then suspected that seasonal variations also have an influence on sediment grains on the Lhoong beach, Aceh Besar. Furthermore, this study also provides an overview of vertical changes to changes in sediment character to a depth of 15 cm.

2 Material dan Method

The research took place on the Lhoong beach, Aceh Besar District. Lhoong Beach is an open sand beach, facing the Indian Ocean. Sediment sampling was carried out at three stations which were 1.2 km apart from each station (Figure 1). The sampling area was defined in the foreshore area, also known as the swash zone, specifically at the low tide boundary.

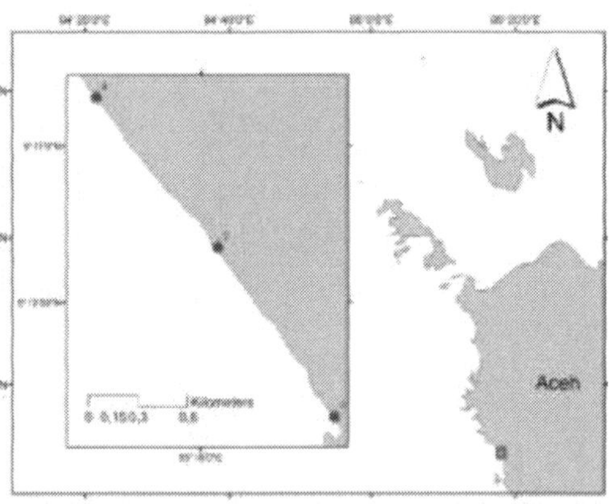

Fig. 1. Sampling location in Lhoong Beach, Aceh Province

Sediment collection was carried out using the coring method to a depth of 15 cm [28], whereupon the sample

* Corresponding author: syahrulpurnawan@unsyiah.ac.id

was further processed by dividing it into three layers, each layer having a thickness of 5 cm: i.e. upper layer (0-5 cm), mid-layer (5-10 cm), and lower layer (10-15 cm). Data collection was carried out in November 2018, January 2019 and March 2019.

The grain size data analysis was carried out using a wet sieve method, using a stratified sieve. A total of approximately 200 grams of sample was separated based on the mesh size of 2 mm; 1mm; 0.5mm; 0.25mm; 0.125mm; 0.063 mm; 0.038mm. Size classification follows the Wentworth scale [29]. The calculation of the percentage weight percent of sediment can be known from the weight of each sediment fraction (eq. 1), and then the mean grain size of the sediment ($\bar{x}_a$) (eq. 2) is calculated using the following equation:

$$Class\ weight(i)\ (\%) = \frac{weight\ size(i)}{total\ sample\ weight} x\ 100\% \quad (1)$$

$$\bar{X}_a = \sum_{i=1}^{n} \frac{(class\ weight(i)\ x\ mesh\ size\ (i))}{n} \quad (2)$$

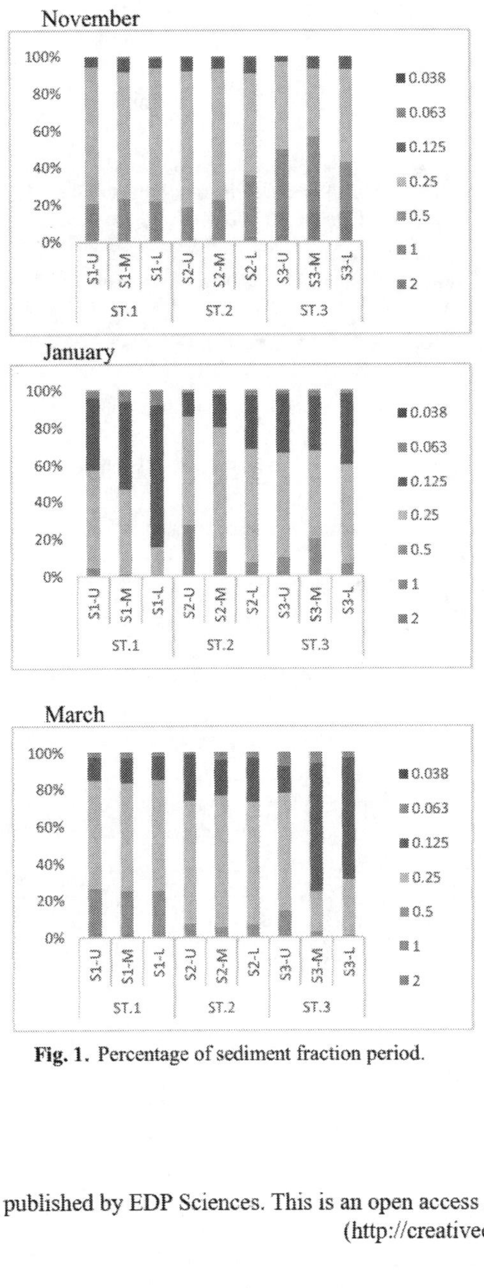

Fig. 1. Percentage of sediment fraction period.

3 Results and Discussion

The percentage of sediment weight for each grain size fraction was obtained from the sediment sieving process. The medium sand fraction has the highest percentage. The percentage of coarse sand was relatively high in November, followed by the percentage of fine sand. In the following period, January and March, the percentage of medium sand was seen to be lower, while the percentage of fine sand was higher than the results obtained in November. The weight percentage of each sediment from the three locations is presented in Figure 2.

The information on the percentage weight per fraction (from Figure 2), was used to calculate the mean grain size (MGS). The MGS value was processed arithmetically based on the layer and sample collection period. From Table 1 it is assumed that there was a spatial variation in the MGS value in Lhoong, which roughly mean that station 3 has coarser grains than stations 1 and 2. MGS variation itself is common in profiles along the beach, which can be caused by beach morphology [30] as well as coastal slope and wave effects [31].

Table 1. Mean grain size value (in millimeter)

Layer	St.	MGS (mm)			Average
		Nov	Jan	Mar	
Upper (0-5 cm)	1	0.29	0.20	0.32	0.27
	2	0.28	0.30	0.24	0.27
	3	0.49	0.23	0.27	0.33
Mid (5-10 cm)	1	0.29	0.18	0.32	0.26
	2	0.30	0.26	0.24	0.27
	3	0.68	0.26	0.16	0.37
Lower (10-15 cm)	1	0.30	0.14	0.31	0.25
	2	0.34	0.22	0.23	0.26
	3	0.52	0.21	0.16	0.30

The variation in MGS values on the Lhoong beach can be assumed to be related to wave activity, as well as the situation on this beach which faces the Indian Ocean. However, this effect can differ between seasons, as previously mentioned [32], that seasons control the variation of wave energy and then affect the MGS value. Moreover, in the swash area, the seasons also make a difference to the sediment transport mechanism that occurs [17], [33]. The variation of MGS at all stations was the highest at November compared to January and March, according to the obtained standard deviation value (Figure3). The highest MGS value in November can be attributed to the seasonal pattern that occurs in the western waters of Aceh. The first data collection was carried out in November and was in the transition season from the southwest to the northeast monsoon [34], so that before data collection was carried out, Lhoong beach had experienced the previous southwest monsoon period. The southwest monsoon in Aceh provides high energy exposure, especially the west coast of Aceh, since the dominant wind comes from the Indian Ocean [27], [35]. We speculated that the presence of high energy during the previous period still has an influence on the grain size during the observation period, and the presence of a high standard deviation is also a consequence of the transitional phase or transitional season that was occurring at the time of data collection. Furthermore, from Figure 3, it can also be seen that the MGS value in January was lower and slightly increased again in March. The January period was more correlated with the northeast monsoon, where the west coast of Aceh experienced calmer conditions than the southwest monsoon. Likewise, the standard deviation value also seemed to follow the trend, where the standard deviation value in January was quite small.

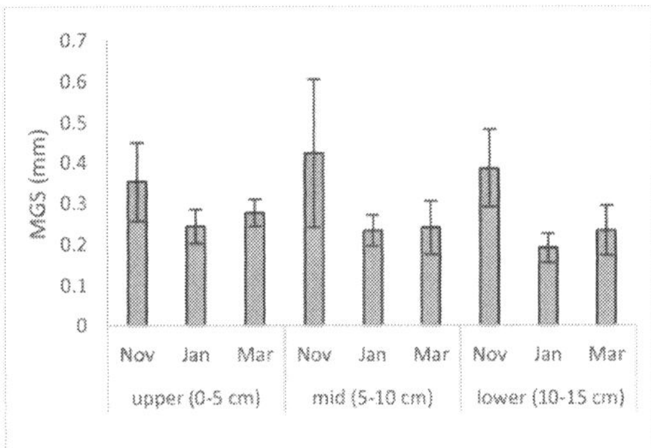

Fig. 3. MGS value based on depth layer

The pattern of sediment transport in the swash area is quite complex [36], since the phenomenon of sediment flow in this area is influenced by the interaction of waves and backwash [37]. Based on vertical layering, we also suspected that there was a seasonal effect. Furthermore, the mean value of standard deviation per layer showed that the highest value was obtained from the middle layer. We could not find out the details of the process that has formed the highest standard deviation values in the middle layer (5-10 cm), as we suspected the greatest variation should occur in the surface layer. However, we assumed that this is related to a transitional process of depositional composition, which at the end of the southwest monsoon would make the finer grains more likely to settle on the seabed. The presence of waves on the beach will more easily carry and stir the fine particles into water column [38]. Furthermore, the stirring process and the deposition of fine sediment that occurred can directly affect the variation in grain size on the seabed even in the deeper layers, as the 10-15 cm layer also has a higher standard deviation value than the surface area.

4 Conclusion

Different observation periods responded to the mean grain size value at three locations on the Lhoong beach. The November period produced the coarsest grains compared to the January and March periods, in line with the standard deviation which was also quite high in November. Vertically, the sediment layering showed an influence from the observation period, where the mid-layer produced sediment with the coarsest grains, along with the highest standard deviation compared to the lower (10-15 cm) and upper (0-5 cm) layers.

Acknowledgments. The writing of this manuscript is supported by the Matching Fund – Kedaireka funded by the Ministry of Education, Culture, Research, and Technology Indonesia awarded to the Research Center for Marine Sciences and Fisheries, Universitas Syiah Kuala.

References

1. J. Obu, H. Lantuit, G. Grosse, F. Günther, T. Sachs, V. Helm, and M. Fritz, Geomorphology **293**, 331 (2017)
2. C. Armaroli and P. Ciavola, Geomorphology **126**, 201 (2011)
3. N. A. Berard, R. P. Mulligan, A. M. F. da Silva, and M. Dibajnia, Coast. Eng. **125**, 70 (2017)
4. G. Incelli, N. Dodd, C. E. Blenkinsopp, F. Zhu, and R. Briganti, Coast. Eng. **115**, 42 (2016)
5. R. T. McCall, G. Masselink, T. G. Poate, J. A. Roelvink, L. P. Almeida, M. Davidson, and P. E. Russell, Coast. Eng. **91**, 231 (2014)
6. N. Cohn and P. Ruggiero, Coast. Eng. **115**, 79 (2016)
7. H. Sawada, H. Saito, K. Adachi, and H. Toyohara, J. Sea Res. **65**, 275 (2011)
8. M. C. Menéndez, C. A. Baleani, M. R. Amodeo, E. M. Acha, and M. C. Piccolo, Estuar. Coast. Shelf Sci. **227**, 106307 (2019)
9. A. Anandkumar, R. Nagarajan, E. Sellappa Gounder, and K. Prabakaran, Chemosphere **287**, 132069 (2022)
10. R. Ranasinghe, M. Larson, and J. Savioli, Coast. Eng. **57**, 1006 (2010)
11. A. H. Reis and C. Gama, Geomorphology **114**, 276 (2010)
12. E. P. Beetham and P. S. Kench, Geomorphology **209**, 98 (2014)
13. K. F. Nordstrom, J. Sediment. Res. **47**, 1287 (1977)
14. G. M. Curtiss, P. D. Osborne, and A. R. Horner-Devine, Mar. Geol. **259**, 73 (2009)
15. A. Ribal, A. V Babanin, S. Zieger, and Q. Liu, Renew. Energy **160**, 1349 (2020)
16. G. Yang, X. H. Wang, Y. Zhong, Z. Cheng, and F. P. Andutta, Estuar. Coast. Shelf Sci. **244**, 106931 (2020)
17. J. M. Alsina, J. van der Zanden, I. Cáceres, and J. S. Ribberink, Coast. Eng. **140**, 23 (2018)
18. T. E. Baldock, J. A. Alsina, I. Caceres, D. Vicinanza, P. Contestabile, H. Power, and A. Sanchez-Arcilla, Coast. Eng. **58**, 214 (2011)
19. N. Senechal and A. Ruiz de Alegría-Arzaburu, in edited by D. W. T. Jackson and A. D. B. T.-S. B. M. Short (Elsevier, 2020), pp. 461–486
20. S. J. Kulkarni, P. G. Deshbhandari, and K. S. Jayappa, Aquat. Procedia **4**, 117 (2015)
21. K. D'Haen, G. Verstraeten, and P. Degryse, Prog. Phys. Geogr. Earth Environ. **36**, 154 (2012)

22. H. E. Wittmeier, J. Bakke, K. Vasskog, and M. Trachsel, Quat. Sci. Rev. **114**, 78 (2015)
23. K. M. Zahid and D. L. Barbeau, Sediment. Geol. **229**, 64 (2010)
24. R. Jagodziński, B. Sternal, W. Szczuciński, C. Chagué-Goff, and D. Sugawara, Sediment. Geol. **282**, 57 (2012)
25. S. Purnawan, N. Ihsan, I. Setiawan, and S. Yuni, Depik **7**, 22 (2018)
26. S. Purnawan, I. Setiawan, H. A. Haridhi, and M. Irham, IOP Conf. Ser. Earth Environ. Sci. **106**, (2018)
27. I. Setiawan, S. Rizal, Y. Haditiar, Y. Ilhamsyah, S. Purnawan, M. Irham, and S. Yuni, in *IOP Conf. Ser. Earth Environ. Sci. 176* (IOP Publishing, Bogor, 2018), p. 012016
28. ASTM, *D4823–95 Standard Guide for Core Sampling Submerged, Unconsolidated Sediments* (ASTM International, West Conshohocken, 2014)
29. C. K. Wentworth, J. Geol. **30**, 377 (1922)
30. A. H. F. Klein, Ó. Ferreira, J. M. A. Dias, M. G. Tessler, L. F. Silveira, L. Benedet, J. T. de Menezes, and J. G. N. de Abreu, Coast. Eng. **57**, 98 (2010)
31. S. Sharma, V. S. Kumar, and R. Gowthaman, Mar. Georesources Geotechnol. **35**, 887 (2017)
32. M. Samsuddin, J. Coast. Res. **5**, 57 (1989)
33. J. M. Alsina, I. Cáceres, M. Brocchini, and T. E. Baldock, Coast. Eng. **68**, 31 (2012)
34. I. Imanullah, S. Sofyan, S. Sugianto, Y. Haditiar, and S. Rizal, IOP Conf. Ser. Earth Environ. Sci. **348**, 12066 (2019)
35. S. Rizal, I. Setiawan, T. Iskandar, Y. Ilhamsyah, A. W. Mulyadi, and M. Musman, Sains Malaysiana **39**, 519 (2010)
36. P. Chardón-Maldonado, J. C. Pintado-Patiño, and J. A. Puleo, Coast. Eng. **115**, 8 (2016)
37. J. van der Zanden, I. Cáceres, S. Eichentopf, J. S. Ribberink, J. J. van der Werf, and J. M. Alsina, Coast. Eng. **152**, 103519 (2019)
38. L. Zuo, D. Roelvink, Y. Lu, and G. Dong, Int. J. Sediment Res. (2021)

Concentration of dissolved heavy metal around the Rupat waters industrial activity, Malacca Strait

Koko Ondara[1], *Ulung* Jantama Wisha[1], *Ruzana* Dhiauddin[1], and *Syahrul* Purnawan[2*]

[1]Research Institute for Coastal Resources and Vulnerability, Ministry of Marine Affairs and Fisheries, Padang 25245, Indonesia
[2]Department of Marine Sciences, Faculty of Marine and Fisheries, Universitas Syiah Kuala, Banda Aceh, Indonesia

Abstract. The Rupat Strait is a water area located between Rupat Island and Dumai City. This area is growing rapidly which is characterized by increasing coastal residential areas, commercial centers, industries, ports and also increasing marine transportation. This increase in activity can lead to a decrease in the water quality of the Rupat Strait. This study aims to analyze the water quality of heavy metals Cadmium (Cd), Cuprum (Cu), Lead (Pb), Zinc (Zn) and Nickel (Ni) in the waters of the Rupat Strait. Sampling of surface water using sample bottles was carried out at 9 observation stations from the waters of West Rupat to South Rupat, then testing the water samples in the laboratory. The results of the analysis show that the lead and nickel values still meet the threshold values set by the Minister of Environment No.51 of 2014. A total of 30% of the sample points for cadmium testing water do not meet the quality standards and have been polluted. Meanwhile, the values of cuprum and zinc have exceeded the specified quality standards and have polluted the waters of the Rupat Strait. There is a need for a mechanism from various aspects to maintain water quality that still meets the threshold and certain efforts are needed to reduce heavy metal pollution, namely cuprum and zinc in the waters of the Rupat Strait and its surroundings.

1 Introduction

Heavy metals are metal elements that have density exceeds 5 g/cm3 which are classified as pollutants and are dangerous because they will accumulate in the water column, sediments, organisms and cannot be degraded naturally [1]. Heavy metals are one of the factors in determining the status of water quality [2]. Heavy metal concentrations that have exceeded the threshold will have an impact on the ecosystem around these waters [3], [4]. Cuprum is one of the heavy metal in the water that is needed by the body in the physiological processes of the organism. However, the excess can interfere with health or cause poisoning [5], [6]. Cadmium (Cd) which is also toxic can cause poisoning, health problems and can kill certain biota, for example, crustacean biota will die on exposure to Cd with a concentration of 0.005 – 0.15 ppm [7].

Rupat Strait is a waters located between Dumai City, Sumatra Island and Rupat Island. The strait is a semi-enclosed water which makes the oceanographic conditions of these waters very dynamic depending on external factors that influence it [8]. The Rupat Strait is used as one of the water transportation routes that are widely used by industries in the coastal areas of Riau Province, especially Dumai City and Bengkalis Regency. The Rupat Strait also has high fishery potential, especially red shrimp (Panaeus monodon) and white shrimp (Pennaeus sp.) to be developed [9]. The high level of human and economic activity in the Rupat Strait has an adverse impact on environmental health, especially the marine environment [10]–[14]. The Rupat Strait area has different levels of environmental sensitivity according to environmental characteristics. The areas of West and South Rupat, Mampu Island and Lubuk Gaung are areas that are very sensitive to oil pollution. The area has relatively good mangrove vegetation and is a catchment area [15]. In addition, the potential for heavy metal oil pollution is also potential in the Rupat Strait because many of these waters have been converted into berths for commercial ships. Therefore, this study aims to determine the suitability of heavy metals, namely Cadmium (Cd), Cuprum (Cu), Lead (Pb), Zinc (Zn) and Nickel (Ni) in the waters of the Rupat Strait.

2 Material and Method

Water sampling of the Rupat Strait was carried out at 9 observation stations (Table 1) from the waters of West Rupat to South Rupat (Figure 1). 5 water quality parameters of dissolved metals, namely cadmium (Cd), Cuprum (Cu), Lead (Pb), Zinc (Zn) and Nickel (Ni). The water quality data analyzed using the quality standards of the Minister of Environment Decree No. 51 of 2004 [16] to see the suitability of the health of marine waters to biota.

Data collection was carried out when the weather conditions were sunny on August 28 to 30, 2020, for the last 3 (three) days there was no rain and the water sample taken was surface water (<1 meter depth). The test was

* Corresponding author: syahrulpurnawan@unsyiah.ac.id

carried out at the Environmental Laboratory of PT. Mutuagung Lestari in Pekanbaru, Riau to obtain water heavy metal values.

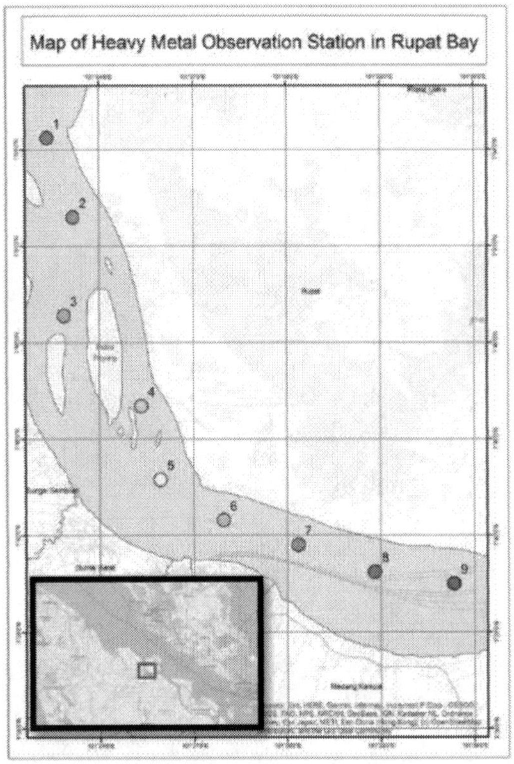

Fig. 1. Observation station map in Rupat Strait.

Table 1. Observation station coordinate.

Station Number	Longitude	Latitude
1	101.372972	1.906833
2	101.386750	1.865472
3	101.381250	1.814917
4	101.422590	1.767140
5	101.432690	1.729470
6	101.466700	1.708330
7	101.506220	1.695470
8	101.547570	1.681680
9	101.589840	1.675250

3 Results and Discussion

The cadmium value in the waters of the Rupat Strait ranges from 0.0001 to 0.021 mg/l. The cadmium value at observation stations 1, 3 and 4 did not meet the marine water quality standard that had been set at 0.001 mg/l. Cadmium values for observation stations 2, 5, 6, 7, 8 and 9 are below 0.0001 mg/l and meet the water quality standards for marine biota (Figure 2).

The main sources of Cd metal contamination are through the earth's layers and human activities [17] such as waste and household waste and fishing activities such as repairing and painting ships [18]. Cd metal can also come from the activity of giving a protective zinc layer to iron and steel to avoid rust [19]. The activity of giving the coating is mostly found in industrial areas located in coastal areas that are prone to rust. The contamination of Cd sourced from various marine activities is potentially absorbed by marine biota in the form of free ions $Cd2+$. This ion is going to be able to compound with chloride ions within pH 7.0, yielding in 51% $CdCl2$, 39 % $CdCl+$, 6 % $CdCl3+$, and $Cd2+$ complex about 2.5% of total concentration.

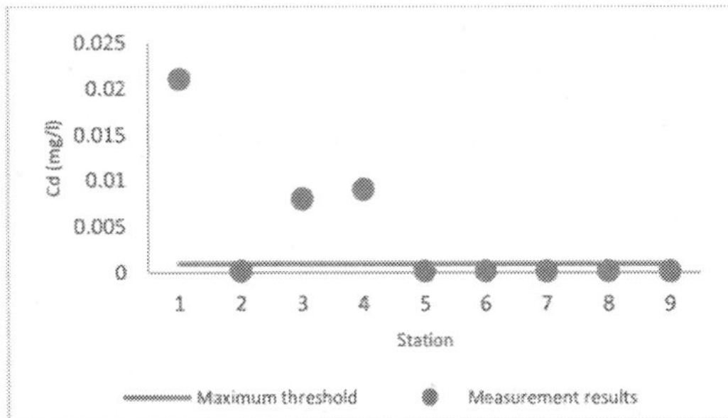

Fig. 1. Cadmium concentration to water quality threshold

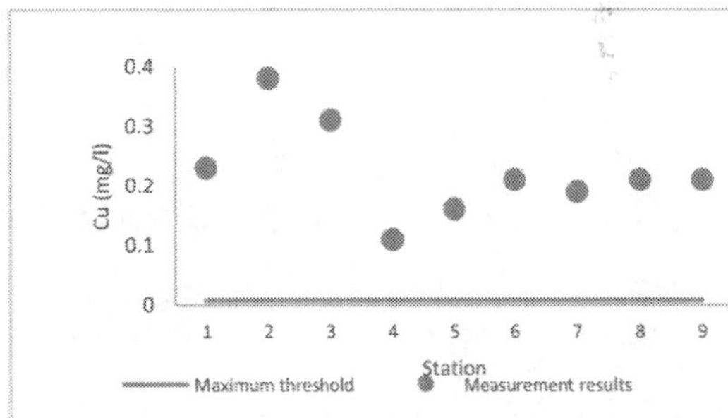

Fig. 2. Cuprum concentration to water quality threshold

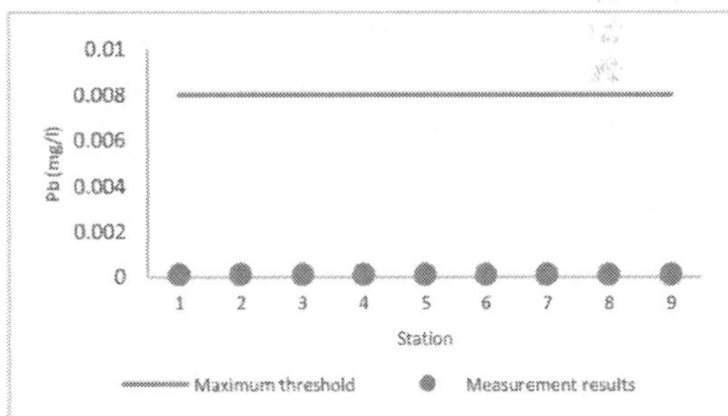

Fig. 3. Lead concentration to water quality threshold

The threshold value for cuprum concentration is 0.008 mg/l. The cuprum value in the Rupat Strait is 0.11 mg/l to 0.38 mg/l, so it exceeds the threshold that has been set for all observation points (Figure 3). The largest values are at observation stations 2 and 3 which are around the northern part of the estuary. These states showed that the Rupat Strait received sufficiently high Cu-contained pollutant. The Rupat Strait, which the center for marine and port activities, has become the source of heavy metals accumulations. Cu is contained within the various paint of the ships serving as an anti-rust paint and antifouling [20]. These materials also contain of Pb compounds such as $PbCO3$, Pb white, and $Pb3O4$.

https://doi.org/10.1051/e3sconf/202233902009

The lead value in the waters of the Rupat Strait is below 0.0001 mg/l and is still below the water quality standard for marine biota of 0.008 mg/l (Figure 4). This result showed that there are no contaminations of lead within the Rupat Strait or perhaps, the device used in the laboratory could not detect the concentration of Pb which the concentration of lead might be tremendously low. Pb is kind of heavy metals that relates to the human health, because it could impact on the large number of mankind toxicities, environmental pollution, and harmful toxic properties [21]. This heavy metal is potentially contained within the marine organisms that often consumed by human [22].

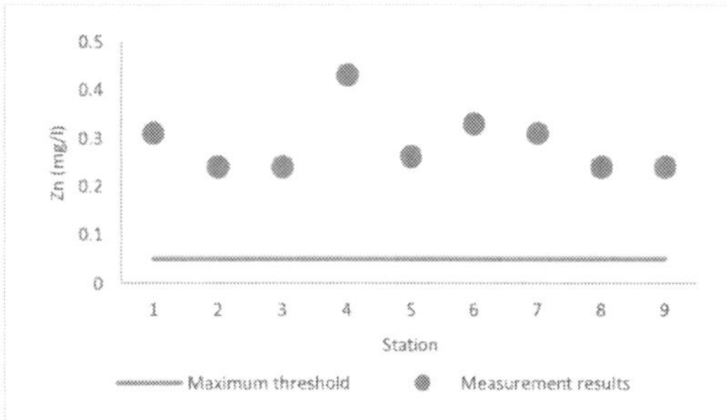

Fig. 4. Zinc concentration to water quality threshold

The zinc concentration in the waters of the Rupat Strait (Figure 5) ranges from 0.24 to 0.43 mg/l. This value exceeds the maximum limit of marine water quality standard of 0.05 mg/l. The high concentration of Zn could be sourced from the presence of Pier, shipping lane, and ship building activities within the Rupat Strait [23]. This metal can threat the survival ability of local marine biota. Metal distribution in the water environment depends on physiochemical alteration in the water and surface bottom. The precipitation of Zn and other heavy metals is caused by the alkaline pH state in the form of insoluble hydroxides, oxides, and carbonates.

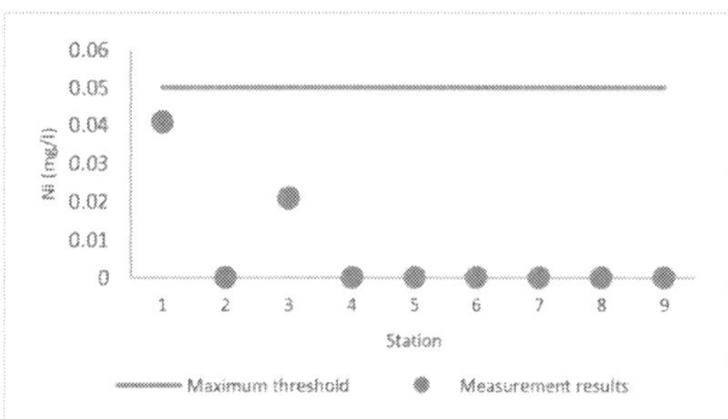

Fig. 5. Nickel concentration to water quality threshold

The nickel concentration in the waters of the Rupat Strait ranges from below 0.0001 to 0.041 mg/l and is still below the maximum water quality standard set at 0.05 mg/l. The highest nickel value is at stations 1 and 3 (Figure 6), while for other observation stations it is below 0.0001 mg/l. These states indicate that the nickel metal does not pollute the Rupat Strait. In the waters, nickels have interacted with organic compounds within the aqueous phase and eventually settled in the sediment, resulting a high accumulation of nickel in the sea bottom sediment.

4 Conclusion

The values of Lead (Pb) and Nickel (Ni) still meet the threshold values set by the Minister of Environment Decree No. 51 of 2014. As many as 30% of the sample locations for Cadmium (Cd) testing waters do not meet the quality standards and have been polluted. Meanwhile, the values of Cuprum (Cu) and Zinc (Zn) have exceeded the specified quality standards and have polluted the waters of the Rupat Strait.

References

1. N. H. Harun, P. M. Tuah, N. Z. Markom, and M. Y. Yusof, in *Int. Conf. Environ. Res. Technol.* (2008)
2. N. Y. Gerhanae and Y. Permanawati, J. Geol. Kelaut. **13**, 99 (2016)
3. Darmono, *Lingkungan Hidup Dan Pencemaran, Hubungan Dengan Toksikologi Senyawa Logam* (Universitas Indonesia Press, Jakarta. Indonesia, 2001)
4. A. H. Kusuma, T. Prartono, A. S. Atmadipoera, and T. Arifin, J. Teknol. Perikan. Dan Kelaut. **6**, 41 (2016)
5. H. Palar, J. Chem. Inf. Model. **53**, 1689 (2004)
6. R.B. Clark, J. Coast. Res. **7**, (2012)
7. S. Wulandari, Y. B, G. Santosa, and K. Suwartimah, Indones. J. Mar. Sci. **14**, 170 (2012)
8. S. Nedi, B. Pramudya, E. Riani, and Manuwoto, J. Enviromental Sci. **1**, 25 (2010)
9. D. Sarianto, S. Asrina Ikhsan, R. bayu Kusuma Haris, and D. Djunaidi, J. Ilmu-Ilmu Perikan. Dan Budid. Perair. **14**, (2019)
10. Y. Badrun, J. Enviromental Sci. **1**, 17 (2008)
11. G. Geureuha, Natadisastra, Z. Hasan, Sriati, and W. Lili, J. Perikan. Kelaut. **9**, 41 (2018)
12. C. E. Larasati, M. Kawaroe, and T. Prartono, Indones. J. Mar. Sci. **20**, 223 (2015)
13. K. Ondara, R. Dhiauddin, and U. J. Wisha, J. Kelaut. Nas. **15**, (2020)
14. S. Nedi, J. Perikan. Dan Kelaut. **17**, 26 (2012)
15. S. Nedi, B. Pramudya.N, E. Riani, and Manuwoto, Model Pengendalian Pencemaran Minyak Di Perairan Selat Rupat Riau, Institut Pertanian Bogor, 2010
16. KLH, *Decree of the State Minister of the Environment Number 51 of 2004, Concerning Seawater Quality Standards.* (2004)
17. Masriadi, Patang, and Ernawati, J. Pendidik. Teknol. Pertan. **5**, 14 (2019)
18. D. Rumahlatu, Ilmu Kelaut. Indones. J. Mar. Sci. **16**, 78 (2012)

19. S. Widowati, A. Sastiono, and R. Jusuf, *Efek Toksik Logam* (Andi, Yogyakarta, 2009)
20. E. Rochyatun, M. T. Kaisupy, and A. Rozak, Makara Seri Sains **10**, 35 (2006)
21. A. Kumar, A. Kumar, M. Cabral-Pinto, A. K. Chaturvedi, A. A. Shabnam, G. Subrahmanyam, R. Mondal, D. K. Gupta, S. K. Malyan, S. S. Kumar, S. A. Khan, and K. K. Yadav, Int. J. Environ. Res. Public Health **17**, (2020)
22. T. Purnomo and M. Muchyiddin, Neptunus **14**, (2007)
23. Meilanty, H. Wahab, and Y. Hala, Mar. Chim. Acta **6**, 12 (2005)

Characterization of hydrological aspects in Kolong Enam Retention Basin, Kijang, Bintan Island, Indonesia

Winny Retna Melani*, *Andi* Zulfikar, *Tri* Apriadi, *Wahyu* Muzammil, and *Deni* Sabriyati

Departmentof Aquatic Resources Management, Faculty of Marine Science and Fisheries, Raja Ali Haji Maritime University, Politeknik Street, Senggarang, Tanjungpinang City 29111, Indonesia

Abstract. Bintan Island in the Riau Islands Province is classified as a small island (±1170 km^2) with limited freshwater resources. The availability of raw water on Bintan Island will certainly be determined by the characteristics of hydrobiological aspects of Bintan Island. Fulfillment of freshwater needs is carried out through the creation of raw water shelters in the form of reservoirs, pond, retention basin, dams, or abandoned bauxite mining pits. One of the raw water sources on Bintan Island is Kolong Enamretention basin. This study aims to examine the hydrological aspects of Kolong Enam retention basin in Bintan Island. The main data used in this study is digital elevation model (DEM) to extract hydrobiological aspects on Bintan Island. In our study found that Bintan Island has nine sub-watersheds. Kolong Enam retention basin is located in the catchment area of Gunung Lengkuas, East Bintan District. The catchment area of Kolong Enam retention basin has an area of 1067,31 ha, which consists of 3 river orders. With regard to vegetation coverage from 2000-2018, it is indicated that there has been a decrease in vegetation coverage in the Kolong Enam retention basin water catchment area.

1. Introduction

Bintan Island is determined by the government as a National Strategic Industrial Estate based on Presidential Regulation No. 87 of 2011, will have implications for increasing the demand for raw water needs. Characteristics of Bintan Island are categorized as small islands where that island is a post-bauxite mining area, [1, 7, 9, 13, 14, 15]is expected to have dynamic hydrological characteristics of existing reservoirs or water reservoirs. A dam or retention basin is a basin used to regulate and accommodate the supply of rainwater and to improve water quality in related water bodies (rivers, lakes). The rainwater storage can be used during the dry season [8]. Small islands also have very limited water catchment capacity and have specific hydrological conditions [5].

Although Bintan Island is categorized as a small island, development on Bintan Island, especially in Kijang City, begins with the rapid development of the mining sector such as granite and bauxite. In early 2000 this sector was no longer a priority due to the diminishing economic potential of granite and bauxite resources. The abandoned bauxite mining pits on Bintan Island has changed the land cover pattern a lot and also left the post-mining tailings pond which is known by the local community as'*kolong*'.

Kolong Enam retention basin is one of the sources of raw water in the City of Kijang, Bintan Island. At the beginning of its existence, it was a holding basin originating from the Lengkuas Mountain water catchment area which was used for the operational needs of the Aneka Tambang (Antam) company. In line with the rapid population growth and regional expansion, Kolong Enam retention basin has become one of the providers of raw water in the City of Kijang, Bintan Island. The existence of Kolong Enam retention basin in Bintan Island is expected to have dynamic hydrological characteristics of the existing reservoir or water reservoir. This study aims to examine the hydrological characteristics of Kolong Enam retention basin on Bintan Island based on DEM (digital elevation model) analysis.

2. Material and Methods

The study was conducted from June to August 2021 at the location of Kolong Enam retention basin in Kijang, East Bintan District, Bintan Regency. Kolong Enam retention basin has a watershed area of 2.29 km^2, an inundation area of 18.73 ha, an inundation volume of 370,000m^3 and a retention basin bottom elevation of +5.75. The main data used in this study is DEM (digital elevation model) to extract hydrobiological aspects on Bintan Island. The DEM data was obtained from the USGS website (https://www.usgs.gov/) which was accessed in July 2021. Changes in vegetation coverage were analyzed from the classification of Landsat 7 imagery in 2000 and Sentinel 2 image in 2018. The

*Corresponding author: winny@umrah.ac.id

 https://doi.org/10.1051/e3sconf/202233902010

image classification process is carried out automatically. Based on spectral patterns that have been determined at the time of the selection of AOI (area of interest). The watershed parameters studied consist of watershed area, main retention basin length, watershed width, watershed length, flow density, flow length,average watershed slope, average slope, average river channel, river order, watershed circumference, total river length, and watershed shape coefficient. Field surveys were conducted to collect secondary data from relevant agencies.

Slope, elevation and catcthment area delineation from dem data were conducted using Qgis version 3.20 [4]with the following steps:

1. Mosaicking dem tiles from downloaded dem to create a single raster dem
2. Reprojecting dem
3. Calculating the flow direction map
4. Fill sinks/remove spikes to make a hydrological correct dem by filling sinks and removing spikes from raw dem
5. Subsetting dem to area of interest
6. Calculating slope (percent) and elevation (meter), polygonizing the output to vector
7. Deriving streams from subsetted dem by calculating the flow direction for each cell
8. Defining outflow point
9. Deriving catchment area and polygonizing the output to vector to extract area extent

Change detection analysis of vegetation cover was carried out using caret package [3]in R programming language[10]. The flow of analysis is presented in Figure 1. Random forest classification method was used for each image (Landsat 7 and Sentinel 2).

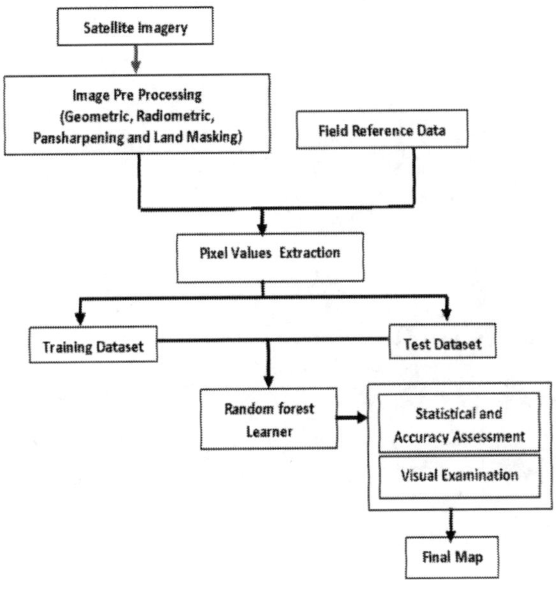

Fig 1. The flow of analysis

3. Result and Discussion

3.1 Geomorphology of Watershed

Bintan Island has a topography which is generally an area with sloping plains on the coast. This island has a varied and undulating topography with slopes ranging from 0-3% to above 40% in mountainous areas. The altitude of the islands ranges from 0-50 m above sea level to a height of 400 m above sea level [2].

Digital elevation model (DEM) data processing is carried out to determine the water catchment path in a watershed.The results of the study from the extraction of DEM data showed that Bintan Island has 9 sub-watersheds (Figure 2). Kolong Enam retention basin as one of the raw water sources in this study is located in the Kalang Tua sub-watershed which is in the southernmost position of Bintan Island. The entire sub-watershed is included in the Bintan Regency area.

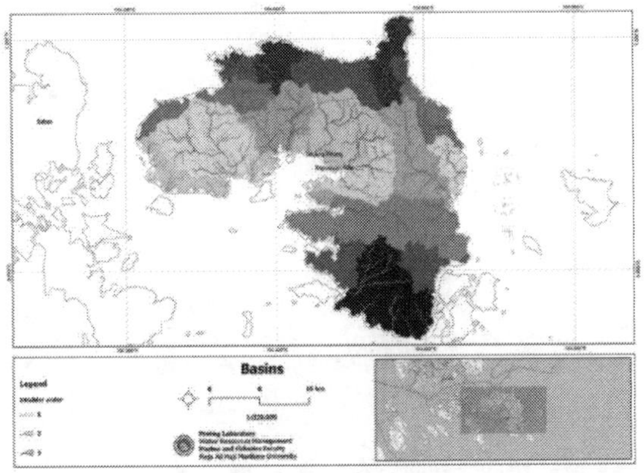

Fig 2. Map of Bintan Island watershed

Bintan Island is the largest island in the Riau Islands Province. In general, Bintan Island has an undulating hills geomorphology with a subtle elevation difference, namely with an altitude between 0 to 372 m above sea level[11]. The landscape of Bintan Island can be classified into several groups (Figure 3) as follows:

1. Lowland units are spread over most of the Bintan Regency area with an altitude of <25 m above sea level and the average slope is relatively flat and slightly sloping.
2. Units of wavy hills with low relief in areas has undulating hills with a height of <50 m above sea level and a rather steep slope.
3. The unit of wavy hills with moderate relief is in the form of a ridge with a height of> 50 m above sea level and a steep slope.

Based on the results of the morphological analysis presented in Figure 3, it is known that the Kolong Enam retention basin, holds water from a high hill area on Bintan Island, specifically in Gunung Lengkuas area. The geographical condition of the sub-watershed shows that the Kolong Enam retention basin catchment area consists of 3 river orders, the sub-watershed name is Kalang Tua, and the catchment area is Kolong Enam retention basin. The location of the retention basin that catches water from steep to flat areas will determine the availability of raw water to meet of water needs by society, especially in the dry season with a catchment area of 1067.31 ha. The results of the morphometric analysis presented in Figure 4 show that the steep upstream slope and the flat downstream slope have great potential to fill water reserves during the rainy season.

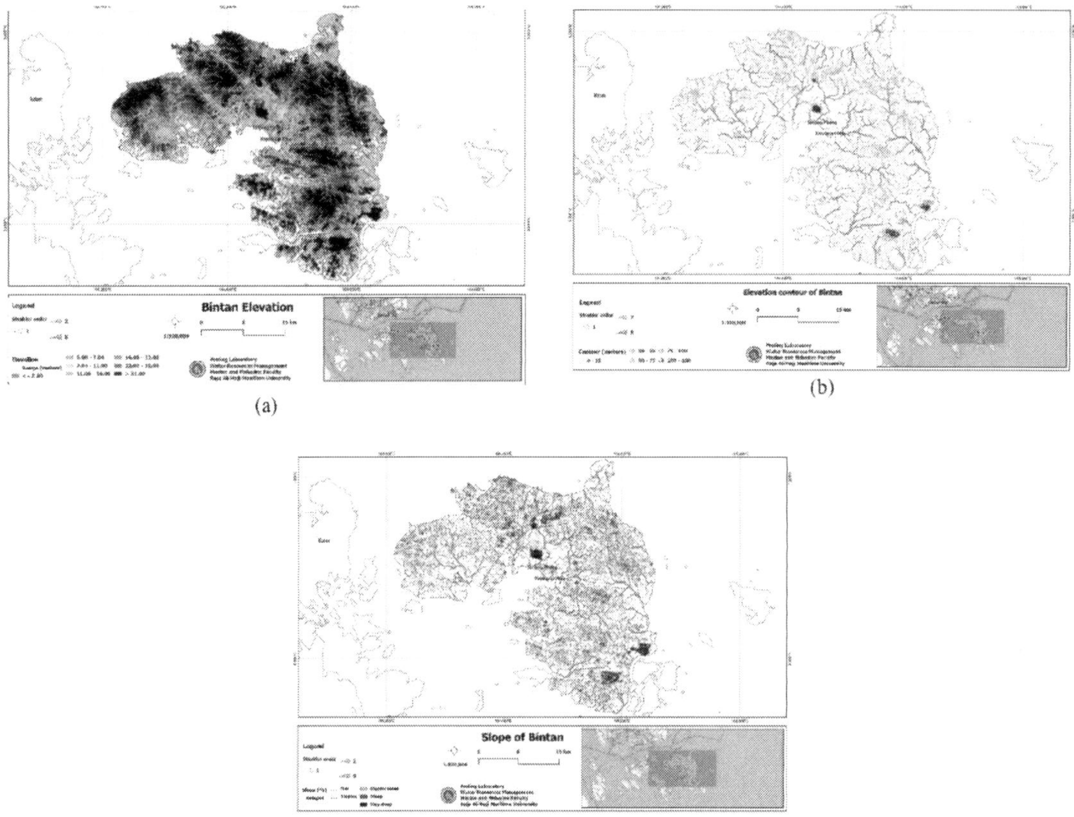

(a) (b) (c)

Fig 3. Bintan landscape:(a) elevation; (b) contour,and (c) slope

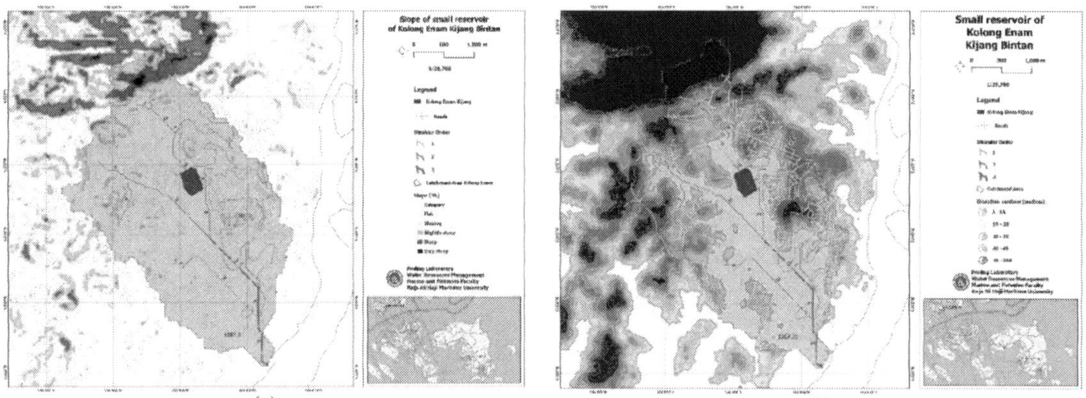

(a) (b)

Fig 4. Kolong Enam retention basin: (a) Location; (b) Elevation and river orders

3.2 Vegetation Coverage

Climate change dynamicsand human activities have various impacts based on environmental conditions in the Kolong Enam retention basin area. As time goes by, the condition of the forest area can be seen from the condition of the vegetation coverage that has undergone rapidly and dynamically changes in accordance with developments that occur. Changes in landcover, especially in vegetation coverage, have a major influence on the availability of water sources in an area [6].

Analysis of vegetation coverage result in the Kolong Enam retention basin area from year of 2000 to 2018 indicate a decrease in vegetation coverage in the water catchment area (Figure 5). In 2000 the vegetation coverage in the Kolong Enam retention basin area 710.18 hectares, until 2018 the vegetation coverage in the Kolong Enam retention basin was 383.24 hectares. There has been a decrease in area to 326.94 hectares for 18 years. Based on the condition of the vegetation coverage in 2000, the vegetation coverage of forest land was still dominant, indicated by vegetation coverage that classified as dense. Based on vegetation coverage condition in 2018 it could be concluded that vegetation coverage is still dense only in the upstream areas, but has begun to turn into open land in the catchment area around the retention basin. The vegetation areasdecreased year by year for 18 years occurred due to the conversion of forest land to other land uses so that it has implications for quality and area of the forest decreased. [12]stated that for more than 30 years, Kijang District was the center of concentration of bauxite mining on Bintan Island.

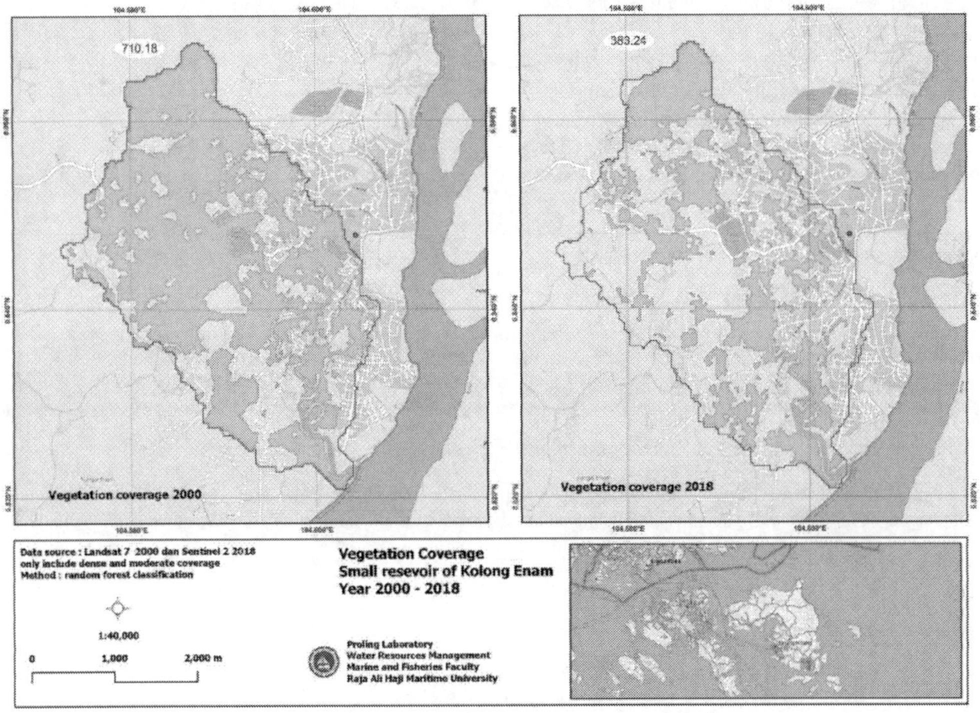

Fig 5. Vegetation coveragefrom 2000 (left) to 2018 (right)

4. Conclusion

Bintan has 9 sub-watersheds. Kolong Enam retention basin is located in the Kalang Tua sub-watershed. The catchment area of Kolong Enam retention basin has an area of 1067.31 ha which consists of 3 river orders. With regard to vegetation coverage, from 2000-2018, it is indicated that there has been a decrease in vegetation coverage in Kolong Enam retention basin water catchment area.

References

1. T. Apriadi, W. Muzammil, W.R. Melani, A. Safitri, Com.Str. of Mac.Org. in Seng. Cre. Bin. Isl, J.D, 9,119 (2020)
2. http://www.sippa.ciptakarya.pu.go.id., Profil Wilayah Kabupaten Bintan. Accessedby 27 Agust2021
3. M. Kuhn, Caret: classification and regression training. Rpackage version 6.0-88, https://CRAN.R-project.org/package=caret, (2021)
4. Kwast HVD, QGIS and Open Data for Hydrological Applications. IHE-DELFT Vitens Evides International (2018)
5. Latuamury, H. Marasabessy, M. Talaohu, W. Imlabla, Small island watershed morphometric and hydrological characteristics in Ambon Region, Maluku Province, in International Conference on Sustainable Utilization of Natural Resources, IOP Conf. Series: Earth and Environmental Science 800 012047, (2021)
6. K. Mane, R.K. Jella, H. Ballu,Lnd.Cgh.Cath.Imp. Cmd. Irrig, J. Sc. R& D, 8, 294 (2021)

7. W.R. Melani, T. Apriadi, W. Muzammil, A. Zulfikar, D. Sabriyati, Strg.Iss. Scrg. of Res. Mng. Prbl. as Raw.Wtr. Scr.in Btn. Isn, J. P. A, 5, 365(2021)
8. Ministry of PUPR., Hydrological analysis modul. Water Resources and Construction Education and Training Center, Jakarta(2017)
9. W. Muzammil, T. Apriadi, W.R. Melani, K.D. Handayani, Leng.Wei. Env. Par. of Macrobrachiummalayanum (J. Roux, 1935) in Sen.Wat.Flo.Tpi. Ria. Isl, J.A.S, 5, 18 (2020)
10. R. Core Team, . R: A language and environment for statisticalcomputing, R Foundation for Statistical Computing, Vienna,Austria, URL https://www.R-project.org/ (2021)
11. E.K. Rohmana, P. Djunaedi, P. Pohan, Inventaritation raw mining in Bintan Island ex-mining areas, Riau Islands Province, in Prosiding of Presentation of the Results of Field and Non-Field Activities of the Geological Resource Center, (2007)
12. R. Rositasari, Bintanamazing island, Oceana XLI, 4, 15(2016)
13. A. Safitri, W.R. Melani, W. Muzammil, Mac.Com.Rel.Seng.Riv.Flo.Wat.Qua.Tpi, J.A.S, 8, 103 (2021)
14. A.B. Supangat, Hyd.Cha.Bas. on Mor.Par. of Mer. Bet. Nat. Par, J. P. H.K. A, 9, 275 (2012)
15. O.A.P.Yolanda, W.R. Melani, W. Muzammil, Sed. of Sei. Car. Wat. Tpi, J. H. A, 1, 11 (2020)

E3S Web of Conferences **339**, 02011 (2022)
10th ICMR-2nd INSAEF 2021
https://doi.org/10.1051/e3sconf/202233902011

Understanding swarm earthquakes in Southeast Aceh, Sumatra

Andrean V. H. Simanjuntak[1,3], *Djati* C. Kuncoro[1], *Irwandi* Irwandi[2,3] and *Umar* Muksin[2,3]

[1]Badan Meteorologi Klimatologi dan Geofisika (BMKG), 22123 Aceh Besar, Aceh, Indonesia
[2]Department of Physics of Universitas Syiah Kuala (USK), 22122 Banda Aceh, Aceh, Indonesia
[3]Tsunami Disaster and Mitigation Hazard Center (TDMRC), Universitas Syiah Kuala (USK), 22122 Banda Aceh, Aceh, Indonesia

Abstract. The cause of the swarm earthquakes occurring from September until November 2020 remains unclear. The swarm activities are close to the Tripa active fault but also close to Mt. Bendahara. This study offers the analysis of seismic P and S wave arrivals and relocate the hypocenter of the swarm. We analysed the hypocentre distribution that recorded by Badan Meteorologi Klimatologi dan Geofisika (BMKG) from July 2020 to January 2021. The arrival of P and S-wave from 103 events was used obtain 1-D seismic velocity model. The earthquake used for this study consisted of at least 6 phases of seismic arrivals with the azimuthal gap angle less than 180 degree. The different velocities occurred in the upper and lower of the crustal thickness of 0 – 40 km. The Vp values are about 5.0 – 7.0 km/s and 3.0 – 5.0 km/s for Vs value while the Vp/Vs ratio are ranging from 1.5 to 1.7. Our preliminary results indicate that the potential swarm activities near to Mt. Bandahara maybe triggered by an unknown fault activity as no volcanic activity was reported.

1. Introduction

The September-October 2020 swarm activities in the northern part of Southeast Aceh (Figure 1) are categorized as a rare event since there was no previous being recorded before. The swarm activities may be undetected due to limited seismic sensor in the Aceh province. The first assumption of the swarm activities in Southeast Aceh was generated by the Tripa segment activities. However, intense swarm activities with a sequence of >100 occurring events might had been influenced by volcanic activities because it is closed to Mt. Bandahara.

Historically, the Tripa fault has generated several earthquakes in 1990 with M 6.8 and in 1996 with M 7.2 [1, 2, 3, 4]. Another major earthquake occurred in the northern part of the Tripa fault on July 2nd, 2013 with magnitude of M 6.2 which was strongly felt by the local people in the Southeast Aceh [3,4,5,6]. Local people reported that they felt the swarm in the I – III MMI which was related to the seismic hazard investigation.

Irwandi et al., [7] proposed an updated seismic hazard model for the Southeast Aceh region in order to understand the probability of seismic hazard with an exceeded limit percentage of 2% and 10% for 50 years without including the swarm activities.

The percentage and the year probability were chosen to calculate the PGA (Peak Ground Acceleration) map on the bedrock structure. The amplification results show that the PGA can reach 0.9 g and 1.5 g for the exceeded limit of 10% and 2% respectively [7].

The amplification is likely to be exceeded for 50 years with the ratio reaching 1.6 times in areas of densely populated people in Kutacane. It is also important to conduct a more suitable seismic hazard model in the future by including the earthquakes from the swarm activities [9, 10, 11, 12]. The detailed study of swarm needs to be conducted in order to understand the cause of the swarm.

In this research, we tried to analyse the earthquake arrival time of the P- and S-wave of the swarm. The arrival time data was used to determine an optimum 1-D seismic velocity model. The earthquakes were simultaneously relocated. The results of the relocation and the seismic velocity model may provide a preliminary explanation of the swarm activities.

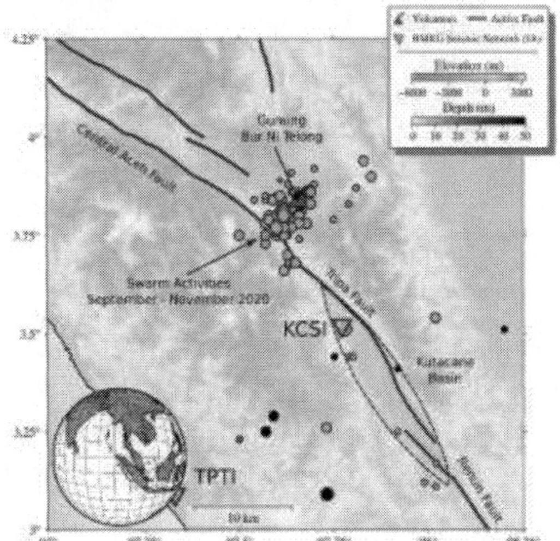

Fig. 1. The tectonic system and the seismic activities … between September to October 2020 in the Southeast Aceh region. The swarm activities were located close to the Tripa fault and beneath Mt. Bandahara. About 103 events were recorded by the seismic network in the northen part of Sumatra with KCSI station has more arrival times data.

https://doi.org/10.1051/e3sconf/202233902011

2. Data and Method

2.1. Data Observation

The Meteorological, Climatological, and Geophysical Agency has operated 20 recording seismometers which have been established since 2009 [8, 10]. BMKG, as the official agency of earthquake monitoring, operates various seismometers to record the earthquake waveform and routinely pick the P and S phase arrivals to determine and disseminate the earthquakes quickly [13, 14].

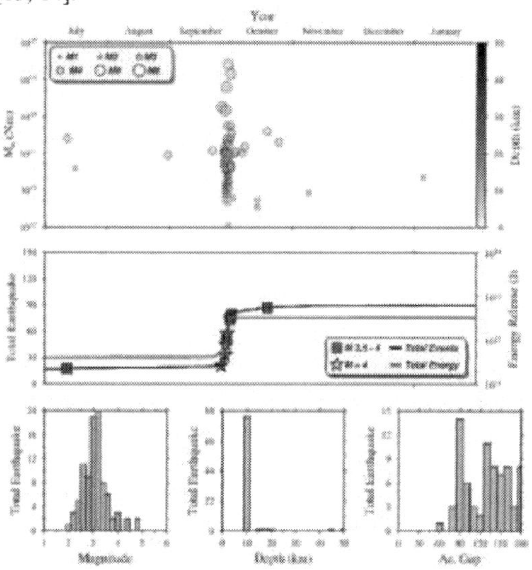

Fig. 2. Temporal graphic of swarm activities in the study area (upper panel). The estimation of seismic energy accumulation that was released from swarm activities around 10^{12} – 10^{13} J (middle panel). The distribution of magnitude, depth and azimuthal gap for all events were selected by optimum criteria to obtain 1-D seismic velocity in the study area (bottom panel).

For this research, we analysed the arrival times of body waves (P and S-waves) recorded by 20 seismic stations in the northern part of Sumatra from July 2020 to January 2021 with the total of 103 events. To initiate a unique solution, in this study we used the data of events with depth of less than 50 km according to the BMKG catalogue This means that all earthquakes location occurred in the crust. Furthermore, we firstly choose events recorded by at least 6 seismic arrivals (P or S waves) with the azimuthal gap angle < 180° and the distance between earthquakes and stations were less than 5° (< 500km).

For statistical purposes, time series analysis (Figure 2) was applied to figure out the temporal variation in the study area. The earthquake distribution underwent a significant change in September 2020 while no activities occurred in the previous month. The total seismic energy release was around 10^{11} – 10^{12} Joule and the accumulation of energy was up to 10^{13} Joule. The total of energy release was not bigger than the common crustal earthquakes release. The reason might be due to the absence of fault activities [15].

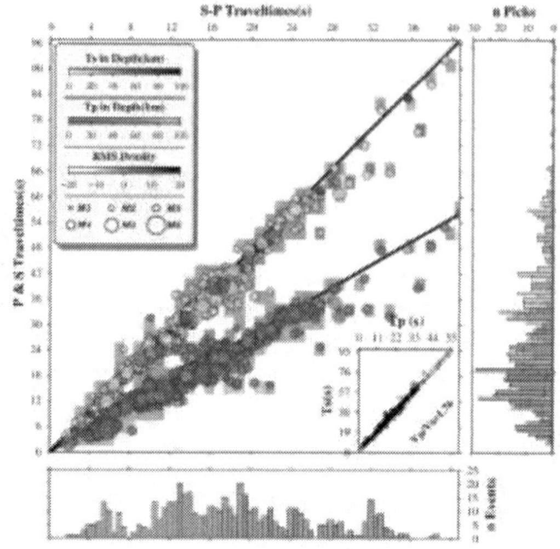

Fig. 3. The interactive of the Wadati diagram describing the distribution of arrival times of the P-phase (grey circle) and the S-phase (purple circle). The rms values were set for both phases in from -20 to 20 seconds which might had been adapted from the global velocity model IASP-91. The ratio of Vp/Vs was 1.76. The total events were 103 with the total of P-phase picks were 687 and S-phase were 321 with the distance of 0° - 5° and the azimuthal gap < 180°.

2.2. Coupled-Velocity Relocation

The hypocenter distribution of the swarm was nonlinearly influenced by velocity model and the distribution of seismic stations [2, 3]. The semi nonlinear approach was among several earthquake relocation that often used. The method of the coupled-velocity relocation was chosen to relocate the events and simultaneously determine the velocity model. [4, 8, 13, 14].

The semi nonlinear approach is useful to relocate a fixed initial regional or local earthquake which trapped on a specific depth. In this study, the initial hypocentres data from BMKG shows a fixed depth solution based on the global velocity model [14].

To improve the accuracy of the hypocentre locations, an appropriate 1-D seismic velocity model was needed. We used the VELEST program [6] to determine a new velocity model that compatible with the hypocentres distribution and the tectonic system in the study area.

A total of 64 different initial velocity models and locations of earthquakes were used in this study. Subsequently, the simultaneous inversion was applied for each initial velocity model. The inversion result, including the new hypocenter locations and the 1D velocity model of each input model was compared to obtain best solution. The result with the smallest RMS value (see Fig. 5) and consistently decrease in each iteration was chosen as the best 1-D velocity model. Furthermore, the new hypocentres were also accepted as the final earthquake locations.

The result was illustrated in a unique 1-D velocity model that is suitable with the new hypocentres location and correct the delayed time in all recording station. The selected data of an event used for the inversion analysis were limited to those with a minimum of six phases and the azimuthal gap angle less than 180°.

3. Result and Discussion

The 81 velocity models (Vs, Vs and Vp/Vs models) used in the inversion are shown in Figure 4.

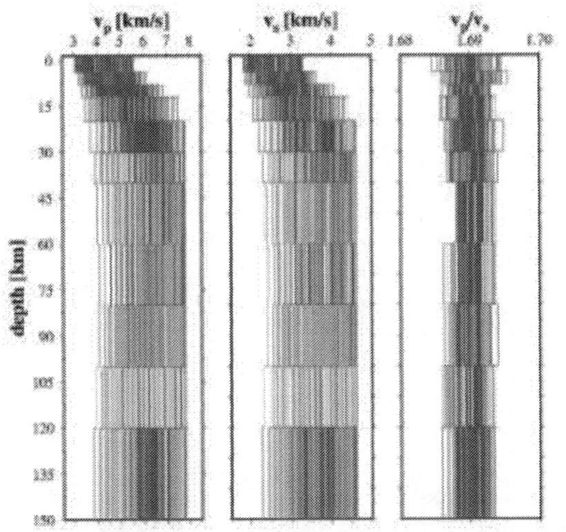

Fig. 4. The total of 81 initial 1D velocity models was appointed as the first model that creates simultaneous inversion (a) Vp model, (b) Vs model and (c) Vp/Vs ratio.

Each model was inverted by 30 iteration to generate the final solution. All initial provided more realistic with models of Vp, Vs and the ratio of Vp/Vs. Simultaneously, the hypocenter locations and the station corrections were also obtained.

Initial models produced a new model that relevant and suitable with the hypocenter distribution. The models were randomly chosen to be within the boundary corridor that indicated by dashed line in Figure 5. All imposed models were convergent to produce the best velocity model with lowest rms values.

The average rms value for all imposed models were smaller than 1.0. The best velocity models are represented by the red lines in Figure 5, which has the lowest rms values and the convergence solution.

The depth variation of the earthquakes in figure 5 is significantly changing from the fixed depth to the different depth. From this research, we found the optimum velocity model that relevant the study area around the Mt. Bandahara.

The velocity model changes in accordance to the increasing depth. It is based on the depth level of each layer. The model should be further modified into the 3-D model for a clear depiction of different seismic velocity model in each layer as well as to explain the lateral heterogeneity beneath an active system [3, 4, 13, 16, 17].

The 1-D model can also be used as the velocity reference to conduct an earthquake analysis in Southeast Aceh, such as the inversion mechanism [6] and double difference relocation. The shallow depth is categorized as the active cluster since most hypocentres shallow. The shallow depth earthquake is considered as higher risk to building damages [7]. Special attention regarding the soil properties in areas where swarm can be felt by the local people should be studied further in order to reduce the earthquake risk [11, 14, 17]. The 1-D velocity model can be a useful parameter to calculate the NDSHA [7], generate the focal mechanism [6, 15], tomography [3], conducting a realistic seismic map for important subsurface information [5, 18] and detecting precise swarm activities [20].

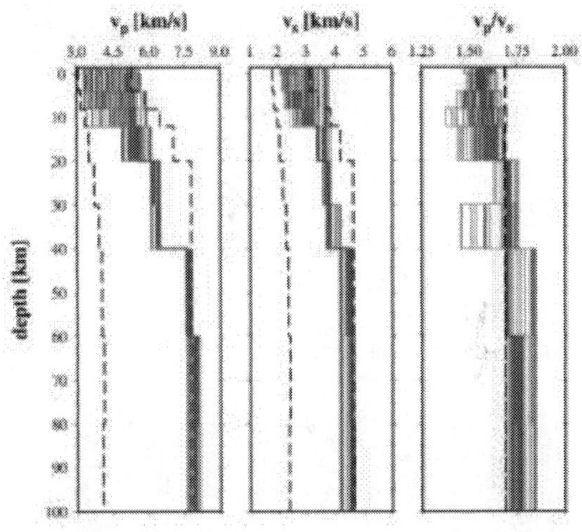

Fig. 5. The 1D velocity model after 30 iterations Vp model, Vs model, and Vp/Vs model. The red lines is the best solution with the smallest RMS values.

4. Conclusion

The swarm activities in 2020 in the northern part of Southeast Aceh may be caused by volcanic activities or shallow tectonic swarm. With the total of P-phase picks of 687 and S-phase 321, we successfully generated the optimum 1D velocity model by applying the simultaneous inversion with 81 initials models that converged to a unique final model. Different velocities are located in the upper to the lower level of the crust area with depth of 0 – 40 km. The Vp values range between 5.0 – 7.0 km/s and 3.0 – 5.0 km/s for Vs whereas the Vp/Vs ratio is 1.5 to 1.7. Our preliminary results figure the swarm activities may be triggered by an unknown active fault since there was no indication of volcanic activities in the area.

References

1. Pasari, S., Simanjuntak, A. V., Mehta, A., Sharma, Y. Pure and Applied Geophysics, 1-18. (2021)
2. Muksin, U., Bauer, K., & Haberland, C. Journal of volcanology and geothermal research, 260, 27-42. (2013).

3. Muksin, U., Bauer, K., Muzli, M., Ryberg, T., Nurdin, I., Masturiyono, M., Weber, M. Journal of Asian Earth Sciences, 171, 20-27. (2019).
4. Muksin U, Haberland C, Nukman M, Bauer K, &Weber M. Journal of Asian Earth Sciences. 2014. 96: 123-131. (2014)
5. Rusydy, I., Idris, Y., Muksin, U., Cummins, P., & Akram, M. N. Geoenvironmental Disasters, 7(1), 1-16. (2020).
6. Qadariah, Q., Simanjuntak, A. V., Umar, M. Journal of Aceh Physics Society, 7(3), 127-132. (2018).
7. Irwandi, I., Muksin, U., Simanjuntak, A. V. In IOP Conference Series: Earth and Environmental Science (Vol. 630, No. 1, p. 012001). IOP Publishing. (2021).
8. Simanjuntak, A. V., Muksin, U., Setiawan, Y. In IOP Conference Series: Earth and Environmental Science (Vol. 273, No. 1, p. 012021). IOP Publishing. (2019, June).
9. Pasari, S., Simanjuntak, A. V., Sharma, Y. Geoscience Letters, 8(1), 1-13. (2021)
10. Pasari, S., Simanjuntak, A. V., Mehta, A., Sharma, Y. Natural Hazards, 1-13. (2021).
11. Julius, A. M., & Widana, I. D. K. K. In IOP Conference Series: Earth and Environmental Science (Vol. 708, No. 1, p. 012106). IOP Publishing. (2021, April).
12. Julius, A. M., Pribadi, S., & Muzli, M. In IOP Conference Series: Earth and Environmental Science (Vol. 132, No. 1, p. 012012). IOP Publishing. (2018, March).
13. Muksin, U., Rusydy, I., Erbas, K., & Ismail, N. In Journal of Physics: conference series (Vol. 1011, No. 1, p. 012031). IOP Publishing. (2018, April).
14. Simanjuntak, A. V., Sihotang, N., Kuncoro, D. C., Irwandi, I. In Journal of Physics: Conference Series (Vol. 1811, No. 1, p. 012123). IOP Publishing. (2021, March).
15. Daniarsyad, G., Sianipar, D., Heryandoko, N., & Priyobudi, P. Pure and Applied Geophysics, 1-13. (2021).
16. Muzli, M., Umar, M., Nugraha, A. D., Bradley, K. E., Widiyantoro, S., Erbas, K, Jousset, P., Wei, S. (2018). Seismological Research Letters, 89(5), 1761-1772.
17. Kissling E., Ellsworth W., Eberhart-Phillips D., Kradolfer U. 1994. Journal of Geophysical Research: Solid Earth (1978).
18. Simanjuntak, A. V., Muksin, U., & Sipayung, R. M. In Journal of Physics: Conference Series (Vol. 1116, No. 3, p. 032033). IOP Publishing. (2018, December).
19. Idris, Y., Cummins, P., Rusydy, I., Muksin, U., Syamsidik, Habibie, M. Y., & Meilianda, E. Journal of Earthquake Engineering, 1-18. (2019).
20. Hendriyana, A., Bauer, K., Muksin, U., & Weber, M. Geophysical Journal International, 213(2), 952-962. (2018).

Body wave dispersion characteristics of regional deep earthquake at Southern Ryukyus subduction zone

Haekal A. Haridhi[1,2,3,*], *Bor-Shouh* Huang[4], *Dimas* Sianipar[5], *Syahrul* Purnawan[1,2], *Ichsan* Setiawan[1,2]

[1] Department of Marine Sciences, Faculty of Marine and Fisheries, Universitas Syiah Kuala, Banda Aceh 23111, Indonesia.
[2] Tsunami and Disaster Mitigation Research Center (TDMRC), Universitas Syiah Kuala, Banda Aceh 23111, Indonesia.
[3] Research Center for Marine Sciences and Fisheries, Universitas Syiah Kuala, Banda Aceh 23111, Indonesia.
[4] Institute of Earth Science, Academia Sinica, Taipei 115, Taiwan.
[5] Agency for Meteorology, Climatology, and Geophysics (BMKG), Jakarta 10610, Indonesia.

Abstract. The convergent plate boundaries known as the subduction zone is the location where the cold, hydrate, and old oceanic plate subduct under the hot, dry, and young continental plate. The top of the oceanic plate, i.e. the oceanic crust usually is less than 6 km thick carries the hydrate material to the hot continental mantle, which is commonly an indication of the occurrence of the deep earthquake. From the recent studies, the deep regional earthquake body wave signal that occurred at the subduction zone was found dispersed and or shown as a waveguide. Thus, by examining these signals, we could further estimate the body wave characteristics at the subduction zone. A narrow bandpass signal and the maximum of Hilbert transform envelope are being used to estimate the delay time of the dominant frequency energy at the Southern Ryukyus subduction zone. It was found that the Southern Ryukyus has a larger delay with the dominant frequency at 2 Hz, thus, the signal travel within the oceanic crust where it acts as a low-velocity layer compared to the surrounding continental and oceanic mantle, inversely it is known as a high-quality material (preserve the high-frequency energy).

1 Introduction

Taiwan is considered the most actives tectonic region. Its geographical location emerges between two distinct convergent systems of the Eurasian and Philippine Sea Plates, where at southern Taiwan the Eurasian Plate plunge under the Philippine Sea Plate while opposing to that situation occurred at its northern [1]. With support from the Ministry of Science and Technology of Taiwan, in recent years, The Institute of Earth Sciences of the Academia Sinica and the Tatun Volcano Observatory deployed a new high-density broadband seismic station at the northern Taiwan area named as Formosa Array (FM Array), with a total of 140 stations being constructed [2]. These seismic stations were aimed to understand the geometry of the magma chamber beneath the Tatun volcanic area, however, we could also utilize them to study the properties of the subducting slab.

One important observation applied to the intermediate-deep earthquake is the body wave dispersion. The body wave dispersion study has recently been used as a tool to estimate the properties of the subducting plate [3–5].

Understanding the subducting plate properties is essential, not only could explain the occurrence of the intermediate-deep seismicity, but also could be used to estimate its effect on ground motion. The study of the 1909 Taipei earthquake particularly indicates the importance to assess the shaking hazard resulting from an intermediate-deep earthquake [6] which is usually underestimated, such as the study of the Inishkin 2016 Mw 7.1 at 111 km depth, Alaska earthquake, indicates the subduction zone structure may contribute to the hazardous strong shaking at the forearc area (Anchorage and the Kenai Peninsula) of the Alaska subduction zone [7].

Therefore, studying the properties of the subducting slab, especially through the characteristics of the body wave signals from the deep regional earthquake are considered urgent and important. It may give a contribution to understanding the anomalous ground shaking distribution along the north eastern coast of Taiwan associated with deep earthquakes [8]. The finding in Taiwan resulting from this study could be applied elsewhere, along the subduction region around the world.

2 Seismotectonic Background

Taiwan is located between two convergent plate boundaries, where at the south of Taiwan, the subduction of the South China Sea (SCS) as part of Eurasian Plate subducted beneath the Philippine Sea Plate (PSP) along the manila trench [9]. Taiwan has been known as one of the very active tectonics activities, as a number of large earthquakes, occurred along this island due to its geologic configurations [1]. As it is

*Corresponding author: haekal.azief.haridhi@unsyiah.ac.id

shown in Figs. 1(a), almost all regions in Taiwan experience earthquake events.

One of many recent destructive earthquakes indicate that a continuous improvement needs to be done to understand the characteristics of the ground movement and the earthquake source [10]. Not only for the shallow destructive earthquake but also the deep regional earthquake, which ever destroy most of the buildings in Taipei, in the northern Taiwan region, and its deemed important to study the characteristics of the regional deep earthquakes, especially with the high density of population [6].

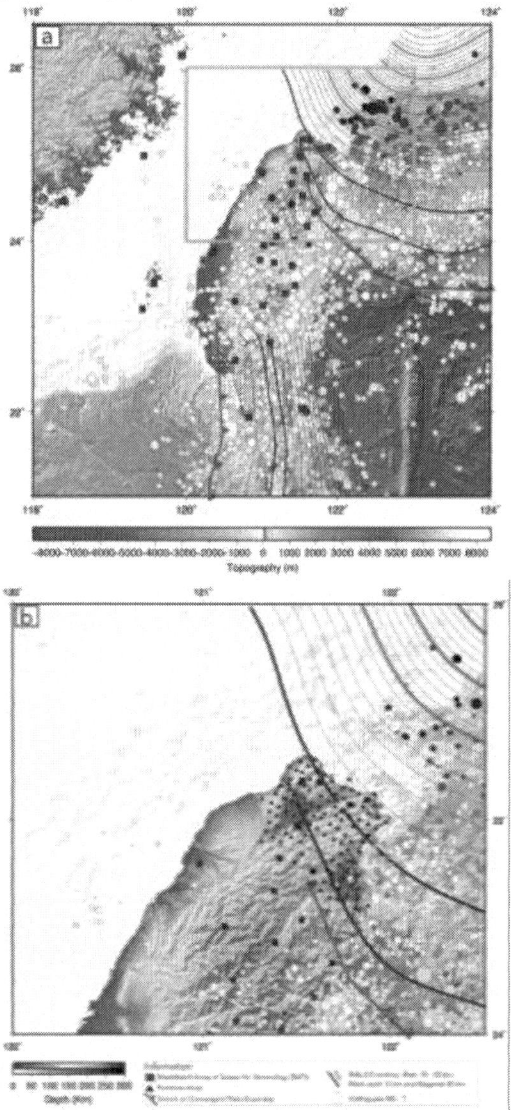

Fig. 1. (a) The topographic map of Taiwan showed the distribution of seismicity along the convergent plate boundary between Eurasian and Philippine Sea Plates in the south and the north. The distinct plate boundaries at the south and north called the Manila and Ryukyu trenches showed as a thick red line with tooth representing their locations, respectively. The distribution of broadband seismic stations, known as Broadband Array in Taiwan for Seismology (BATS) covered the whole island and its outskirt, is shown as a blue square and the study location is shown by the orange rectangle at northern Taiwan region shown in (b). (b) The northern Taiwan region with the location of Formosa Array (FM Array) seismic stations is shown as a black triangle. The grayscale colored circles indicate the earthquake event retrieved from the ISC earthquake catalogue (http://www.isc.ac.uk/), while the hot colored circle indicates the intermediate – deep regional events analyzed in this study. Slab 2.0 [11] contour is shown by different colors indicate different depths as it is shown in the information.

The Regional deep earthquake in northern Taiwan has been known to produce an anomalous ground movement distributed at the east – north coast of Taiwan [8]. The presence of the subducting slab is considered as the major factor in contributing to the enhanced ground movement, where such observation is true along the Alaska subduction zone [7]. However, northern Taiwan region due to its complicated tectonics, the north eastern region, which is located at the edge of the Ryukyu subduction, hence it is difficult to quantify the slab contribution to the ground movement. Thus, in recent year, the Institute of Earth Science, Academia Sinica, and the Tatun Volcano Observatory has been installed a high-density seismic station in the northern Taiwan region called as Formosa Array (FM Array), as is shown in Fig. 1 (b) [2]. Had the advantage of high seismic stations density, the characteristic of the slab estimated by the regional deep events could be assessed.

3 Material and Method

3.1 Regional deep earthquake catalogue

The regional deep earthquakes data that we analyse in this study are listed in Table 1.

Table 1. The Taiwan Central Weather Bureau intermediate-deep (60 – 152 km depth) earthquake catalogue during 2018 – 2019 located at the north – north eastern Taiwan with magnitude between 3 – 5. DT: date time, LON: longitude, LAT: latitude, M: magnitude, Z: depth in km.

No	DT	LON	LAT	M	Z
1	2018/01/17 13:59:11.11	121.72	25.06	5.5	140.2
2	2018/03/01 21:42:26.26	123.92	24.42	5.8	68.3
3	2018/03/06 17:42:10.10	121.86	24.7	4.9	81.7
4	2018/03/28 18:29:58.58	121.98	24.65	4.9	63.7
5	2018/04/20 13:25:06.06	122.37	24.37	4.7	70.6
6	2018/05/13 05:18:44.44	121.95	24.73	4.3	78.8
7	2018/06/02 19:25:15.15	121.91	24.5	4.2	60.4

8	2018/06/18 23:06:15.15	122.63	24.81	4.7	126.4
9	2018/06/16 09:08:58.58	121.83	24.68	3.8	63.2
10	2018/06/16 04:53:56.56	121.84	24.67	5.3	63.2
11	2018/06/26 03:07:31.31	122.06	24.75	4.1	68.2
12	2018/07/14 19:46:07.07	121.8	24.56	4.3	67.4
13	2018/07/20 09:39:15.15	121.51	20.78	5.6	99.8
14	2018/08/04 09:49:20.20	121.79	24.66	4.1	76.4
15	2018/08/06 05:30:22.22	122.65	24.6	4.6	97.4
16	2018/08/09 16:01:32.32	120.18	20.99	4.9	86.2
17	2018/08/11 10:24:20.20	121.55	22.31	4.2	116.2
18	2018/08/31 04:14:24.24	122.31	24.72	4.6	100.5
19	2018/09/04 00:37:10.10	121.99	24.83	4.3	89.8
20	2018/10/21 07:10:42.42	121.41	24.08	4.3	63
21	2018/11/03 03:23:54.54	121.73	24.58	3.7	64.4
22	2018/11/09 14:59:17.17	121.93	24.76	3.9	87.2
23	2018/11/16 10:01:35.35	121.81	24.79	4.4	82.3
24	2018/11/16 08:34:19.19	122.69	24.31	4.9	69.1
25	2018/11/27 19:25:49.49	121.83	24.7	4	83.6
26	2019/01/02 20:27:21.21	121.91	24.72	3.9	80.1
27	2019/03/06 09:33:29.29	121.79	24.76	4.1	86.7
28	2019/03/13 04:19:15.15	121.86	25.02	5.3	138
29	2019/03/19 12:01:34.34	121.93	24.75	4.9	83.2
30	2019/03/22 20:05:48.48	122.45	24.49	4.9	88.8
31	2019/03/24 15:56:50.50	122.43	24.7	4.7	109.3
32	2019/03/31 00:07:08.08	121.59	24.53	4	66.3
33	2019/04/01 10:54:35.35	122.24	24.93	5.1	128.6
34	2019/04/07 02:22:56.56	121.6	24.66	3.8	60.5
35	2019/05/08 23:48:47.47	122.49	24.82	5.1	122.7
36	2019/05/03 05:47:43.43	121.91	24.83	4.3	101.1
37	2019/05/16 12:16:34.34	121.56	24.67	3.8	60.7
38	2019/06/02 15:51:23.23	121.94	24.71	4.2	64.9
39	2019/06/30 00:20:06.06	122.18	25.07	5	152.1
40	2019/07/16 19:12:30.30	122.43	24.22	4.7	64.2
41	2019/07/26 14:39:57.57	121.84	24.75	4.7	62.7
42	2019/08/08 07:02:22.22	121.82	24.61	4.6	68
43	2019/09/19 22:50:21.21	121.8	24.78	3.9	80.5
44	2019/10/13 13:20:50.50	122.01	24.8	3.9	86.6
45	2019/11/09 07:47:05.05	121.83	24.81	3.7	83.3
46	2019/11/12 02:11:33.33	121.92	24.88	4.5	109
47	2019/12/07 03:32:33.33	122.66	24.54	4.7	100.5

3.2 Method

To study the body wave dispersion, the continuous record signals from FM Array were cut based on the catalogue shown in Table 2. The seismogram then processed to remove its instrument response by using

 https://doi.org/10.1051/e3sconf/202233902012

the transfer function in the Seismic Analysis Code (SAC) [12,13] to displacement.

The horizontal components of the East-West and North-South, were then rotated to Great Circle Path (GCP) as the Radial and Transverse components. Then manual picking of P- and S-wave throughout all events are performed. After the picking, 10 seconds signal with 5 seconds before and after P- and S-wave arrivals were cut. A tapper of 0.05 was applied at both end of the 10 second signals. A narrow bandpass of 0.25 Hz is applied to the seismogram until its corner frequency, then Hilbert transform was performed. The peak amplitude of the Hilbert transforms envelope function for each narrow bandpass signal was then used to calculate the lag time relative to the manual pick and 0.25 Hz.

The narrow band signal, is following this form exp{-a(f-fi)2/fin)}, where fi are the narrow band frequency, a and n is constant. Following this procedure, it gives a smooth dispersion curve. This approach are intensively being used to estimate the body wave dispersion curve at the subduction zone around the world [3–5,14] and it is similar to that being used in measuring the surface-wave group velocity [15].

4 Result

4.1 Waveform characteristics

The characteristics of the body wave that travel within the subducting slab have distinctive characteristics. The body wave arrival showed a small amplitude onset, and a large amplitude arrived soon after the onset with a couple of milliseconds intermittent between them, which is known as a waveguide (focused energy) or dispersed signals. Figs. 2 a-b and c-d indicate a waveform characteristic of a station that records the signal that does travel within the slab and with the one that did not.

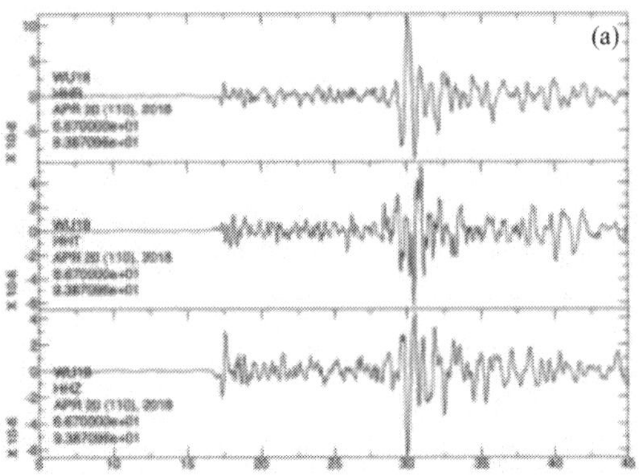

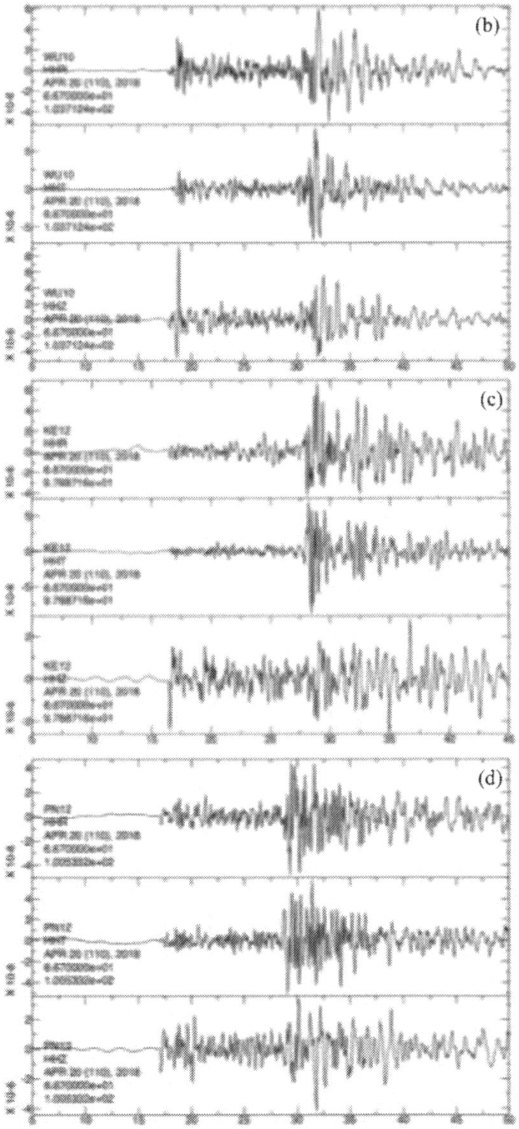

Fig. 2. Seismogram records of 4 stations with a similar distance between one and another, that showed a distinctive characteristic of P- and S-wave arrivals. (a) and (b): non-impulsive arrival of P-wave on HHZ component as well as for the S-Wave on HHT component, thus sampled the slab (waveguide). (c) and (d): impulsive arrival of P-wave on HHZ component as well as for the S-wave on HHT, thus non-sampled the slab (non-waveguide).

The waveguide or dispersed signal is observed only at the station that sampled the slab, which means the seismic wave trapped within a lower velocity medium to its surroundings which is able to trap the energy.

Differ to the non-guided signal which has the dominant frequency mostly depends on the soil type on the receiver or stations, i.e. soil or rock sites, the guided signal has its dominant frequency depending on the thickness of the low-velocity layer, appropriately within a certain range of frequency which depends on the thickness the layer, i.e. the subducted oceanic crust.

 https://doi.org/10.1051/e3sconf/202233902012

4.2 Body wave arrival time characteristics

The characteristics of the dispersed body wave at northern Taiwan, the southern Ryukyus subduction zone observed delayed at some particular station locations, especially the stations that located between 10 - 20 km from the subducted slab interface. For example, Figs. 3a showed an event located at a depth of 66.7 km with M 4.7 on 20 April 2018 (the same event shown in Figs. 2 a-d). The WU18 stations showed a distinctive delay in P-wave arrivals compared to the other recording stations (Figs. 3b).

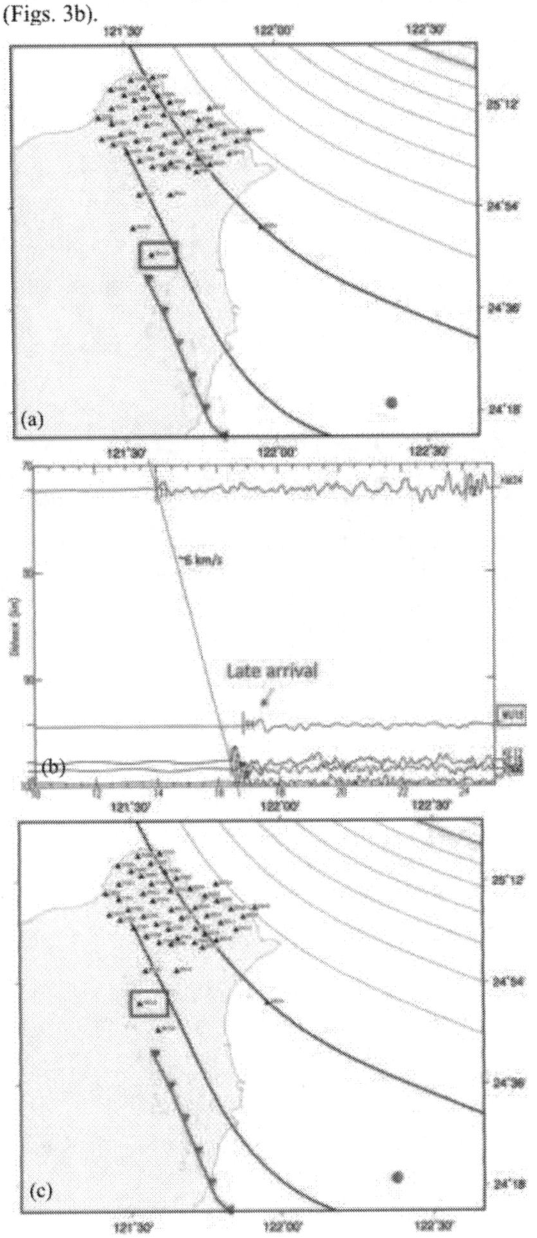

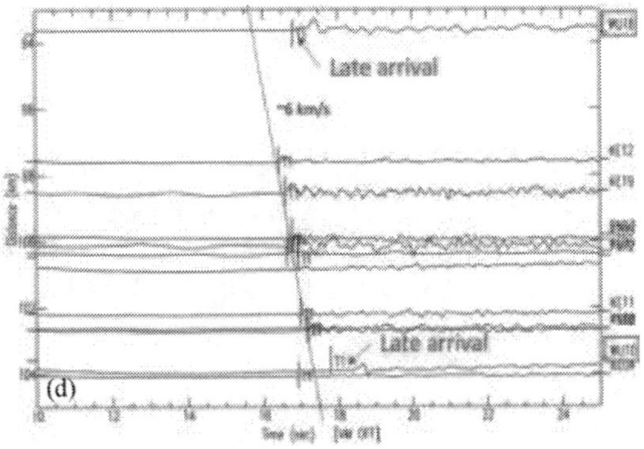

Fig. 3. (a) and (c) The northern Taiwan region and southern Ryukyus subduction zone. The land is shown as a cream color and the location of FM Array stations is shown by black triangle with text indicating its station name. The trench is indicated by the red toothy lines, and its interface contoured for each 10 km depth, with 10-20 km depth colored in blue, 100 km depth colored in magenta, while the rest colored in a thin solid black line. The event of DT 20 April 2018, M 4.7, Z 66.7 km indicated by a red dot. The red rectangle indicates the location of WU18 and WU10 seismic stations for (a) and (c), respectively. (b) and (d) The Signal Stacking Subprocesses (SSS) with an apparent velocity normal moveout (VNMO) of 6 km/s shown as the solid blue line. The WU18 and WU10 seismic stations indicated an obvious delay compared to the other recording stations.

As it is shown in Figs. 3 a-d, the arrival of P- and S-wave were delayed only at the stations that located 10 – 20 km above the slab interface. The delayed usually correspond positively with the depth of event, where a larger delay being observed at a deeper event.

4.3 Spectrogram characteristic

The spectrogram characteristics of the waveguide or dispersed signals indicate that the signal recorded at the station 10-20 km above the slab interface displays enriched high-frequency amplitudes. The arrival of each frequency energy showed a delay of high-frequency signal about 1.5 – 2 seconds, as is shown in Figs. 4 a and b. Focusing on high-frequency energy was one of the characteristics of a waveguide. On the other hand, the non-sampled slab stations showed no delay between the low-high frequency on the spectrogram, as it is shown in Figs. 4 c and d.

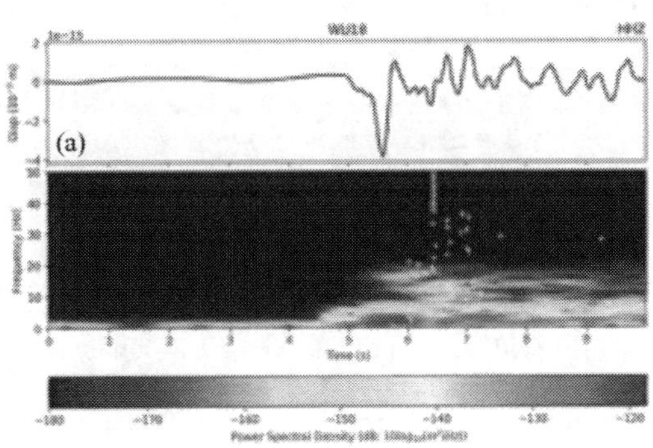

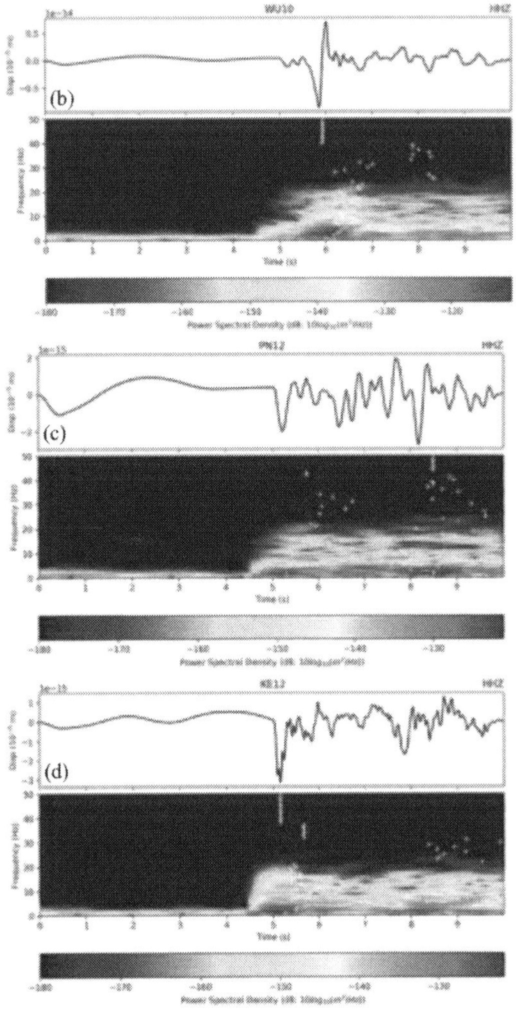

Fig. 4. The 10 seconds seismogram (P-wave are set at 5 second) and its corresponding spectrogram of the same event shown in Figs. 2 a-d. (a) and (b) A typical wave guide or dispersed body wave. (c) and (d) A typical non-wave guide or a typical body wave.

5 Conclusion

The characteristics of body wave observed in the northern Taiwan region showed as a guided wave at the particular station locations, especially those located at between 10 - 20 km above the subducting plate interface. The distinct characteristics of the seismogram, P- and S-wave arrival time as well as its spectrogram are distinguishable between the seismic wave that travels within the oceanic crust and the other that does not. The waveguide is recently being used as a tool to estimate the subducting slab properties, as well as a mitigation effort. The focused energy in a certain frequency has a significant effect on the building, hence a higher risk of disaster and or damage to the building should be anticipated. This observation is only possible with a dense seismic array such as the FM Array, beside the tectonic setting of the region, such that, the Indonesian subduction zone, thought, limited stations are available, this phenomenon is also observed at the forearc islands [16].

Acknowledgment: We would like to send our greatest gratitude to the Institute of Earth Sciences (IES), Academia Sinica and the Tatun Volcano Observatory (TVO) at Taipei, Taiwan for the availability FM Array dataset. The FM Array dataset is available upon request to IES and TVO. This study was supported by the Ministry of Education and Culture of the Republic of Indonesia under the Institute of Research and Community Service (LPPM) of Universitas Syiah Kuala within the H-Index Scheme (grant number 169/UN11/SPK/PNBP/2021). The Generic Mapping Tools software package was used to draw map figures [17].

References

1. Y.-B. Tsai, Tectonophysics **125**, 17 (1986)
2. Academia Sinica, Institute of Earth Science, and Tatun Volcano Observatory, Int. Fed. Digit. Seismogr. Networks. Dataset/Seismic Netw. (2017)
3. G. A. Abers and G. Sarker, Geophys. Res. Lett. **23**, 1171 (1996)
4. G. A. Abers, Phys. Earth Planet. Inter. **149**, 7 (2005)
5. G. A. Abers, Earth Planet. Sci. Lett. **176**, 323 (2000)
6. H. Kanamori, W. H. K. Lee, and K. F. Ma, Geophys. J. Int. **191**, 126 (2012)
7. M. E. Mann and G. A. Abers, Seismol. Res. Lett. **91**, 85 (2019)
8. K. H. Chen, B. L. N. Kennett, and T. Furumura, J. Geophys. Res. Solid Earth **118**, 665 (2013)
9. C. H. Lin, B. S. Huang, and R. J. Rau, Earth Planet. Sci. Lett. **174**, 231 (1999)
10. H. Kanamori, L. Ye, B. S. Huang, H. H. Huang, S. J. Lee, W. T. Liang, Y. Y. Lin, K. F. Ma, Y. M. Wu, and T. Y. Yeh, Terr. Atmos. Ocean. Sci. **28**, 637 (2017)
11. G. P. Hayes, G. L. Moore, D. E. Portner, M. Hearne, H. Flamme, M. Furtney, and G. M. Smoczyk, Science (80-.). **362**, 58 (2018)
12. P. Goldstein and A. Snoke, Inc. Institutions Seismol. Data Manag. Cent. Electron. Newsl. (2005)
13. P. Goldstein, D. Dodge, M. Firpo, and L. Minner, Int. Geophys. **81**, 1613 (2003)
14. T. Garth and A. Rietbrock, Earth Planet. Sci. Lett. **474**, 237 (2017)
15. A. Dziewonski, S. Bloch, and M. Landisman, Bull. Seismol. Soc. Am. **59**, 427 (1969)
16. H. A. Haridhi and D. Sianipar, IOP Conf. Ser. Earth Environ. Sci. **674**, (2021)
17. P. Wessel and W. H. F. Smith, Eos, Trans. Am. Geophys. Union **72**, 441 (1991)

Community structure of invasive aquatic invertebrates in coral reef ecosystems in the KKPD PISISI Simeulue

Chitra Octavina[1,2,3*], *Maria* Ulfah[1,2,3], *Sri* Agustina[1,2,3], *Yayang* Rifka Putri[1,4], *Indra* Indra[5]

[1]Departement of Marine Sciences, Faculty of Marine and Fisheries, Universitas Syiah Kuala, Indonesia
[2]Research Center for Marine and Fisheries, Universitas Syiah Kuala, Indonesia
[3]Laboratory of Marine Biology, Faculty of Marine and Fisheries, Universitas Syiah Kuala, Indonesia
[4]Fauna & Flora Internasional - Aceh Programme, Aceh, Indonesia
[5]Ocean Diving Club, Faculty of Marine and Fisheries, Universitas Syiah Kuala, Indonesia

Abstract. This study aimed to determine the structure of invasive aquatic invertebrates through the approach, diversity, similarity, dominance, distribution, and comparison of invasive aquatic invertebrates with coral cover in coral reef ecosystems in PISISI KKPD. This research was conducted in September 2019 using the purposive sampling method to research points, collecting invasive aquatic data used the belt transect method and collecting coral reef data used the Point Intercept Transect (PIT) method. The results of the study found that the structure of invasive aquatic invertebrates in KKPD PISISI was in the good category indicated by the density invasive aquatic invertebrates of 39.67 individuals/m^2, the biodiversity index was 1.50 with the medium category, the similarity index is 1.06 in the high category and the dominance index is 0.44 in the low category. The distribution of invasive aquatic invertebrates in the PISISI KKPD includes a cluster distribution. The highest value of invasive invertebrate water was 12.83 ind/m^2 and the proportion of coral cover was 29.83% and the lowest was 3.67 ind/m^2 and the proportion of coral cover was 54.67% showing the effect of invasive aquatic invertebrates on the proportion of coral cover.

1 Introduction

The Pinang Island, Siumat Island, and Simanaha Island Regional Water Conservation Area (KKPD PISISI) is a marine protected area located in Simeulue Regency which has been established through Aceh Governor Decree Number: 523/1297/2018, which Aan area of 44,404.10 Ha. The PISISI KKPD has enormous marine and fishery potential and is the main source of livelihood for coastal communities living in the PISISI KKPD area. One of the fishing grounds for fish and other marine products is located in the coral reef ecosystem in the PISISIS KKPD.

Coral reefs are a habitat for various marine biota to grow and reproduce [1]. There are biota associated with coral reefs that can cause damage to coral reefs, especially invasive aquatic invertebrates that act as coral predators, thus having an impact on damage to coral community structures or on the abundance of coral species [2]. Invasive species are species that are able to survive in their new environment, reproduce, and spread very widely in their distribution [3]. In addition, according to [4] Invasive species can cause drastic changes to an ecosystem by eradicating native organisms as they compete.

Predation is one of the main ways for invasive species to affect aquatic communities [5]. Coral predators according to [6] cause damage and or loss of coral tissue so that it has a direct impact on the growth of coral colonies and coral survival. Coral predators are a natural part of a healthy coral reef ecosystem, but overcrowding of some predators that are *corallivores* can result in drastic and widespread reductions in coral cover.

Invasive species studies have been carried out by [7] on invasive marine species that threaten marine biodiversity and industry. A similar study has also been conducted by [3] which discusses the interactions of aquatic invasive species with the natural and anthropogenic environment. Other studies have also been carried out by [8] regarding the distribution of potential invasive tunicates of *Didemnum vexillum* species. Research on invasive species in Indonesia has been carried out by [9] regarding the distribution of benthic invasion on coral reef ecosystems in the Thousand Islands, Jakarta and by [10] regarding the relationship between coral reef ecosystem conditions and invertebrates that can kill corals in the Thousand Islands, Jakarta.

Aquatic invertebrates act as a balancer in coastal ecosystems including coral reef ecosystems, but if their abundance is high, it can have a negative impact on the sustainability of the coral reef ecosystem. Assessment of invasive aquatic invertebrates on coral reefs still needs to be done, especially in areas that have the potential as conservation areas. Information from this study will be useful in determining the location of conservation area zoning which has a lower level of vulnerability to

* Corresponding author: chitraoctavina@unsyiah.ac.id

attacks and threats from invasive aquatic invertebrates. Therefore, researchers feel the need to study invasive aquatic invertebrates, especially in the PISISI KKPD.

2 Materials and Methods

2.1 Site and time

Data collection was carried out in September 2019 at the KKPD PISISI Simeulue. The research was conducted at seven stations, namely Air Manis Islet, Batee Dua Islet, Putih Islet, Ujung Babang Islet, Gabui Island, Babi Island, and Limau Island (Figure 1).

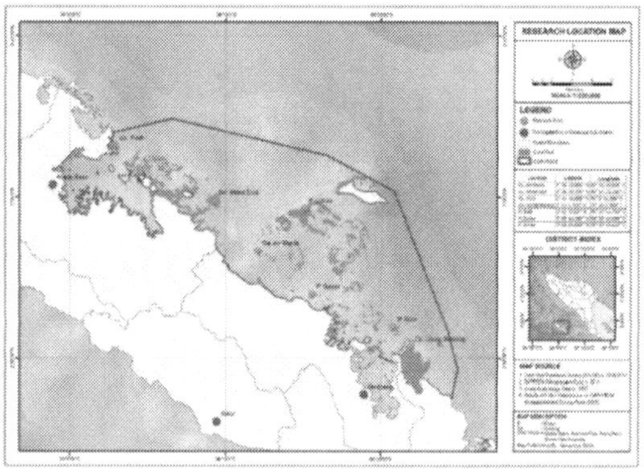

Fig. 1. Map of the research location.

3 Data Analysis

3.1 Station determination

Determination of the research location using purposive sampling method. The purposive sampling method determines the location chosen based on the sample requirements to suit the research objectives [11]. This research's location was determined based on the representation of the coral reef ecosystem and the affordability of the area for research observations.

3.2 Invasive aquatic invertebrate data collection

Invasive aquatic invertebrates were sampled using the belt transect method (Figure 2) by stretching roll meter along 50 m following the water bed contours with an observation area of 2.5 m on the left and 2.5 m on the right. Data were collected at two depths, namely 5 m and 10 m. Observations were made three times at each station. The recording of invasive aquatic invertebrates was carried out by species and number. Identification of invasive aquatic invertebrate species using the book Tropical Pacific Invertebrates [12].

Fig. 2. Belt transect method.

3.3 Coral reef data collection

Coral reef data was collected using the PIT (Point Intercept Transect) method using the same roll meter as the invasive aquatic invertebrates. Data was collected for three replications at each station at a depth of 5 m and 10 m. Coral reef data was recorded based on the form of growth (lifeform) and coral genus.

3.4 Water quality data collection

Aquatic environmental parameter data collection includes chemical and physical parameters at each observation station that has been previously determined. Measurement of the aquatic environment parameters was carried out directly at the research location (in situ) and was carried out three times at each station. According to [13] environmental parameters such as salinity, temperature, pH, and flow velocity can affect the high and low abundance of organisms.

4 Data analysis

4.1 Abundance of invasive aquatic invertebrates

The Abundance, according to [14], can be calculated using the following formula, namely:

$$D = \frac{Ni}{A} \quad (1)$$

Where, D is an abundance of species (individuals /m^2); Ni is the total number of individual types (individuals); A is the area sampled (m^2).

4.2 Diversity of invasive aquatic invertebrates

The diversity of invasive aquatic invertebrates can be calculated using the Shannon-Winner formula [14] is:

$$H' = \sum_{i=1}^{s} Pi \, Log2 \, Pi \quad (2)$$

Where, H' is diversity Index (ind/m^2); Pi is ni/N, number of individuals of species i to total individuals (ind); N is total number of individuals of all species (ind), S is number of taxa / species (ind). [14] divides the criteria for diversity with low diversity if the value is less than 1, moderate diversity if the value is between 1 – 3, and high diversity if the value is more than 3.

 https://doi.org/10.1051/e3sconf/202233903001

4.3 Evenness of invasive aquatic invertebrates

The similarityof invasive aquatic invertebrates can be determined using [14]:

$$E = \frac{H'}{H\ maks} \quad (3)$$

Where, E is Evenness Index (ind/m^2); H' is diversity index; H maks is maximum species diversity; S is number of species found.
The uniformity of invasive aquatic invertebrates can be categorized with an index value of E<0.4 is categorized as low E 0.4-0.6 is categorized as moderate, and E>0.6 is categorized as high [14].

4.4 Dominance of invasive aquatic invertebrates

The dominance of invasive aquatic invertebrates can be calculated using the Simpson Dominance Index [14]:

$$C = \sum \left(\frac{ni}{N}\right)^2 \quad (4)$$

Where, C is dominance index (ind/m^2); ni is number of individuals of the I species (ind); N is total number of individuals (ind).
According to [14], the dominance criterion is a low category with values ranging from 0.00 - 0.50, medium category values ranging from 0.50 - 0.75, and high category values ranging from 0.75 - 1.00.

4.5 Distribution of invasive aquatic invertebrates

The distribution of invasive aquatic invertebrates can be calculated using the Morisita Index [15] with equations:

$$Id = n \times \frac{\sum X^2 - N}{N\,(N-1)} \quad (5)$$

Where, I_d is species distribution index (ind/m^2); n is Number of plots; N is total number of individuals acquired (ind); Σx^2 is multiplication of the number of species in each plot (x) and the frequency found in the number of plots f(x). The criteria of Morisita index is Id = 1, distribution pattern is random; Id < 1, is uniform, Id > 1 is cluster.

4.6 Percentage of coral reef cover

The calculation of the percentage of coral reef cover is carried out by comparing each category's total length with the total length of the transect using the following formula [16]:

$$Percent\ Cover = \frac{Number\ found}{Number\ of\ dots} \times 100\% \quad (6)$$

The standard criteria for coral reef damage according to based on [17] the percentage of the area of live coral cover can be categorized as follows: 0 – 24,9% is bad; 25 – 49,9% is moderate; 50 – 74,9% is good; 75 – 100% is very good.

5 Results and Discussion

5.1 Community structure of invasive aquatic invertebrates

The observations the study found five invasive aquatic invertebrates from 5 different families in the coral reef ecosystem in the KKPD PISISI; namely *Acanthaster planci, Culcita novaeguinaea, Diadema antillarum, Drupella* sp., and *Spirobranchus giganteus*. The total abundance value of invasive aquatic invertebrates in the KKPD PISISI was 39.67 ind/m^2 (Figure 3).

The highest abundance of invasive aquatic invertebrates was found at station Batee Dua Islet, with a total number of 12.83 ind/m^2 and the lowest abundance of invasive aquatic invertebrates, was found at the Limau Island station as much as 3.17 ind/m^2. This study's results are different from the results of research conducted by [18] in Morowali where the highest abundance value of aquatic invertebrates was 2.37 ind/m^2, and the lowest was 2.13 ind/m^2. Other research conducted by [19] in Ambon Bay waters also had different results where the highest abundance value of aquatic invertebrates was 5.59 ind/m^2, and the lowest was 0.04 ind/m^2. Causes of the high abundance of invasive aquatic invertebrates in Batee Dua Islet are thought to be caused by physical and chemical factors in the waters that support invasive aquatic invertebrates growth. According to [19], the high abundance of aquatic invertebrates in the research location is due to the suitability of their habitat for aquatic invertebrates. The low abundance of invasive aquatic invertebrates in Limau Island is thought to be related to the percentage of coral cover in that location. According to [18], the abundance of aquatic invertebrates at the study site was influenced by the coral cover percentage.

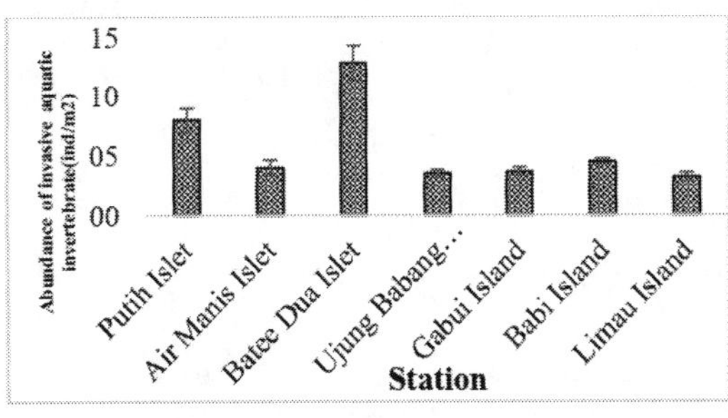

Fig. 3. The abundance of invasive aquatic invertebrates.

The difference in the abundance of invasive aquatic invertebrates at each station in the KKPD PISISI is thought caused by differences in the substrate from each station and other chemical and physical factors. It is supported by the opinion of [20], which statedthat differences in the abundance of aquatic invertebrates can be caused by the suitability of the habitats occupied by the organisms. High species abundance was found at Batee Dua Islet is *Diadema antillarum* as much as 6.17

ind/m^2 and *Acanthaster planci* as much as 5.83 ind/m^2. The high abundance of *Diadema antillarum* is thought to be caused by the substrate at the Batee Dua Islet location in the form of coral fragments and sand. According to [21], the substrate in the form of coral fragments, sand, and muddy sand is suitable for sea urchin life because coral fragments can grow algae. Sea urchins use seagrass and algae as food sources [22].

The high abundance of *Acanthaster planci* is caused by food availability in the Batee Dua Islet location, namely coral reefs, which are the types of coral that *Acanthaster planci* likes. [23] stated that certain coral species contributed to the high abundance of *Acanthaster planci*. According to [24], *Acanthaster planci* does not prey on all types of corals but generally chooses dominant coral colonies such as *Acropora, Montipora, Seriatopora*, and *Pocillopora*. Another factor that causes the high species of *Diadema antillarum* and *Acanthaster planci* is assumed due to the appropriate quality of the waters for these biota's growth. The brightness at Batee Dua Islet is relatively high, namely 5 - 10 m, which causes sunlight to still penetrate to the bottom of the water. Therefore, photosynthesis still occurs in these waters. The deeper the waters, the less light can penetrate the bottom of the water. [25] stated that if the deeper the waters, the waters' productivity decreases, and the organic matter contained in the waters is small so that the density of organisms is low. Apart from that, the salinity and temperature at the Batee Dua Islet supports the growth of invasive aquatic invertebrates with a salinity of 32.7‰ and a temperature of 29.7 C. This is consistent with the statement of [26] that good salinity for the growth of marine biota is 25-40 ‰ and the normal temperature of water for marine life ranges from 20-30^{0}C.

The low abundance of invasive aquatic invertebrates at Limau Island station is thought not to be influenced by water parameters. The brightness value of Limau Island waters reaches a depth of 10 m, a salinity of 31.9 ‰ and a temperature of 29.7°C, which supports the growth of invasive aquatic invertebrates. The cause of the abundance of species at this station is related to the live coral cover at that location. A low abundance was found in the species *Achantaster planci, Culcita novaeguineae,* and *Drupella* sp. with a value of 0.17 ind/m^2 where these three species make live corals as food (Figure 4).

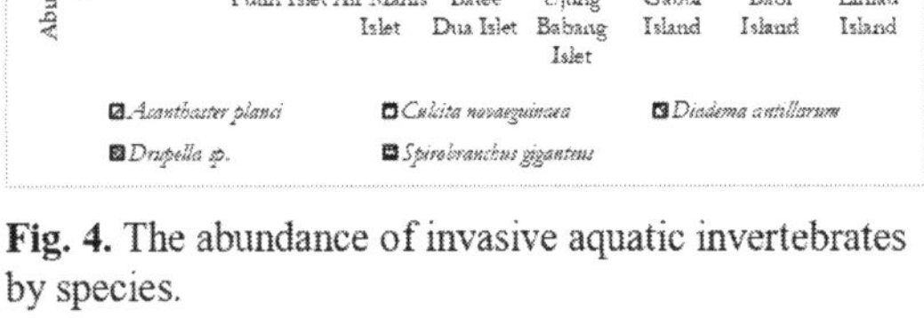

Fig. 4. The abundance of invasive aquatic invertebrates by species.

Live coral cover at Limau Island station (Figure 5) has a 41.67% value, which is categorized as moderate. Although the percentage of live coral cover is moderate, coral predators are thought to have the preferred coral species. [27] *Drupella* sp. does not prey on all types of coral but also selects the form of coral colony growth, the ease of taking live coral tissue, the production of mucus from the coral, the nutritional value of coral tissue as well as the ability of stinging cell defense from the coral. Based on the results of research conducted by [28] regarding the food preferences of *Drupella* sp. indicates that *Drupella* sp has a high degree of preference for certain coral species.

Achantaster planci also has a food preference for certain types of corals. Based on [29] research, this starfish has shown a preference for food in experiments conducted in the laboratory where *Acropora* corals are preferred over *Fungia* corals. Other research conducted by [30] regarding the predation rate of *Achantaster planci* showed that the coral's growth form also determines the food choices of this starfish species, where the tabulate growth form has a higher predation rate than the *submassive* growth form.

Sea star *Culcita novaeguineae* also shows a preference for the type of coral eaten. Based on research conducted by [31], observations made in the field showed that Culcita novaeguineae preferred Acropora and Monopora corals to Porites corals. Another thing that is thought to be low in Culcita novaeguineae abundance is water depth. According to [32] several individuals of Culcita novaeguineae were more frequently found at depths of 1 m - 2 m. Besides, [32] also stated that Culcita novaeguineae is the dominant type found in sandy seagrass microhabitates.

5.2 Biological index of invasive aquatic invertebrates

The biotic index of invasive aquatic invertebrates in the KKPD PISISI can be shown through the diversity index (H '), Evenness (E), and dominance (C). These three indices indicate whether a community has diversity or species domination. The diversity value (H ') of invasive aquatic invertebrates in the KKPD PISISI can generally be classified into the moderate category with a value of 1.50 (Table 1). The H 'value of invasive aquatic invertebrates was highest at station Babi Island with a value of 1.89, and the lowest was at Air Manis Islet station with a value of 1.06. Both of these values are still medium categories. Where according to [14], if the value of H 'ranges from 1 - 3, then the diversity is classified as moderate.

The diversity value (H ') which belongs to the medium category at the research location is thought to be caused by the small number of invasive aquatic invertebrates that are found in the research location and have different abundances. According to [15], a community is said to have high species diversity if it is composed of many species with the same or nearly the same species abundance and a community has low species diversity if the community is composed of very few species and only a few dominant species. The higher the H 'value, the more variety of invasive aquatic invertebrates in these waters.

Table 1. The biological index of invasive aquatic invertebrates.

Station	H'	Cat.	E	Cat.	C	Cat.
Putih Islet	1.32	Moderate	0.95	High	0.49	Low
Air Manis Islet	1.06	Moderate	0.97	High	0.59	Moderate
Batee Dua Islet	1.33	Moderate	0.96	High	0.44	Low
Ujung Babang Islet	1.71	Moderate	1.23	High	0.36	Low
Gabui Island	1.71	Moderate	1.06	High	0.38	Low
Babi Island	1.89	Moderate	1.36	High	0.29	Low
Limau Island	1.47	Moderate	0.91	High	0.50	Low
Average	1.50	Moderate	1.06	High	0.44	Low

Note. Cat. Is Category

The diversity value (H'), which belongs to the medium category at the research location, is thought to be caused by the small number of invasive aquatic invertebrates found in the research location and have different abundances. According to [15], a community is said to have high species diversity if it comprises many species with the same or nearly the same species abundance. A community has low species diversity if the community comprises very few species and only a few dominant species. The higher the H ' value, the more variety of invasive aquatic invertebrates in these waters.

The Evennes value (E) of invasive aquatic invertebrates in the KKPD PISISI can be categorized as high, namely 1.06. The highest E value was found at Babi Island station, namely 1.36 ind/m^2, and the lowest was found at Limau Island station with a value of 0.91. Both values are classified as high uniformity, where according to [14], if the value of E> 0.6, then the evenness is high and is classified as a stable community.

Evenness shows the stability of a community through high and low values of the evenness index. According to [33], if the uniformity index value approaches 1, the community can be said to be stable. The high uniformity index value shows the even distribution of invasive aquatic invertebrates. As stated by [34], if the evenness is close to 1, then the uniformity between species is classified as evenly distributed.

The dominance value (C) of invasive aquatic invertebrates in the KKPD PISISI is generally in the low category, with an average value of 0.44. According to [14], if the C value ranges from 0.00 - 0.50, it is in a low category. This indicates that there is no dominance of certain invasive aquatic invertebrates in the KKPD PISISI. According to [35], if the value of C is getting smaller, the dominance of species in an area will decrease. The low C value is thought to be due to the even distribution of species shown by the high E value of invasive aquatic invertebrates. A species dominance with other species occurs when the C value approaches 1 [14].

5.3 Distribution of invasive aquatic invertebrates

The distribution shows the distribution pattern of an organism in a community. Each species has a unique distribution according to the type of habitat. Distribution occurs due to differences in response to habitat [36]. The distribution of invasive aquatic invertebrates in the KKPD PISISI is generally classified as a group, except for *Culcita novaeguinaea*, which has a uniform distribution (Table 2).

The distribution of invasive aquatic invertebrates in the KKPD PISISI is generally classified as a cluster, except for *Culcita novaeguinaea*, which has a uniform distribution. Similar distribution occurs due to antagonistic interactions between individuals due to competition for resources. According [14], similar distribution can occur if the competition among individuals is robust, which encourages space division.

The clustered distribution occurs due to the high availability of organic matter in the waters [38]. Flow has a relationship in the availability of organic matter in waters. According to Muqsit *et al.* (2016), currents play a role in the movement of nutrients, larvae, sediment, and oxygen. The current speed supports this in the KKPD PISISI, ranging from 0.07 m/s - 0.19 m/s. The clustered distribution of invasive aquatic invertebrates in the KKPD PISISI is thought to be related to the substrate shape. In general, the substrate form in the KKPD PISISI consists of sandy substrate and coral fragments. Based on the research conducted by [39], cluster distribution is related to the shape of the substrate in the form of chunks of coral, coral rubble, and a little sand and coral area reef flats. According to [36] the most common distribution is group distribution because individuals in the population tend to form groups of various sizes.

Table 2. The distribution of invasive aquatic invertebrates.

No	Species	Distribution	Category
1.	*Acanthaster planci*	1.67	Clustered
2.	*Culcita novaeguinaea*	0.62	Uniform
3.	*Diadema antillarum*	1.22	Clustered
4.	*Spirobranchus giganteus*	1.66	Clustered
5.	*Drupella* sp.	1.88	Clustered

5.4 Comparison of Abundance of Invasive Aquatic Invertebrates to the Percentage of Coral Reef Cover

The abundance of invasive aquatic invertebrates and the percentage of coral reef cover have different values at each KKPD PISISI station. The results showed a difference between the abundance of invasive aquatic invertebrates and the percentage of coral cover where the low coral cover had a high abundance of invasive aquatic invertebrates, namely at station Batee Dua Islet. In contrast, high coral cover has an abundance of less invasive aquatic invertebrates at the Gabui Island station (Figure 5).

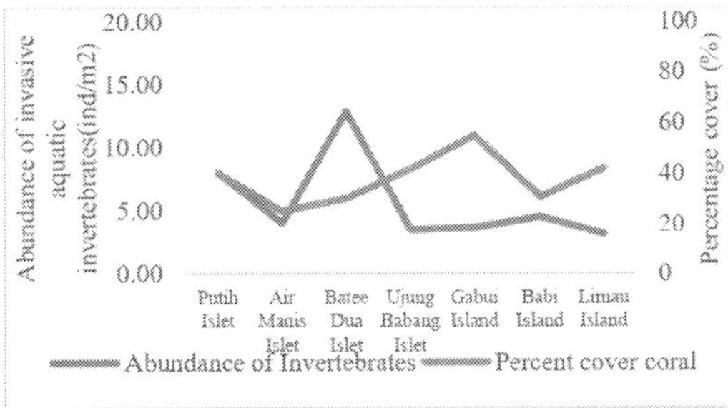

Fig. 5. The abundance of invasive aquatic invertebrates with live coral cover.

The high and low abundance of invasive aquatic invertebrates is thought to be related to the coral cover percentage. A study conducted by [40] high percentage of coral cover is associated with a low abundance of invasive aquatic invertebrates, so that predation of coral reefs is not too high. Another statement is supported by [41], which stated that invertebrate predation activities would increase if there were a decrease in coral cover. Another study conducted by [42] showed that a decrease in the percentage of live coral cover has high invasive aquatic invertebrates.

Other causes that affect the abundance of invasive aquatic invertebrates are thought to be caused by human activities such as fishing activities. The KKPD PISISI area is an area used by fishermen to catch fish and other marine biotas, so there is a possibility that the high and low abundance of invasive aquatic invertebrates can be caused by fishing. This is consistent with [43] which states that intensive fishing activities on coral reefs can impact ecosystems because they can change the composition of organisms in the ecosystem and change the processes that occur in coral reef ecosystems.

6. Conclusion

The community structure of the invasive aquatic invertebrates in the KKPD PISISI is in a suitable category, as indicated by the total abundance of invasive aquatic invertebrates of 39.67 ind/m^2. The diversity index of invasive aquatic invertebrates of 1.50 with moderate category, index evenness of 1.06 with a high category, and dominance index of 0.44 with a low category. Distribution of invasive aquatic invertebrates in the KKPD PISISI is classified into clustered distribution. The value of the highest abundance of invasive aquatic invertebrates was 12.83 ind/m^2, and the percentage of coral cover was 29.83%. The lowest abundance of invasive aquatic invertebrates was 3.67 ind/m^2. The percentage of coral cover was 54.67%., indicating an influence between the abundance of aquatic invertebrates invasive to the coral cover percentage.

Acknowledgement. The authors would like to thank Fauna and Flora International (FFI) Aceh for facilitating this research activity and Ocean Diving Club (ODC) team and ODC who have helped during the research

References

1. S. Hasma, B. Sadarun, R.D. Palupi, Sapa Laut **4**, 2 (2019)
2. P.W. Glynn, I.C. Enochs, *Invertebrates and their roles in coral reef ecosystems. In Coral reefs: an ecosystem in transition* 273- 325 (2011)
3. J.E. Havel, K.E. Kovalenko, S.M. Thomaz, S. Amalfitano, L.B. Kats, Hydrobiologia **750** (2015)
4. İ. E. Büyüktahtakın, Z. Feng, G. Frisvold, F. Szidarovszky, App. Math **4** (2013)
5. D.W. Schindler, B.R. Parker, Wat. Ai. So. Poll. Foc **2** (2002)
6. R.D. Rotjan, S. M. Lewis. Mar. Eco. Prog. Ser. **367** (2008)
7. N.A. Bax, Williamsons, M. Aguero, E. Gonzalez, W. Geeves. Mar. Poli **27** (2003)
8. L.M. Herborg, P. O'Hara, T.W. Therriault, J. of App. Ecol **46** (2009)
9. M.Y. Sastri, *Percentage of cover and distribution of benthic invasion of invertebrates in coral reef ecosystems* (Thesis, Bogor Agricultural University, 2016)
10. R.T. Utami. *Relationship of coral reef ecosystem conditions to the distribution of coral killer invertebrates in the waters of Pulau Dua and Pulau Tikus, Bengkulu as well as the Island of the Netherlands and Pulau Dapur, Kepulauan Seribu, Jakarta* (Thesis, Bogor Agricultural University, 2019)
11. S. Nasution. *Research Methods: Scientific Research* (Bumi Aksara, Jakarta, 2004)
12. L. Collin, A.C. Arneson, *Tropical pacific invertebrates* (Coral Reef Press, California, 1995)
13. T. Rahmasari, T. Purnomo, R. Ambarawati, Biosaintifika, **7**, 1 (2015)
14. E.P. Odum, *Basic Ecology* (Saunders College Publishing, 1983)
15. A. Soegianto, *Quantitative ecology* (Indonesian National Business, 1994)
16. S. English, C. Wilkinson, V. Baker, *Survey manual for tropical marine resources* 2nd edd, Townsville: Australian Institute of Marine Science (1997)
17. Keputusan Menteri Lingkungan Hidup Nomor 04/ 02/2001 (2001)
18. S. Bangapadang, E. Emiyarti, W. Nurgayah, Sapa Laut **4**, 2 (2019)
19. H.A.W. Cappenberg, Ocea. Limnno **37**, 2 (2011)
20. D. Alwi, S.H. Muhammad, S. Bisi, J. Archip. Mar. Sci **1**, 1(2018)
21. N.P.E. Noviana, P.G.S. Julyantoro, D.A.A. Priestess, Curr. Trends in Aquat Sci **II**, 1 (2019)

22. S. Suryanti, W. Wahyuni, J. Fish. Sci **10,** 1 (2014)
23. S. Alustco, Y. Wardiatno, I. Setyobudiandi, JIAFS **17**, 1 (2011)
24. N. P. Zamani, Oseana **18**, 2 (2015)
25. J. Souhoka, S. Patty, Scien. Jour **1**, 3 (2013)
26. G. Gabbi. *Shells: guide to the jewels of the sea*, Periplus, Turin (1999)
27. B. Morton, G. Blackmore, C.T. Kwok, JMS **68**, 3 (2002)
28. A. Aziz, Oseana **XXI,** 3 (1996)
29. C. Rani, S. Yusuf, F. D. S. Benedikta, J. Fac. Mar. Fish. Sci. (2007)
30. J. Bell, 2008. *Feeding preferences of the cushion star Culcita novaeguineae in the presence of the crown of thorns starfish Acanthaster planci. UCB Moorea Class: Biology and Geomorphology of Tropical Islands,* Student Research Paper (2008)
31. D.A. Setyowati, Supriharyono, W.T. Taufani, J. Maquares **6,** 4 (2017)
32. I. Irawan. *Structure of mollusc communities (gastropods and bivalves) and their distribution in Pulau Burung and Pulau Tikus, Pari Island Cluster, Kepulauan Seribu* (IPB University, 2008)
33. A.M. Pirzan, Utojo, M. Atmomarso, M. Tjaronge, A.M. Tangko, and Hasnawi, Indones. Fish. Res **J11**, 5 (2005)
34. A. Indarjo, H. Retno, I. Samidjan, S. Anwar, J. Mar. Sci. (2007)
35. R. Werdiningsih. *Crab Community Structure in Mangrove Habitat, Tanjung Pasir Beach,* (Tanggerang, Banten, IPB University, 2005)
36. M.C. Molles, *Ecology: concepts and applications (5th ed.)* (NY: McGraw-Hill, 2010)
37. N. Yuniarti. 2012. *Diversity and Distribution of Bivalves and Gastropods (Mollusks) in the Glayem Juntinyuat Coast of Indramayu, West Java* (IPB University, 2012)
38. S. Steven, S. Nasution, Thamrin, SOJ **1,** 2 (2014)
39. R. Riska, B. Sadarun, L.O.M.Y. Haya, JMLI **2, 6** (2013)
40. L. Lalang, B. Sadarun, L.O.M.Y. Haya, JMLI **1**, 1 (2013)
41. R. Mauliza, D.J. Prihadi, M.L. Syamsuddin, JMF **7**, 2 (2016)
42. C. Rani, Rev. Bionatura **5,** 2 (2003)

Biodiversity of the Indonesian marine sponge genus *Aaptos* with molecular networking approach

Viqqi Kurnianda[1], *Ulung* Jantama Wisha[2], and *Musri* Musman[3*]

[1]Department of Marine Science, Faculty of Marine and Fisheries, Syiah Kuala University, Banda Aceh 23111, Indonesia
[2]Research Institute for Coastal Resources and Vulnerability, Ministry of Marine Affairs and Fisheries, Padang 25245, Indonesia
[3]Department of Chemistry Education, Faculty of Teacher Training and Education, Syiah Kuala University, Banda Aceh 23111, Indonesia

Abstract. Many natural products have been isolated from the cryptic species of the Indonesian marine sponge. Despite this, it remains unclear how sponges produce such a diverse array of metabolites and their relationship to sponge diversity. During our expedition in 2017-2019, we collected the Indonesian marine sponge of the genus Aaptos, which was found surrounding Weh and Aceh Islands, the northwesternmost part of the Indonesian archipelago. All specimens were subjected to a feature-based molecular networking approach to understand their relationship. The molecular phylogenetic clades of genus *Aaptos* were examined by cytochrome oxide I (COI) while the presence of their chemotypes were confirmed by chemical analysis based on bioassay-guided separation with several steps of chromatography. As results, the presence *Aaptos suberitoides, Aaptos aaptos, Aaptos nigra, Aaptos lobata,* and *Aaptos sp.* were confirmed by the molecular phylogenetic. On the other hand, chemical analysis showed the presence of alkaloid-class molecules on all specimens. Interestingly, feature-based molecular networking revealed the whole figure of cryptic species were formed a clear and distinct group together with their chemotypes. Consequently, it confirmed that the relationship between the presence of each species as well as its chemotype can be distinguished clearly. In conclusion, we suspect the chemical analysis of the genus *Aaptos* have unknowingly conflated between different cryptic species, resulting in the seemingly idiosyncratic chemical variations.

1 Introduction

Metabolites found in marine ecosystems exhibit wide structural diversity and bioactive properties [1]. Sponge, which belongs to the phylum Porifera, is known to contain a variety of secondary metabolites [2]. Interestingly, compounds from sponges have often been found to be promising pharmaceutical agents [3]. To date, FDA has been approved 3 marine drugs derived from sponges such as anticancer cytarabine (ara-C) and anticancer Eribulin mesylate, and antiviral vidarabine (ara-A) [4-7].

Sponge genus *Aaptos* is found in a wide range of ecosystems, including the South China Sea, Japanese, Indonesian, and Caribbean waters [8-11]. The first discovery was conducted by Nakamura and co-workers from an Okinawan sponge *Aaptos aaptos* which contain aaptamine, an alkaloidal molecule, in 1981 [12]. To date, many other research groups throughout the world have also carried out chemical analyses on different *Aaptos* species to investigate the structural diversity [13-15].

During our expedition around Weh and Aceh Islands in the early 2017, we discovered the marine sponge of the genus *Aaptos*. Since there are no reported papers on the *Aaptos* sponges around that area to date, we examined both genetic and chemical contents of them. In this paper, we revealed the whole figure of genetic and chemical diversity on *Aaptos* sponges around Weh and Aceh Islands using the metabolomics approach.

2 Materials and methods

2.1 Materials

Nuclear Magnetic Resonance (NMR) 500 MHz spectroscopy (Bruker Biospin, Billerica, MA, USA) in $CDCl_3$ (Cambridge Isotope Labs). All extracts were injected into LC-MS/MS data using a reversed-phase analytical HPLC column (5μm, 250mm x 4.6mm). Peaks were detected by positive ion mode of high-resolution MS/MS system.

2.2 Extraction and isolation

Specimens were collected by hand using SCUBA at 10 to 30 m depth off surrounding Weh and Aceh Islands from 2017 to 2019 (Fig. 1). The specimens were kept in the freezer during the trip. Furthermore, all specimens consist of 209 specimens from Weh and 347 specimens from Aceh Islands were extracted by acetone. After partitioning and concentrating the extracts, the organic layer was used for purification and metabolomics purposes. On the other hand, all extracts which contain

* Corresponding author: musrimusman@unsyiah.ac.id

alkaloidal markers were purified to obtain pure material used for the chemical standard.

2.3 Phylogenetic analysis

All crude extracts (10.0 mg) were extracted using Qiagens' DNeasy Blood & Tissue Kit (Qiagen, Hilden, Germany), followed by PCR amplification using dmaW forward (5′-TTGATCTGGAGTGGTTCCGC-3′) and reverse (5′-CGTTCATGCCGAAGGTTGTG-3′) under the standard condition. Amplicons were analysed by AB3130 with ABI PRISM BigDye Terminator Cycle Sequence Kit v1.1, and sequence alignment used MEGA v5.1.

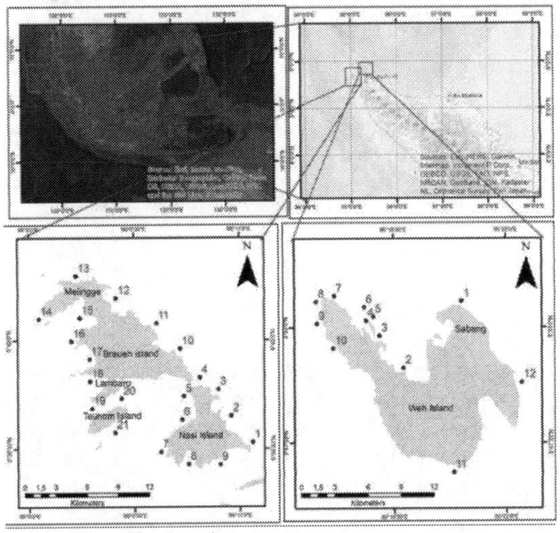

Fig. 1. The collection sites.

2.4 Metabolomics approach

The presence of the marker alkaloidal was confirmed by a ¹H NMR spectra comparison. On the other hand, For LC-MS/MS analysis, MeOH solution (20 μL) containing each crude extract was prepared. We applied a flow rate of 0.5 mL/min under positive ion mode of high resolution to samples ranging from 50 to 1000 m/z. Furthermore, all raw data were converted to mzXML format, and it was uploaded online with reference number MSV000088027 (https://massive.ucsd.edu) [16]. To further analysis, we subjected them to MZmine2 software. On the GNPS website (http://gnps.ucsd.edu), quantitative and MS/MS tables have been uploaded for use in a feature-based molecular networking study [17]. A cosine score of 0.6 was applied to the parameters with a maximum of 10 matched peaks. Then, by using Cytoscape 3.7.2 version, the results were visualized [18].

2.5 Spatial analysis of the Sponge genus *Aaptos*

The specimen data collected at each observation station were then spatially analysed using an interpolation method. Inverse Distance Weighted (IDW) interpolation was employed to perform the distribution model of the detected individual. This method corresponds to the Indonesian Standard (SNI 7644:2010) for oceanographic-spatial data provider. IDW method can calculate faster than the other methods such as kriging, even though the calculation quality is categorised to "acceptable" [19]. IDW assumes the value resulted from interpolation process as similar as the closer sampled data compared to the distant ones. The distance among the interpolated points toward the reference is going to influence the weight of calculation [20]. The interpolation results were limited to the initial data so that the product will not be exceeded or extrapolated.

3 Results and discussion

All specimens both from Weh and Aceh Islands were examined by ¹H NMR comparison with those reported [21-25]. As results, all extracts were identified as known compounds. However, since there was a methoxy moiety was introduced with those reported, it was likely a typical solvolysis product during chemical works. Careful analysis by LC-MS/MS revealed the presence of phenolic groups on all extracts (Fig. 2). It also confirmed that those likely reported to be artifacts by solvolysis reaction to yield methoxy derivatives [26].

Since there is quite similar morphology on the marine sponge of the genus *Aaptos*, we examined the whole diversity figure using molecular networking approach. After obtaining the MS/MS data, all raw data were subjected to MZmine2 for feature detection, grouping, and alignment. By uploading the results on the GNPS web platform, the feature-based on molecular networking may reveal the whole figure of diversity in the sponge of the genus *Aaptos* relating to their chemotypes. As results, the presence of some clades on the sponge of the genus *Aaptos* were confirmed by the phylogenetic tree. Moreover, feature-based on molecular networking revealed the whole figure of their diversity (Fig. 3).

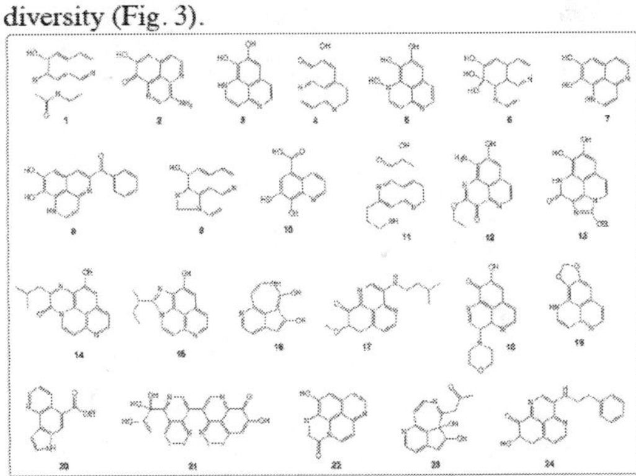

Fig. 2. Chemotypes of sponge of the genus Aaptos around Weh and Aceh islands.

To understand how the sponge was spread, we used the distribution analysis to examine their spread (Figs. 4-5). As results, *A. aaptos, A. lobata, A. nigra,* and *A. suberitoides* were spread in the northern part while *Aaptos sp.* was spread in the western part of Weh Island (Fig. 4). The characteristic of geography in Aceh Island may introduce unique sponge spread in the area. The presence of *A. nigra*, and *A. suberitoides* were discovered in the northern part of the collection site in

https://doi.org/10.1051/e3sconf/202233903002

Nasi Island, meanwhile, *Aaptos sp*. was found in the southern part of Nasi Island. On the other hand, the sponge of the genus *A. lobata* was spread in both northern and southern parts of Breueh Island (Fig. 5). We suspected that *A. lobata* is a more tolerant sponge of big waves and strong currents from the Indian Ocean in the western part of Aceh Island. As the sponges contain the potential of secondary metabolite, it also gives us an opportunity to discover the resources for the pharmaceutical sciences purposes.

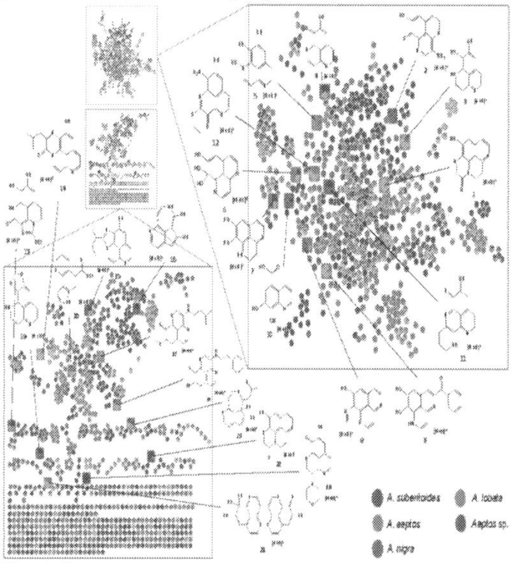

Fig. 3. Feature-based on molecular networking.

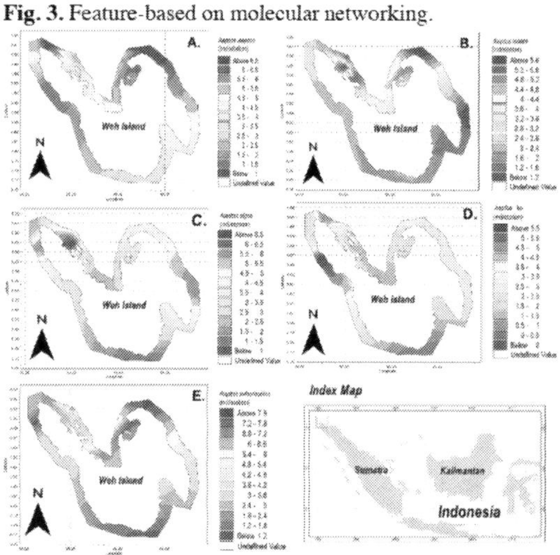

Fig. 4. Distribution of the genus *Aaptos* in Weh Island; (a) *Aaptos aaptos*, (b) *Aaptos lobata*, (c) *Aaptos nigra*, (d) *Aaptos* sp. and (e) *Aaptos suberitoides*.

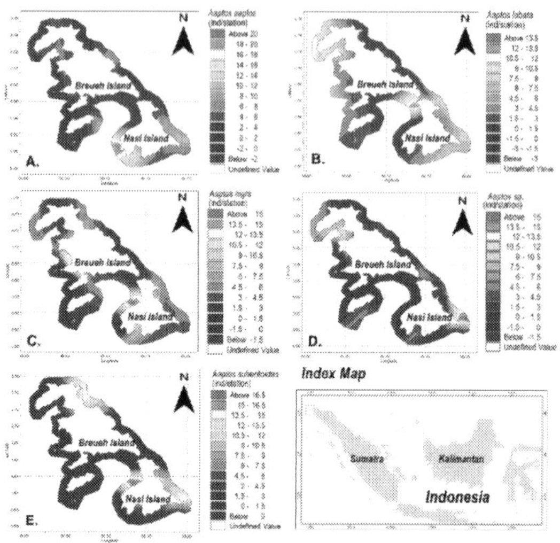

Fig. 5. Distribution of the genus *Aaptos* in Aceh Island; (a) *Aaptos aaptos*, (b) *Aaptos lobata*, (c) *Aaptos nigra*, (d) *Aaptos* sp., and (e) *Aaptos suberitoides*.

4 Conclusion

The presence of *Aaptos* suberitoides, *Aaptos aaptos, Aaptos nigra, Aaptos lobata,* and *Aaptos sp*. from the Indonesian marine sponge of the genus *Aaptos* were confirmed by the molecular phylogenetic. The chemical analysis showed the presence of alkaloid-class molecules on all specimens. Interestingly, feature-based molecular networking revealed the whole figure of cryptic species were formed a clear and distinct group together with their chemotypes. Further analysis revealed that, *A. aaptos, A. lobata, A. nigra,* and *A. suberitoides* were spread in the northern part while *Aaptos sp*. was spread in the western part of Weh Island. Moreover, the presence of *A. nigra,* and *A. suberitoides* were discovered in the northern part of the collection site in Nasi Island, meanwhile, *Aaptos sp*. was found in the southern part of Nasi Island. On the other hand, the sponge of the genus *A. lobata* was spread in both northern and southern parts of Breueh Island. As the sponges contain the potential of secondary metabolite, it also gives us an opportunity to discover the resources for the pharmaceutical sciences purposes.

References

1. Nadar, V.M. S. Manivannan., R. Chinnaiyan., M. Govarthanan., and Ponnuchamy, K. Chem. Biol. Drug Des. **2021** (in press).
2. He, Q., S. Miao., N. Ni., Y. Man., K. Gong. Nat. Prod. Commun. **2020**, 15, 1934578X20951439.
3. Pham, C.D., R. Hartmann., W.E. Müller., N.J. de Voogd., D. Lai,. P.J. Proksch. J. Nat. Prod. **2013**, 76, 103-106.
4. Calcabrini, C., E. Catanzaro., A. Bishayee., E. Turrini., C. Fimognari. Mar. drugs **2017**, 15, 310.
5. Sagar, S., M. Kaur., K.P. Minneman. Mar drugs **2010**, 8, 2619-2638.

6. Gyawali, B., A.S. Kesselheim. Nat. Rev. Clin. Oncol. **2021**, 1-2.
7. Del Paggio, J.C., I.F, Tannock. Lancet Oncol. **2019**, 20, 1065-1069.
8. Elissawy, A.M., D.E. Soleiman., N. Mehdinezhad., M.L. Ashour., P.P. Mohammadi. Biomolecules **2021**. 11, 258.
9. Trang, D.T., B.H. Tai., D.T.T. Hang., P.H. Yen., N.X. Nhiem., P.V. Kiem. Nat. Prod. Res. **2021**, 1-10.
10. Fristiohady, A., B. Sadarun., W. Wahyuni., M.H. Malaka., F. Ahmad., F. Malik., L.O.M.J. Purnama., I. Sahidin. J Appl Pharm Sci. **2020**, 10, 081-089.
11. Anjum, K., S.Q. Abbas., S.A.A. Shah., N. Akhter., S. Batool., S.S. ul Hassan. Biomol. Ther. 2016, 24, 347.
12. Nakamura, H., J.I. Kobayashi., Y. Ohizumi., Y. Hirata. Tetrahedron Lett. **1982**, 23, 5555-5558.
13. Nguyen, T.L.A., T.H.N. Doan., D.H. Truong., N.T.A. Nhung., D.T. Quang., D. Khiri., S. Taamalli., F. Louis., A. El Bakali., and D.Q. Dao. RSC Adv. **2021**, 11, 21433-21446.
14. Paliwal, D., S. Srivastava., P.K. Sharma., I. Ahmad. Curr. Tradit. Med. **2021**, 7, 93-106.
15. Chabowska, G., E. Barg., A. Wójcicka. Molecules **2021**, 26, 4324.
16. MassIVE **2021**. https://massive.ucsd.edu/. Accessed on August 27, 2021
17. Nothias, L.F., D. Petras., R. Schmid., K. Dührkop., J. Rainer., A. Sarvepalli., I. Protsyuk., M. Ernst., H. Tsugawa., M. Fleischauer., F. Aicheler. Nat. Methods **2020**, 17, 905-908.
18. Aron, A.T., E.C. Gentry., K.L. McPhail., L.F. Nothias., M. Nothias-Esposito., A. Bouslimani., D. Petras., J.M. Gauglitz., N. Sikora., F. Vargas., J.J. van Der Hooft. Nat. protocols **2020**, 15, 1954-1991.
19. Qin, R., L. Lin. Computers and Géosciences **2017**, 103, 70-79.
20. Childs, C. *Interpolating Surface in ArcGIS Spatial Analysis*. ESRI Education Services **2004**.
21. Jang, K.H., S.C. Chung., J. Shin., S.H. Lee., T.I. Kim., H.S. Lee., K.B. Oh. Bioor. Med. Chem. Lett. **2007**, 17, 5366-5369.
22. Pham, C.D., R. Hartmann., W.E. Müller., N.J. de Voogd., D. Lai,. P.J. Proksch. J. Nat. Prod. **2013**, 76, 103-106.
23. Yu, H.B., F. Yang., F. Sun., G.Y. Ma., J.H. Gan., W.Z. Hu., B.N. Han., W.H. Jiao., H.W. Lin. J. Nat. Prod. **2014**, 77, 2124-2129.
24. Wu, C.F., M.G. Lee., M. El-Shazly., K.H. Lai., S.C. Ke., C.W. Su., S.P. Shih., P.J. Sung., M.C. Hong., Z.H. Wen., M.C. Lu. Mar. Drugs **2018**, 16, 18.
25. He, Q., S. Miao., N. Ni., Y. Man., K. Gong. Nat. Prod. Commun. **2020**, 15, 1934578X20951439.
26. Kurnianda, V., H. Fujimura., Y. Kanna., J. Tanaka. Chem. Lett. **2021**, 50, 220-222.

Diversity, distribution, and species status of the fish in Banten Bay, Indonesia

Sulistiono Sulistiono[1*], *Dudi* Muhamad Wildan[1], *Ayu* Ervinia[1], *Nur* Rohim[2], *Dimas* Angga Hedianto[2], *Faqih* Baihaqi[2], *Didit* Abdillah[2], *I Nyoman* Yoga Parawangsa[3], *Gema* Wahyudewantoro[4], and *Masashi* Yokota[5]

[1]Department of Aquatic Resource Management, Faculty of Fisheries and Marine Science, IPB University, Bogor, 16680, Indonesia.
[2]Study Program of Aquatic Resource Management, Graduate School, IPB University, Bogor, 16680, Indonesia.
[3]Indonesian Ichthyological Society, Bogor. 16004. Indonesia
[4]Center for Biological Research, Indonesia Institute of Sciences, Cibinong, Bogor, 16004, Indonesia
[5]Department of Aquatic Biosciences, Tokyo University of Marine Sceince andTechnology, Minato-Ku, Kounan 4-5-7. Tokyo, 108-8477, Japan.

Abstract. This study aims to analyze the species diversity, distribution and status of fish found in the coastal waters of Banten Bay. Samplings were carried out 7 times at three locations namely Cengkok coastal waters (on February-March 2019), Bojonegara coastal waters (January-February, August-September 2020), and Kemayungan-Linduk coastal waters (February-April 2021) of the Banten Bay, using gill nets, trammel net, cast net, hand-line, beach seine, and mini trawl. Results of this study showed that the fish caught in the Banten Bay coastal waters consisted of 111 species, distributed at the three areas, namely Cengkok (32 species), Bojonegara (58 species), and Kemayungan-Linduk coastal waters (51 species). The diversity, eveness, and dominance indices were 2.185 - 3.763, 0.630 - 0.781, and 0.072 - 0.233 respectively, indicating the medium-high species diversity. These fish were spreading in marine water, fresh water, estuary, freshwater and estuary, marine and estuary, and marine, estuary and freshwater. The common species found at the study location were Ellochelon vaigiensis, Planiliza subviridis, and Leiognathus equula. In general, most of the fish found in Banten Bay were classified as least concern (LC), while two species were vulnerable (VU), one species was near threatened (NT), and ones species was endangered species (EN).

1 Introduction

Banten Bay is located in the west of Java Island in the northern part of Serang City [1]. The bay covers area about 150 km^2 [2], which is part of the Java Sea, with relatively shallow water conditions. The oceanographic conditions of these waters are influenced by a number of freshwater inputs, originating from rivers that empty into the waters of the bay [3], including the Cibeureum River, Cibanten River, and Cikadueun River [4]. Banten Bay has experienced various environmental pressures from anthropogenic activities [5-6], which includes various activities around the area, such as industry, shipping and fishing ports, housing, and aquaculture [4]. Various anthropogenic activities of the surrounding community may have an impact on the existence of fish resources [7].

The coastal waters of Cengkok, Bojonegara, and Kemayungan-Linduk are three areas in Banten Bay, which have quite large potential of fishery resources. Several fishing ground were found in these areas, either in the fish, shellfish or shrimp resources, and location of green mussel cultivation. Cengkok Beach, is a coastal area located in the middle of Banten Bay. This beach is adjacent to Gopek beach (in the west) as the centre of port and tourism activities in the region. Bojonegara Beach is a coastal area located in the western part of Banten Bay. This coastal area is quite dense with industry and large ship ports. There is also a fishing port (on the Wadas River and the Terate River), but it is not as big as the fishing port in Karangantu (Gopek beach). Meanwhile, Kemayungan-Linduk beach is one of the beaches located in the eastern region of Banten Bay. Many aquaculture and agriculture of local community are found in this coastal area.

Several study activities in Banten Bay waters were carried out and published, including study on water quality [2, 8-9], hydrodynamics [5], fish biology [10-11], and heavy metal content in aquatic biota [12-13]; [14-15]. However, to date there have been few publications or study activities related to the status of fish caught in this area. Therefore, this study aims to examine the diversity and status of fish caught in the coastal waters of Banten Bay. This study is necesary as a basis information for better management of the fisheries resources.

2 Materials and Methods

* Corresponding author: onosulistiono@gmail.com

2.1 Time and sampling location

Sampling stations were chosen based on the character of the ecosystem, namely rivers, estuaries, and seas. Fish sampling was carried out nine times, Mar-Apr 2019 (Cengkok, C1-C5), Jan-Feb and Aug-Sep 2020 (Bojonegara, B1-B6), and Feb-Apr 2021 (in Kemayungan-Linduk, KL1-KL6), of Banten Bay, (Figure 1).

2.2 Data collection

2.2.1 Fish sampling

Fish were collected using gill nets (1-3 inches), trammel nets (1.5; 6 inches), hand-lines, beach seines, mini trawls (>1 inches), and cast nets. The gill nets and trammel nets used were operated as fixed gillnets or set gillnets. The fishing gear was operated in the coastal areas by placing it along the surface of coastal waters and rivers in accordance with the sampling point. The gill and trammel net were left for about 30 minutes to 1 hour. In shallow water, gill nets and cast net were usually used. Fish samples were also obtained from fishermen who caught in the area around the study location. Fish samples were sorted and preserved with 10% formalin, and brought to the Bio-Macro Laboratory of IPB University for identification.

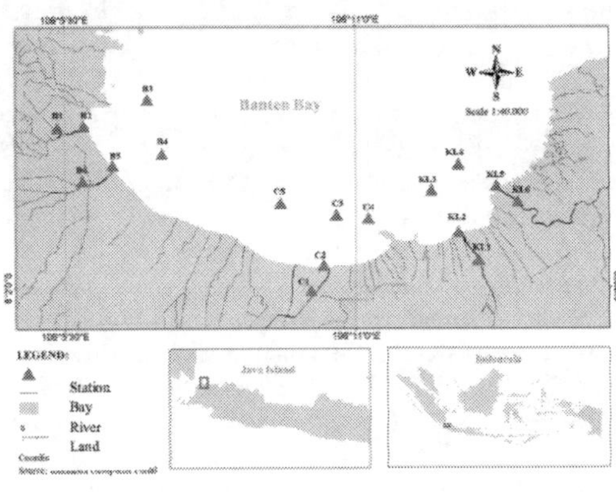

Fig. 1. Location of fish sampling in the Cengkok (C1-C5), Bojonegara (B1-B6), and Kemayungan-Linduk Coastal Waters (KL1- KL6)

Note: B1 (5.984096S, 106.090157E), B2 (5.983031S, 106.110909E), B3 (5.97762S, 106.12730E), B4 (5.991025S, 106.126765E), B5 (5.99523S, 106.10731E), B6 (6.00047S, 106.09703E), C1 (6.031318S, 106.173411E), C2 (6.02659S, 106.17492E), C3 (6.013604S, 106.177051E), C4 (6.01174S, 106.18898E), C5 (5.99988S, 106.15621E), KL1 (6.02402S, 106.22164E), KL2 (6.015710S, 106.216530E), KL3 (6.001726S, 106.208904E), KL 4 (5.994793S, 106.218296E), KL5 (6.002112S, 106.227292E), KL6 (6.00593S, 106.23437E)

2.2.2 Fish observation and fish conservation status (IUCN)

Fish samples of each station were distinguished by species, and identified to the lowest taxon, using the books of [16-20]. Fish status was based on the status criteria from International Union for Conservation of Nature (IUCN).

2.3 Data analysis

Fish data were analyzed using diversity, evenness and dominance indices, and fitted descrptively based on conservation status.

2.3.1 Diversity Index

Calculations of the diversity index (H) can be made using the Shannon-Wiener diversity index [21].

$$H' = \sum_{i=1}^{s} (p_i)(log_2\, p_i)^k ; p_i/N$$
$$H' = \sum_{i=1}^{s} (p_i)(log_2\, p_i)^k ; p_i/N \quad (1)$$

Information:
H = diversity index
p_i = i-th type ratio
n_i = number of i-th individual species
N = number of individuals of all types
S = number of taxa (type)

2.3.2 Evenness Index

Calculations of the evenness index (E) can be done using the evenness index formula, and the magnitude of the species evenness index ranges from 0 to 1 [21]:

$$E = \frac{H'}{Hmax} \quad (2)$$

Information:
E = evenness index
H = diversity index
H_{max} = Log_2 S
S = number of taxa (species)

2.3.3 Dominance Index

The calculation of the dominance index uses the Simpson dominance index formula [22]. The value of the dominance index (C) ranges from 0 to 1. If the value of C is close to 0, then no individual dominates, whereas if the value of C is close to 1, then there is an individual who dominates in a community [23].

$$C = \sum_{i=1}^{s} \left(\frac{n_i}{N}\right)^2 \quad (3)$$

Information:
C = dominance index
n_i = number of i-th individual species
N = total number of individuals

3 Results and Discussion

3.1 Results

3.1.1 Fish Diversity

The total number of fish caught during the research activity was 1277, consisting of 111 species, and divided into 51 families and 29 orders (Table 1). These fish species caught in the Cengkok coastal waters consisted of 32 species (and 349 individuals), in the Bojonegara coastal waters consisted of 58 species (and 473 individuals), and in Kemayungan-Linduk coastal waters consisted of 51 species (and 455 individuals). The fish caught in Cengkok Beach consisted of 23 families (18 orders), with the most numerous families being Mugillidae, followed by Sciaenidae, Mulidae, Gobiidae, Gerreidae, and Carangidae. In the Bojonegara, the fish consisted of 37 families (22 orders) where the Clupeidae family was the most common family found on the beach, followed by the Mugilidae, Leiognathidae, Carangidae, and Ariidae. Meanwhile, 34 families (21 orders) were found in the Kemayungan-Linduk beach, where the Sciaenidae is the most common family found on this beach, followed by the Clupeidae, Carangidae, Butidae, and Gobiidae. Results from the three sampling locations showed that the Sciaenidae and Clupeidae families (with 4 members each) are the families that have the most members found in the waters of Banten Bay, followed by Carangidae, Mugilidae, Leiognathidae, Butidae and Gobiidae.

In general, the most common species of fish found in those location were Ellochelon vaigiensis, Planiliza subviridis, and Leiognathus equula, followed by Trichopodus trichopterus, Ambassis cf kopsii, Eleutheronema tetradactylum and Scatophagus argus (Table 1). Based on the calculation of the biological indices, the diversity index ranged from 2,185 to 3,763, the evenness index ranged from 0.630 to 0.781, and the dominance index ranged from 0.072 to 0.233. These values indicated that the diversity of fish in this area was classified as medium-high, uniform, and there is no dominant fish.

3.1.2 Fish Distribution

The distribution of various species of fish was found in river, estuary, and marine waterss. In addition, there were several species of fish found in the freshwater and estuary, sea and estuary, as well as fresh, estuary and marine locations. Three commonly fishes found in those areas are presented in Figure 2. In Cengkok coastal waters, *Oxuderces dentatus*, *Glossogobius giuris* and *Planiliza planiceps* were group of fish that are quite commonly found in rivers (C1). In the estuary (C2), the most common species of fish found were *Ellochelon vaigiensis*. *Glossogobius giuris*, and *Atule mate*. Meanwhile, in marine waters, the most common fish species found were *Ambassis* cf *kopsii* (C3), *Planiliza subviridis* and *Ellochelon vaigiensis* (C4), and *Ellochelon vaigiensis*, and *Ambassis* cf *kopsii* (C5). In this study, fish caught in fresh water and estuaries were also found, namely *Ambassis* cf *kopsii*, *Escualosa thoracata*, and *Oxuderces dentatus*. *Ellochelon vaigenesis* was found in seawater and estuarie, while *Butis butis*, *Hexanemalichthys sagor*, and *Glossogobius giuris* were found in marine, estuary and freshwater areas.

Sampling in the Bojonegara coastal waters was carried out in the two river watershed areas (Wadas and Terate Rivers), including the estuary and the adjacent sea (Figure 2). At the freshwater station of the Wadas River (B1), this study found *Ellochelon vaigiensis*, *Planiliza subviridis*, and *Mystus gulio*. At the estuary station (B2), the fish species commonly found were *Leiognathus equula*, *Scomberoides commersonianus* and *Pomadasys argenteus*. While at the sea station (B3), the species of fish that are commonly found were *Leiognathus equula*, *Nuchequula gerroides* and *Siganus javus*. At the freshwater station in the Terate River (B6), *Trichopodus tricopterus*, *Ellochelon vaigiensis*, and *Planiliza subviridis* were the most common fish found in that location. At the estuary station (B5), the species of fish commonly found were *Chanos chanos*, *Siganus javus* and *Nibea saldado*. While at the sea station (B4), the species of fish that are commonly found were *Leiognathus equula*, *Siganus javus*, and *Anodontosoma chachunda*. In these waters were found several species that inhabit more than one waters. *Hexanmalichthys sagor* and *Barbonymus gonionotus* were found in freshwater and estuary areas. *Johnius belangeri* and *Tetraodon lineatus* were found in marine and estuary areas. While *Arius maculatus*, *Planiliza subviridis*, and *Scatophagus argus* were found in river, estuary and marine areas.

Observations made in the Kemayungan-Linduk coastal waters were also carried out in two watersheds, namely the Kemayungan River and the Linduk River (Figure 2). In the Kemayungan watershed, the common fish species found at the freshwater station (KL1), the most common fish species found were *Planiliza subviridis*, *Barbonimus gonionatus* and *Chana striata*. At the estuary (KL2), the common fish species found were *Thryssa mystax*, *Scatophagus argus*, *Planaliza subviridis*, and *Eleutheronemaan tetradactylum*. Meanwhile, in the sea (KL3), the common fish found were *Eleutheronema tetradactylum*, *Ostemugil cunnesius*, *Johnius belangeri*, and *Scomberomerus commerson* (Figure 2). In observations in this area, *Glossogobius giuris* was found in freshwater and estuary areas, while *Eleutheronema tetradactylum*, *Hyporhamphus dussumieri*, *Johnius belangeri*, *Osteomugil cunnesius* and several other species were found in marine and estuary areas, while *Planiliza subviridis* were found in fresh water, estuaries and seas. Observation conducted in Linduk River (KL6), the most common species of fish found were *Glossogobius giuris*, and *Planiliza subviridis*. At the estuary (KL5), the fish species commonly found were *Arius maculatus*, *Planiliza subviridis*, *Scatophagus argus* and *Johnius belangeri*. While at the sea (KL4), the species of fish that are commonly found were *Thryssa mystax*, *Eleutheronema tetradactylum* and *Hypoatherina temminckii*. In this area, fishes of *Butis butis* and *Hexanomachthys sagor* were found in fresh water and estuaries, *Johnius belangeri* and *Tetraodon lineatus* were found in sea and estuaries, while *Arius maculatus*, *Planiliza subviridis* and *Scatophagus argus* were found in sea, estuaries and fresh water.

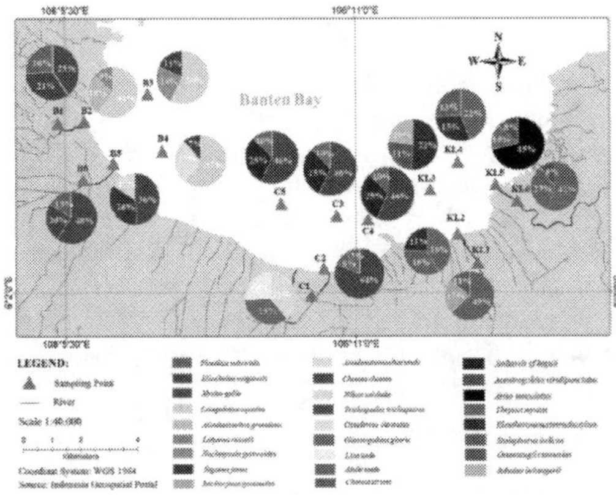

Fig. 2. Distribution of three commonly fish found in Banten

3.1.3 Species status.

The International Union for the Conservation of Nature and Natural Resources (IUCN) performs categorization based on the IUCN Red List conservation status in classifying the status of various species of living organisms that are threatened with extinction. From this conservation status, the IUCN issued the IUCN Red List of Threatened Species or abbreviated as IUCN Red List, which is a list of the rare status of a species. Based on the IUCN Redlist version 3.1, the conservation category consists of Extinct (EX; Extinct); Extinct in the Wild (EW; Extinct in the Wild); Critically Endangered (CR; Critical), Endangered (EN; Critical or Endangered), Vulnerable (VU; Vulnerable), Near Threatened (NT; Near Threatened), Least Concern (LC; Low Risk), Data Deficient (DD; Lack of Information), and Not Evaluated (NE; Not yet evaluated).

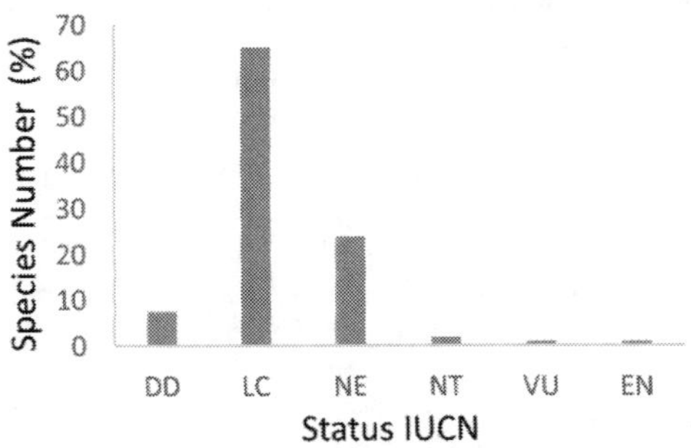

Fig. 3. Status of fish found on the coastal waters of Banten Bay according to IUCN. Note: EW: Extinct, CR: Critical, EN: Endangered, VU: Vulnerable, NT: Near Threatened, LC: least concern, DD: Data deficient, NE: Not Evaluated.

Based on checking the status of these fish at the IUCN, it was found that almost 64% of fish in the coastal waters of Banten Bay are classified as least concern-low risk (LC), followed by 26% not evaluated-not yet evaluated (NE), 10% data deficient-less information (DD), 2% near threatened-endangered (NT), and 2% vurnerable-extinct (VU). The results of the grouping the species status based on the IUCN is presented in Figure 3. From these results, there were four species that are quite worrying, namely *Brevitrygon cf walga*, *Scomberoides commerson* (NT), *Oreochromis mossambicus,* (VU) and *Maculabatis gerrardi* (EN)(Figure 3).

3.2 Discussion

Banten Bay, as one of the areas that has potential for fish resources, has currently developed into an industrial area. Such conditions became challenges for regional managers to be able to carry out activities that are not contradictory to one another. Industries that are estimated to produce waste may certainly affect the fish resources found in the area around the industry. Therefore, important scientific information is needed as a basis for carrying out development policies in the region. The watershed (DAS-Indonesian) that leads to the waters of Banten Bay has a variety of designations. In this study, samples were taken which were estimated to have designation characteristics in the surrounding area, namely the Cengkok (center), Bojonegara (west), and Kemayungan-Linduk coastal waters (east).

Based on the results of observations in the three coastal areas of Banten Bay, it was found about 111 species, which were spread on the cengkok beach (32 species), Bojonegara coastal waters (58 species), and Kemayungan-Linduk beach (51 species). The number of fish species found on the Cengkok, Bojoneraga and Kemayungan-Landik beaches was greater, compared with number of species found in the mangrove ecosystem of Ujung Kulon National Park, Pandeglang (Banten Province) [24], in several estuaries in Banten Bay [4], in Kumbe River estuary, Merauka Regency (Papua Province) [25], in Sembilang River estuary, Banyuasin (South Sumatra Province) [26], in Bojong Langkap River estuary, Segara Anakan (Central Java Province) [27], in the mouth of the Musi River (South Sumatra Province) [28], and in the Sembilang River estuary (South Sumatra Province) [29]. However, the results of this study are similar to those conducted in the Mayangan estuary of West Java [30].

Several species found in the coastal area of the Banten Bay are also similar to those found in the mangrove ecosystem of the Ujung Kulon National Park such as *Planiliza subviridis*, *Planiliza planticeps*, *Mystus gulio*, *Leiognathus equula*, *Gerres oyena*, *Terapon jarbia*, *Butis butis*, *Scatophagus argus*, *Strongylura leiura*, *Chelonodon patoca*, *Lutjanus russeli*, and *Caranx sexfasciatus* [24]. The species of fish from the results of this study are also similar to the results of previous research conducted in 2013 in several estuaries of the Wadas, Cibanten, Cengkok, and Pamong Rivers of Banten Bay. In that previous study, there were 17 species of fish (Estuary of Wadas), 20 species of fish (Estuary of Cibanten), 15 species of fish (Estuary of Cengkok), and 1 species of fish (Estuary of Pamong). Some of these species of fish are also similar to the fish found in this study, i.e. *Scatophagus argus,*

Scomberomorus commersoni, *Mugil cephalus*, *Sardinella fimbriata*, *Eleutheronema tetradactylum*, *Rastrelliger* sp., *Leiognathus* sp., and *Ambssis* sp [11].

In the current study, it appears that the fish species found in the Cengkok coastal waters (32 species) were fewer than in other locations (51 species in the Kemayungan-Linduk coastal waters, and 58 species in the Bojonegara coastal waters). This condition is thought to be due to the sampling frequency which is less than the other locations and requires further observations. The species of fish found in the three locations were similar, but there were several species of fish that were only caught in certain locations. Species of *Ambassis cf kopsii*, *Atherinomorus lacunosus*, *Atule mate* were only found at the Cengkok coastal waters. *Albula argentea*, *Allenbatrachus grunniens*, *Boleophthalmus boddarti*, *Brevitrygon cf wedga*, *Hyporhampus quoyi*, and *Lagocephalus lunaris* were only found in the Bojonegara coastal waters. Meanwhile, *Chirocentrus dorab*, *Platycephalus indicus*, *Pseudotriacanthus strigilifer*, *Tetraodon lineatus*, and *Trichiurus lepturus* were only found in Kemayungan-Landik coastal waters. In these observations, it is not known with certainty the parameters that distinguish the diversity of species found, but if sampling activities can be carried out with the same frequency, it is estimated that the species of fish found are more similar. In this study, no invasive species were found, but introduced species of fish (Chiclid) was found. According to the biological indices, similar condition was also conveyed [11], who conducted research in estuaries, and showed that the diversity of fish in some of these estuaries (Wadas, Cibanten, Cengkok) was moderate, and study conveyed the diversity of fish in the mangrove ecosystem of Ujung Kulon National Park, Pandeglang (Banten) [24]. However, the conditions on the Bojonegara coast and the Kemayungan-Linduk coast are somewhat different from the observations made show higher diversity [11]. The evenness index value also shows a medium-high value (0.630-0.781), and there is no particular species that dominates (0.072-0.233) in the area.

The status of fish species found in Banten Bay waters is generally least concern (LC), and only a few are found in near threatened (NT), vulnerable (VU), and endangered (EN) status. Two species of fish belonging to the NT species are *Himantura cf walga* and *Scomberoides commerson*, one species of fish that classified as vulnerable is *Oreochromis mossambicus* and one endangered species (EN) is *Maculabastis gerrardi*. Such conditions are similar to those found in several coastal areas in Indonesia. This condition is a common condition in estuarine and coastal areas, as is the case with estuarine fish species in the Kayan-Sembakung Delta which are generally also in LC status [31].

4 Conclusion

Based on observations made in three areas of Banten Bay (Cengkok, Bojonegara, and Kemayungan-Linduk coastal waters), the diversity of fish species consists of 111 species, including 29 orders and 51 families. Common species found include *Ellochelon vaigiensis*, *Planiliza subviridis*, and *Leignathus equula*.. The diversity, evenness, and dominance index values ranged from 2.185 to 3.763, 0.630 to 0.781, 0.072 to 0.233, respectively which indicated moderate-high diversity and none was dominant. Based on the IUCN status, generally the fish found were least concern (LC), but there were also one species included in the vulnerable (VU) (*Oreachromis mossambicus*, one species included Endangered (EN) *Maculabastis gerrardi*), and two species in the near threatened (NT) (*Brevitrygon cf walga*, *Scomberomorus commerson*). Therefore, fishing activities in this area are also important to maintain these species, in the context of conserving fishery resources

References

1. Rustam, N.S. Adi, E. Mustikasari, T.L Kepel, M.A. Kusumaningtyas. *Jurnal Segara* **14**, 3 (2018)
2. Afdal S.H., Riyono. *Oseanologi dan Limnnologi.* **33** (2007)
3. M. Simanjuntak. *Ilmu Kelautan* **12**, 3 (2007)
4. H.S. Sugiarti, S.H. Nasution. *Hubungan panjang berat ikan belanak (Mugil cephalus) di tiga muara sungai di Teluk Banten.* Prosiding Seminar Ikan ke 8. Bogor, Indonesia (2013)
5. J.W. Wisha, S. Husrin, J. Prihantono. *Ilmu Kelautan* **20**, 2 (2015)
6. S.S. Thayib, D.H. Kunarso. *Penelitian bakteri indikator pencemar di perairan Banten dan sekitarnya.* Prosiding Pemantauan Pencemaran Laut. 07-09 Februari 1994, Jakarta, Indonesia (1994)
7. M.C. Mainassy. *Jurnal Perikanan Universitas Gadjah Mada.* **19**, 2 (2017)
8. N.K. Wardani, T. Prartono, Sulistiono. Jurnal Pendidikan IPA Indonesia. **9**, 4 (2020)
9. E.P. Surbakti, A. Iswantari, H. Effendi, Sulistiono. *Distribution of dissolved heavy metals Hg, Pb, Cd, and as in Bojonegara Coastal Waters, Banten Bay.* IOP Conf. Series: Earth and Environmental Science. Bogor, Indonesia (2021)
10. N. Fitriah, M.F. Rahardjo, R. Affandi, Sulistiono, M. Wildan, C.P.H. Simanjuntak. *Reproductive aspects of greenback mullet Planiliza subviridis (Valenciennes, 1836) in Cengkok coastal waters, Banten Bay Indonesia.* IOP Conf. Series: Earth and Environmental Science. Jakarta, Indonesia (2021)
11. H.S. Sugiarti, S.H. Nasution, Sulistiono. *Estuarine fish community structure in Banten Bay Indonesia.* IOP Conf. Series: Earth and Environmental Science, Bogor, Indonesia (2021)
12. N. Febrianessa, Sulistiono, A.M. Samosir, M. Yokota. 2020. Indonesian Journal of Marine Sciences, **25**, 4 (2020)
13. E. Noviani, Sulistiono, A.M. Samosir. Omni-Akuatika, **16**, 2 (2020)

14. T. Melinda, A.M. Samosir, Sulistiono, C.P.H. Simanjuntak. *Bioaccumulation of lead (Pb) and mercury (Hg) in green mussel Perna viridis (Linnaeus, 1758) in Cengkok Coastal Waters, Banten Bay, Indonesia.* IOP Conf. Series: Earth and Environmental Science. Bogor, Indonesian (2021)
15. A. Dinulislam, Sulistiono, D.T.F. Lumbanbatu, R. Affandi. *Heavy metals (Pb, Hg) in blood cockle (Anadara granosa) in Cengkok Waters, Banten Bay, Indonesia.* IOP Conf. Series: Earth and Environmental Science. Bogor. Indonesia (2021)
16. G. Allen. Marine Fishes of South East Asia. Perth (AU): Western Australian Museum (1997)
17. M. Kottelat, A.J. Whitten, Kartikasari S.N., Wirjoatmo S. Frehswater fishes of western Indonesia and Sulawesi. *Periplus Edition Protek EMDI.* Jakarta. 377 (1993)
18. [FAO] Food and Agriculture Organization of the United Nations. FAO Species Identification Guide for Fishery Purposes. The Living Marine Resources of the Western Central Pacific. Carpenter KE, Niem VH, editor. Volume III-VI. Roma (IT): FAO. (2001)
19. R.H. Kuiter, T. Tonozuka. Pictorial guide to Indonesian reef fishes. Part 1-3. Australia: Zoonetics (2001)
20. G.R. Allen, M.V. Erdmann. Reef Fishes of the East Indies. Volume 1. Tropical Reef Research, Perth, Australia. p 1-424 (2012).
21. C.J. Krebs. Ecological Methodology. Harper Collins Publisher. New York. 649p (1989)
22. E.P. Odum. Dasar-dasar Ekologi. Terjemahan Tjahjono Samingan. Edisi Ketiga. Yogyakarta: Gadjah Mada University Press (1993)
23. E.P. Odum. Dasar-Dasar Ekologi. Diterjemahkan oleh Tjahjono Samingan. Edisi Ketiga. Gadjah Mada University Press. Yogyakarta. (1998)
24. G. Wahyudewantoro. Zoo Indonesia. **18**, 2 (2009)
25. N. Mote, S. Ayarau, S. Elviana. Aquatic Science. **1**,1 (2019)
26. K. Fatah, A.K. Gaffar, Rupawan. Jurnal Protein, **14**, 2(2006).
27. G. Wahyudewantoro. Zoo Indonesia. **21**, 1(2012)
28. E. Prianto, N.K. Suryati. Jurnal Penelitian Perikanan Indonesia. **16**, 1 (2009)
29. S.A. Wardoyo, M. Iqbal. Jurnal Ilmu Ilmu Perikanan dan Budidaya Perairan (JIPBP). **1**, 1(2003)
30. A. Zahid, C.P.H. Simanjuntak, M.F. Rahardjo, Sulistiono. Jurnal Iktiologi Indonesia. **11**, 1 (2011)
31. Sulistiono, T. Nugroho, I. Suyatna, A. Laga. Status keanekaragaman ikan estuari dan upaya pengelolaanya di area Delta Kayan-Sembakung, Kalimantan Timur. Departemen Manajemen Sumberdaya Perairan, Fakultas Perikanan dan Ilmu Kelautan, Institut Pertanian Bogor. Bogor (2021)

Table 1. Diversity of fish found in Banten Bay

No	Ordo	Family	Species	Cengkok (C)	Bojonegara (B)	Kemayungan-Linduk (KL)
1	Acanthuriformes	Scatophagidae	*Scatophagus argus*	1	13	41
2		Siganidae	*Siganus javus*		31	
3	Albuliformes	Albulidae	*Albula argentea*		1	
4	Anabantiformes	Channidae	*Channa striata*			4
5			*Channa gachua*		3	
6	Anabantiformes	Osphronemidae	*Trichopodus trichopterus*	2	60	
7	Atheriniformes	Atherinidae	*Atherinomorus lacunosus*	2		
8			*Hypoatherina temminckii*			15
9	Aulopiformes	Synodontidae	*Synodus indicus*		1	
10	Batrachoidiformes	Batrachoididae	*Allenbatrachus grunniens*		4	
11	Beloniformes	Belonidae	*Strongylura incisa*			11
12			*Strongylura leiura*		3	
13		Hemiramphidae	*Hyporhamphus dussumieri*			5
14			*Hyporhamphus quoyi*		1	
15	Carangiformes	Carangidae	*Alepes kleinii*			4
16			*Atule mate*	4		
17			*Caranx sexfasciatus*	2	2	
18			*Scomberoides commersonnianus*		5	
19			*Scomberoides tala*		12	
20			*Carangoides malabaricus*			2
21			*Carangoides praeustus*			3
22	Cichliformes	Cichlidae	*Oreochromis mossambicus*		3	
23			*Oreochromis niloticus*	2	11	1
24	Clupeiformes	Chirocentridae.	*Chirocentrus dorab*			2
25		Clupeidae	*Anodonstoma chacunda*		35	1
26			*Dussumieria elopsoides*		1	
27			*Hilsa keele*		1	1
28			*Sardinella albella*			6
29			*Sardinella fimbriata*		3	
30			*Sardinella gibbosa*		1	
31			*Escualosa thoracata*	6		
32		Engraulidae	*Thryssa hamiltoni*		3	
33			*Thryssa mystax*			53
34	Cypriniformes	Cyprinidae	*Barbonymus gonionotus*			6
35			*Osteochilus vittatus*			1
36	Cyprinodontiformes	Poeciliidae	*Poecilia latipinna*		4	
37			*Xiphoporus helleri*		1	
38	Elopiformes	Elopidae	*Elops hawaiensis*	1		1
39		Megalopidae	*Megalops cyprinoides*			2
40	Gobiiformes	Butidae	*Butis butis*	6		4
41			*Oxyeleotris marmorata*			3
42		Eleotridae	*Ophiocara porocephala*			1
43		Gobiidae	*Acentrogobius caninus*	4		

44			*Acentrogobius viridipunctatus*			2
45			*Glossogobius giuris*	16	1	16
46			*Odontamblyopus rubicundus*			1
47			*Parapocryptes serperaster*			3
48			*Oxuderces dentatus*	10		
49		Oxudercidae	*Boleophthalmus boddarti*		1	
50	Gonorynchiformes	Chanidae	*Chanos chanos*	2	16	1
51	Mugiliformes	Mugilidae	*Planiliza planiceps*	6		
52			*Mugil cephalus*		3	
53			*Osteomugil cunnesius*			23
54			*Planiliza subviridis*	26	27	63
55			*Ellochelon vaigiensis*	153	42	
56	Mulliformes	Mullidae	*Upeneus quadrilineatus*	1		
57			*Upeneus tragula*	3		
58	Myliobatiformes	Dasyatidae	*Brevitrygon* cf *walga*		7	
59			*Maculabastis gerrardi*		1	
60	Perciformes	Ambassidae	*Ambassis cf kopsii*	61		
61		Gerreidae	*Gerres* cf *limbatus*		1	
62			*Gerres filamentosus*		7	
63			*Gerres longirostris*		2	
64			*Gerres oyena*	6	1	
65			*Pentaprion longimanus*	1		
66		Haemulidae	*Pomadasys argenteus*	3	6	
67			*Pomadays kaakan*		1	
68		Lactariidae	*Lactarius lactarius*			1
69		Latidae	*Lates calcalifer*			7
70		Lutjanidae	*Lutjanus ehrenbergii*	1		
71			*Lutjanus russelli*		5	
72		Polynemidae	*Eleutheronema tetradactylum*		3	53
73			*Polydactylus plebeius*	9		
74		Serranidae	*Ephinephelus coioides*			2
75		Sphyraenidae	*Sphyraena barracuda*		4	
76			*Sphyraena jello*			2
77		Sciaenidae	*Dendrophysa ruselli*		2	
78			*Johnius amblycephalus*	3		
79			*Johnius belangerii*			19
80			*Johnius heterolepis*	2		
81			*Johnius trachycephalus*		1	
82			*Otolithes ruber*			1
83			*Nibea soldado*		15	2
84	Acanturiformes	Leiognathidae	*Eubleekeria splendens*			3
85			*Nuchequula gerreoides*		12	
86			*Leiognathus equula*	4	101	
87	Scombriformes	Trichiuridae	*Trichiurus lepturus*			1
88	Centrarchiformes	Terapontidae	*Terapon jarbua*	1	4	

https://doi.org/10.1051/e3sconf/202233903003

89			*Tetraodon lineatus*			6
90			*Lagocephalus lunaris*		1	
91	Tetraodontiformes	Tetraodontidae	*Chelonodon patoca*	1	1	
92			*Takifugu oblongus*		4	
93	Anabantiformes	Anabantidae	*Anabas testudineus*		34	2
94		Psettodidae	*Psettodes erumei*		1	
95	Pleuronectiformes		*Cynoglossus bilineatus*		2	
96		Cynoglossidae	*Cynoglossus* cf *lida*	2		
97			*Cynoglossus cynoglossus*			1
98			*Rastrelliger brachysoma*			2
99		Scombridae	*Scomberomorus commerson*			24
100	Scombriformes		*Scomberomorus guttatus*		1	
101		Stromateidae	*Pampus argenteus*			1
102			*Pampus chinensis*			4
103	Scorpaeniformes.	Platycephalidae	*Platycephalus indicus*			1
104			*Arius arius*		1	
105		Ariidae	*Arius maculatus*			31
106	Siluriformes		*Plicofollis cf argyropleuron*		1	
107			*Hexanematichthys sagor*	6	1	2
108		Loricariidae	*Pterygoplichthys pardalis*	3	1	
109		Bagridae	*Mystus gulio*		12	10
110	Tetraodontiformes	Triacanthidae	*Pseudotriacanthus strigilifer*			1
111			*Triacanthus biaculeatus*		1	
			Taxon Number	**32**	**58**	**51**
			Abundance	**350**	**523**	**452**
			Diversity Index	**2.185**	**3.763**	**3.055**
			Evenness Index	**0.630**	**0.763**	**0.781**
			Dominance Index	**0.233**	**0.078**	**0.072**

https://doi.org/10.1051/e3sconf/202233903004

Diversity of marine plankton in coral reef ecosystems at Gosong Island, Southwest Aceh

Nurul Najmi[1], *Mai* Suriani[2], *Mira* M Rahmi[1], and *Ananingtyas* S Darmarini[1*]

[1]Department of Aquatic Resources, Faculty of Fisheries and Marine Science, Teuku Umar University, Jl. Alue Peunyareng, Meulaboh 23681, West Aceh, Indonesia.
[2]Department of Marine Science, Faculty of Fisheries and Marine Science Teuku Umar University, Jl. Alue Peunyareng, Meulaboh 23681, West Aceh, Indonesia

Abstract. Plankton is a fundamental part of the waters, especially in coral reef ecosystems. This study aims to determine the abundance and diversity of plankton species, both phytoplankton, and zooplankton in the waters of Gosong Island, Southwest Aceh. This research was conducted in May 2021. Samples of plankton were collected from 4 sampling points with 4 replications. Phytoplankton consists of 51 species which are in 3 families, namely Cyanophyceae, Bacillariophyceae, and Dinophyceae. There are 17 species of zooplankton consisting of 7 families, namely Protozoa, Crustaceans, Urochordata, Polychaeta, Gastropods, Pelecypods, and Nematodes. The total abundance of phytoplankton ranged from 292,826 to 2,661,738 cells/m^3, while the total abundance of zooplankton ranged from 4,957 to 32,870 ind/m^3. The composition of the presence found at the study site was found to be higher in Cyanophyceae than Bacillariophyceae and the lowest number was found in Dinophyceae. Zooplankton is found in large to low numbers in the Crustacea, Protozoa, Urochordata, Gastropod, Pelecypoda, Polychaeta, and Nematoda. This research is able to provide an initial description of diversity, abundance, and several biological indices for follow-up research. Gosong Island is an island that has coral reefs, so it is necessary to carry out routine monitoring related to the type and abundance of plankton regularly.

1 Introduction

Plankton is the base of marine food webs [1,2] and is a staple food source in coral reef ecosystems [1, 3]. In addition, the abundance of plankton has an important role in the trophodynamics and metabolism of reefs [4]. This role is very important for the sustainability of fisheries and marine life [2]. Waters rich in resources have a high fertility rate. This can be seen from the productivity of the waters, including the abundance and diversity of plankton [5].

The role of phytoplankton is very important because it carries out the process of photosynthesis to produce oxygen in the waters [6, 7]. Phytoplankton biodiversity is very important for ecosystem stability and marine biogeochemistry [8], they are involved in trophic and biogeochemical dynamic and recruitment processes [3]. Phytoplankton grazing is an important component in benthic-pelagic connections on coral reefs [9]. In addition, phytoplankton is also a biological carbon pump, this is the physiological walk-on-part of phytoplankton and community structure [10]. The major concern is that phytoplankton is the basis of marine food webs and a fundamental food source in coral reef ecosystems [1].

In addition to phytoplankton, zooplankton also plays a role in coral reef ecosystems. Its important assignment is as part of the dynamic trophic [11]. According to [12] zooplankton has a close relationship with coral reef food webs and is a source of food for corals. Even [13] stated that most of the zooplankton live on coral reefs. As a food source, zooplankton is consumed by corals as they pass through coral reefs [14, 11].

Gosong Island is located in Susoh District, Southwest Aceh Regency, and is one part of the KKPD (Regional Water Conservation Area) of Southwest Aceh. Southwest Aceh Regency was declared a regional waters conservation area through the Aceh Governor's Decree Number 523/1297/2018 in the form of Water Nature Reserve (SAP). Based on Government Regulation No. 60 of 2007, Water Nature Reserve is a water conservation area with certain characteristics to protect the diversity of fish species and their ecosystems. As a marine conservation area, Gosong Island has a large coral reef ecosystem potential that can be found almost around the island.

The waters of Southeast Asia have very diverse habitats, including coral reefs, so that research on the abundance and composition of zooplankton is very likely to be carried out for extensive studies [15]. In addition, little is known about the diversity of plankton in the water column around coral reefs [16]. This reason strengthens the initial studies related to plankton in this area. Research about the diversity, abundance, and composition of plankton on Gosong Island has never been done before. Therefore, it is necessary to conduct

*Corresponding author: ananingtyas.s@gmail.com

 https://doi.org/10.1051/e3sconf/202233903004

preliminary research on the diversity of marine plankton on Gosong Island, Southwest Aceh.

2 Material and Methods

2.1 Study area

Samples were collected from four stations (Figure 1) on Gosong Island, Southwest Aceh. Station 1 (St 1) was located in 096°49'39.2"E and 03°41'48.5"N, station 2 (St 2) in 096°49'29.33"E and 03°41'52.87"N, station 3 (St 3) 096°49'37.3"E and 03°41'45.5"N, and station 4 (St 4) were situated in 096°49'33.09" E and 03°4'54.91" N.

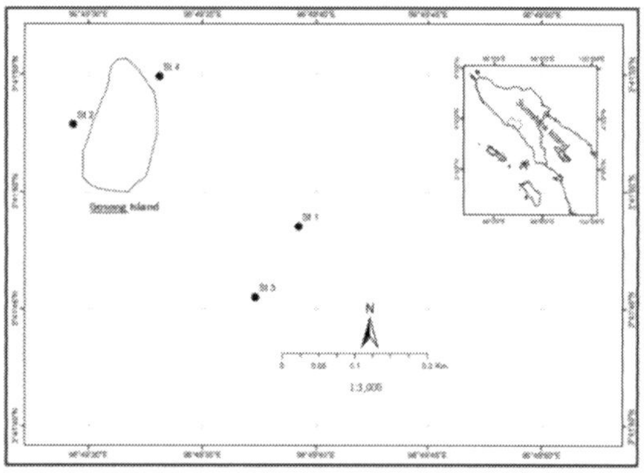

Fig. 1. Sampling station in Gosong Island, Southwest Aceh

2.2 Plankton sampling and identification

The sample was collected in May 2021. Plankton collection is carried out once at each station with four replications. One hundred litters of seawater from each station were filtered by using a plankton net with a 30 µm mesh size. Samples were preserved in a 100 ml plastic bottle by 1 ml of Lugol. 1 ml of preserved samples was used for identification and counting on a Sedgwick Rafter Counting using a binocular microscope Nikon Eclipse 100. The species were identified in Aquatic Productivity Laboratory, Department of Aquatic Resources, Faculty of Fisheries and Marine Science, Teuku Umar University. Identification of plankton based on [17]. An abundance of plankton refers to [18]. The diversity index (H') and evenness (E) of marine plankton in Gosong Island were determined based on the Shannon-Weaner Index and the dominance (D) index was approach with Simpson's index.

3 Result and Discussion

3.1 Plankton Compossition

The results of the analysis showed that the study site had 51 phytoplankton species consisting of 3 classes, namely Cyanophyceae (2 species), Bacillariophyceae (38 species), and Dinophyceae (11 species) (Table 1). The three classes found in the research location are lower than the studies conducted in the eastern waters of Aceh, namely 5 classes with 56 species [19]. Of the 3 classes found at the research site, Bacillariophyceae is a class with a diverse number of species, namely 38 species. This shows that this class is abundant based on its type. While the Dinophyceae and Cyanophyceae classes appear less.

Based on the type, there were 19 species found at all stations, namely: *Trichodesmium* sp., *Amphora* sp., *Asterionella* sp., *Bacillaria* sp., *Bacteriastrum* sp., *Biddulphia* sp., *Campylodiscus* sp., *Chaetoceros* sp., *Climacosphenia* sp., *Coscinodiscus* sp., *Fragilaria* sp., *Licmophora* sp., *Navicula* sp., *Nitzschia* sp., *Rhabdonema* sp., *Ceratium* sp., *Dinophysis* sp., *Orinthocercus* sp., and *Peridinium* sp. On the other hand, some species were only found at one particular station, including *Richelia* sp., *Achnanthes* sp., *Amphiprora* sp., *Asteromphalus* sp., *Cyclotella* sp., *Ditylum* sp., *Hyalodiscus* sp., *Lauderia* sp., *Mostogloia* sp., *Skeletonema* sp., *Gymnodinium* sp., *Gonyodoma* sp., and *Pyrocystis* sp.

Table 1. Phytoplankton diversity in Gosong Island, Southwest, Aceh

No	Genus/Station	1	2	3	4
	CYANOPHYCEAE				
1	*Richelia* sp.	+	-	-	-
2	*Trichodesmium* sp.	+	+	+	+
	BACILLARIOPHYCEAE				
3	*Achnanthes* sp.	-	-	+	-
4	*Amphiprora* sp.	-	+	-	-
5	*Amphora* sp.	+	+	+	+
6	*Asterionella* sp.	+	+	+	+
7	*Asterolampra* sp.	-	-	+	+
8	*Asteromphalus* sp.	-	+	-	-
9	*Bacillaria* sp.	+	+	+	+
10	*Bacteriastrum* sp.	+	+	+	+
11	*Biddulphia* sp.	+	+	+	+
12	*Campylodiscus* sp.	+	+	+	+
13	*Chaetoceros* sp.	+	+	+	+
14	*Climacosphenia* sp.	+	+	+	+
15	*Cocconeis* sp.	+	+	-	+
16	*Coscinodiscus* sp.	+	+	+	+
17	*Cyclotella* sp.	-	-	-	+
18	*Diatom* sp.	+	-	+	+
19	*Diploneis* sp.	-	+	+	+
20	*Ditylum* sp.	+	-	-	-
21	*Fragilaria* sp.	+	+	+	+
22	*Hemiaulus* sp.	-	-	+	+
23	*Hyalodiscus* sp.	-	-	-	+
24	*Lauderia* sp.	-	+	-	-
25	*Leptocylindrus* sp.	+	+	-	-
26	*Licmophora* sp.	+	+	+	+
27	*Melosira* sp.	-	+	+	+
28	*Mostogloia* sp.	-	-	-	+
29	*Navicula* sp.	+	+	+	+
30	*Nitzschia* sp.	+	+	+	+
31	*Pinnularia* sp.	+	+	-	-
32	*Pleurosigma* sp.	-	+	+	+
33	*Podocystis* sp.	-	+	-	+

34	*Rhabdonema* sp.	+	+	+	+
35	*Rhizosolenia* sp.	+	+	-	+
36	*Skeletonema* sp.	+	-	-	-
37	*Surirella* sp.	-	+	+	+
38	*Thalassionema* sp.	-	+	-	+
39	*Thalassiothrix* sp.	+	+	+	-
40	*Triceratium* sp.	+	+	+	+
DINOPHYCEAE					
41	*Ceratium* sp.	+	+	+	+
42	*Dinophysis* sp.	+	+	+	+
43	*Diplopsalis* sp.	+	+	+	-
44	*Gymnodinium* sp.	-	+	-	-
45	*Gonyodoma* sp.	+	-	-	-
46	*Orinthocercus* sp.	+	+	+	+
47	*Peridinium* sp.	+	+	+	+
48	*Podolampas* sp.	-	+	+	+
49	*Prorocentrum* sp.	+	+	+	-
50	*Pyrophacus* sp.	+	+	-	-
51	*Pyrocystis* sp.	-	+	-	-

Cyanophyceae class is a type of class that is only composed of 2 species and one of them is found in all stations, namely Trichodesmium sp., although the number of species in the research location is only a few, it needs attention. Trichodesmium sp. including Cyanobacteria species based on [20] have a role in photosynthesis and nitrogen fixation in lakes and seas. In addition, the genus Trichodesmium is a major source of marine nitrogen and the nitrogen cycle [21] but has the potential to cause blooms [20].

Bacillariophyceae (Diatom) is the class with the highest number of species found in the study area. Several studies have shown that this class is indeed found with various species in marine [22,23], has an important ecological role [21], and contributes remarkably to total global primary [24, 25]. The most common species found from several reports illustrate that this species is dominantly found [26, 27]. The result of the new research the phytoplankton diversity of this community is influenced by geographical factors that do not depend on environmental conditions [28].

The types of zooplankton found in Gosong Island were 26 species consisting of 7 Groups, namely Protozoa (11 species), Crustacea (10 species), Urochordata, Gastropods, Pelecypoda, Polychaeta, and Nematoda 1 species each (Table 2). The number of species found in the study area was less than the results of research in aquaculture areas [29] and seagrass habitats [30]. Table 2 shows some types of zooplankton found at all stations are *Stenosemella* sp., *Nauplius* (stadia), *Oithona* sp., and *Paracalanus* sp.. Some types of zooplankton are also found only at certain stations including *Didinium* sp., *Favella* sp., *Leprotintinnus* sp., *Protorha bdonella* sp., *Corycaeus* sp., *Temora* sp. Larva Pelecypoda, Larva Polychaeta, and Nematoda.

Several types of zooplankton from the Protozoa group found at the research site can be key species in an ecosystem. This is in line with the opinion [31] which states that *Stenosemella nivalis*, *S. ventricosa*, and *Tintinnopsis* sp. are key species in the Northern Adriatic Sea and the loss of this species is a sign of structural changes in the planktonic system.

Table 2. Zooplankton diversity in Gosong Island Southwest, Aceh.

No	Genus/Station	I	II	III	IV
PROTOZOA					
1	*Codonellopsis* sp.	+	+	-	+
2	*Didinium* sp.	+	-	-	-
3	*Favella* sp.	+	-	-	-
4	*Globigerinoides* sp.	-	-	-	+
5	*Globorotalia* sp.	-	+	+	+
6	*Leprotintinnus* sp.	-	+	-	-
7	*Pleurspis* sp.	+	+	-	-
8	*Protorhabdonella* sp.	+	-	-	-
9	*Rhabdonella* sp.	+	+	-	-
10	*Stenosemella* sp.	+	+	+	+
11	*Tintinnopsis* sp.	+	+	+	-
CRUSTACEAE					
12	*Acrocalanus* sp.	+	+	-	-
13	*Corycaeus* sp.	-	+	-	-
14	*Euterpina* sp.	+	-	+	+
15	*Microsetella* sp.	+	+	-	+
16	*Nauplius* (stadia)	+	+	+	+
17	*Oithona* sp.	+	+	+	+
18	*Oncaea* sp.	+	+	-	-
19	*Paracalanus* sp.	+	+	+	+
20	*Temora* sp.	-	+	-	-
21	*Tigriopus* sp.	-	-	+	+
UROCHORDATA					
22	*Oikopleura* sp.	+	+	-	-
GASTROPODA					
23	Larva Gastropoda (sp1)	+	+	-	+
PELECYPODA					
24	Larva Pelecypoda (sp1)	+	-	-	-
POLYCHAETA					
25	Larva Polychaeta (sp1)	-	+	-	-
NEMATODA					
26	Nematoda (sp1)	-	-	-	+

3.2 Abundance and Structure Community

The results of the analysis of community structure (Table 3) show conditions of abundance, diversity index, uniformity, and dominance in the waters around Gosong Island. The highest abundance of phytoplankton was found at station IV, which was 2661738 cells/m^3. The highest abundance by species at the study site was Cyanophyceae: *Trichodesmium* sp. (1436706.5 cell/m^3), Bacillariophyceae: *Rhabdonema* sp. (101217.3 cell/m^3), *Chaetoceros* sp. (18510.74 cell/m^3), and *Nitzchia* sp. (17888.6 cells/m^3), *Biddulphia* sp. (9152.2 cell/m^3), *Licmophora* sp. (8413.1 cell/m^3) and Dinophyceae: *Ceratium* sp. (13554.2 cell/m^3) and,

 https://doi.org/10.1051/e3sconf/202233903004

and *Peridinium* sp. (5489.1 cells/m^3). Forty-three other species have lower abundances. High abundance of *Trichodesmium* sp. presumably because the time of sampling is approaching the transition season so that its presence dominates. This condition is reinforced by the opinion of [32] which states that the abundance and distributionof Trichodesmium thiebautii and Trichodesmium eryhtraeum dominates the transition season. In addition, the abundance of phytoplankton is also influenced by the availability of nutrients, and nitrate enrichment can have an impact on the growth potential and large biomass accumulation [33].

Rhabdonema sp. is the second most abundant species, including the Tablelariaceae family of the Bacilliaophyceae class. This species is an epilithic diatom, apart from this type, there are also other epilithic diatoms with high waste, namely *Licmophora* sp. Both can be found abundantly in shallow areas or deep sites [34]. Other species found in abundance were *Biddulphia* sp., and *Chaetoceros* sp. this species is often found in abundance in several locations [35, 19]. The abundance, Diversity, Evenness, and Dominance index of Plankton in Gosong Island is presented in Table 3. Phytoplankton have an average diversity index is 1.03, the evenness index is 0.33 and the dominance index is 0.64. This value description shows an indication of the dominance of the presence of certain species. The diversity index at the study site is higher when compared to the Gerupuk Bay area, West Nusa Tenggara, Indonesia [35].

The highest abundance of zooplankton reached a value of 32.870 ind/m3 with an average diversity index of 1.4, evenness 0.64, and dominance 0.48. Zooplankton abundance was higher than zooplankton abundance in Lawas seagrass habitat, Sarawak, Malaysia [30]. However, it is lower than research in East Yapen Papisir water, Papua [36], Gulf of Aqaba, and Northern Red Sea [37]. The abundance and distribution of zooplankton are strongly influenced by currents [38]. The zooplankton diversity index on Gosong Island is lower than the 2018 study by [34] which reached 3.35 and [39], which reached 2.4-2.6. The index of uniformity and dominance of the zooplankton community describes a fairly good condition.

The relationship between phytoplankton and zooplankton is seen descriptively based on the results of the analysis, so there is a relationship between the presence of phytoplankton and zooplankton at each station. The linkage relationship is illustrated by if the number of phytoplankton is low, the zooplankton is also low. This shows that the abundance of zooplankton is influenced by the abundance of phytoplankton. In addition, the low number of zooplankton species can be caused by a lack of food sources in the waters. The adaptation made by zooplankton when there is a shortage of food is to look for other food alternatives. For example, Copepoda mentioned will consume terrigenous detritus as an additional food source when the number of phytoplankton is limited [40].

Table 3. Abundance, Diversity, Evenness and Dominance index of Plankton in Gosong Island Southwest, Aceh.

St	Plankton	Abundance	Diversity (H')	Evennes s (E)	Dominance (C)
I	Phytoplankton (sel/m^3)	2545782	0.53	0.18	0.82
II		1052172	1.32	0.41	0.55
III		292826	1.50	0.50	0.45
IV		2661738	0.75	0.23	0.74
I	Zooplankton (ind/m^3)	29,544	1.39	0.63	0.40
II		16,695	1.36	0.66	0.44
III		4,957	0.93	0.80	0.48
IV		32,870	0.87	0.48	0.59

The composition of the presence of plankton is presented in figures 2 a and b. Figure 2a illustrates that the abundance of phytoplankton in the Cyanophyceae class is higher than that of Bacillariophyceae and Dinophyceae. This composition is different from some published research results, such as the results of research by [41] in the Bali Strait, it was found that the composition of Bacillariophyceae offshore reached 85% and nearshore reached 78%. Likewise, the results of research by [19] in the coastal water of Lubuk Damar Aceh Tamiang found that the composition of Bacillariophyceae reached 85%.

The percentage of the various classes sorted largest to smallest: Cyanophyceae> Bacillariophyceae> Dinophyceae. The most common class of phytoplankton was Cyanophyceae with 87.89% (Fig. 2a). *Trichodesmium* sp. from Cyanophyceae was recorded as a dominating species at each station. Phytoplankton dominance by *Trichodesmium* sp. need to be careful because it can cause blooms of *Trichodesmium* sp. which is harmful to other organisms and disturbs the balance of the ecosystem.

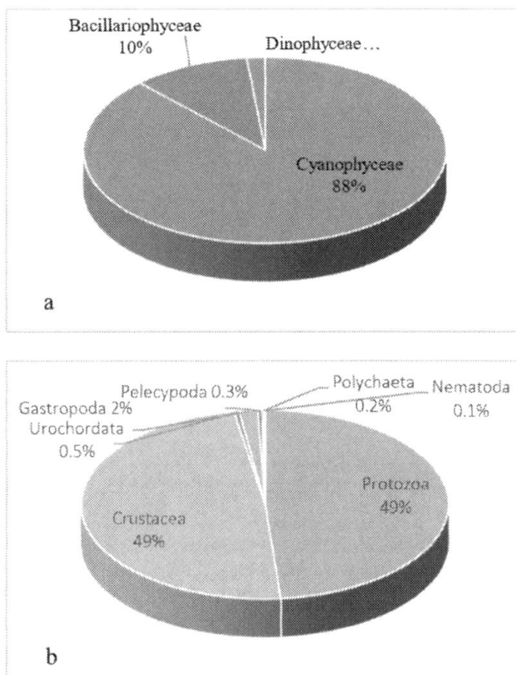

Fig. 2. Abundance of phytoplankton (a) and zooplankton (b) in Gosong Island water

The difference in class composition at different locations is thought to be due to different environmental conditions or the availability of nutrients to support the survival of certain phytoplankton. This is in line with the opinion of [42] which states that high abundance is due to greater adaptation to the environment compared to other classes. Figure 2b shows the composition of zooplankton with a balanced composition of Crustaceans and Protozoa while other species are less than 2%. The large composition of Crustacea and Protozoa classes is caused by the abundance of Nauplius (stadia) and *Codonellopsis* sp. Zooplankton composition comprised Crustacea> Protozoa> Urochordata> Gastropoda> Pelecypoda> Polychaeta> Nematoda. The most common groups found were Crustacea (48.67%) and Protozoa (48.62%). [43] said that overall, crustaceans dominate marine zooplankton. The abundance of crustacean groups is due to the species constituting crustaceans having a wide distribution of species because they can adapt to extreme environmental conditions [44]. The large abundance of crustacean species in the study area is the same as the results of research in Est Yapen Papisir water, Papua [36].

The different composition of each type of plankton is influenced by several factors, including the availability of food sources and environmental conditions. Because the availability of resources, in this case, is nutrition will explain the variation in biomass, productivity, and species richness, this is because each species has a monopoly on the availability of resources in the food web and the environment in this case is food quality and toxicity [45]. The composition of plankton species formed according to [46] plays a key role in the ongoing functioning of the ecosystem. This is due to the ecological and biological differences of phytoplankton characteristics that can be used as an assessment of ecosystem conditions [1]. Protecting the ecosystem is due to the dynamic and trophic relationship between different groups in this case phytoplankton-zooplankton-fish [47].

4 Conclusion

Phytoplankton composition consists of Cyanophyceae, Bacillariophyceae, and Dinophyceae. The most common class of phytoplankton was Cyanophyceae with 88%. Trichodesmium sp. was dominant at each station. Zooplankton consists of Crustacea, Protozoa, Urochordata, Gastropods, Pelecypoda, Polychaeta, and Nematoda. The most common groups of zooplankton found were Crustacea and Protozoa. The abundance of phytoplankton was higher than Zooplankton. The results of the study of marine plankton in the waters of Gosong Island which is a very important coral reef area because it is a food source. The abundance of phytoplankton in the study area will support the existence of zooplankton. Both can support the sustainability of coral in the area. So that it becomes a dynamic relationship between phytoplankton-zooplankton-coral reef. This initial research is the basis for conducting a more in-depth study of the relationship between the diversity of plankton species and the coral reef communities.

Acknowledgments. The author is to express for Pusong Diving Club (PDC) for its excellent facilities and to thank the anonymous reviewers for the worthwhile comments on the manuscript. The authors expresses gratitude to the Institute for Research and Community Service (LPPM) of Teuku Umar University which has funded research through an internal grant scheme. In addition, the authors also thank the Faculty of Fisheries and Marine Science, Teuku Umar University which has supported research and publication activities.

References

1. M. Racault, D.E. Raitsos, M.L. Berumen, R.J.W. Brewin, T. Platt, S. Sathyendranath, I. Hoteit, Remote Sensing of Environment, **160**, 222-234 (2015)
2. S.D. Batten, R. Abu-Alhaija, S. Chiba, M. Edwards, G. Graham, R. Jythibabu, J.A. Kitchener, P. Koubbi, A. McQuatters-Gollop, E. Muxagata, C. ostle, A.J. Richardson, K.V. Robinson, K.T. Takahashi, H.M. Verheye, W. Wilson, frontiers in Marine Science, **6** (2019)
3. A. Carrillo-Baltodano, A. Morales-Ramirez, Revista de Biología Tropical, **64**, 3 (2016)
4. Y.I. Sorokin, Plankton in Coral-Reef Waters. In: Coral Reef Ecology. Ecological Studies (Analysis and Synthesis). Springer Berlin (1995)
5. J.W. Nybakken, Biologi Laut suatu pendekatan ekologis. (Jakarta 1992)
6. F. Not, R. Siano, W.H.C.F. Kooistra, N. Simon, D. Vaulot, I. Probert, Advances in Botanical Research, **64** (2012)
7. F.J. Vincent, S. Colin, E. Scalco, L. Bitter, Y. Garcia, R.M.M. Lopes, J.R. Dolan, A. Zingone, C. de Vargas, C. Bowler, The ISME Journal, **12**,1094-1108 (2018)
8. S. Dutkiewicz, P. Cermeno, O. Jahn, M.J. Follows, A.E. Hickman, D.A.A. Taniguchi, B.A. Ward, Biogeosciences, **17**, 609-634 (2020)
9. G. Yahel, A.F. Post, K. Fabricius, D. Marie, D. Vaulot, A. Genin, Limnol. Oceanogr, **43**, 4 (1998)
10. S. Basu, K.R.M. Mackey, Sustainability, **10**, 869 (2018)
11. Heidelberg KB, O'Neil K.L, Bythell J C, Sebens KP. Journal of Plankton Research. **32**, 1 (2010)
12. O. Esqual-Garrote, A. Morales-Ramirez, Revista de Biologia Tropical **68** (suppl. 1) S261-S270 (2020)
13. A.L. Alldredge, J.M. King, Marine Biology, **41** (1977)
14. Hammer WM, Colin PL, Hamner PP. Marine Ecology Progress Series. **334**, 83-92
15. S. Nishida, N. Miyazaki, in Nishida S, Fortes MD, Miyazaki N eds. Coastal Marine Science in Southeast Asia-Synthesis Report of the Core University Program of the Japan. Society for the Promotion of Science: Coastal Marine Science (2001-2010) (2011)
16. J.K. Pearman, M.M. El-Sherbiny, A. Lanzen, A.M. Al Aidaroos, X. Irigoien, Frontiers in Marine Science **1**, 27 (2014) Congress on Advances in

Nuclear Power Plants, ICAPP, 2-5 May 2011, Nice, France (2011)
17. I. Yamaji, *Illustration of the Marine Plankton of Japan* Hoikusha Publishing Co. Ltd. Japan (1979)
18. E.W. Rice, R.B. Baird, A.D. Eaton, L.S. Clesceri, APHA: Standart Method for The Examination of Water and Wastewater 22th ed. AWWA (American Water Works Association) and WEF (water Environment Federation). Washington DC (US) (2012)
19. A.S. Darmarini, Y. Wardiatno, T. Prartono, K. Soewardi, D. Ardania, IOP Conf. Series, **674**, 012023 (2021)
20. B.A. Whitton, M. Potts, Ecology of Cyanobacteria II: Their Diversity in Space and time (Springer Science (2007)
21. L. Hoffman, Eur. J. Phycol, **34**: 371-379 (1999)
22. S. Malvia, E. Scalco, S. Audic, F. Vincent, A. Veluchamy, J. Poulain, P. Wincker, D. Iudicine, C. de Vargas, L. Bitter, A. Zingone, C. Bowler, PNAS, **29** (2016)
23. X. Lu, Y. Liu, Y. Fan, Water, **12** (2020)
24. A. Yool, T. Tyrell, Global Biogeochemical Cycles, **17**, 4 (2002)
25. S. Licea, J.L. Moreno-Ruiz, R.J. Luna, Biodivers Endanger Species, **4**,174 (2016)
26. D.Y. Wulandari, N.T.M. Pratiwi, E.M. Adiwilaga, Jurnal Ilmu Pertanian Indonesia, **19**, 3 (2014)
27. H. Wiyarsih, Endrawati, S. Sedjati, Buletin Oseanografi Marina, **8**, 1 (2019)
28. P. Vanormelingan, E. Verleyen, W. Vyverman, Biodiversity and Conservation, **17**, 2 (2009)
29. O.Z. Ahmad, H. Endrawati, N. Taufiq, Journal of Marine Research, **3**, 3 (2014)
30. J. Ismail, A.H.M. Kamal, M.H. Idris, S.M.N. Amin, H. Hamli, L.S. Sien, A. Al-Asil, M.H. Abualreesh, Biodiversity Data Journal, **9**, e67449 (2021)
31. M. Monti-Birkenmeier, T. Diociaitui, S.F. Umani, Nature Conservation, **34**, (2019)
32. A. Sediadi, Makara Sains, **8**, 1 (2004)
33. R.M. Kudela, Dugdale, Deep-Sea Research Part II: Topical Studies Oceanography, **47**, 5-6 (2000)
34. D. Hafner, A. Car, N. Jasprica, T. Kapetanovic, I.D. Radic, Medit Mar. Sci, **19**, 2 (2018)
35. Purnamaningtyas SE., Mujiyanto, and Riswanto. Jurnal Akuatika Indonesia **4**, 1 (2019)
36. K. Paiki, J.D. Kalor, E. Indrayani, L. Dimara, Maspari Journal, **10**, 2 (2018)
37. Mantha G, Al-Sofyani A.A, M. Ali A, Crosby M.P. Acta Oceanologica Sinica. **38**, 59-72 (2019)
38. R.A. Morales, M.M. Murillo, Rev. Biol. Trop, **44**, 2 (1996)
39. I. Noventalia, H. Endrawati, M. Zainuri, Journal of Marine research, **1(1)**:19-23 (2012)
40. S. Hu, Z. Guo, C. Xu, H. Huang, S. Liu, J. Lin. Plankton Res, **37**, 2 (2015)
41. N.T.M Pratiwi, D.Y. Wulandari, I.P. Ayu, A. Iswantari, IOP Conf. Series: Earth and Environmental Science, **54**, 012090 (2020)
42. E.P. Odum, *Fundamental of Ecology. Philadelphia*: Ed. W.B. Saunders (1971)
43. J.W. Nybakken, M.D. Bertness, *Marine Biology* Pearson/Benjamin Cummings Science (2005)
44. K. Romimohtarto, S. Juwana, Biota Laut: Ilmu Pengetahuan Tentang Biota Laut, Jakarta (2004)
45. S. Lehtinen, T. Tamminen, R. Ptacnik, T. Andersen, Lomnol. Oceanogr, **62**:1393-1408 (2017)
46. J. Otero, X.A. Alvarez-Salgado, A. Bode, Front. Mar. Sci. (2020)
47. S. Lomartire, J.C. Marques, A.M.M. Goncalves, Ecological Indicator, **129**, 11:107867 (2021)

Community structure of mangrove in Lambadeuk village, Aceh Besar regency, Indonesia

Chitra Octavina[1,2,3*], *Maria* Ulfah[1,2,3], *Rauzatul* Sakinah[1], *Sayed* Abdul Azis[4,5], *Nanda* Muhammad Razi[4,5], *Muhammad* Rizki Fazillah[1,4], and *Muhammad* Agustiar[1,4]

[1]Department of Marine Sciences, Faculty of Marine and Fisheries, Universitas Syiah Kuala, Indonesia
[2]Research Center for Marine and Fisheries, Universitas Syiah Kuala, Indonesia
[3]Marine Biology Laboratory, Faculty of Marine and Fisheries, Universitas Syiah Kuala, Indonesia
[4]Ocean Diving Club, Faculty of Marine and Fisheries, Universitas Syiah Kuala, Indonesia
[5]Masters Program in Integrated Coastal Resource Management, Postgraduate Program, Universitas Syiah Kuala, Indonesia

Abstract. Mangroves are plants that can live in tidal areas. The study aims to determine the condition of the mangrove ecosystem through a mangrove community structure approach in Lambadeuk Village, Peukan Bada District, Aceh Besar District. This research was conducted in November 2017 using the stratified sampling method. The results showed that there were only 2 types of mangrove species found in Lambadeuk Village, Peukan Bada District, Aceh Besar District, the first is *Rhizophora apiculata* as many as 36 individuals and the second one is *Rhizophora mucronata* as many as 116 individuals. The important value index (INP) tree and sapling of *Rhizophora apiculata* at station 1 and station 3 of 253.655% and station 2 of 183.789%. While seedling of *Rhizophora apiculata* at station 1 and station 3 of 250.04% and station 2 of 175.02%. The important value index (INP) tree and sapling of *Rhizophora mucronata* at station 1 and station 3 of 253.655%. While seedling of *Rhizophora mucronata* at station 1 and station 3 of 250.04%. The diversity (H') obtained is low where H' 2.0, dominance (C) is low where 0 < C 0.5 and evenness (E) is also low where 0 < E 0.5.

1 Introduction

The mangrove ecosystem is a vegetation in the tropics invalue as a unique ecosystem, which can live in water with high salinity, and are usually found along the tidal areas [7]. Mangrove forests are the main ecosystem that supports life in coastal and marine areas. The mangrove ecosystem has important ecological function as a place habitat (feeding ground, nursey ground), (spawning ground) for a variety of marine organisms, as well as a carbon sink [16]. Ecosystem also has a physical function as anchoring coastal erosion, protecting against wind and tsunami, absorbing waste, preventing water intrusion sea. In addition, the mangrove forests also contribute economically high as providers of wood, medicines, tools and fishing locations [6].

Aceh Besar District is an area located at the northern tip of Aceh Province with a coastline of 344 km which is rich in biodiversity [5], one of which is mangroves. The potential of mangrove resources can be founded in Lambadeuk Village, Peukan Bada District, Aceh Besar Regency. But most of the mangrove area in the village have been damaged due to the impact and Tsunami disaster in 2004, so that rehabilitation efforts are managed by the government. According to [8], the rehabilitation program is a solution in re-functioning the damaged mangrove ecosystem. In fact, the rehabilitation efforts carried out have not been maximal.

Awareness of local communities in conserving the coastal environment is still lacking, waste disposal and pond waste are not managed properly. In addition, there are no regulations to prohibit community activities that can damage mangrove plants, such as fishpond in the mangrove rehabilitation area. The research reporting on the condition of mangroves in Lambadeuk Village, Peukan Bada District is still limited. A related study reported by [2] who studied the spatial distribution of mangroves using a geographic information system. However, research has never been done at the location. There are some observations in peukan bada district. Therefore, this study aims to determine the condition of the ecosystem mangrove through a community structure approach in Lambadeuk Village, Peukan Bada District, Aceh Besar District.

2 Materials and Methods

2.1 Time and Place

The research was conducted in November 2017 in Lambadeuk Village, Peukan Bada District, Aceh Besar District. Sample identification was carried out at the

* Corresponding author: chitraoctavina@unsyiah.ac.id

https://doi.org/10.1051/e3sconf/202233903005

Marine Biology Laboratory, Faculty of Marine and Fisheries, USK. Figure 1.

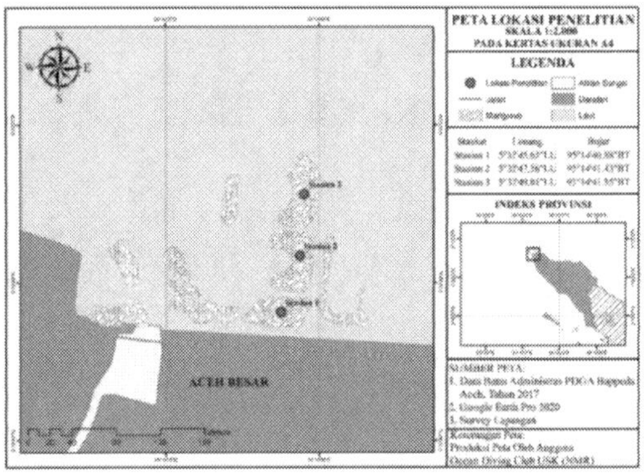

Fig. 1. Map of research locations

2.2 Determination of research Locations

The research locations were determined using the stratified sampling method, based on the needs and representation of the research area as a whole. Designated study area grouped into three observation in which stations are Station 1 area of mangrove vegetation was seaward Station 2 and Station 3 landward. Taking coordinates were marked by using GPS (Global Positioning System).

2.3 Data Collection

The stations were transected perpendicularly from sea to land. Stations are grouped into three plots. Mangrove vegetation is observed and recorded the number and type of stem diameter measured by each vegetation category. Observations on 10x10 m^2 plots for mangrove vegetation trees with measuring high categories> 1.5 m and diameter> 10 cm. Plot 5x5 m^2 for the mangrove tree category with a height of 1.5 m and a diameter of <10 cm. While plot 1x1 m^2 for the category of mangrove seeds with high stem <1.5 m [11].

2.4. Data Analysis

2.4.1 Species Density (Di) and Relative Density (RDi)

Density and relative density of mangroves were calculated based on the equations of [9] as follows:

$$\mathrm{Di} = \frac{Ni}{A} \quad (1)$$

Description:
Di : Density of species i (ind/ m^2)
Ni : Total number of individuals of species i (ind)
A : Total area of sampling (m^2)

$$\mathrm{RDi} = \frac{ni}{\Sigma n} \mathrm{x}100\ \% \quad (2)$$

Information :
RDi: Relative Density (%)
n : Total number of individuals (ind)

2.4.2 Species Frequency (Fi) and Relative Species(RFi)

The Frequency of species and the frequency of mangrove species is calculated based on the equation [9] as follows:

$$\mathrm{Fi} = \frac{Pi}{\Sigma p} \quad (3)$$

Description:
Fi :Frequency Type i (ind)
pi :Number of sample plots where the i-th type is found
p :Total number of sample plots made

$$RFi = \frac{Fi}{\Sigma F} \mathrm{x}100\ \% \quad (4)$$

Note:
RFi : Relative Frequency (%)
p : Total frequency for all species (ind)

2.4.3 Species Coverage (Ci) and Relative Species (RCi) Species

Cover and relative mangrove species cover are calculated based on the [9] equation as follows:

$$Ci = \frac{\Sigma BA}{A} \quad (5)$$

Description:
Ci: The extent of the type of to-i (m^2)

$$BA = \frac{\mathrm{DBH2}}{4} \quad (6)$$

π: 3.14

$$RCi = \frac{Fi}{\Sigma F} \mathrm{x}100\ \% \quad (7)$$

Description:
RCI: Percentage of mangrove cover (%)
C : Total area covered by all species (m^2)

2.3.4 Importance Value Index (INP)

The mangrove significance index is calculated based on the equation of [9] as follows:

$$\mathrm{INP} = \mathrm{RDi} + \mathrm{RFi} + \mathrm{RCi} \quad (8)$$

Note:
INP: Significant Value Index (%)
RDi: Relative Density (%)
RFi : Relative Frequency (%)
RCi : Relatively Important Closure (%)

2.4.5 Species Diversity Index (H ')

The mangrove species diversity index was calculated based on the equation of [24] as follows:

$$H' = \sum_{i=1}^{s} Pi \log2\ Pi \quad (9)$$

Note:
H' : Diversity Index
Pi : Proportion of the number of individuals of the species (ni) to the total number of individuals (n)
S : Number of taxa/species (ind)

2.4.6 Species Evenness Index (E)

The mangrove species uniformity index is calculated based on the [17] equation as follows:

$$E = \frac{H'}{log2\ S} \quad (10)$$

Description:
E : Evenness Index
H :diversity index
Log2S: 3.3219 Log S
S : Number of species (ind)

2.4.7Species Dominance Index (D)

The dominance index of mangrove species is calculated based on the equation [3] as follows:

$$D = \sum_{i=1}^{s} (\frac{ni}{n})^2 \quad (11)$$

Description:
D : Dominance Index, Ni: Number of individuals of the species
N : Total number of individuals (ind)
S : Number of taxa/species (ind)

3 Results and Discussion

The Peukan Bada area is one of the natural mangrove forest areas. In this study, mangrove vegetation was divided into 3 categories, namely trees, saplings and seedlings. Based on observations at the research site, there are 2 types of mangroves located in Lambadeuk village, Peukan Bada sub-district, Aceh Besar district, namely of species *Rhizophora apiculata* and *Rhizophora mucronata*. At station 1, there were two types of mangroves, namely *Rhizophora mucronata* and *Rhizophora apiculata* species. The number of each tree category, saplings and seedlings found at Station 1 is 61 individuals at station 2 there is only one type of mangrove, from *Rhizophora Apiculata*. The number of each tree category, sapling and seedling contained in plot 2 of 20 stems. While at station 3 there were two species of mangroves of the same genus. The number of each category of trees, saplings and seedlings contained at station 3 is 61 stems. This research was conducted at 3 different stations with 10 x 10 m^2 tree category, 5 x 5 m^2 sapling category and 1 x 1 m^2 seedling.

The overall density value for the growth rate of trees and saplings was higher than the seedling level. The results of the study stated that the value of species density (Di) at station 1 *Rhizophora apiculata* of 0.08 ind/m^2. While *Rhizophora mucronata* of 0.53 ind/m^2. Station 2 there is only 1 type, namely *Rhizophora apiculata* with a value of 0.2 ind/m^2. Then station 3 there were 2 types, for *Rhizophora apiculata* the value of 0.08 ind/m^2. while for *Rhizophora mucronata* of 0.53 ind/m^2 (Table 1). The two types of mangroves *Rhizophora apiculata* and *Rhizophora mucronata* were included in the true mangrove category (Table 2) [4,16].

The greater the density of a species in Lambadeuk, the more individuals of that species per unit area. The density value can be used to see the magnitude of the disturbance to a habitat [23]. Plant habitats are considered damaged, if the value of the density of plant species is low or small.

Species frequency is one of the vegetation parameters that can show the distribution pattern or distribution of plant species in the ecosystem or show the distribution pattern of plants. The frequency value is influenced by the value of the plot where mangrove species are found. The greater the number of squares found, the higher the frequency of mangrove species presence [10]. Species frequency values (Fi) at stations 1 and 3 for *Rhizophora apiculata* and *Rhizophora mucronata* was 1 ind. The value of the species frequency Fi at station 2 for *Rhizophora apiculata* was 100 ind. The results of the analysis of the relative frequency (RFi) of mangroves that have been carried out in Lambadeuk Village, Peukan Bada District, Aceh Besar District at the tree level at stations 1 and 3 for *Rhizophora apiculata* and *Rhizophora mucronata was* 75.02%. Species frequency value (Fi) at station 2 for *Rhizophora apiculata was* 75.02 ind. The large number of species was *Rhizophora apiculata* due to the condition of the substrate at the research site in the form of muddy sand. According to [14] that locations not far from the beach generally have a muddy sand type of sediment. Meanwhile, the location not far from the mainland has a sandy mud sediment type.

Table 1. Mangrove biological indeks value.

Mangrove	Number of Stands	Transect Area (A) ha	Specific Density (Di)	Relative Density	Relative Frequency	Relative Type Frequency	Closure Type	Relative Closure Type	Important Value Index (INP)	
			Ind/ha	(RDi)	(Fi)	(RFi)	(Ci)	(Rci)	Tree and sapling	Seeding
Rhizophora apiculata	36	0.01								

Station 1	8		0.08	13.115	1	75.02	2.26	1.81	253.655	250 .04
Station 2	20		0.2	100	100	75.02	10.96	8.77	183.789	175 .02
Station 3	8		0.08	13.115	1	75.02	2.26	1.81	253.655	250 .04
Rhizophora mucronata	116	0.01								
Station 1	58		0.53	86.89	1	75.02	2.26	1.81	253.655	250 .04
Station 2	0		0	0	0	0	0	0	0	0
Station 3	58		0.53	86.89	1	75.02	2.26	1.81	253.655	250 .04
Quantity			1.42	300	104	375.1	20	16.01		

Table 2. Diversity, dominance, evenness and temperature.

Diversity (H')			Dominance (C)			Evenness (E)			Temperature
Tree	Sapling	Seeding	Tree	Sapling	Seeding	Tree	Sapling	Seeding	
0.09	0.04	0.02	0	0	0	0.01	0.01	0	29,51 °C
3	3	3	1	1	1	1	1	1	
0.09	0.04	0.02	0	0	0	0.01	0.01	0	
2.31	2.66	2.79	1	1	1	0.35	0.4	0.42	
0	0	0	0	0	0	0	0	0	
2.31	2.66	2.79	1	1	1	0.35	0.4	0.42	
1.35	1.45	1.49	0.5	0.5	0.5	0.2867	0.303	0.307	

Species dominance is used to determine the concentration and distribution of dominant species. If the dominance is more than one species, then the dominance index value will increase. Furthermore, if several species dominate simultaneously, the dominance index value obtained will be low. [13]. Specific closure (Ci) is the area of the one type of closure in an area. The species closure value (Ci) at stations 1 and 3 for *Rhizophora apiculata* and *Rhizophora mucronata was* 2.26 m^2. The species closure value (Ci) at station 2 for Rhizophora *apiculata* was 10.96 m^2. The relative cover (RCi) of mangroves at stations 1 and 3 for *Rhizophora apiculata* and *Rhizophora mucronata was* 1.81 % and at station 2 for *Rhizophora apiculata* was 8.77%. The high relative cover of Rhizophora apiculata indicates that the presence of the mangrove species Rhizophora apiculata dominates the mangrove species in Lambadeuk Village, Peukan Bada District, Aceh Besar District. This is caused by the condition of the sandy mud substrate at the research site. Sandy mud substrate is suitable for the mangrove species *Rhizophora apiculata* because it has a high fertility rate, so that the mangrove species Rhizophora apiculata dominates in the Lambadeuk Village area and has relatively high cover values in almost all categories [21]. Saplings are an indicator of the sustainability of each mangrove species in the Lambadeuk area, where mangroves will grow and develop naturally without human intervention.

The Important Value Index for mangrove vegetation trees and seedlings can be obtained from the sum of the relative frequency, relative density, and relative cover of a vegetation expressed in percent (%) [13]. From the results Table 1, it can be seen that there are differences in the value of the Important Value Index in each category. The important value index (INP) tree and sapling of *Rhizophora apiculata* at station 1 and station 3 of 253.655% and station 2 of 183.789%. While seedling of *Rhizophora apiculata* at station 1 and station 3 of 250.04% and station 2 of 175.02%. The important value index (INP) tree and sapling of *Rhizophora mucronata* at station 1 and station 3 of 253.655%. While seedling of *Rhizophora mucronata* at station 1 and station 3 of 250.04%. The high Importance Value Index in the seedling category of *Rhizophora Apiculata was* caused by the condition of the muddy sand substrate [21]. The Important Value Index (INP) shows the range of the index that describes the community structure and distribution pattern of mangroves [21]. Furthermore, [21] stated that if the INP of the species obtained is high, then there are species that dominate a habitat. And this species is better able to adapt to its environment.

The high and low index of mangrove vegetation importance is caused by competition between species for nutrients and sunlight at the research site. Nutrients are very important for the stability of mangrove vegetation such as nitrate (NO_3) and phosphate (PO_4) [25]. Nitrate (NO_3) and phosphate (PO_4) are the most important nutrients that can determine the stability of vegetation growth. According to [12] stated that nutrient elements are absolute for organisms such as N and P because they cannot be replaced by other elements.

The value of H' is a value that states how diverse a species is in a certain area, an area can be said to have moderate species diversity if the value of H' exceeds 1 or equals 1, if the value of H' exceeds 3, the species diversity obtained is high. However, if the H' value found exceeds 0 but is less than 1, it can be said that a community in the area has low species diversity.

4 Conclusion

The results of the study found 2 types of mangroves from the genus Rhyzophora. Which consists of 2 species Rhyzophora mucronata and Rhyzophora apiculata found in Lambadeuk village, Peukan Bada Regency, Aceh Besar district. RFi stations 1,2, and 3 with a value of 100, RFi stations 1 and 3 with a value of 150.04, and RFi for station 2 with a value of 75.02. For RCi stations 1 and 3 with a value of 36.2, and RCi for station 2 with a value of 8.77. The highest INP is trees and saplings with a value of 253,655, while the lowest INP is trees and saplings 183,789. The highest INP of seedlings was 250.04 and the lowest INP of seedlings was 175.02. The highest value (H') from the category of trees, saplings, and seedlings was at station 2 with a value of 3 ind from the species Rhizophora apiculata. The highest value (C) was found at stations 1 and 3 species Rhizophora mucronata and Rhizophora apiculata with a value of 1 ind. The highest value (E) obtained from the categories of trees, saplings, and seedlings was at station 2 with a value of 1 ind from the species Rhizophora apiculata.

Acknowledgement. The writing of this manuscript is supported by the Matching Fund – Kedaireka funded by the Ministry of Education, Culture, Research, and Technology Indonesia awarded to the Research Center for Marine Sciences and Fisheries, Universitas Syiah Kuala.

References

1. A. Arief, Hakekat hutan dan pengaruhnya terhadap lingkungan.Yayasan Obor Indonesia. Jakarta (1994).
2. R. Azhar, M. Rusdi., M. Irham, A. Fuadi. Spatial distribution of mangrove using a geographic information system in Aceh Besar. IOP Conference Series: Earth and Environmental Science. International and National Symposium on Aquatic Environment and Fisheries. Banda Aceh. Indonesia. **674** (2021)
3. J. Z. Brower, C. Jerroid, V. Ende. Field and Laboratory Methods for General. Ecology. Mc Grow-Hill Science, New York (1997).
4. T. Cahyanto, R. Kuraesin, Jurnal Istek, **7**, 2 (2013)
5. DKP Aceh Besar. Statistik Dinas Kelautan dan Perikanan Kabupaten Aceh Besar Tahun 2010. Dinas Kelautan dan Perikanan Kabupaten Aceh Besar. Jantho (2011)
6. R. Dahuri, J. Rias, S. P. Ginting, M. J. Sitepu, Pengelolaan Sumberdaya Wilayah Pesisir dan Lautan Secara Terpadu. Pradnya Paramita. Jakarta (2001).
7. Departemen Kehutanan. Statistik Kehutanan Indonesia, Frorestry Statistics of Indonesia 2003. Badan Planologi Kehutanan, Departemen Kehutanan, Jakarta (2004).
8. I. Dewiyanti, Identifikasi dan kelimpahan hama penyebab ketidakberhasilan rehabilitasi ekosistem mangrove Di Sekitar kawasan Aceh Besar dan Banda Aceh. Prosiding Seminar Nasional XXI PBI Aceh. PBI Aceh, Banda Aceh (2011).
9. S. English, C. Wilkinson, V. Baker, Survey manual for tropical marine resource. Townsville, Autralian Institute of Marin Science (1994)
10. M. F. Fachrul, Metode Sampling Bioekologi. Bumi Aksara. Jakarta (2007).
11. M. H. Ghufran, K. Kordi, K. Ekosistem Mangrove: Potensi, Fungsi dan Pengelolaan. Jakarta. Rineka Cipta (2012)
12. A. P. Hartoko, Soedarsono, A. Indrawati. Water Resour. Manag. **2**, 2 (2013)
13. Indriyanto. Ekologi Hutan. Bumi Aksara. Jakarta. 138 (2006).
14. M. Irham, S. Adhla, C. Octavina, Depik **9**, 1 (2020).
15. F. F. Muhsoni, JST **13**, 3 (2020).
16. Y. R. Noor, M. Khazali, I. N. N. Suryadiputra, I. N. N. Panduan Pengenalan Mangrove di Indonesia: PHKA/WI-IP, Bogor (2006).
17. E. P. Odum, Dasar-Dasar Ekologi. Edisi Ketiga. Gadjah Mada University Press. Yogyakarta. 697 (1996)
18. E. Rochana, Ekosistem Mangrove Dan Pengelolaanya di Indonesia. IPB. Bogor (2001)
19. G. Raymond, N. Harahap, Soenarno. Agritek, **18**, 2 (2010)
20. Y. P. Sari, M. L. S. Pessy, I. Lidiawati, Persepsi Masyarakat Pesisir dalam Pengelolaan Ekosistem Hutan Mangrove di Muara Gembong Bekasi Jawa Barat. Perenial, **14**, (2) 78-85p (2018)
21. Supriharyono. Konservasi Ekosistem Sumberdaya Hayati di Wilayah Pesisir dan Laut Tropis. Pustaka Pelajar. Yogyakarta (2007)
22. Supriadi, A. Romadhon, A. Farid, J. Kelautan, **8**, 1 (2015).
23. M. Warpur, Struktur Vegetasi Hutan Mangrove di Kampung Kunef Distrik Supiori Selatan Kabupaten Supiori. Paper presented at the Seminar Nasional Edusainstek (2018).
24. J. T. Wilhem, Dorris. Fundamental of Ecology. New York: Drenker Inc. (1986)
25. S. Yahra, Z. A. Harahap, E. Yusni, E., R. Leidonald, Jurnal Enggano, **5**, 3 (2020).

Mahseer, the history of the king of the river

Zainal A. Muchlisin[1*], *Firman* M. Nur[2], *Siti* Maulida[2], *Luvi* Syahfrida Handayani[3] and *Sri* Riska Rahayu[3]

[1] Department of Aquaculture, Faculty of Marine and Fisheries, Universitas Syiah Kuala, Banda Aceh 23111, Indonesia
[2] Graduate School of Mathematics and Applied Sciences, Universitas Syiah Kuala, Banda Aceh 23111, Indonesia
[3] Integrated Coastal Management Postgraduate Program, Universitas Syiah Kuala, Banda Aceh 23111, Indonesia

Abstract. Mahseers (Cyprinidae) is one of the most important freshwater fish in Southeast Asia. This fish is being a consumption fish, ornamental and sport fishing. Mahseers can grow up to more than 30 kg. Therefore, this species is called The King of The River. This paper reviewed some aspects of this interesting fish. A total of 54 papers have been reviewed where some of the papers come from our Ichthyos Research Group, Universitas Syiah Kuala. There are 50 species of mahseer throughout the world, where 20 species of which are found in Asia. Indonesia has four species of mahseer, namely; *Tor tambra*, *T. tambroides*, *T. duorenensis*, and *T. soro*, of which two species, T. tambra and T. tambroides are found in Aceh province. Between the two species, *T. tambra* has a wide distribution. The original habitat of the mahseer fish waters with speed currents, clear water, high dissolved oxygen, and low water temperatures. In the wild, mahseers migrate upstream for spawning during periods of high flooding in the early rainy season. The spawning occurs one or two months at the place that is inundated during floods. Mahseer is omnivorous, they fed on mollusks, aquatic plants, small fish, insects to fruits that grow on the banks of rivers. Mahseer fish in rearing ponds are often attacked by *Lernea* sp., *Argulus* sp., *Trichodina* sp., and *Bothriocephalus acheilognathi* worms

1 Introduction

Mahseers (Cyprinidae) is one of the most valuable river fish in Southeast Asia. Apart from being a consumption fish, this fish is also popular as an ornamental fish and for sport fishing [1-3]. This fish has a high nutritional value, good meat texture, and taste that makes this fish much sought after, hence making it one of the most expensive freshwater fish [3, 4]. Generally, these fish lives in waters that have relatively cold temperatures, clean and speed currents with rocky bottoms [4, 5].

In Aceh, this fish is a favorite for fishermen because it has a high selling price so that this fish becomes the main target for local fishermen [6]. Mahseers are one of the largest freshwater fish in Indonesia which can grow up to 30-45 kg. However, human activities such as dam construction and unfriendly fishing practices such as electric fishing, poison, and dynamite cause the population the fish have decreasing significantly [6]. Presently, the mahseers are rarely to catch, and the fish size is also smaller as an indication that overfishing. Therefore, currently, mahseer is listed in the IUCN Red List as an endangered species [7-9]; Therefore, it is necessary to conduct intensive research on various of bio-ecology, aquaculture development, and conservation plan [10].

Mahseer has the potential to be used as a cultured fish in Indonesia [11]; therefore catching mahseer from the wild has to be reduced or avoided, for that fishermen need to shift from fishing activity to cultivation [6]. Therefore, it is necessary to develop mahseer fish breeding technology both for cultivation (economic) purposes and for restocking (conservation), so that the supply of this fish can be met without disturbing its wild population. For example, the technique of domestication of mahseer broodstock has been developed by Muchlisin [11], and the breeding technique has been developed by Ingram et al. [12]. This article reviews several aspects of mahseer fish that have been studied and reported by several researchers including the results of our study.

2 Morphology, diversity, distribution and conservation

The mahseer has a total of four dorsal spines, and eight dorsal soft fin rays (D.IV.8), three anal spines, and five anal soft rays (A.III.5), and 39–41 vertebrae [13]. The upper lip is curved upwards and backward, and with the median lobe pointing upwards. The fins are yellow during younger and turn black in adults. Adult fish can grow to one meter a maximum [13]. The shape, size, and length of the median lobe are common morphology used to distinguish species within the Tor genus [14], besides, the color of the scales and mouth structure are also frequently used to differentiate the mahseers species [15].

Mahseer is a common name for *Tor* group of fishes, this group comprises 50 species worldwide, and 20 species are found in Southeast Asia. The fish had the symbolic name "The king of the river" and "Fish of

*Corresponding author: muchlisinza@unsyiah.ac.id

God" [16]. There are several local name for mahseer in Indonesia, namely; tambra (Jawa), sapaan (Kalimantan), semah (Sumatera), jurung (Sumatera), kelah (Melayu), garing (West Sumatra), and keureling (Aceh). Mahseer or keureling is one of the popular fish for consumption, and sport fishing [17]. The fish can grow up to above 30 kg [14, 18, 19], and therefore the mahseer is one of the bigger fish in the river.

Indonesia has at least four species of mahseer, namely; *Tor soro, T. tambra, T. douronensis* and *T. tambroides* [20, 21], while in Aceh province has two valid species of mahseers, namely; *T. tambra* dan T. *tambroides* (Figure 1). Presently morphological based observation especially on the presence or absence of lobes on the lower lip and the size of the lobes were used to distinguish between species of mahseers in Indonesia [7, 22]; but this method is considered less accurate to distinguish to the species level and therefore other more accurate methods have been developed by using the *COXI* marker gene or commonly known as DNA barcoding [23-27]. For example, based on morphological observation, there are three presumed species of *Tor* occurring in Aceh water, but after clarified by DNA barcoding there are only two valid species (*T. tambra* and *T. tambroides*) (Muchlisin et al., submitted paper).

In Aceh, the mahseers is distributing over the regions including Aceh Barat, Nagan Raya, Aceh Selatan, Aceh Tenggara, Aceh Besar, Pidie, and Gayo Lues [6, 28-32], and Aceh Jaya [33]. Mahseers conservation program have been carried out in several areas, for example in Jambi, West and North Sumatra Provinces by establishing a reservation area called "lubuk larangan", for instance; Lubuk Larangan Lobu Sangkunur in North Sumatra, Lubuk Larangan Sikucur in Padang Pariaman, West Sumatra, and Lubuk Larangan Beringin, Jambi. But unfortunately in Aceh, the conservation of this fish has not been initiated, while the population decline is more intense over the years.

Lubuk larangan is established and managed by indigenous peoples. The community is prohibited from catching fish in that locations for a certain period of time, and if there are violations of the rule, they will be subject to customary sanctions in the form of fines, such as having to pay a certain amount of food, a goat or even a buffalo depending on the scale of the violations. As one of the provinces that has the big potential for mahseer, Aceh should also develop rules, both community-based or those initiated by the government to establish river conservation areas in Aceh, so that not only mahseers would be protected but also other fish in the areas.

3 Habitat, reproduction and breeding

The original habitat of mahseer is in the upstream area of the river in mountain areas which is characterized by strong currents, clear water, high dissolved oxygen content, and relatively low water temperature (<25 °C) with rock bottom [16, 18, 34]. Mahseers fed on both plant and animal-based foods, and therefore they are omnivores [35, 36]. In the wild, mahseers migrate upstream for spawning during periods of high flooding, and the spawning season is limited to one or two months [12]. This is an agreement with Desai [37] who stated that freshwater fish generally spawn during the rainy season. Mahseers are similar to other freshwater fish, they spawn in places that are inundated during floods [38, 39].

Fig. 1. Mahseer fish occurs in Aceh waters; (a) *Tor tambra*, (b) *Tor tambroides.*

Study on the reproduction of mahseers over the last two decades has developed an effective and productive spawning method [3]. Ingram et al. [12] is a pioneer in mahseer spawning research, who has conducted a three-year trial of spawning using hormone treatments and resulted in significant findings and insights on mahseer breeding. The best hormonal treatment was Ovaprim with a dose of 0.5 ml kg^{-1}, the success rate of spawning induction also increased dramatically with the administration of Ovaplant at a dose of 28-68 8 g kg^{-1} with pre-treatment two to seven months before treatment [12]. In addition, immersing eggs in formalin solution is very important to prevent fungus infection and increases egg hatchability [12]. Ingram et al. [40] also reported that the best interval time for breeding was 30 days, but if it was shorter than 30 days, it was suspected that it would have an impact on sperm quality decreasing.

According to Gaffar et al. [41] and Sabar and Rachmatika [42], the fecundity of mahseers ranges from 9,180 to 100,000 eggs. Meanwhile, Rupawan and Husnah [43] stated that female mahseer had a fecundity of 11,689-14,433 eggs. However, fish fecundity, in general, is highly dependent on the size and age of the fish, where the bigger fish is produced more eggs than the smaller ones [44]. Mahseer eggs will hatch 48 hours after fertilization [42]. The age of the first sexually mature mahseer fish is over two years, the size of the male fish at the same age is smaller than the female [16].

4 Feeding and nutrition requirements

Mahseers are omnivores, they fed on molluscs, aquatic plants, algae, worms, small fish, and insects [3, 6]. In addition, mahseers have a unique food preference for fruits with high protein energy content such as *Shorea macrophylla* and *Ficus* sp. [6, 45]. Mahseer fish harvested from the west coast of Aceh most preferred algae and worms as food [6]. Tan [39] also reported that in addition to liking brown and green filamentous algae, *T. tambroides* in certain seasons, they also fed ripe fruit and seeds of *Ficus vareigata*, *Dipterocarpus oblongifolius*, and *Eugenia* sp.

In some areas including Aceh, mahseer has been reared in ponds and floating cages in rivers. According to Halver [46], fish require protein as their main energy source. The research results of Muchlisin et al. [47] also showed that fingerling mahseer requires a minimum of 30% protein for optimal growth. The growth of mahseer T. tambra in rearing ponds can be increased by the addition of vitamin E as much as 300 mg kg-1 of feed [48] and papain enzyme as much as 27.5 mg kg-1 [49] and the addition of probiotics in the feed as much as 10 ml kg^{-1} [47].

5 Parasites

Fish diseases are generally divided into two types, namely infectious and non-infectious diseases. Infectious diseases are caused by pathogens, such as bacteria, fungi, and viruses. While non-infectious diseases are caused by chemical and physical environmental factors for example; poisoning and malnutrition [50]. Lernea sp. is common exto-parasite attack mahseer in cultivated ponds [51]. Apart from *Lernea, Argulus* and *Trichodina* are also frequently attack mahseers in aquaculture ponds [51]. In addition to exto-parasites, mahseer fish in rearing ponds was also often attacked by endo-parasites especially the Asian fish tapeworm, *Bothriocephalus acheilognathi* [52]. To avoid parasitic infection on fish is to keep the pond environment clean and eradicate parasitic hosts, such as snails and wild fish. In addition, filtration of water entering the rearing pond, isolation of infected fish, and drying of the pond can also reduce the risk of cultured fish infecting parasites and disease [53, 54].

6 Conclusion

Mahseer is one of the bigger freshwater fish that lives in rivers. This fish has high economic value and its distribution is mostly in the Asian region. Indonesia has four species of mahseer, namely *Tor soro*, *T. tambra*, *T. douronensis* and *T. tambroides*. While in the waters of Aceh there are two valid species, namely *T. tambra* and *T. tambroides*. This fish spawns in the early of the rainy season and this fish is an omnivore. Mahseer fish in rearing aquaculture ponds were often attacked by ectoparasites *Lernea*, *Argulus*, and *Trichodina*, while endoparasites from the worm *Bothriocephalus acheilognathi* was also frequently found in the alimentary tract of mahseer. In rearing ponds, mahseer fingerling fish require feed with a minimum protein content of 30%, and their growth can be increased by adding vitamin E, papain enzymes, and probiotics to the feeds.

References

1. V. Mohindra, P. Khare, K. Lal, P. Punia, R. K. Singh, A. Barman,W. S. Lakra. Acta Zool Sin. (2007)
2. N. Shahi, S. K. Mallik,D. Sarma. Aquac Asia. **14,** 22 (2014)
3. M. M. L. Lau, L. W. K. Lim, S. D. Ishak, A. B. Abol-Munafi,H. H. Chung. Proc Zool Soc. **74,** 227 (2021)
4. L. K. Soon, S. Lihan, F. F. G. Dasthagir, K. M. Mikal, F. Collick,N. K. Hua. Int J Sci Technol Res. **3,** 285 (2014)
5. T. K. Shrestha. The mahseer in the rivers of Nepal disrupted by dams and ranching strategies (1997)
6. Z. A. Muchlisin, A. S. Batubara, M. N. S. Azizah, M. Adlim, A. Hendri, N. Fadli, A. A. Muhammadar,S. Sugianto. Biodiversitas. **16,** (2015)
7. M. Kottelat, A. Whitten, S. Katikasari,S. Wirjoatmodjo. Ikan air tawar Indonesia bagian Barat dan Sulawesi (1993)
8. U. Sarkar, B. Mahapatra, S. R. Saxena,A. Singh. J Ecophysiol Occup Health. **15,** 45 (2015)
9. M. Kottelat, A. Pinder,A. Harrison. Tor tambra. The IUCN Red List of Threatened Species 2018: e.T188012A89801879 31 October 2021.(2018)
10. C. Kulkarni,S. Ogale. The present status of mahseer (fish) and artificial propagation of Tor khudree (Sykes) 651 (1978)
11. Z. A. Muchlisin. JII. **13,** 91 (2013)
12. B. Ingram, S. Sungan, G. Gooley, S. Y. Sim, D. Tinggi,S. S. De Silva. Aquac Res. **36,** 983 (2005)
13. Bleeker. *Tor tambroides* 1854)
14. Z. Wei,G.-H. Cui. Ichthyol Explor Freshw. **7,** 131 (1996)
15. S. Siraj, Y. Esa, B. Keong,S. Daud. Malays Appl Biol. **36,** 23 (2007)
16. N. C. Kiat. Kings of the rivers : Mahseer in Malaysia and the region (2004)
17. V. Desai. Synopsis of biological data on the tor mahseer *Tor tor* (Hamilton, 1822) (2003)
18. S. J. Haryono. Biodiversitas. **9,** 306 (2008)
19. H. M. Smith. Bull US Natl Mus. **188,** 9 pls. (1945)
20. S. S. De Silva, B. Ingram, S. Sungan, D. Tinggi, G. Gooley,S. Y. Sim. Aqua Asia. **9,** 15 (2004)
21. H. Haryono. Biodiversitas. **7,** (2006)
22. T. R. Roberts. Raffles Bull Zool. **47,** 225 (1999)
23. R. Floyd, E. Abebe, A. Papert,M. Blaxter. Mol Ecol. **11,** 839 (2002)
24. S. N. Farhana, Z. A. Muchlisin, T. Y. Duong, S. Tanyaros, L. M. Page, Y. Zhao, E. A. Adamson, M. Z. Khaironizam, M. de Bruyn,M. N. S. Azizah. Sci Rep. **8,** 1 (2018)
25. P. D. N. Hebert, A. Cywinska, S. L. Ball,J. R. deWaard. Proc Biol Sci. **270,** 313 (2003)

https://doi.org/10.1051/e3sconf/202233903006

26. Z. A. Muchlisin, Z. Thomy, N. Fadli, M. A. Sarong,M. N. Siti-Azizah. Acta Ichthyol Piscat. **43,** (2013)
27. Z. A. Muchlisin, A. S. Batubara, N. Fadli, A. A. Muhammadar, A. I. Utami, N. Farhana,M. N. Siti-Azizah. F1000Research. **6,** (2017)
28. Z. Muchlisin,M. S. Azizah. Int J Zool. **5,** 62 (2009)
29. Y. Akmal, Y. Dhamayanti,E. Paujiah. Biodiversitas. **21,** (2020)
30. T. F. Haser, M. S. Nurdin, E. Supriyono, D. Radona, F. Azmi, K. Nirmala, T. H. P. Widanarni, T. Budiardi,R. Y. Valentine. (2021)
31. A. Hendri,S. Wahyuni. JPTropis. **3,** (2016)
32. S. Zuraidah, B. Budiman,E. Safutra. J Akua. **2,** (2018)
33. M. R. M. Radhi, I. Zulfahm,Y. Akmal. Prosiding Biotik. **8,** (2021)
34. Haryono. (1994)
35. S. Sulastri, I. Rachmatika,D. Hartoto. Berita Biol. **3,** 84 (1985)
36. Haryono. Perikanan dan aspek budidaya pada masyarakat Dayak di sekitar Cagar Alam Kayan Mentarang Kalimantan Timur; 1992; Kota Kinabalu.
37. V. Desai. Indian Nat Sci Acad. **39,** 228 (1973)
38. K. Alikhuni,S. Rao. Jerdon Rec Indian Mus. **49,** 157 (1951)
39. E. S. Tan. Aquaculture. **20,** 281 (1980)
40. B. Ingram, S. Sungan, D. Tinggi, S. Y. Sim,S. S. De Silva. Aquac Res. **38,** 809 (2007)
41. A. Gaffar, A. Utomo,S. Adjie. BPPD. **10,** 17 (1991)
42. F. Sabar,I. Rachmatika. Zoo Indonesia. (1983)
43. A. Rupawan,G. Husnah. JPII. **5,** 1 (1999)
44. Z. Muchlisin, M. Musman,M. S. Azizah. Reprod Biol Endocrinol. **8,** 1 (2010)
45. M. S. Kamarudin, M. L. Bami, A. Arshad, C. R. Saad,M. Ebrahimi. Fish Sci. **84,** 385 (2018)
46. J. Halver. Fish nutrition (2013)
47. Z. A. Muchlisin, T. Murda, C. Yulvizar, I. Dewiyanti, N. Fadli, F. Afrido, M. N. Siti-Azizah,A. A. Muhammadar. F1000Research. **6,** 1 (2017)
48. Z. A. Muchlisin, A. A. Arisa, A. A. Muhammadar, N. Fadli, I. I. Arisa,M. N. Siti-Azizah. Fish Aquatic Sci. **24,** 47 (2016)
49. Z. A. Muchlisin, F. Afrido, T. Murda, N. Fadli, A. A. Muhammadar, Z. Jalil,C. Yulvizar. Biosaintifika. **8,** 172 (2016)
50. Z. Muchlisin. Pengantar akuakultur (2019)
51. Z. A. Muchlisin, A. M. Munazir, Z. Fuady, W. Winaruddin, S. Sugianto, M. Adlim, N. Fadli,A. Hendri. Hum Vet Med. **6,** 148 (2014)
52. Z. Muchlisin, Z. Fuadi, A. Munazir, N. Fadli, W. Winaruddin, C. N. Defira,A. Hendri. Bulg J Vet Med. **18,** 361 (2015)
53. Z. Kabata. Parasites and diseases of fish cultured in the tropics (1985)
54. E. Soulsby. Helminths, Arthropods and Protozoa of domesticated animals (1982)

https://doi.org/10.1051/e3sconf/202233903007

Length-weight relationships and condition factors of *Cephalopholis argus* harvested in the northern coast of Aceh, Indonesia

Nur Fadli[1*], *Adrian* Damora[1], *Zainal* Abidin Muchlisin[1], *Irma* Dewiyanti[1], *Mutia* Ramadhaniaty[1], *Rianjuanda* Djamani [1], *Alfis* Syahril[1], *Firman* M. Nur[2], *Agung* S. Batubara[3], *Nanda* Muhammad Razi[1], *Edison* D. Macusi[4] and *Mohd* N. Siti-Azizah[5]

[1]Faculty of Marine and Fisheries, Universitas Syiah Kuala, Banda Aceh, Indonesia
[2]Graduate School of Mathematics and Applied Sciences, Universitas Syiah Kuala, Banda Aceh 23111, Indonesia
[3]Faculty of Mathematics and Natural Sciences, Universitas Negeri Medan, North Sumatra 20221, Indonesia
[4]Institute of Agriculture and Life Sciences (IALS), Davao Oriental State University (DOrSU), Mati City, Davao Oriental, Philippines
[5]Institute of Marine Biotechnology, Universiti Malaysia Terengganu, Terengganu, Malaysia

Abstract. The peacock hind grouper (*Cephalopholis argus*) is among several highly exploited grouper fishes globally. Categorized as least concern (LC) in the IUCN, biological data on the species is limited, especially in the Aceh region. This research targeted to investigate the biological features of the *C. argus* fished on the northern coast of Aceh. The fish samples were collected from Pelabuhan Perikanan Samudera (PPS) Lampulo, Kuta Alam District, Banda Aceh, Aceh from June - September 2020. Overall, 30 specimens *C. argus* were compiled in this study. The total length (TL) varied between 136.0-284.0 mm, while weight (W) ranged between 46.5-460.0 g. In addition, Fulton's condition (K) ranged from 1.45-4.61, and relative weight (Wr) ranged from 81.75-128.53 g. The LWRs analysis revealed a b value of 3.1329 with a correlation coefficient of 0.94, indicating a positive allometric growth pattern. This study provides length-weight relationships (LWR) and condition factors value for the peacock hind grouper that will assist in building a feasible management of the fish.

1 Introduction

One of the groupers species fished in Aceh northern coast is the peacock hind grouper (*Cephalopholis argus*). The fish is categorized as least concern (LC) in the IUCN List [1]. *Cephalopholis argus* is distributed in the Indo-Pacific from the Red Sea to South Africa and eastward to French Polynesia and the Pitcairn group, Ryukyus Islands and Ogasawara islands in the north, northern Australia coast and Lord Howe Island in the south. The fish is considered reef-associated fishes and mainly feed on fishes and crustaceans [2, 3]. Despite its commercial importance, there is still deficient information on their biological characteristics, including length-weight relationships (LWR) and condition factors.

The LWRs information can be utilized to examine the growth shape between fish sexes [4-7], seasonal variation [4, 8, 9], the population from different locations [5, 9, 10], and different environmental conditions [9]. On the other hand, the condition factor value can determine the state of a fish population in their habitat [11]. This information is essential in fisheries management.

There have been limited extensive studies on length-weight relationships (LWR) and condition factors of groupers in Aceh. For example, the study on the growth pattern of the leopard coralgrouper (*Plectropomus leopardus*), blacksaddled coralgrouper (*P. laevis*), duskytail grouper (*E. bleekeri*), and brown-marbled grouper (*Epinephelus fuscoguttatus*) [12]; the golden hind grouper (*Cephalopholis aurantia*) [13]; the orange-spotted grouper (*Epinephelus coioides*) [14] and white-edged lyretail (*Variola albimarginata*) [15]. However, similar information is not available for *Cephalopholis argus* in the northern Aceh region. Hence, the recent research's purpose was to expand further the biological investigations of *C. argus* harvested in the region.

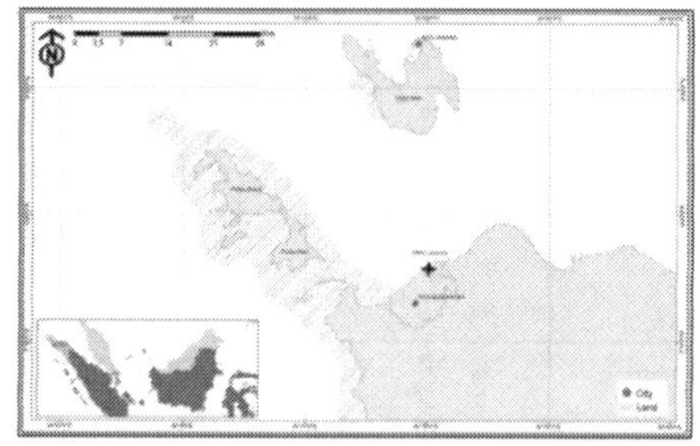

Fig. 1. Location of PPS Lampulo in the northern Aceh coast.

* Corresponding author: nurfadli@unsyiah.ac.id

2 Materials and Method

The study was conducted from June - September 2020 in Pelabuhan Perikanan Samudera (PPS) Lampulo, Kuta Alam District, Banda Aceh, Aceh (Figure 1). For further analysis and measurements, the collected *Cephalopholis argus* samples were then transported to Genetics and Biodiversity Laboratory in the Faculty of Marine and Fisheries, Universitas Syiah Kuala.

The length-weight relationship (LWRs) was evaluated applying the linear allometric model (LAM) following De Robertis and Williams [16] and Muchlisin, Musman and Siti Azizah [17]. In addition, two condition factors were assessed in this study: Fulton's condition (K) [18] and Relative weight (Wr) [19].

Table 1. Values of several biological parameters, length, and weight relationships, and condition factors for *Cephalopholis argus* harvested on the northern coast of Aceh.

Parameters	mean ± SD
Total length (mm)	136.0-284.0 (237.7 ± 29.5)
Body weight (g)	46.5-460.0 (253.6 ± 90.8)
Fulton's condition (K)	1.45-4.61 (2.02 ± 0.59)
Prediction weight (Ws) (g)	41.77-419.52 (251.50 ± 81.95)
Relative weight (Wr) (g)	81.75-128.53 (100.59 ± 11.28)
b Value	3.1329

3 Results and Discussion

Overall, 30 specimens of *C. argus* were compiled in this research. The observed total length (TL) of the fish samples varied from: 136.0-284.0 mm (mean 237.7 ± 29.5) with weight (W): 46.5-460.0 g (mean 253.6 ± 90.8). Additionally, the Fulton (K) condition factor value ranged from 1.45 to 4.61 (mean 2.02 ± 0.59). The Ws values (predicted weight) ranged from 41.77 to 419.52 g (mean 251.50 ± 81.95). Furthermore, the relative weight (Wr) for *C. argus* fish ranged from 81.75 to 128.53 g (mean 100.59 ± 11.28) (Table 1).

The linear allometric model (LAM) generated a *b* value of 3.1329 with a correlation coefficient value of 0.94. These results indicate that *C. argus* has a positive allometric growth pattern (Figure 2a). In addition, the regression model for the species showed a equal growth pattern between the observation and the prediction (Figure 2b).

The examined Fulton's condition factor (K) was > 1 (2.02 ± 0.59) and parallel with earlier grouper studies [6, 12, 20]. In addition, the average relative weight (Wr) was greater than 100 (100.59 ± 11.28 g). The value corresponds with previous grouper studies worldwide [13, 14]. Similar findings were also reported in several marine fishes surveys [4, 21]. The K and Wr values observed in this study showed that the populations of *C. argus* harvested in this region are still in good condition. Morton and Routledge [11] stated that a fish population is good when the K value is higher than 1, while a Wr value is> 100. Furthermore, the growth pattern of *C. argus* observed (positive allometry) in this study is similar to several other grouper species in northern Aceh [12, 13]. This positive allometric growth pattern was also noted in many other marine fishes studies [4, 22].

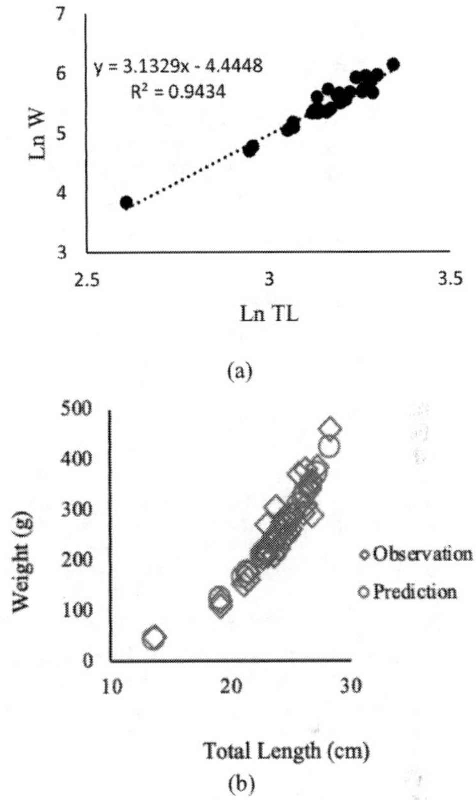

Fig. 2. The length-weight relationships of *C. argus* (a) and comparison of observation and prediction growth (b).

Some abiotic factors were identified as the major factors affecting fish condition factors and growth patterns, such as environmental condition [9, 21] and temperature [23]. In addition to abiotic factors, biotic factors such as food accessibility [4] and predator density [21] also identified affected the fish condition factors and growth patterns.

This study provides length-weight relationships (LWR) and condition factors value for the peacock hind grouper that will assist in building a feasible fisheries management of the fish.

Acknowledgments

The study was funded by Universitas Syiah Kuala (Penelitian Calon Professor Research Scheme, contract number: 58/UN11.2.1/PT.01.03/PNBP/2021) and Universiti Malaysia Terengganu (Addressing Fisheries and Aquaculture Challenges in the Grouper (Family Epinephelidae) through Molecular Genetics with

Special Focus on *E. coioides* - UMT/RMIC/STRA/53280).

References

1. IUCN. The IUCN Red List of Threatened Species. Version 2019-1. ; 2019.
2. R. Froese,D. Pauly. (2019)
3. M.T. Craig, Y. Sadovy de Mitcheson,P.C. Heemstra. (2011)
4. D. Yulianto, I. Indra, A. Batubara, D. Efizon, F. Nur, S. Rizal, R. Elvyra,Z. Muchlisin. F1000Research **9**, 259 (2020)
5. U.R. Zargar, A. Yousuf, B. Mushtaq,D. Jan. Turkish Journal of Fisheries and Aquatic Sciences **12**, (2012)
6. U. Bulanin, M. Masrizal,Z.A. Muchlisin. Aceh Journal of Animal Science **2**, 1 (2017)
7. M.S. Alavi-Yeganeh, S.J. Seyfabadi, Y. Keivany, B. Kazemi,G.P. Wallis. Journal of Applied Ichthyology **27**, 6 (2011)
8. E.M. Kembenya, E.O. Ogello, C.M. Githukia, C.N. Aera, R. Omondi, J.M. Munguti. International Journal of Sciences: Basic and Applied Research **14**, 2 (2014)
9. E.Ö. Özbek, T. Kebapçıoğlu,M. Çardak. Journal of the Black Sea/Mediterranean Environment **19**, 1 (2013)
10. S. Munilkumar, Y. Rameshori, W.R. Mangang, M. Kamei, Y. Chinglemba, A. Pavan-Kumar, B.K. Mahapatra, A. Das, M.A. Hanif, M.A.B. Siddik. Journal of Applied Ichthyology **37**, 2 (2021)
11. A. Morton,R.D. Routledge. North American Journal of Fisheries Management **26**, 1 (2006)
12. A. Ramadhani, Z.A. Muchlisin, M.A. Sarong,A.S. Batubara. DEPIK Jurnal Ilmu-Ilmu Perairan, Pesisir dan Perikanan **6**, 2 (2017)
13. N. Fadli, A. Damora, Z.A. Muchlisin, I. Dewiyanti, M. Ramadhaniaty, A. Hanif, C.S. Zahara, M. Rusdi, F.M. Nur, A.S. Batubara, M.N. Siti-Azizah. IOP Conference Series: Earth and Environmental Science **674**, 1 (2021)
14. N. Fadli, A. Damora, Z.A. Muchlisin, I. Dewiyanti, M. Ramadhaniaty, Z. Zhelfi, N.F. Roka, F. Fitriani, M. Rusdi, F.M. Nur, A.S. Batubara,M.N. Siti-Azizah. IOP Conference Series: Earth and Environmental Science **674**, 1 (2021)
15. A. Damora, N. Fadli, Z.A. Muchlisin, I. Dewiyanti, A.S. Batubara, F.M. Nur, Z. Zhelfi, N.F. Roka, F. Fitriani,M.N. Siti-Azizah. IOP Conference Series: Earth and Environmental Science **674**, 1 (2021)
16. A. De Robertis,K. Williams. Transactions of the American Fisheries Society **137**, 3 (2008)
17. Z.A. Muchlisin, M. Musman,M.N. Siti Azizah. Journal of Applied Ichthyology **26**, 6 (2010)
18. B.G. Blackwell, M.L. Brown,D.W. Willis. Reviews in Fisheries Science **8**, 1 (2000)
19. A.L. Rypel, T.J. Richter. North American Journal of Fisheries Management **28**, 6 (2008)
20. W. Ndiaye, K. Diouf, O. Samba, P. Ndiaye,J. Panfili. International Journal of Advanced Research **3**, 3 (2015)
21. Z.A. Muchlisin, V. Fransiska, A.A. Muhammadar, M. Fauzi,A.S. Batubara. Croatian Journal of Fisheries **75**, 3 (2017)
22. G. Mahadevan, S.M. Gosavi,P. Murugesan. Thalassas: An International Journal of Marine Sciences **36**, 2 (2020)
23. W. Sabbir, M.Y. Hossain, M.A. Rahman, M.R. Hasan, Z. Mawa, S. Tanjin, H. Ul-Hassan,J. Ohtomi. Egyptian Journal of Aquatic Biology and Fisheries **24**, 2 (2020)

Community structure of plankton in Aneuk Laot Lake Sabang Pulau Weh

Nurfadillah Nurfadillah[1,3,4], *Desrita* Desrita[5*], *Irma* Dewiyanti[2,3,4], *Sri* Riska Rahayu[1], *Seri* Maulidawati[1], *Siswani* Sari[6], *Surikawati* Surikawati[6], and *Indri* Karina[6]

[1] Department of Aquaculture, Faculty of Marine and Fisheries Universitas Syiah Kuala, Banda Aceh, Indonesia
[2] Department of Marine Science, Faculty of Marine and Fisheries Universitas Syiah Kuala, Banda Aceh
[3] Laboratory of Marine Biology, Faculty of Marine and Fisheries, Universitas Syiah Kuala, Banda Aceh
[4] Marine and Fishery Research Center, Universitas Syiah Kuala, Banda Aceh, Indonesia
[5]Aquatic Resources Management Study Program, Agriculture Faculty, Universitas Sumatera Utara, Medan
[6]Local environmental agency, Sabang

Abstract. The role of plankton as natural food affects the potential of fish in the waters. The structure of the plankton community in the Aneuk Laot Lake can provide an overview of the stability of these waters. This study aims to determine the plankton biomass and the structure of the plankton community of Aneuk Laot Lake Sabang. This research was conducted in September 2019 in the waters of Aneuk Laot Lake. Determination of observation stations using the stratified random sampling method, identification of plankton using the census method and analysis of chlorophyll-a using the Trichomatic method. The results showed that the plankton composition consisted of 27 phytoplankton and 4 zooplankton genera which were divided into Bacillarophyceae (36%), Chlorophyceae (19%), Cyanophyceae (16%), Zygnematophyceae (10%), Dinophyceae (3%), Euglenophyceae (3%) and Crustacean class (13%). The abundance of plankton was 21,333 ind/l, the most abundant type is Navicula sp. Diversity index value (2.97) is category high, eveness index (0.87) and low dominance index (0.07). Chlorophyll-a values ranged from 2.84 to 4.84 mg/l including the medium nutrient category. Aneuk Laot Lake has a high level of community stability with trophic status of mesotrophic-eutrophic.

1 Introduction

Aneuk Laot Lake is one of the lakes where is located in Sabang, Pulau Weh. The lake has produced clean water for communities of Sabang to fulfill their needs. Besides the sources of life, the lake also becomes a magnificent place for tourism, agriculture, farming, and fishing spot. Edyanto [1] said if the appearance of Aneuk Laot Lake in Sabang as the freshwater resources importantly for supporting plenty of community activities on the island. Referring to Yusifa *et al* [2] said Aneuk Laot Lake has not equipped the inlet and outlet of the water flow.

The exceeded exploitation of the resources is expected will corrupt the stability of the ecosystem in Anak Laot Lake. The high of community activities surrounding the waters tend to affect the contamination and damage of sustainability in the waters, with no exception of the plankton [3-5]. Plankton is one of the biotas that can be observed and measured for evaluating of quality of the waters [6-8]. Their appearance in the waters is an important function due to for delivering in food chain cycle. The phytoplankton has the ecology function as the primer producer in the food chain, thus making them as indicated the stability of the waters [9 – 12].

The research regarding the waters Aneuk Laot Lake had been launched minimally. Some of the research that had been done namely the quality of the waters of Aneuk Laot Lake in Pulau Weh [1], the number quantity of waters of Aneuk Laot Lake[13], Morphometric the five of fish species which dominantly caught in Aneuk Laot Lake [14]. The chemical parameters and microbiology of Aneuk Laot Lake[15], and Investigation of shrinkage of water in Aneuk Laot Lake [2]. However, the structure of plankton in the waters of Aneuk Laot Lake has not been reported. One of the biotic components that can be the main factor in the waters is the plankton or the bioindicator.

Wirabumi et al [16] argued the existence of plankton particularly phytoplankton in the water importantly due to their characters as the autotroph which transformed the inorganic to be the organic through the photosynthesis process. This research had been conducted as the initial data for evaluating in managing the waters of Aneuk Laot Lake. This research was aimed for observing the abundance and structure of plankton in the waters of Aneuk Laot Lake Sabang, Pulau Weh.

2 Methodology

* Corresponding author: desrita@usu.ac.id

2.1 Sampling Area

The research was conducted in Aneuk Laot Lake, Sabang, Province of Aceh. The sampling was tackled in September 2019. The map of research was portrayed in Figure 1.

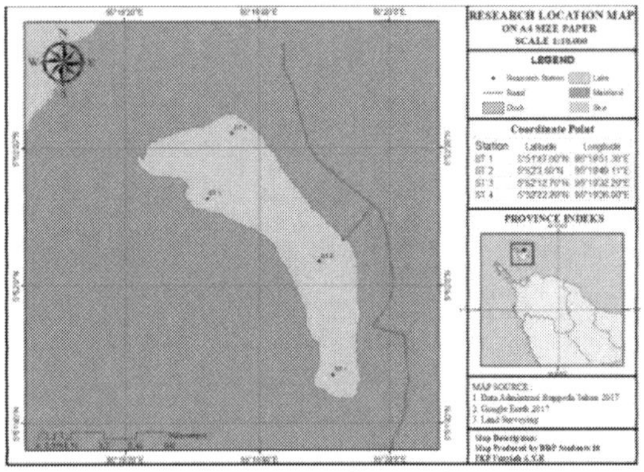

Fig. 1. Map of research

The determination of sampling spot was used the stratified random sampling. This research was divided into 4 (four) stations of observing. Namely, station 1 (The protected forest around of edge of the lake). Station 2 (The area of dock and tourism). Station 3 (The location which there was the intake of PDAM as the area for obtaining the clear water directly. Station 4 (the area of settlement, agriculture, and planting).

2.2 Research procedure

The sampling process had been conducted quantitatively through the filtering of plankton with equipment of the plankton net no. 25. The taking of water used the Van Dorn water sampler. The observing of the plankton used the census method. Identification of the plankton was addressed with the book of plankton [17], identifying marine diatoms and dinoflagellates [18], and identifying marine phytoplankton [18]. The parameters that had been measured comprised physic, chemical, and biology. Namely temperature, visibility, Dissolved Oxygen, pH, nitrate-nitrogen (NO_3-N), a total of phosphate dan chlorophyll-a.

2.3 Data analysis

2.3.1 The abundance of phytoplankton

The abundance of phytoplankton can be observed with the equation [19]:

$$N = n\,x\frac{a}{A}\;x\;\frac{v}{vc}x\frac{1}{V} \qquad (1)$$

Note: N = amount of plankton per litre (ind/l); n = amount of phytoplankton were identified (ind); a = wide of cover glass (400 mm^2); A= wide of far view (400 mm^2); v= volume of water concentrated (50 ml); vc= volume of water dropped under of cover glass (0.05 ml); V= volume of water filtered (9 L).

2.3.2 Diversity of plankton

The index that was used for evaluating the diversity of species-level was used the diversity index [20]

$$H = -\sum_{i=0}^{s} \text{pi ln pi} \qquad (2)$$

Note: H'= Index of species diversity, pi= probability of each part in overall (ni/N), ni= number of species individual to-i N= number of the total individual.

2.3.3 Evenness Index

The evenness index of plankton is used for evaluating on how the huge similarity of the spread of individual each species in the communities. The evenness index is based on the equation [20]

$$e = \frac{H'}{Ln\,S} \qquad (3)$$

Note: e = Evenness Index, H'= Diversity index, S = Number of species.
The range of evenness index between 0 -1. If the value of e close to 1, the spreading of species evenly. If the value of e close to 0 indicated the spreading of species unequally, or there was the particular species dominantly [20].

2.3.4 Dominance index

The dominance index was obtained through the Simpson Index (Odum, 1971).

$$C = \sum_{i=1}^{s} (n_i\,/N)^2 \qquad (4)$$

Note: C= Simpson Dominancy Index, ni= number of species individual i, N = number of total individual. The range of dominancy between 0 – 1.

3 Result and Discussion

3.1 Composition of species plankton

Composition of plankton were discovered in the waters of Aneuk Laot Lake consisted of Bacillariophyceae (36%; 11 genera), Cyanophyceae (16%; 5 genera), Chlorophyceae (19%; 6 genera), Crustacea (13%; 4 genera), Dinophyceae (3%; 1 genera), Euglenophyceae (3%; 1 genera), dan Zygnematophyceae (10%; 3 genera). Based on the composition, Bacillariophyceae was the various dominantly found. It was related to Andriani *et al* [21] said if Bacillariophyceae was the species in generally was discovered in the waters due to adapted in their environment. The data composition of plankton in the waters of Aneuk Laot Lake is served in figure 2.

https://doi.org/10.1051/e3sconf/202233903008

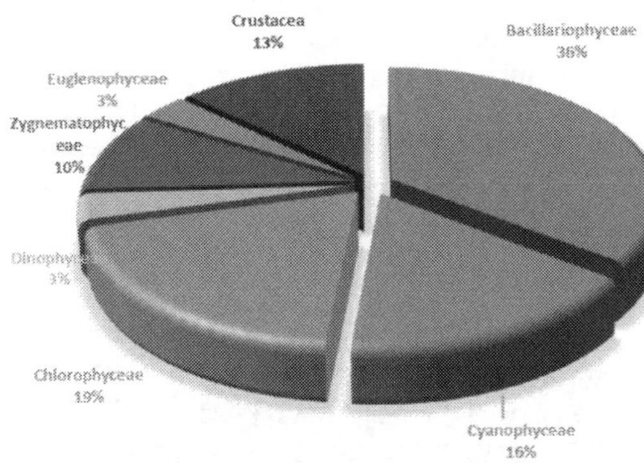

Fig. 2. Composition of plankton in Aneuk Laot Lake

Navicula sp is the biggest genus from Bacillariophyceae were found in the composition. The huge of *Navicula* sp. was depicted if this genus has the adapted level and high tolerance in the environment of waters such as DO, CO_2, visibility, and pH [22]. Martin and Fernandez [23] said *Navicula* sp. has the slimy stick which was used for patch tool on the substrate, both of in current waters and without the current entirely. Referring to Aryawati and Thoha [24] argued the Bacillariophyceae was cosmopolite which high adapted into the waters. The increase of phytoplankton diversity was influenced by the eutrophication process.

3.2 Abundance of Plankton

The abundance of plankton is the calculation amount of individual per volume of water (ind/l). The abundance of plankton in the waters of Aneuk Laot Lake can be found in Table 1.

Table 1. Abundance of Plankton in Aneuk Laot Lake

No	Plankton Species	Station			
		1	2	3	4
	Class Bacillariophyceae				
1	*Aulacoseira sp.*	+	-	+	-
2	*Amphipleura sp*	-	-	-	+
3	*Cocconeis* sp.	+	-	-	+
4	*Cymbella sp*	+	-	-	-
5	*Cyclotella* sp.	-	-	-	+
6	*Fragilaria* sp.	+	-	-	+
7	*Gomphonema* sp.	-	-	-	+
8	*Navicula* sp.	+	-	-	+
9	*Pinnularia* sp.	+	-	-	-
10	*Synedra* sp.	+	+	+	+
11	*Skeletonema sp*	+	-	-	-
	Class Cyanophyceae				
12	*Agmenellum* sp.	-	-	-	+
13	*Cuspidothrix* sp	-	+	-	-
14	*Cuspidothrix sp*	+	-	-	-
15	*Merismopedia sp.*	-	-	+	+
16	*Oscilatoria* sp	-	-	-	+
	Class Chlorophyceae				
17	*Ankistrodesmus*	-	-	-	+
18	*Closterium* sp.	+	-	+	-
19	*Chlorella* sp.	-	+	+	-
20	*Cosmarium* sp.	-	+	-	+
21	*Cylindrospermopsis sp.*	+	+	-	+
22	*Spirogyra sp*	-	-	-	+
	Class Zygnematophyceae				
23	*Desmidium sp.*	-	-	-	+
24	*Meugeotia* sp.	-	-	+	+
25	*Zygnema* sp.	+	-	+	-
	Class Dinophyceae				
26	*Ceratium sp*	+	-	-	+
	Class Eugleonaphyceae				
27	*Euglena* sp	+	+	-	-
	Class Crustacea				
28	*Acartia sp*	-	-	-	+
29	*Calanus* sp.	-	-	+	-
30	*Cyclops sp*	+	-	-	+
31	*Oithona* sp	-	-	-	+
Abundance of Plankton (ind/l)		11,111	1,667	2,889	5,556

The abundance of plankton was observed in four stations have an average of around 5.306 ind/l. The highest phytoplankton was found in station 1 (11,111 ind/l) and the lowest in station 2 (1,667 ind/l). The abundance indicates of each station has a high nutrient and is suitable for plankton living. Based on Barus [25] the exceeding of plankton in the waters is closely related to the content of nutrients namely phosphate, nitrate, silicate, and other nutrients.

The highest abundance was in station 1, wherein that station there was plenty of nutrients such as nitrate and phosphate which derived from the activities of the settlement around the lake. Linus and Irawati [26] said the huge concentration of nitrogen in the waters was caused by the waste of people's activities and also by the metabolism of sewage which sedimented in the bottom of waters. The highest abundance of Bacillariophyceae was expected due to the plenty of nitrogen. Besides that, Bacillariophyceae is one of diatom which tolerance through the temperature and can adapt in the extreme waters.

3.3. Biology Index

The structure of plankton communities is determined based on the several biology index, namely diversity index, evenness index, and dominancy. The diversity of plankton in the waters showed a value of around 2,97. Based on [20] if the diversity index was above 2,3, indicated the waters were categorized as the middle stability. The evenness index of plankton showed averages 0,87 which illustrated the uniformity of plankton in the waters of Aneuk Laot Lake spread equally.

The dominancy of plankton in Aneuk Laot Lake was around 0,07. The value indicated there were no plankton species dominantly in the location. Based on [27] argued if the range of dominancy index 0 - 0,50, it showed the waters dominancy was low. Overall, the structure of plankton communities in Aneuk Laot Lake was categorized stably, where the diversity index in the middle value, the evenness index equally, and dominancy lowly. However, this condition can transform anytime based on the environment changes drastically. There are several factors that can change the plankton communities, namely sampling time, the appearance of nutrients, and climate change.

The quality of waters in all of the stations fluctuated and the range of value is still optimum for the living of plankton (Table 2).

Table 2. Parameters of water quality

No.	Parameter	Station			
		1	2	3	4
1	Temperature (°C)	29,44	29,42	30,37	30,36
2	Visibility (cm)	99,2	128,5	82	75,5
3	Depth (m)	12	16,82	12,3	8,15
4	Current (m/s)	0,02	0,31	0,18	0,36
5	pH	7,8	8	8	8,2
6	DO (mg/l)	7,15	6,95	6,85	7,1
7	Total of phosphate (mg/l)	0,28	0,43	0,38	0,36
8	Total of nitrogen (mg/l)	3,45	3,34	1,44	1,26
9	Chlorophyll-a (μg/l)	4,84	4,66	2,84	2,88

The nutrient which affected the abundance of plankton is the total of phosphate and nitrogen. Phosphate and nitrogen are the nutrients that are absorbed by the plankton directly. Based on Indrayani *et al* [28] the P-totally in the waters depending on the particle which derived from the land into the waters caused by the erosion and the activities of humans and disturbed the living of plankton in the waters. Meanwhile, the relation of temperature and pH are the components that increased each other. The temperature experienced a slump along with the increasing deeply of waters, pH, and DO as well.

4 Conclusion

The composition of plankton was discovered in Aneuk Laot Lake were 31 genera namely Bacillariophyceae 11 genera, Cyanophyceae 5 genera, Chlorophyceae 6 genera, Crustacea 4 genera, Dinophyceae 1 genera, Euglenophyceae 1 genera, and Zygnematophyceae 3 genera. The composition of plankton showed the class of Bacillariophyceae was dominantly in their species. The highest abundance was in station 1 around 11.111 ind/l. Overall, the structure of communities in Aneuk Laot Lake showed the diversity index (2,97), the evenness index was equally (0,87), and the dominancy index was low (0,07).

References

1. C. B. H. Edyanto. Jurnal Teknik Lingkungan **1**:115-124 (2006)
2. N.A. Yusifa, S. Yanto, Sairiyyah dan I. Muhammad. Jurnal Aceh Physic Social **8**, 2: 47-54 (2019)
3. E. A. Tyas, S. Hutabarat, and C. Ain. Management of Aquatic Resources Journal (MAQUARES) **6**, 2 (2018)
4. N. Nurfadillah, E. M. Adiwilaga, A. Damar J. Depik **1**, 2: 93-98 (2012)
5. B. Suharto, L. Dewi, A. N. Mustaqiman, T.R.A.K. Marjo. Indonesian Journal of Urban and Environmental Technology **2**, 2 (2019)
6. A. Muhtadi, Y. Yunasfi, F. F. Rais, N. Azmi, D. Ariska. Acta Aquatica: Aquatic Sciences Journal, **2**, 2 (2015)
7. Z. Hasan, I. N. Syawalludin and W. Lili. Jurnal Akuatika **4**, 1 (2013)
8. N. Nurfadillah, S. Tarina, E. Miswar, I Dewiyanti and S. Agustina (IOP Conference Series: Materials Science and Engineering, 567, 2019)
9. Gazali, Widiatmono, Rahadi and R. Wirosoedarmo. Jurnal Keteknikan Pertanian Tropis dan Biosistem **1**, 2: 1-8 (2013)
10. A. Rahman and H. Satria. LIMNOTEK, 23 (1):17-25 (2016)
11. F.F. Rais, D. Yunasfi and A. Muhtadi Aquacoastmarine. **9,** 4:34-42 (2015)
12. E.P. Sagala. LIMNOTEK, 20 (2):151-158 (2013)
13. A. Riyad. Jurnal Teknik Lingkungan PTL-BPPT. **7,** 2: 166 – 172 (2006)
14. A. Khayra, Z.A. Muchlisin and M. A. Sarong. J. Depik, **5,** 2: 57-66 (2016)
15. T. Widiyanto. Jurnal Limnotek Perairan Darat Tropis, **24,** 2: 83-92 (2017)
16. P. Wirabumi, Sudarsono and Suhartini. Jurnal Prodi Biologi **6,** 3: 174-184 (2017)
17. I.M. Shuthers and D. Rissik. Plankton. A Guide to Their Ecology and Monitoring for Water Quality (CSIRO Publish, Melbourne, 2009)
18. G. Hassle, E. Syvertsein, K. Streidinger, K. Tangen and C. Tomas. Marine diatom: identifying marine diatoms and dinoflagellates (Academic Press, 1996)
19. American Public Health Association [APHA]. Standard methods for the examination of water and wastewater. Amer. Publ. 17th Edition (New York Health Assosiation, 2005)

20. E.P. Odum. Fundamental of Ecology Edisi III. Sounders College Publishing (Philadelphia, 1971)
21. S. Andriani, T.R. Setyawati, and I. Lovadi. Protobiont **4**, 1 (2015)
22. T.R.S. Andriansyah and I. Lovadi. Protobiont **3**, 1 (2014)
23. G. Martin and M.R. Fernandez. Diatoms as Indicators of Water Quality and Ecological Status: Sampling, Analysis and Some Ecological Remarks (InTech Europe, Croatia, 2012)
24. R. Aryawati and H. Thoha. Maspari Journal: Marine Science Research **2**, 1: 89-94 (2011)
25. T.A. Barus. Jurnal Manusia dan Lingkungan **11**, 2: 64-72 (2004)
26. Y. Linus & N. Irawati. Jurnal Manajemen Sumber Daya Perairan **2**,1 (2016)
27. J.A. Ludwig, L. Quartet, J.F. Reynolds. Statistical ecology: a primer in methods and computing (Vol. 1) (John Wiley & Sons, 1988)
28. E. Indrayani, K.H. Nitimulyo, S. Hadisusanto & R. Rustadi. Jurnal Manusia dan Lingkungan, **22**, 2 (2015)

Comparative anatomy of the caudal fin (*pinna caudalis*) *Tor douronensis* (Valenciennes, 1842) and *Tor soro* (Valenciennes, 1842)

Yusrizal Akmal[1], *Ilham* Zulfahmi[2,]*, *Muliari* Muliari[1], *Andi* Iqbal Burhanuddin[3], *Budimawan* Budimawan[3], *Yeni* Dhamayanti[4], *Rindhira* Humairani[1], *Irfannur* Irfannur[1], *Rinaldi* Rinaldi[1]

[1]Department of Aquaculture, Faculty of Agriculture, Universitas Almuslim, Bireuen, Indonesia
[2]Department of Fisheries Resources Utilization, Faculty of Marine and Fisheries, Universitas Syiah Kuala, Banda Aceh, Indonesia
[3]Faculty of Marine Science and Fisheries, Hasanuddin University, Sulawesi Selatan, Indonesia
[4]Department of Veterinary Anatomy, Faculty of Veterinary Medicine, Universitas Airlangga, Indonesia

Abstract. *Tor douronensis* and *Tor soro* were two of the four species of the Tor genus that live in Indonesian waters. However, studies related to the skeleton of these two fish are still rarely disclosed. The aim of this study compared the morphology of the caudal-fin (*pinna caudalis*) *T. douronensis* and *T. soro*. The research stages include sample preparation, making skeleton preparations, image analysis, and identification of skeleton terminology. *T. douronensis* fish were collected from the waters of the Pagar Alam area, Lahat Regency, South Sumatra, while *T. soro* was collected from the waters of Bukit Lawang, Bohorok District, Langkat Regency, North Sumatra Province. The caudal-fin (*pinna caudalis*) is part of the ossa urostylus which produces optimal hydrodynamic propulsion. The caudal-fin (*pinna caudalis*) Genus Tor is part of the *ossa urostylus* which is composed of 31 caudal-fin rays (*pinnae*), six *hypural* bones, *parhypural*, *pleurostylus*, *epural*, and *uroneuralis*. The ventral part of *T. douronensis* and *T. soro* is composed of the *parhypural*, and the 1st and 2nd *hypural* bones. *T. douronensis* had a *parhypural* bone that was more prominent and separates from the *spina hemalis* compared to *T. soro*. The dorsal part was composed of the 3rd *hypural* bones to 6th *hypural*, in *T. soro* the *hypural* bone was fused with cartilage. *Os pleurostylus T. douronensis* had a more prominent shape than *T. soro* and there was enlargement in the posterior part. The 3rd and 4th *hypural* bones on *T. soro* had the largest size. The *T. soro* had three *spina neuralis* and three *spina hemalis* to support the rays of the tail. The bones that composed the caudal-fin of *T. douronensis* and *T. soro* were relatively similar to those of some fish from the family Cyprinidae. The results of this study could be used as an alternative to identify *T. douronensis* and *T. soro* from the skeleton.

1 Introduction

Although many studies of the fish skeleton morphology had been carried out, the nomenclature of the bones that make up the skeleton, especially the head, still shows a high level of inconsistency so that it had an impact on the bias of the bone constituent data for different fish species (Bogutskaya et al. 2008) [1]. Fish of the same genus generally have almost similar external morphological characteristics so that it was sometimes difficult to distinguish, therefore additional analysis was needed to be able to identify differences between these species, including through skeleton comparisons. Several studies on skeleton comparisons of several fish families had been carried out previously, including fish from the Nemacheilidae family [2], Characidae family [1], Cichlidae family [3], and Zoarcoidei family [4].

The limb bones of fish were composed of fins supported by skeletal and controlled by muscles [5]. According to Standen, fish depend 20% of their movement and propulsion on their fins [6]. Generally, fish have two paired fins and three single fins. Paired fins consist of pectoral (*pinna pectoralis*) and pelvic fins (*pinna pelvis*), while single fins consist of dorsal (*pinna dorsalis*), anal (*pinna analis*), and caudal fin (*pinna caudalis*) [7][8]. The caudal-fin had optimal hydrodynamic propulsion and the shape could also indicate the swimming habits of fish [9]. In some fish species, the shape of the fins and the rays of the fins are modified from the general form and had several other additional functions, such as sperm distribution and attachment tools [9].

Tor douronensis and *Tor soro* were two of the four species of the Tor genus that live in Indonesian waters. However, the study related to the osteology of these two fish are still rarely disclosed. The caudal-fin was one of the body parts of the fish with striking differences between species in the same genus in addition to weber bones [10]. The aims of this study was to compare the caudal-fin (*pinna caudalis*) of *T. douronensis* and *T. soro*.

Corresponding author: Ilham.Zulfahmi@unsyiah.ac.id

2 Materials and Methods

2.1 Location and Sample Preparation

The research stages include sample preparation, making skeleton preparations, image analysis, and identification of the skeleton terminology. Making skeleton preparations at the Aquaculture Laboratory, Aquaculture Study Program, Faculty of Agriculture, Almuslim University. The fish was used in this study were obtained by direct capture from fish collectors. Sample of the fish have a minimum weight, at least 2 kg with a length 30-50 cm. *T. douronensis* were collected from the waters of the Pagar Alam area, Lahat Regency, South Sumatra, while *T. soro* was collected from the waters of Bukit Lawang, Bohorok District, Langkat Regency, North Sumatra Province. 10 sample for each type of the fish were collected. The fish are then preserved using ice and transported to the laboratory.

2.2 Caudal fin preparation

The procedure of the preparation tail was based on Zulfahmi et al. 2020[11]. Tail's fish was cutting located at the posterior aspect of the last centrum of the *ossa vertebrae caudales*. The separated fish fins are then soaked in hot water at a temperature range of 80–90°C until the muscle and connective tissue are blistered and white. Soaking in hot water was done carefully so that the bones did not become brittle. Preparation of fish skeleton preparations was done physically and chemically. Physically, fish scales are removed using a knife or tweezers and then doused with hot water until the muscles scald and turn white. The chemical stage begins with immersing the skeleton sample into formalin 10%. Specimens were transferred to 70% ethanol solution to remove stains on the skeleton [12].

2.3 Image editing and Identification of Skeleton Terminology

The digital image of the fish was taken from the lateral direction with *Canon EOS 750D*. After that, it was analyzed with the *Adobe Photoshop CS6* software and then presented with a table and image.

3 Results and Discussion

The cauda fin (*pinna caudalis*) of the Tor's Genus was part of the urostylus bones. These bones had 31 of the cauda fin (*pinnae*), six of the hypural bones, pleurostylus, epural, and uroneuralis bone. The ventral side of the *T. douronensis* and *T. soro* is composed of parhypural bone, and 1st – 2nd hypural bones. The parhypural bone of the *T. douronensis* more prominent than T. *soro*. The parhypural bone of the *T. soro* fused with cartilage bone and enlargement at the posterior part.

The dorsal part was composed of the 3rd to 6th hypural bones, and in *T. soro* the hypural bone was fused with cartilage. Os pleurostylus *T. douronensis* was a more prominent shape than *T. soro*, and there was an enlargement in the posterior part. The 1st, 3rd, and 4th hypural bones of the *T. soro* had the largest size, while the 5th and 6th hypural was the smallest size. The 2nd hypural *bone of T. douronensis* was the largest one, while the 6th hypural was the smallest size. *T. soro* had three neural spines and three haemal spines to support the rays of the caudal fin (Figure 1.)
The bones that make up the fins of the tail of *T. douronensis* and *T. soro* are relatively the same as some fish from other Cyprinidae families, such as *Ischikauia steenackeri*, *Alburnus amirkabiri*, *Cyprinion milesi*, and *Barbus cyri* [14][15][16][17]. However, the Barbus cyri fish had a larger hypural bone size than the parhypural bone, compared to that T. *soro* fish.

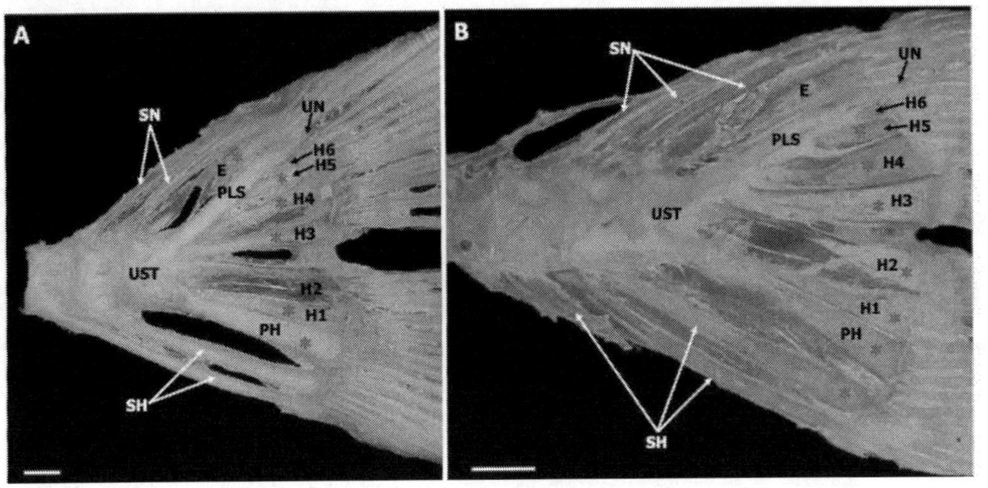

Fig. 1. *Pinna caudalis* lateral view of *T. douronensis* (A) and *T. soro* (B). Description: *arcus neuralis* (AN); *Centrum* (C); *Os epural* (E); *Os hypural* (H); *Os parhypural* (PH); *Os pleurostylus* (PLS); *Os preural urostylus* (PU); *spina haemalis* (SH); *spina neuralis* (SN); *Os uroneuralis* (UN); *Ossa urostylus* (UST). Bar scale: 0.5 cm.

Table 1. Comparative morphology of the caudal fin (*pinna caudalis*) *T. douronensis* and *T. soro*

No	Bone name	view	*T. douronensis*	*T. soro*
1	*Os parhypural*	Lateral	The shape more prominent and slim, separated from *spina hemalis*	The shape thinner, enlargement at the posterior part, and fused to *spina hemalis*
2	*Os hypural 1*	Lateral	Slim and sturdy	Enlargement at the posterior part
3	*Os hypural 2*	Lateral	Slim and thin	Slim and short
4	*Os hypural 3*	Lateral	Slim and sturdy	Wide and enlargement at the posterior part
5	*Os hypural 4*	Leteral	Had a clear boundary with the 5th *hypural* bone	Had the largest size with indistinct boundaries with the 5th hypural O
6	*Os hypural 5*	Leteral	Not wide and short	Wide and long
7	*Os hypural 6*	Lateral	Clearly boundry with the 5th hyoural bone	Indistinct boundry with the 5th hypural bone
8	Neural spines	Lateral	2 spines	3 spines where the last one bigger than others
9	Haemal spines	Lateral	2 spines	3 spines

T. douronensis had more tail fin rays than *Petrocephalus simus* (family Mormyridae) and lower than *Oreochromis lorenzoi* fish (family Cichlidae) [18][19]. The caudal fin rays of *T. douronensis* and *T. soro* fish were divided into two parts, namely the ventral fin rays and the dorsal fin rays with sharp angles. According to Lauder, the function of the ventral caudal fin rays was more dynamic than the dorsal [20]. The shape of the pointed tail fin with long fingers on the *Keureling* fish indicated that the fish was capable of swimming at high speeds [21]. The caudal fin was one of the bones that had a high variation between species compared to the bones that make up other body parts.

4 Conclusion

The caudal fin (*pinna caudalis*) of *T. douronensis* and *T. soro* fish was part of the ossa urostylus. The urostylus bone was composed of 31 caudal-fin rays (*pinnae*), six hypural bones, parhypural, pleurostylus, epural, and uroneuralis bones. The ventral parts of *T. douronensis* and *T. soro* were composed of the parhypural bone and the 1st and 2nd hypural bones. *T. douronensis* had a parhypural bone that was more prominent and separated from the *haemalis spine* than *T. soro*. The dorsal part was composed of the 3rd to 6th hypural bones. The hypural bone of the *T. soro* fused with cartilage. The shape pleurostylus *bone T. douronensis* was more prominent than *T. soro,* and there was an enlargement in the posterior part. The size of the 3rd and 4th hypural bones of the *T. soro* was the largest. The *T. soro* had three neural spines and three haemal spines to support the rays of the caudal fin. The results of this study could be used as an alternative to identify *T. douronensis* and *T. soro* by osteology.

Acknowledgments. We would like to thank the Ministry of Education, Culture, Research, and Technology Indonesia for funding this research through the Penelitian Kerjasama Antar Perguruan Tinggi/PKPT scheme (Number: B/112/E3/RA.00/2021 and Agreement/Contract Number: 173/ E4.1/AK.04.PT/2021, 035/LL13/AKA/LT /2021, 140/LPPM-Umuslim/KP/2021).

References

1. Y. Akmal, Y Dhamayanti, & E. Paujiah. Osteocranium of Tor tambroides (Cypriniformes: Cyprinidae) from Tangse River, Aceh, Indonesia. *Biodiversitas Journal of Biological Diversity*, *21*(2) (2020).
2. P. Mafakheri, S. Eagderi, H. Farahmand Mousavi-Sabet H. Osteological structure of Kiabi loach, *Oxynoemacheilus kiabii* (Actinopterygii: Nemacheilidae). *Iranian Journal of Ichthyology*, 1(3): 197 – 205 (2015).
3. K. Dierickx, W. Wouters, & W. Van Neer. Comparative osteological study of three species of distinct genera of *Haplotilapiini* (Cichlidae). *Cybium*, 41(3): 223-235 (2017).
4. EJ. Hilton & NJ. Kley. Osteology of the Quillfish, *Ptilichthys goodei* (Perciformes: Zoarcoidei: Ptilichthyidae). *Copeia, 2005*(3): 571-585 (2005).
5. EJ. Hilton. The Skeleton Bony Fish Skeleton. Elsevier Inc. All rights reserved, Virginia Institute of Marine Science, Gloucester Point, VA, USA. 434-436p (2011).
6. EM. Standen. *Buoyancy, Locomotion, and Movement in Fishes, Paired Fin Swimming*. Elsevier Inc, McGill University, Canada. 564p (2011).
7. P. Lőw, K. Molnár, G. Kriska. Atlas Of Animal Anatomy And Histology. Springer, pp17 (2016).
8. J. Cardeira, R. Valles, G. Dionisio, A. Estevez, E. Gisbert, P0.00000 Pousao-Ferreira, ML. Cancela, PJ. Gavaia. Osteology of the axial and appendicular skeletons of the meagre *Argyrosomus regius* (Sciaenidae) and early skeletal development at two rearing facilities. *Journal of Applied Ichthyology*, 28(12): 464–470 (2012).
9. Y. Akmal, & MF. Rahardjo. Morphology of appendicular skeleton of the Thai mahseer's Tor tambroides (Bleeker, 1854). *Jurnal Iktiologi Indonesia, 18*(3), 261-274 (2018).

https://doi.org/10.1051/e3sconf/202233903009

10. I. Zulfahmi, Y. Akmal, M. Radhi, M. Hidayat, & M. Muliari. Comparative osteology of Tor tambroides (Bleeker, 1854) and Tor tambra (Valenciennes 1842) vertebral column (ossa vertebrae). *Jurnal Iktiologi Indonesia, 20*(3), 235-249 (2020).
11. I. Zulfahmi, & Y. Akmal. Komparatif Osteologi Tor tambroides & Tor tambra. PT Penerbit IPB Press (2020).
12. WR. Taylor, & CC. Van Dyke. Revised procedures for staining and clearing small fishes and other vertebrates for bone and cartilage study. *Cybium*, 9(2): 107–119 (1985).
13. I. Zulfahmi, Y. Akmal, Muliari. *Osteologi Ikan Keureling (Tor tambroides)*. IPB Press, Bogor. Pp. 107 (2019).
14. H. Takeuchi K. Hosoya. Osteology of *Ischikauia steenackeri* (Teleostei: Cypriniformes) with comments on its systematic position. *Ichthyological Research,* 58(1): 10–18 (2011).
15. P Jalili, S. Eagderi, M. Nasri, H. Mousavi-Sabet. Descriptive osteology study of *Alburnus amirkabiri* (Cypriniformes: Cyprinidae), a newly described species from namak lake basin, central of Iran. *Bulletin of the Iraq Natural History Museum*, 13(4): 51–62 (2015a).
16. P. Jalili, S. Eagderi, N. Nikmehr, Y. Keivany. Descriptive osteology of *Barbus cyri* (Teleostei: Cyprinidae) from southern Caspian Sea basin. *Iranian Journal of Ichthyology*, 2(2): 105–112 (2015b).
17. M. Nasri, S. Eagderi, H. Farahmand. Descriptive and comparative osteology of Bighead Lotak, *Cyprinion milesi* (Cyprinidae: Cypriniformes) from southeastern Iran. *Vertebrate-zoology*, 66(3): 251–260 (2016).
18. EJ. Hilton. Comparative osteology and phylogenetic systematics of fossil and living bony-tongue fishes (Actinopterygii, Teleostei, Osteoglossomorpha). *Zoological Journal of the Linnean Society*, 137(1): 1–100 (2002).
19. G. Carnevale, C. Sorbini, W. Landini. *Oreochromis lorenzoi*, a new species of tilapiine cichlid from the late Miocene of central Italy. *Journal of Vertebrate Paleontology*, 23(3): 508–516 (2003).
20. GV. Lauder. Function of the caudal fin during locomotion in fishes: kinematics, flow visualization, and evolutionary patterns. *Integrative and Comparative Biology*,40(1):101–122 (2000).
21. JA. Walker, MW. Westneat. Performance limits of labriform propulsion and correlates with fin shape and motion. *The Journal of Experimental Biology*, 205: 177–187 (2002).

Sustainable management of fisheries resources in Java Sea: Utilization status of longtail tuna (*Thunnus tonggol*) in Indramayu waters West Java case study

Gatut Bintoro*, *Daduk* Setyohadi, *Tri* Djoko L, *Rofiiqoh* Affif N, *Almira* Syawli, and *Muammar* Kadhafi

Fisheries and Marine Science Faculty, Universitas Brawijaya, Veteran Street 65145 Malang, Indonesia

Abstract. Longtail tuna (*Thunnus tonggol*) is a pelagic fish which is generally found in tropical waters. Research on fishery resource sustainability by applying bioeconomic and length weight assessments is necessary to be done in order to determine status of longtail tuna. This research was conducted in south part of Java Sea namely Indramayu waters West Java from January 2020 to April 2020 which aims to estimate utilization status of longtail tuna (*T. tonggol*) namely exploitation rate, Maximum Sustainable Yield (MSY), Total Allowable Catch (TAC), Maximum Economic Yield (MEY), and Open Access (OA), and to determine length and weight relationship. The research used quantitative descriptive method based on catch and effort data from 2010 to 2019 as well as length weight primary data as much as 987 fish. Result showed that catch and effort at MSY level was 8,453 tons/year and 2,974 trips/year respectively with value of TAC as much as 6,762 ton/year. While the estimated values of catch and effort at MEY level was 8,168 tons/year and 2,427 trips/year with optimum profit as much as 88.98 billion Indonesia dollar rupiah (IDR). Another analysis indicated that OA level was achieved when effort reached at 4,855 trips/year and whilst exploitation rate was 93%. Meanwhile length weight relationship analysis showed that fish had a balance length weight growth which b value = 3.039 with equation $W = 0.0147L^{3.0393}$. In conclusion, the fish has an isometric length weight relationship and the utilization status of fish is in fully exploited condition hence additional fishing effort must be prohibited in order to maintain stock sustainability.

1 Introduction

Java Sea, including Indramayu waters, has high potency of marine fisheries resource. Indramayu Regency has three small islands, namely Biawak Island, Gosong Island and Candikian Island. Karangsong Fish Landing Base (FLB), in Indramayu, is one of the fish landing baes for pelagic fish catches and one of the busiest fisheries landings in Indramayu [1].

Longtail tuna (*Thunnus tonggol*) is a type of pelagic fish that is oceanodromous and its population is mainly in clear neritic waters. This fish is spread throughout the Indo-West Pacific region from northern Australia, Papua New Guinea and Indonesia, while from the northwest through Malaysia and Thailand, continues to spread to the northeast and south of Japan, northwest of Iran to the Red Sea. Longtail tuna (*T. tonggol*) is the second smallest of the eight species *Thunnus*, growing to a maximum size of 142 cm and a weight of 35.9 kg. This type of longtail tuna (*T. tonggol*) is generally the target of fishing in purse seines, gill nets, and line fishing [2].

The bioeconomic approach used in this research is surplus production. While there is no information about economic benefits regarding with fisheries resource utilization, it hopes that through this approach, the estimation of biological values can be known and achieve sustainable welfare. Utilization of fishery resources must be based on socio-economic aspects as well as fish biology, sustainability and environmental conditions to support sustainable utilization activities. Management of fishery resources is how the activities of utilizing these resources produce high economic benefits for users, but their sustainability is maintained [3].

The analysis of the length and weight relationship is intended to measure the variation in expected weight for a certain length of fish individually or in groups of individuals as an indicator of obesity, health, gonad development, and so on. Another usefulness of length-weight relationship analysis is that it can be used to estimate condition factors or often referred to as the index of plumpness, which is one of the important derivatives of growth to compare the relative health condition or condition of a particular fish population or individual regarding with resource sustainability [4].

2 Methodology

The material used in this research is longtail tuna (*T. tonggol*) caught in Java Sea and landed at the Karangsong FLB, Indramayu, West Java. The observations made were regarding with the length and weight relationship and fish samples were taken as many as 987 fish. The bioeconomic aspect uses tuna resource

* Corresponding author: gbintoro@ub.ac.id

data from 2010-2019 which were obtained from statistical data from the Karangsong FLB, Indramayu Regency, West Java. The study also used direct observation and descriptive measurements of length and weight of fish. The data used are longtail tuna production, fishing effort in trip units and catch per unit Effort (CPUE). The length and weight sizes were measured directly, the data obtained were processed using Microsoft Word and Microsoft Excel software.

2.1 Data analysis of length weight relationship

Relationship of length and weight is estimated based on the cubic model [5-7] in the exponential equation:

$$W = a\,Lb(1) \quad (1)$$

Where:
W = fish weight (g);
L = length of fish (TL = mm, or cm);
a = constant, and
b = allometric constant;
Estimation of constants a and b is done by transforming the cubic equation above into a linear equation:

$$Ln\,(W) = Ln\,(a) + b\,Ln\,(L) \quad (2)$$

The estimation of the values a and b is based on the linear regression approach (value intercept and regression direction coefficient). The increase in fish body length is equal to the increase in weight so it is called an isometric growth model (b = 3), a deviation of the value of b from 3 is called an allometric growth model. Value b> 3, then the shape of the fish tends to be fat (an indication of the availability of natural food), while the value of b <3 indicates the opposite. The length weight relationship is part of fish biology data to illustrate fish stock sustainability [8, 9].

2.2 Fishing effort standardization

The formula for standardization calculation of fishing gear [10] is as follows:

$$Pj = Uj\,/\,Us \quad (3)$$
$$Ejt = Pj\,.\,Tjt\,.\,Vjt \quad (4)$$

Where:
Pj = Average fishing power of fishing gear j, with constant value every year;
Uj = Average vessel productivity using fishing gear j;
Us = Average productivity of fishing gear s;
Ejt = Fishing effort of fishing gear j year t;
Tjt = Average fishing days of vessel j year t; and
Vjt = Amount of vessel j year t.

2.3 Maximum sustainable yield (MSY) calculation

2.3.1 Schaefer (1954) model

Optimum fishing effort rate (fMSY) and the maximum sustainable catch (YMSY) can be known by the following equation:

a. Relationship between catch (Y) and fishing effort (f)

$$Y = af + bf2 \quad (5)$$

$$Y/f = U = a + bf \quad (6)$$

b. The maximum sustainable effort (f_{MSY}) is obtained from equation (7) as follows:

$$Y/f = a + 2bf \quad (7)$$
$$0 = a + 2bf \quad (8)$$
$$a = 2bf \quad (9)$$
$$f_{MSY} = -a/2b \quad (10)$$

c. The maximum sustainable catch (Y_{MSY}) is obtained by substituting the optimum effort value (f_{MSY}) to equation (5),

$$YMSY = af_{MSY} + bf_{MSY}^2 \quad (11)$$
$$YMSY = -a^2/4b \quad (12)$$

The calculation of the catch per fishing effort (UMSY) is obtained by the formula equation:

$$U_{MSY} = Y_{MSY}/f_{MSY} \quad (13)$$

Estimation of the Total allowable catch value (Y_{TAC}) based on Schaefer model using the following equation:

$$Y_{TAC} = -a^2/4b \times 80\% \quad (14)$$

$$Y_{TAC} = Y_{MSY} \times 80\% \quad (15)$$

The estimated value of the allowable fishing effort (f_{TAC}) can be found using the abc math formula with the following equation:

$$TAC = \frac{a \pm \sqrt{(a^2 - 4b \times c))}}{2b} \quad (16)$$

Where:
Y = Yield or catch (tonnes)
f = Effort or number of attempts to catch (trip)
a = intercept
b = slope
c = Total allowable catch (ton)
Y_{MSY} = Maximum sustainable catch (ton)
f_{MSY} = Sustainable effort (ton)
U_{MSY} = Maximum CPUE (ton/trip)
f_{TAC} = Allowable effort (trip)
Y_{TAC} = Allowable catch (tonnes)

2.3.2 Fox (1970) model

The assumption of Fox's (1970) exponential model is that the population is considered not to be extinct [5, 7, 11]. The Fox method uses an exponential pattern based on equation:

$$\frac{y(i)}{f(i)} = e^{(c+d*f(i))} \quad (17)$$
$$Y(i) = f(i) * e^{(c+d*f(i))} \quad (18)$$

Where:
c = intercept
d = slope

From this equation, a derivative equation is obtained to work out the utmost sustainable catch the maximum sustainable fishing effort as follows:

$$f_{MSY} = -\frac{1}{d} \quad (19)$$
$$Y_{MSY} = -\frac{1}{d*e^{(c-1)}} \quad (20)$$

2.4 Maximum economic yield (MEY) calculation

Calculation of possible maximum economic yield can be gained is often obtained using the Gordon-Schaefer method of calculation, which uses the principle of total revenue (TR) minus total costs (TC):

$$\pi = TR - TC \quad (21)$$
$$\pi = \boldsymbol{p.Y - C.f} \quad (22)$$

Where:
π = profit
p = average fish price (IDR/kg)
y = volume of production (kg)
c = monetary value per trip (IDR/trip)
f = catch effort (trip)

Analysis of maximum economic yield (MEY) can be calculated based on the equation of [12]:

$$f_{FAE} = \left(\left(\frac{c}{p}\right) - a\right) \quad (23)$$
$$Y_{FAE} = \frac{(c*p)}{f_{FAE}} \quad (24)$$
$$f_{MEY} = \frac{1}{2} f_{FAE} \quad (25)$$

Where:
f_{FAE} = Effort to catch at the equilibrium point
Y_{FAE} = The catch at the purpose of balance
f_{MEY} = Sustainable economic fishing effort
a = intercept
b = slope

Then to determine the value of sustainable economic catch (Y_{MEY}) it can be calculated by substituting the value of f_{MEY} into the equation as follows:

$$Y_{MEY} = a\,xf_{MEY} + bxf_{MEY}^{\;2} \quad (26)$$

2.5 Utilization Rate (UR)

The calculation of the percentage value in biology that has been used will be known using the following formula:

$$UR = (f_i / f_{TAC}) \times 100\ \% \quad (27)$$

The calculation of the percentage value economically that has been utilized can be found using the following formula:

$$UR = f_i/f_{MEY} \times 100\% \quad (28)$$

Where :
f_i =Last year's catch
f_{TAC} =Total allowable catch
f_{MEY} = Economically sustainable effort
UR = Utilization rate

3 Results

Based on fishery statistics data at Karangsong FLB, Indramayu from 2010 to 2019, longtail tuna production has fluctuated. (Fig 1).

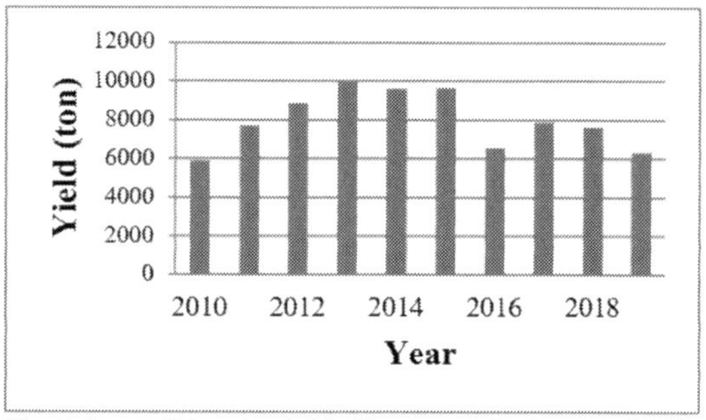

Fig. 1. Longtail tuna landed in Karangsong Indramayu (2010-2019).

Figure 1 above illustrates the catch of longtail tuna for a period of 10 years (2010-2019) at Karangsong FLB, Indramayu. The highest production occurred in 2013 with a volume of 9,968 tons and the lowest production occurred in 2010 with a volume of 5,877 tons.

The nature of fish growth can be divided into 2, namely isometric (b = 3) or length growth equivalent to weight growth, and allometric (b ≠ 3) or long growth does not equal weight growth. Growth Allometric can be divided into 2, namely positive allometric (b> 3 or growth in weight faster than length growth) and negative allometric (b <3 or growth in length faster than weight growth). Analysis of the length and weight relationship was used to determine the growth patterns of fish. Analyzing the relationship between length and weight of fish can use linear regression analysis with the model W = aLb, where the values of a and b are obtained from linear regression analysis. The length-weight relationship of longtail tuna (*T. tonggol*) is analized based on total data obtained by 987 fish.

3.1 Length weight relationship

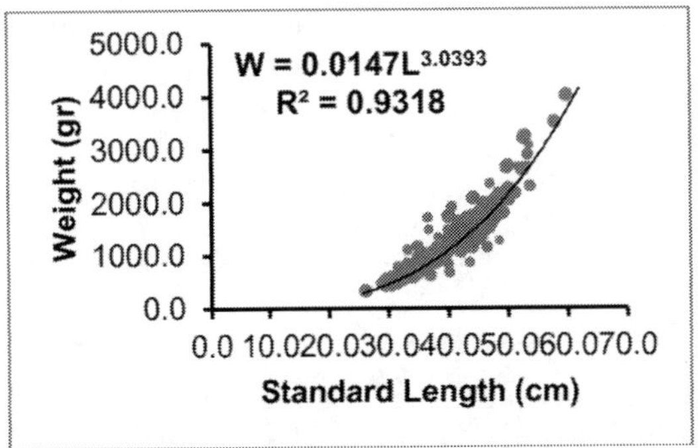

Fig. 2. Length weight relationship of longtail tuna in Karangsong, Indramayu.

Data analysis results for samples, heavy longtail tuna (*T. tonggol*) shows that it is W = 0.0147L3039 (Fig. 2). It shows that seen from the growth of the weight gain of longtail tuna (*Thunnus tonggol*) is faster than the length of the fish (b> 3). It is known that the b value is 3.0393, the R square is 0.9318, the t value is 1.49 and the t table value is 1.96. This shows that the value of b> 3 and the value of t count <t table which means the weight and length of longtail tuna (*T. tonggol*) is isometric or balance length and weight growth.

3.2 Fishing gear Standardization

Standardization of fishing gear can be calculated after knowing the value of fishing power index (FPI). The FPI is the capability level of a fishing gear to catch a type of fish within a certain period of time. Calculating the FPI is necessary if a type of fish is caught using several fishing gears (Table 1).

Table 1. Productivity and FPI in Karangsong FLB, Indramayu (2010-2019).

Fishing gears	Produktivity (kg/trip)	FPI	Rasio
Millenium	1,497.63	1.00	1.00
Gill net	971.21	0.65	1.54
Seine net	973.75	0.65	1.54
Hand line	977.62	0.65	1.53
Trammel net	827.39	0.55	1.81

Standardized fishing gear is used to know which fishing gear is the most dominant in catching tuna, this is used to determine unit of effort the standard before calculating the potential for sustainable fishing (MSY).

3.3 Maximum sustainable yield (MSY) analysis

Table 2. Effort, catch, CPUE,and Ln CPUE of longtail tuna in Indramayu West Java.

Year	Effort (Trip)	Catch (Ton)	CPUE	Ln CPUE
2010	1623.0	5877.1	3.62	1.29
2011	1906.0	7732.4	4.06	1.40
2012	2356.0	8878.5	3.77	1.33
2013	2883.0	9967.8	3.46	1.24
2014	2562.0	9632.4	3.76	1.32
2015	2802.0	9666.8	3.45	1.24
2016	2371.0	6572.5	2.77	1.02
2017	2659.0	7868.0	2.96	1.08
2018	3282.0	7632.4	2.33	0.84
2019	3665.0	6313.3	1.72	0.54

3.3.1 Schaefer (1954) Model

The highest CpUE value was obtained in 2011 at 4.06 tonnes/trip, while the lowest value was in 2019 at 1.72 tonnes/trip. This CpUE catch has experienced changes that fluctuate from year to year (Table 2). Regression analysis of CpUE against effort indicated that value of intercept (a) and slope (b) was 5.6854 and -0.000955995 respectively (Fig. 3). Hence non linear equation of MSY based on Schaefer model was $Y = 5.6854f - 0.000955995f^2$. With the result of sustainable catch and effort 8453 ton/year and 2974 trip/year respectively (Fig. 4).

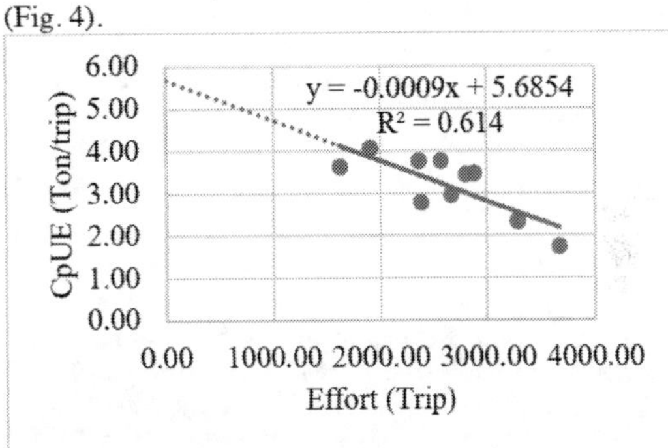

Fig. 3. Relationship between CpUE and effort of Schaefer model for Longtail tuna in Indramayu.

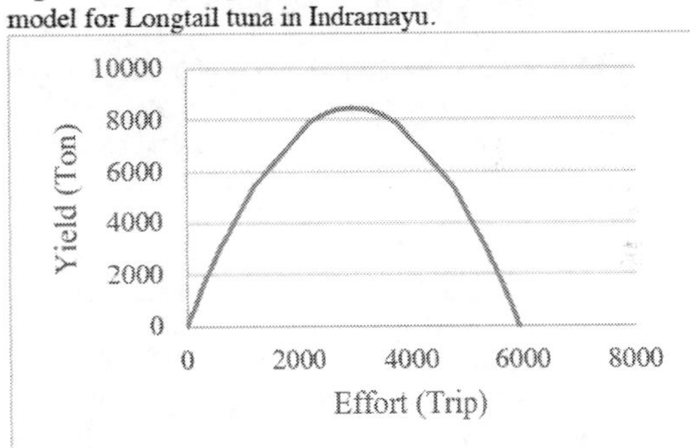

Fig. 4. Maximum sustainable yield (MSY) of Schaefer model for longtail tuna in Indramayu.

3.3.2 Fox (1970) Model

Calculation the maximum sustainable yield (MSY) using Fox model required effort and ln CpUE data to apply regression analysis (Table 2). It was found that intercept (c) and slope (d) values were 2.040591946 and -0.0003484 respectively (Fig. 5). The Fox model uses the Gompertz growth function to analyze a surplus production model which results in an exponential relationship between fishing effort and population size and an asymmetrical production curve. Hence MSY calculation based on Fox model was 12,650 ton/year and 7,959 trip/year respectively.

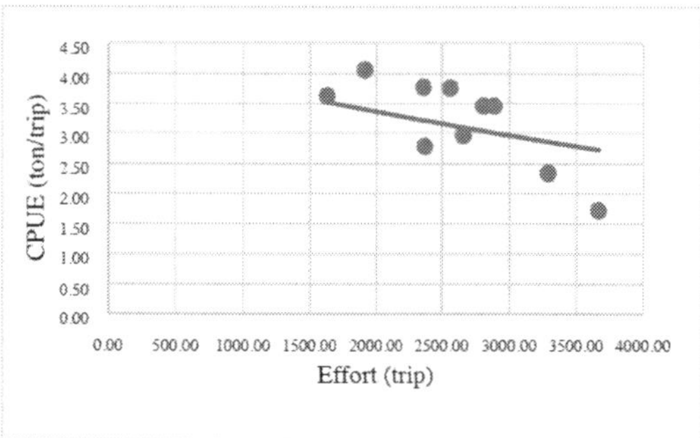

Fig. 5. Relationship between Ln CpUE and effort of Fox model for longtail tuna in Indramayu.

3.4 Bioeconomic analysis of Gordon-Schaefer model

Determining the MEY calculation that must be done is knowing the price of fish and the operational costs of fishing for tuna.

3.4.1 Fish price

Calculation of fish price can be done by identifying fish landed and production value in local fish landing place It was found that the fish price of longtail tuna at FLB Karangsong Indramayu in year 2019 was IDR 15.798 Million/ton.

3.4.2 Operational costs

The calculation of operational costs is obtained from the sum of fixed costs (fixed costs) andvariable costs. Fixed costs include depreciation costs of vessel, machinery, fishing gear and permit fees. The depreciation is obtained by applying calculation of fishing gear and equipment prices divided by their technical age. The variable costs include fishing accomodation, namely diesel, oil, consumption, ice block, and bait.

The proportion of catch of millenium fishing gear for long tail tuna based on production value is 31%. The operational cost of millenium for fishing activity is 53.225 million IDR per trip. Hence, operational cost of the millenium to catch longtail tuna is of 16,499,750 IDR/trip.

3.4.3 MEY calculation

After knowing operational costs and the price of tuna, then a sustainable economic catch potential will be calculated based on the formula equation of [12] as follow:

$f_{MEY} = ap\text{-}c/2bp$
$= (73,315,906/30,204)$
$= 2,427$ trip/year

$Y_{MEY} = a^2/4b - c^2/4bp^2$
$= 8,453 - 285 = 8,168$ ton/year

3.4.4 Bioeconomic Analysis

The catch shows that the production of longtail tuna is different at certain levels of effort. When fishing is still low, the increase in costs will follow the increase in business revenue so as to achieve economic balance (Table 3 and Fig. 6).

Table 3. Calculation of total revenues and total costs.

Variable	Existing (2019)	MSY	MEY	OA
Catch (ton)	6,313	8,453	8,168	5,070
Effort (trip)	3,665	2,974	2,427	4,855
Revenue (Billion IDR)	99.7	133.5	129.1	80
Cost (Billion IDR)	60.5	49.1	40.1	80
Profit (Billion IDR)	39.3	84.5	89	0

Total revenue earned is greater than the fishing costs incurred so that fishermen will get a large profit or profit until with the point of F_{MEY}.

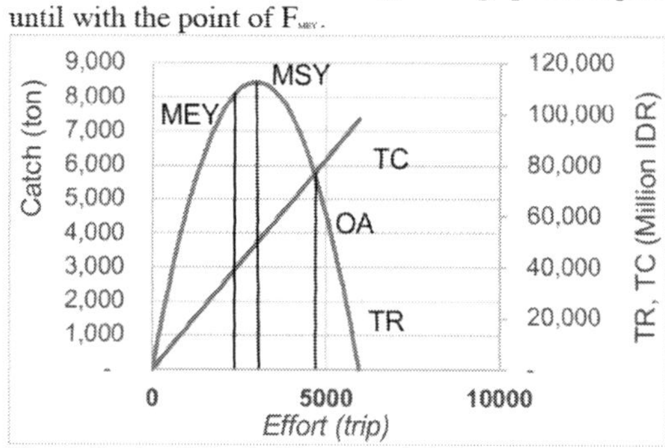

Fig. 6. Gordon-Schaefer bioeconomic balance.

3.5 Estimation of utilization rate

Estimation of the utilization rate of the Schaefer model using the equation UR = (f_i / f_{TAC}) x 100%, where in the past 10 years, an average catch of 8,014 tonnes was obtained with a total allowable catch (TAC) value of 6,762 tonnes, so that the results of the Utilization Rate (UR) was 119%. These results showed that the exploitation level of longtail tuna Indramayu waters in the period of 2010-2019 was in over exploited which means that the fisheries resources have been exploited over MSY level. So, it is recommended to reduce fishing effort because the stock has been disturbed.

Estimation of Fox model utilization rate (Table 4) using the equation UR = (f_i / f_{TAC}) x 100%, where in the last 10 years an average catch was obtained of 8,014.12 tons with a value total allowable catch (TAC) is 10,120 tonnes, so the results of Utilization rate (UR) of 79% show that the utilization rate of tuna landed in Indramayu waters in the period 2010-2019 is fully exploited, which means fish resources which have been

https://doi.org/10.1051/e3sconf/202233904001

exploited are close to the equivalent of 79% of MSY's value. Increasing the number of fishing efforts is not recommended although the number of catches can still increase but it can disrupt the preservation of longtail tuna resources.

Based on the economy, the number of sustainable fishing efforts (f_{MEY}) is 2,427 trips/year, while according to social, sustainable catch efforts (f_{OA}) of 4,855 trips/ year. Number of fishing gear MEY (f_{MEY} = 2,427 trips/ year) and social/open access (f_{OA} = 4,855 trips/ year). The number of fishing efforts will then be used as input data to estimate the level of exploitation of tuna resources in Indramayu waters (Table 5).

Table 4. Results of Analysis of Schaefer and Fox Model.

Variabel	Analysis	
	Schaefer	Fox
Intercept	5.68539	1.46334
Slope	-0.00096	-0.00013
R2	0.6147	0.01889
F_{MSY} (trip/year)	2974	7959
Y_{MSY} (ton/year)	8453	12650
U_{MSY} (ton/trip)	2.84	1.59
Y_{TAC} (ton/year)	6762	10120
UR	93%	62%
Status Utilization	*Fully Exploited*	*Moderate exploited*

Table 5. Results of Analysis of the exploitation level and status.

Variabel	MEY	OA
F	2.427	4.855
Y	8.168	5.070
Y/F	0.30	1.04
UR	98%	158%
Status	*Fully exploited*	*Depleted*

The results of the economic analysis show that the level of exploitation of tuna resources is 98%. This value indicates that the status of the tuna resources exploitation (fully exploited) is approaching and leads toover-exploitation. Results of social analysis or open access to get that level of exploitation of 158% indicates thestatus of utilizationdepletion(depleted) as well as fish resources have been exploited.

4 DISCUSSION

4.1 Length Weight Relationship

The results of the analysis of the relationship between length and weight obtained in this study indicate that the b value is 3.0393, the R square is 0.9318, the t count value is 1.49 and the t table value is 1.96. This shows that the value of b> 3 and the value of t count <t table which means the weight and length of longtail tuna, (*T. tonggol*) is isometric or length and weight are in balance growth, and the value of the coefficient of determination (R2) indicates that length growth affects fish weight growth. The multiple r value is 0.9652 which indicates the degree of closeness between the length and weight variables which are closely related.

The length weight relationship of fish generally is influenced by gonad maturity. While isometric growth is proportional development between length and weight. There are differences in the nature of this growth because fish growth can be influenced by heredity, sex, age, disease, food availability and water conditions [13].

4.2 MSY and MEY calculation

Schafer and Fox models is applied to find out which one is the most appropriate to estimate the status of longtail tuna resource utilization. It can be seen from the R square value, if the R square value approaching the value of 1, the model is getting better. R square in the Schaefer model has a value of 0.614 while the Fox model has a value of 0.0188, so it can be concluded that the appropriate and better model is the Schaefer model.

The coefficient of determination (R^2) aims to determine how much influence the independent variable has on the dependent variable. The value of the coefficient of determination is in the range of $0 \leq R^2 \leq 1$, which means that if the value of R^2 approaches a value of 1, it can be said to be independent variables on the dependent variable models used large means good and appropriate to clarify the effect of these variables [14].

Fishing effort will increase with increasing level of fishing effort. The total revenue is obtained by multiplying the price by the catch, while the total operational cost per trip is obtained by multiplying the fishing effort by the fishing cost per trip. Economic gain is the difference between the total revenue and the total cost for conducting a trip fishing.

Sustainable economic potential is the maximum value of the catch that can provide maximum benefits. MEY needs to be calculated so that the exploitation of fishery resources can run healthy and efficiently to get maximum benefits without disturbing the regeneration process or the recovery of these resources [7].

4.3 Bioeconomics

The total income earned is greater than the fishing costs incurred so that fishermen will get a large profit or profit up to the F_{MEY}. If the fishing business continues until F_{MSY}, then physically there will be a higher production but economically the profit will decrease. In this condition, it can be said that the fishing does not experience a loss because the total revenue is the same as the total expenditure, meaning that the tuna fishing effort has broken even.

In principle, the catch is the output of the fishing activity, while the effort required is the input of the fishing activity. In a production economy the comparison between output and input reflects the level of technical efficiency of each input use. Increased effort can also increase fishing costs, which means that increasing the use of fish resources beyond the maximum effort level in MEY can reduce the value of rent. The decline in rent value will not stop until it reaches a balance of open access [15].

The results of the economic analysis show that the level of exploitation of tuna resources is 95%. This value indicates that the status of the tuna resources exploitation (fully exploited) is approaching and leads to over-exploitation. The results of the social analysis or open access show that the level of exploitation of 158% indicates the status of depleted utilization as well as exploited fish resources.

The catch of longtail tuna resources in Indramayu waters has been categorized as over exploited, fish resources have decreased because they are exploited more than MSY and MEY levels. This is in accordance with the Ministerial Decree NO. 45/MEN/2011 concerning the estimated potential of fish resources which states that fish resources have been categorized as over exploited in 2019. Fishing efforts must be reduced because the sustainability of fish resources has been disrupted.

5 CONCLUSION

Length weight relationship of longtail tuna (*T. tonggol*) is isometric or balance length and weight growth with equation $W = 0.0147L$[illegible]. Value of maximum sustainable yield (MSY) takes place at 8,453 ton/year and 2,974 trip/year respectively with amount of TAC as big as 6,762 ton/year. While value of maximum economic yield (MEY) occurs at 8,168 ton/year and 2,427 trip/year respectively. In addition, utilization rate (UR) is 93% so that the status of the stock is fully exploited. Hence additional fishing effort of longtail tuna in Indramayu waters must be prohibited in order to maintain stock sustainability.

Acknowledgements: The authors sincerely thank the Fisheries and Marine Science Faculty, Universitas Brawijaya for providing the opportunity to carry out this piece of research and the Karangsong Fish Landing Base (FLB) staff members, Indramayu, West Java who have given permission to obtain field data. Last but not least, we would like to thank Prof Ali Suman (Research Institute for Marine Fisheries, Jakarta) for his nice review in this article.

References

1. G. Yulianto, S. Kadarwan, A. Lucky, Machfud. 2016. Demersal Fisheries Status in Indramayu adjacent coast West Java. Fisheries and Marine Faculty, IPB University, Bogor **10** (2016).
2. S.P. Griffiths, J Pepperell, M. Tonks, W. Sawynok, L. Olyott, S. Tickell, M. Zischke, K. Moyle. Biology,fisheries, and status of longtail tuna (*Thunnus tonggol*),with special reference to recreational fisheries in Australian waters. Final Report. Fish. Res. and Dev. Corp. and CSIRO Mar. and Atmosph. Res., **105** (2010).
3. A. Fauzi, and S. Anna. Fisheries and Marine Resource Modelling for Policy Analysis. Gramedia Pustaka Utama, Jakarta, **162** (2005).
4. A. Wujdi and Suwarso. Catch Fluctuation and Composition of neritic tuna caught by gill net in South China Sea. Tuna Fishery Research Institute. Benoa Bali, **8** (2014).
5. P. Sparre and S.C. Venema. Introduction to tropical fish stock assessment. Part I. Manual. FAO *Fish. Tech. Paper*. No.**306** (1), Rev.2, 407 (1998).
6. D.G.S. Wiadnya. 1992. Analysis of Catch and Effort Data on Marine Capture Fisheries in East Java. Indonesia. Verslag Nummer 1376. Landbouw. Universiteit Wageningen. The Netherlands. Unpublished (1992).
7. G. Bintoro. Sustainable utilization of fringescale sardine *(Sardinella fimbriata* valenciennes, 1847*)* fishery in Madura Strait waters East Java. Disertation IPB University Bogor, **302** Unpublished (2005).
8. J.A. Gulland. Manual of the method for fish stock assessment part 1 fish population analysis. Fish. Science. FAO. 154 pp (1969)
9. D. Pauly. A Selection of sample method of the stock assessment of tropical fish stock. FAO. Journal of Fish. Science 727 pp (1980).
10. S.Y. Tai, and T. Heaps. Effort Dynamics and Alternative Management Policies For The Small Pelagic Fisheries of Northwest Peninsular Malaysia. Mar. Res. Ec. 11. USA. 85-103 (1996).
11. Fauziyah, A.I.S. Purwiyanto, F. Agustriyani, W.A.E. Putri, Ermatita, and A Putra. Determining the stock status of snapper (*Lutjanus* sp.) using surplus production model: a case study in Banyuasin coastal waters, South Sumatra, Indonesia. IOP Conf. Series: Earth and Env. Science **404.** 012009. 12 (2020).
12. M. King. *Fisheries Biology, Assessment and Management*. Fishing News Books Ltd. Oxford, **341** (1995).
13. M.I. Effendie, 2002. Fisheries Biology. Yayasan Pustaka Nusantara. Bogor, **163** (2002).
14. Y. M. Harahap, F. Bu'ulolo, H. R. Sitepu. Several factors that influence the demand of fresh water consumption in local fresh water company (PDAM Indonesia) Tirtanadi Medan. J. Saintia Math. **1** (4), 325-336 (2013).
15. A. Fauzi. Fisheries Economics : Policy and Management Theory. Gramedia Pustaka Utama. Jakarta, **224** (2010).

Study of determining the fishing season for tiger grouper (*Epinephelus fuscoguttatus*) landed at Kutaraja Fishing Port, Banda Aceh

Alvi Rahmah[1,3*], *Julyanur* Sahputri[1], *Ratna* Mutia Aprilla[1,3], *Makwiyah* Chaliluddin[1,3], and *Sayyid* Afdhal Elrahimi[2,3]

[1] Department of Fisheries Resources Utilization, Faculty of Marine and Fisheries, Universitas Syiah Kuala, Banda Aceh 23111, Indonesia
[2] Department of Marine Sciences, Faculty of Marine and Fisheries, Universitas Syiah Kuala, Banda Aceh 23111, Indonesia
[3] Research Center for Marine and Fisheries, Universitas Syiah Kuala, Banda Aceh 23111, Indonesia

Abstract. The tiger grouper (*Epinephelus fuscoguttatus*) is a type of fish landed at the Kutaraja Fishing Port using handline. The production of tiger grouper from 2014 to 2019 fluctuated, so it is necessary to determine the right fishing season to make it easier for fishermen to carry out fishing operations for tiger grouper effectively. This research was conducted in July-August 2020, at Kutaraja Fishing Port. The primary data consists fishing trip, location of fishing ground, and total catch, the secondary data (time series data of tiger grouper catches on 2014-2019, as well as research-related literature). The data analysis consists of catch per unit effort (CPUE) analysis and time series analysis using the moving average method.The catch per unit effort (CPUE) of handline at Kutaraja Fishing Port tends to decrease every year. The highest CPUE occurred in 2015 (26.06 kg/trip) and the lowest in 2018 (0.25 kg/trip). The fishing season index of tiger grouper (*Epinephelus fuscoguttatus*) showed that fish can be caught throughout the year. The results showed that the peak season for tiger grouper landed at Kutaraja Fishing Port occurred in February (201.0%), March (127.2%), June (161.0%), and July (146.1%). Meanwhile, the fishing season index of 50% to <100% includes the moderate season in January (75.0%), August (91.2%) and December (84.3%), and the low season occurs in May (33, 6%), September (31.3%), October (19.9%) and November (41.8%) with the index value less than 50%. In conclusion, the catch per unit effort (CPUE) of tiger grouper (*Epinephelus fuscoguttatus*) at Kutaraja Fishery Port in 2014-2019 fluctuated but tends to experience a significant decline, where the value ranged from 0.25 kg/trip to 26.06 kg/trip. The peak season for fishing of tiger grouper occurs in February, March, June, and July; moderate season in January, August, and December; and low season occurs in May, September, October, and November.

1 Introduction

The Kutaraja Fishing Port is the largest fishing port in Aceh Province. The fishing gear used at Kutaraja fishing port to catch tiger grouper is handline and longline. Tiger grouper (*Epinephelus fuscoguttatus*) is one of the important economic commodities and include a group of coral fish with a wide distribution throughout Indonesia [1]. Annual production of tiger grouper (*Epinephelus fuscoguttatus*) tends to decrease significantly. This can be seen from the total production of tiger grouper (*Epinephelus fuscoguttatus*) at Kutaraja Fishing Port in 2014 which was 3,075 kg/year and decreased in 2019 by 56 kg/year. This condition is thought to be influenced by the fishing methods, fishing effort (trip), damage aquatic environment, and fishing season [2,3].

Fishing season information is used to determine the right time for carrying out fishing operations, where fishing activities without a clear fishing plan and target will lead to non-optimal profits [4]. Determining the appropriate fishing season for tiger grouper (*Epinephelus fuscoguttatus*) is important to know so that fishermen can obtain optimal catches, minimize fishing operation costs, and maintain the availability of tiger grouper [5-7].

Research on the fishing season for tiger grouper (*Epinephelus fuscoguttatus*) at Kutaraja fishing port has never been carried out. This study discusses the period of the fishing season for tiger grouper (*Epinephelus fuscoguttatus*) which landed in Kutaraja Fishing Port. Based on the catch and the number of fishing trips per month, it is necessary to determine the appropriate fishing season for tiger grouper (*Epinephelus fuscoguttatus*) in order to reduce time losses in fishing, so that the catches obtained by fishermen can be more optimal.

* Corresponding author: alvi_rahmah@unsyiah.ac.id

https://doi.org/10.1051/e3sconf/202233904002

2 Materials and methods

2.1 Site and time

This research was conducted in August 2020 at the Kutaraja Fishing Port, Banda Aceh, Aceh Province.

2.2 Sampling and measurements

The data collection method in this study used purposive sampling. Determination of the number of samples as much as 10% of the total population studied [8]. The total population of fishermen is 231 people with 10% taking based on the size of fishing vessels at the Kutaraja Fishing Port, so the sample of respondents per ship is 22 respondents. Data was collected through interviews using a questionnaire.

This study uses primary data and secondary data. Primary data consists of the number of fishing trips, time of capture (in months), number of catches, and distance from fishing grounds. Meanwhile, secondary data regarding data production of tiger grouper during 2014-2019 and data on fishing trips for 2014-2019 were obtained from Marine and Fishery Resources Supervision Unit of Kutaraja Fishing Port.

2.3 Data and analysis

The analysis of the fishing season index conducted in this study aims to determine the time series analysis of monthly data on tiger grouper catches for 6 years followed by the calculation of the moving average. With the following formula [9] :

a) Compile monthly CPUE series (2014-2019)

$$CPUE_i = \frac{ci}{fi} \quad (1)$$

Where :
CPUE = Catch per fishing effort year i (kg/trip)
i = 1,2,3,..., n
ci = Catch year i (kg)
fi = Attempt to catch year i (trip)

b) Compile a 12 month CPUEi moving average (RG)

$$\text{RGi} = \frac{1}{12}(\textstyle\sum_{i=i-6}^{i+5} CPUE) \quad (2)$$

Where:
RGi = 12-month moving average i
CPUE = CPUE of order i
i = 6,7,8 -, n-5

c) Compile the i-month centered CPUE moving average value (RGPi)

$$\text{RGPi} = \frac{1}{2}(\textstyle\sum_{i=i}^{i=1} RGi) \quad (3)$$

Where:
RGPi =i-month centered CPUE moving average
RGi = 12-month moving average i

d) Compile the average ratio of the first month (RBi)

$$\text{RBi} = \frac{\text{CPUEi}}{\text{RGPi}} \quad (4)$$

Where:
Rbi = Average ratio of the i-month
CPUEi = CPUE month i
i = month 6,7,8…., n-5

e) Compile the average value in an ixj matrix which is compiled for each month starting from July 2014 to June 2019, then the calculation is carried out:

1. Average ratio for the first month (RBBi)

$$\text{RBBi} = \frac{1}{n}(\textstyle\sum_{i=1}^{n} RBij) \quad (5)$$

Where:
RBBi = Average Rbij for month i
Rbij : Ratio of monthly average for matrix size ixj

2. Total monthly average ratio (JRB)

$$\text{JRBB} = \textstyle\sum_{i=i}^{12} RBBi \quad (6)$$

Where:
JRBB = Total monthly average ratio
RBBi = Average Rbij for month i
i = 1,2,3,……,12

3. Correction Factor (FK)

$$\text{FK} = \frac{1200}{\text{JRBB}} \quad (7)$$

Where:
FK = Correction Factor Value
JRBB = Total monthly average ratio

4. 1st Month Catching Season Index (IMPi)

$$IMPi = RBBixFK \quad (8)$$

Where:
IMPI = Fishing Season Index value i month
RBBi = Average ratio for month i

Criteria for determining the fishing season :
IMP Value = 100% : Normal (Moderate season)
IMP Value > 100% : Catching Season (Peak Season)
IMP < 100% : Not fishing season (low season)

3 Results and Disscussion

3.1 Trends in Catch Per Unit Effort (CPUE) Tiger Grouper (*Epinephelus fuscoguttatus*)

The catch per unit effort (CPUE) can provide an overview of the abundance of fish resources in a waters. CPUE analysis can be considered as an indicator of whether fish stocks in these waters are still in good condition or have run out. CPUE analysis can be obtained by way of dividing the tiger grouper catches by fishing effort (fishing trip). The dominant of fishing gear used by fishermen at Kutaraja Fishing Portto catch tiger grouper is handline. The highest catch per effort (CPUE) was found in 2015 which was 26.06 kg/trip and the lowest in 2018 was 0.25 kg/trip which landed at Kutaraja Fishing Port tends to decrease every year, it can be seen in Table 1 below:

Table 1. The CPUE value of tiger grouper during 2014-2019

Year	Catch (Kg)	Effort (Trip)	CPUE(kg/trip)
2014	3.075	428	7.18
2015	10.372	398	26.06
2016	3.008	1.030	2.92
2017	1.399	1,124	1.24
2018	282	1.119	0.25
2019	1.161	1,131	1.03

The CPUE trend which shows a downward trend is an indirect indication of the occurrence of excessive utilization, on the contrary the CPUE trend which tends to increase indicates that resource utilization is still within safe limits and has the potential for further development [10]. Based on the results of research for the last 6 years, the production value of tiger grouper in Kutaraja Fishing Port tends to fluctuate, where in 2014-2015 fishing production increased by 237%, then in 2015-2016 capture production decreased by 71%, in 2015-2016 in 2016-2017 the catch production decreased by 53%, in 2017-2018 the catch production decreased by 80% and in 2018-2019 the catch production increased by 312%.

Catching efforts have increased where the highest fishing effort occurred in 2019 as many as 1,131 trips and the lowest effort occurred in 2015 which only reached 398 trips. The average fishing effort in 2014-2019 was 5,230 trips. This is because fishermen who increase their fishing effort in the form of fishing trips occur because fishermen assume that the more fishing efforts they make will increase the production of their catch, but in fact, after research, the production of the catch has decreased. [11] states that several characteristics that can be used as benchmarks for a fishery towards a condition of more fishing effort are the time at sea being longer than usual, the size of the net being smaller than usual, followed by a decrease in catch per unit effort (CPUE), smaller fish sizes, and usually higher fishing costs.

The effort used to calculate the CPUE value is the number of fishing trips operating from the number of ships landing the catch of tiger grouper (*Epinephelus fuscoguttatus*) at Kutaraja Fishing Port for the last 6 years (2014-2019). The result shows that the CPUE of tiger grouper landed at Kutaraja Fishing Port for the last 6 years tends to decreasebecause there was a very high fishing effort in that year and the previous year (in 2018) thus reducing the catch per unit effort (CPUE).

The changes in CPUE value every year is due to a decrease in catches and the large burden of production costs compared to the profits obtained. This is in accordance with the research of [12] that a decrease in effort in the same year is not always followed by an increase in production and vice versa. This shows that the increase in the number of fishing efforts is not the only factor causing the decline in catches, but may be influenced by various factors such as environmental changes that can affect fish abundance. This happened because the number of fishing efforts made in those years could increase the level of competition between fishermen so that the catch also decreased. [13] states that the catch is influenced by the number and efficiency of the fishing unit, the length of operation and the availability of fish caught in addition to being influenced by internal and external factors.

The changes in the CPUE value each year are influenced by the addition or reduction of the number of trips (effort). The CPUE value is inversely proportional to the effort value , where each additional effort will reduce the catch per unit of effort (CPUE). This is because the resources will tend to decrease if the fishing effort carried out continues to increase [14]. The increase in the CPUE value indicates that with a decrease in the number of fishing efforts, there is an increase in catch per trip [15]. [16] stated that the addition of fishing effort that is not followed by an increase in the number of catches will reduce the CPUE value, this is supported by the research of [17] that the size of the trip does not guarantee the amount of production of the catch obtained is large. This can be an indicator that the utilization of fish resources in these waters is already high. [18] states that increasing efforts will lead to a decrease in fish biomass. This is also similar to the results of research by [19] in PPN Prigi which shows that CPUE continues to decline along with the increase in fishing efforts made.Resources will tend to decrease if the fishing effort carried out continues to increase. It is well known that CPUE is negatively correlated with catch effort.When the number of capture attempts is low, the CPUE is high, and vice versa when the number of trips is high, the CPUE decreases. This relationship forms the basis of a popular surplus production model in the discipline of fish stock studies, the Schaefer model assumes that CPUE decreases linearly, as fishing effort increases, therefore, the moving average method only considers changes in the actual CPUE during certain months. the number of boat trips differs consistently from other months. The correlation between CPUE and effort shows a negative relationship, namely the higher the effort, the lower the CPUE value.The negative correlation between CPUE and effort indicates that the productivity of fishing gear will decrease if effort increases [20].

3.2 Fishing Season Index (FSI) of Tiger Grouper (*Epinephelus fuscoguttatus*)

The results show that the peak season for fishing of tiger grouper occurs in February with an FSI of 201%, and the low season in October with FSI of 19.9 %, it can be seen in Figure 1.

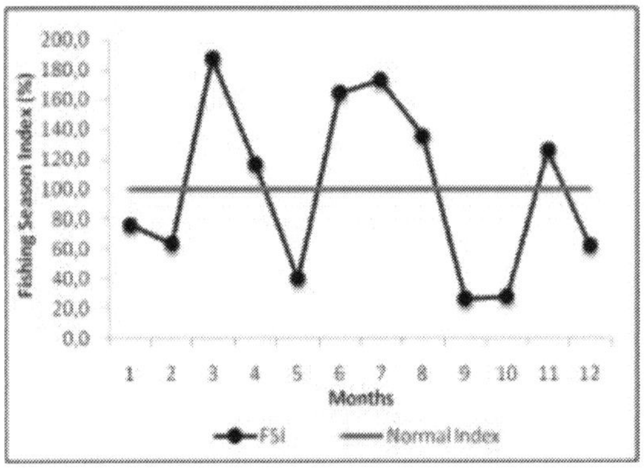

Fig. 1. The fishing season index (FSI) of tiger grouper at Kutaraja Fishing Port

Fishing Season Index (FSI) can be used in determining the right time to carry out fishing operations. Based on the result of FSI, it can be seen the trend of the fishing season so that the right time for fishing can be determined [5]. The FSI of tiger grouper (*Epinephelus fuscoguttatus*) based on research results for the last 6 years shows that tiger grouper can be caught throughout the year. [21] states that the value of the fishing season index can be used in determining the right time to carry out fishing operations.

Based on the criteria for determining the fishing season, the FSI that have a value greater than 100% are included in the peak season . The results of the FSIshow that the peak season for tiger grouper fishing occurs in February (201%), March (127%), June (161%), and July (146%); moderate fishing season occurs in January (75%), August (91%) and December (84%), and the low season occurs in May (34%), September (31%), October (20%) and November (42%) with the FSI of less than 50%. This is in accordance with research from [22] which found that the fishing season of grouper for ten months starts from January, February, March, May, June, July, August, September, November, and December with the peak season in March. March is the peak season for fishing of grouper, possibly due to low rainfall. This increases the opportunities for fishermen to carry out fishing operations. This is in accordance with the research of [23] which states that the fishing season index for reef fish shows that each species has its own fishing season and replaces each other every month.The results of interviews with fishermen said that the tiger grouper (*Epinephelus fuscoguttatus*) in Kutaraja Fishing Port is a fish whose existence is difficult to predict by fishermen, so that the catch of tiger grouper (*Epinephelus fuscoguttatus*) is relatively uncertain and fish that exist every year.

The caught of tiger grouper throughout the year can be related to the variability of oceanographic factors in the waters. The two most commonly associated oceanographic parameters are sea surface temperature and chlorophyll-a concentration [24]. According to [25] oceanographic parameters are one of the factors that greatly influence the variability of fish catches because temperature greatly affects the biological metabolism of fish.Judging from the physical effect, the surface temperature can cause upwelling (reversal of water mass) which brings nutrients to the surface and makes foraging places for fish. While chlorophyll-a is an indicator of primary productivity for fish, where the peak season for fishing of grouper is in February because in this month the rainfall is low, making it easier for fishing. It is suspected that high chlorophyll-a is caused by an upwelling process that causes an abundance of phytoplankton which resulted in the gathering of groupers.In general, there is a linear relationship between chlorophyll-a and grouper catches. When the amount of chlorophyll-a increases, it will increase the number of phytoplankton on the sea surface. That way the fish will look for places or locations that store a lot of phytoplankton, it is also suspected that there will be a decrease in sea surface temperature.

4 Conclusions

The catch per unit effort (CPUE) value of tiger grouper (*Epinephelus fuscoguttatus*) landed at Kutaraja Fishing Port based on 2014-2019 fluctuated but tended to experience a significant decline, where the highest CPUE occurred in 2015 with a value of 26.06 kg/trip and CPUE the lowest occurred in 2018 with a value of 0.25 kg/trip.

The Fishing Season Index (FSI) for tiger grouper (*Epinephelus fuscoguttatus*) landed in Kutaraja Fishing Port based on data for 2014 to 2019 ranging between 19.9% to 201.0%. The peak season for fishing of tiger grouper occurs in February, March, June, and July; moderate season occurs in January, August and December, and the low season occurs in May, September, October and November with the FSI of less than 50%.

References

1. S. Made, S. Fakhriyyah, A. Darawelalangi, ECSOFiM, **04**, 9 (2017)
2. R. Kurnia, K. Suwardi, I. Muchsin, M. Boer, Buletin PSP, **19**, 7 (2011)
3. Fauziyah, F. Agustriani, D.M. Situmorang, Y. Suteja, *Fishing seasons of fish landed at Sungailiat Archipelago Fishing Port in Bangka Regency*, in Environment, Energy, and Earth Sciences, E3S Web of Conferences 7-9 May 2018, Purwokerto, Indonesia (2018)
4. E. Hamka, M. Rais, J. Science and Technology Utilization of Fishery Resources, **3**, 6 (2017)
5. M. Syahrir, M.S. Baskoro, Darmawan, E. Lubis, E.S. Wiyono, J. Ilmu Per. Trop, **13** (2010)
6. D. Simbolon, B. Wiryawan, P.I. Wahyuningrum, H. Wahyudi, Bul. PSP, **19** (2011)
7. J. S. Kekenusa, V.N.R. Warung, D. Hatidja, J. Ilm. Sains, **12** (2012)
8. S. Arikunto, Research Methodology A Proposal Approach (PT. Rineka Cipta, 2002)
9. A. Dajan, Introduction to Statistical Methods (Research Institute for Social Economic Information, 1983)

10. Ministry of Marine Affairs and Fisheries, Attachment to the decision of the Director General of Capture Fisheries Number18/kep-djpt/2014 concerning Technical Guidelines for Indicator Assessment for Fisheries Management with an Ecosystem Approach (Ministry of Marine Affairs and Fisheries, 2014)
11. J. Widodo, Suadi, Management of Marine Fisheries Resources (Gadjahmada University, 2018)
12. B. S. Mujais, J. Agrikan, **9**, 2 (2016)
13. D. Simanjuntak, Usman, T.E.Y. Sari, JOM UNRI, **3**, 1 (2018)
14. B. Suswanto, Z. Anna, I. Gumilar, J. Perikanan Kelautan, **2**, 1 (2015)
15. R. I. Wahju, Zulkarnain, K.P.S. Mara, PSP Bulletin Journal, **19**, 1 (2011)
16. E. Nugraha, B. Koswara,Yuniarti, J. of Fisheries and Marine Affairs, **3**, 1 (2012)
17. R. M. Aprilla, Mustaruddin, E.S. Wiyono, N. Zulbainarni. JTPK, **4**, 1 (2013)
18. I. Nahib, Bioeconomic analysis of the impact of the presence of FADs on the sustainability of small tuna fisheries resources (a case study in the waters of Palabuhanratu Bay, Sukabumi Regency). (Thesis, Faculty of Fisheries and Marine Sciences, Bogor Agricultural University, 2008)
19. E. Nurdin, A. A. Taurusman, R. Yusfiandayani, J. Lit Fishing. Eng, **8**, 1 (2012)
20. S. Aminah, J. of Fish Scientiae. **1**, 2 (2011)
21. Ihsan, E. S. Wiyono, S. H. Wisudo, J. Haluan, J. of Mar. Fis, **5**, 2 (2014)
22. Andriyeni, Zulkhasyni, Fishing Season for Grouper (*Epinephelus* sp.), (Research Results Report, Faculty of Agriculture, Universitas Prof. Dr. Hazairin, S.H, 2015)
23. E. Yuliana, M. Boer, A. Fahrudin, M.M. Kamal, E. Muttaqin, Indonesian Journal of Fisheries Research, **22**, 1 (2016)
24. W. Kurniawan, J. Oseana, **40**, 4 (2015)
25. Adnan, J. Amanisal PSP FPIK Unpati, **6**, 1 (2010)

The abundance of herbivorous and coralivorous fishes during and after coral bleaching in Krueng Raya Waters, Aceh Besar District

Maria Ulfah[1,2*], *Khairunnas* Khairunnas[1,4], *Syahrul* Purnawan[1,3] and *Afrita* Ida Utami[4]

[1]Marine Science Department, Faculty of Marine and Fisheries, Universitas Syiah Kuala, Banda Aceh – Indonesia, 23111
[2]Laboratory of Marine Biology, Faculty of Marine and Fisheries, Universitas Syiah Kuala, Banda Aceh – Indonesia, 23111
[3]Marine Acoustic Laboratory, Faculty of Marine and Fisheries, Universitas Syiah Kuala, Banda Aceh – Indonesia, 2311
[4]Ocean Diving Club, Faculty of Marine and Fisheries, Universitas Syiah Kuala, Banda Aceh – Indonesia, 23111

Abstract. This study aims to determine the community structure changes and to determine the abundance of herbivorous fish and coralivorous fish in Krueng Raya waters during and after the occurrence of coral bleaching. This research was conducted in three stages: June 2016, October 2016, and February 2017. There are 3 locations for data collection, including Benteng Inong Balee, Ahmad Rhang Manyang, and Lhok Mee in Mesjid Raya Sub-District, Aceh Besar District. Reef fish abundance data were collected at a depth of 3-6 meters using the visual census technique (VCT) - belt transect method. The study results show that there were six families, 17 genera, and 47 species belonging to the herbivorous and coralivorous fish groups. The most abundant family was the Chaetodontidae family, while the least fish family was the Kyphosidae family. The highest abundance of corallivorous fish was found at station 2 in October 2016, with a value of 509 ind/ha. The lowest was found at station 3 in October 2016, with a value of 182 ind/ha. The highest abundance of herbivorous fish was found at station 3 in February 2017, with a value of 498 ind/ha. The lowest was found at station 3 in October 2016, with a value of 150 ind/ha.

1 Introduction

Coral reefs are a typical ecosystem found in shallow tropical waters, with primary productivity and high diversity. An indication of good ecosystem health can be seen from the recovery level, namely the ability of the ecosystem to repair itself after experiencing damage [1]. One factor affecting the recovery rate of coral reef ecosystems is the availability of hard substrates as a place for coral larvae to attach [1,2]. Especially for coral reef ecosystems, the level of recovery of coral reef ecosystems is an essential indication of its management. This information is needed to assess the vulnerability of coral reefs to be managed and estimate the length of the ecosystem repair process, whether through restoration or rehabilitation [3,4].

Reef fish are essential resources in coral reef ecosystems and are the most abundant organisms in coral reef ecosystems. Therefore, reef fish have an essential role as a supporter of existing relationships in coral reef ecosystems. The existence of reef fish is strongly influenced by the condition and health of coral reefs, which can be assessed from the percentage of live coral cover [5]. The diversity of reef fish will decrease if coral reefs are damaged. [6], stated that of the estimated 12,000 marine fish species in the world, approximately 7,000 species (58.3%) are fish that live in coral reef areas.

Herbivorous and coralivorous fish can control the dynamics of macroalgae on coral reefs, which is one of the bioindicators of coral reef ecosystems during and after coral bleaching. Therefore, consider their important ecological and economic functions, herbivorous fish and reef fish are required to continuously monitor and assess their condition. Furthermore, monitoring these fish and their habitats must always be carried out correctly and appropriately. So that, appropriate conclusions can be drawn to take policies and strategic steps, especially for managers and other related stakeholders.

2 Research Methodology

2.1 Time and Place

This research was conducted in three stages: June 2016, October 2016, and February 2017. There are 3 locations for data collection: Benteng Inong Inong Balee, Ahmad Rhang Manyang, and Lhok Mee in Krueng Raya District, Aceh Besar District.

2.2 Research Methods

Reef fish data were collected at a depth of 3-6 meters using the visual census technique (VCT) - belt transect

* Corresponding author: mariaulfah@unsyiah.ac.id

[7]. The observation area is 2.5 meters on the right and left along the transect for fish with a size above 10 cm, and 1 meter on the right and left along the transect for fish with a size below 10 cm. Identification of reef fish refers to the Reef Fish Identification Guide Book [5].

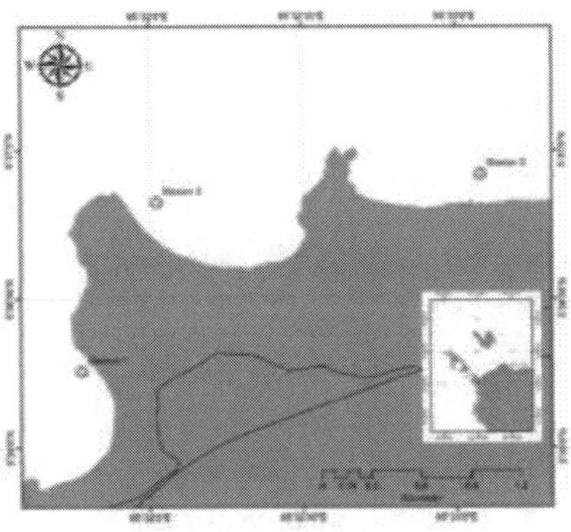

Fig. 1. Map of research location.

3 Results and Discussion

Based on the results of research that have been carried out, there are six families, 17 genera, and 47 species found to belong to the herbivorous and coralivorous fish groups (Appendix 1). The most commonly found family is the Chaetodontidae family, a family of the corallivorous fish group. Meanwhile, the fish family that was the least found was the Kyphosidae family, a herbivorous fish family. Only one family of fish was included in the corallivore category, namely Chaetodontidae. While five families of fish belonging to the herbivorous category were found, namely Acanthuridae Kyphosidae, Pomacanthidae, Scaridae, and Siganidae (Table 1).

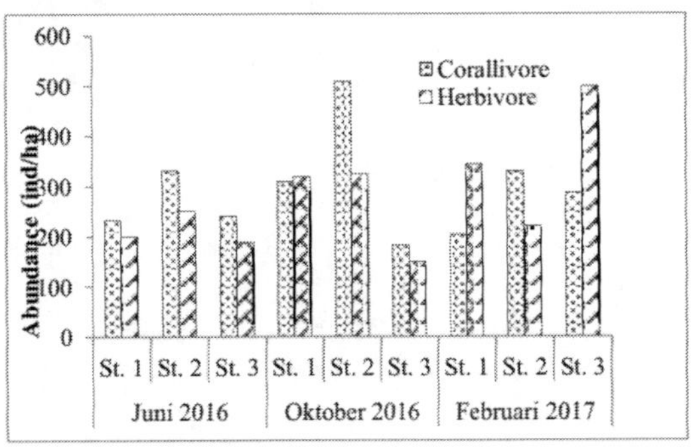

Fig. 2. The abundance of Coralivorous and Herbivorous Fish in Mesjid Raya Sub-District

The abundance value of fish shown in Table 1 shows that there are only two families of fish that are always found at each station and time of observation. The fish are from the Acanthuridae and Chaetodontidae families. The highest average fish abundance value came from the Chaetodontidae family, with 292 ind/ha values. In contrast, the lowest average fish abundance value came from the Kyphosidae family, which was 41 ind/ha. Other fish families also have different values, such as families Acanthuridae and Pomacentridae with values of 282 ind/ha and 270 ind/ha, while Scaridae and Siganidae only have an abundance of 89 ind/ha and 73 ind/ha, respectively. According to Madduppa [8], based on its role, fish from the family Acanthuridae are major fish whose habitat is in coral reef ecosystems and has schooling characteristics.

3.1 The abundance of Herbivorous and Coralivorous Fish

You are free to use colour illustrations for the online version of the proceedings but any print version will be printed in black and white unless special arrangements have been made with the conference organiser. Please check whether or not this is the case. If the print version will be black and white only, you should check your figure captions carefully and remove any reference to colour in the illustration and text. In addition, some colour figures will degrade or suffer loss of information when converted to black and white, and this should be taken into account when preparing them.

The abundance of corallivorous and herbivorous showed a fairly varied value at each research location. The abundance of corallivorous fish species was most abundantly found at station 2 in October 2016, reaching 509 ind/ha. In contrast, the lowest abundance was found at station 3 in October 2016 with 182 ind/ha. The high abundance of corallivorous fish indicates that the condition of coral reefs at station 2 is quite good, around 50% - 74.9% [10]. While at Station 3, the condition of coral reefs is in the moderate category, which is around 25% - 49.9%. This condition shows that coral reefs have a very close relationship with biota and have an important role in their lives [11].

In addition, there was a natural phenomenon, namely the warming of sea surface temperatures in June which caused mass coral bleaching in several locations in the Aceh Besar area, including the research site [12]. Allen [13] stated that when environmental changes threaten coral reef ecosystems, fish with high mobility will tend to move to look for coral reef locations that are still good. At the same time, coral reefs cannot avoid threats that originate from coral reef's environmental changes.

The abundance of herbivorous fish had different numbers at each station, where the highest average number of fish was found at station 3 in February 2017 with a value of 498 ind/ha. In contrast, the lowest value was found at station 3, with an average abundance of herbivorous fish of 100 ind/ha. In general, it can be seen that the herbivorous fish abundance graph has the same model as the coralivorous fish abundance graph but has a different value. This condition indicates that the presence of herbivorous and corallivorous fish are related to each other. This is following the functional group where the number of individuals, types, and sizes of herbivorous fish play a crucial role in coral reef resilience, which functions as a controller of algae growth. The same is true for fish of the Coralivorous group. There is a positive correlation between the functional groups of obligate coral indicator fish and the percentage of coral reef cover [12, 15,16,].

Overall, the most dominant herbivorous fish group came from the Acanthuridae family, with an average abundance value of 282 ind/ha. Acanthuridae is a major fish that lives in groups (schooling) with the primary habitat in coral reef ecosystems [18, 19].

4 Conclusion

Based on the research results conducted in the waters of Krueng Raya, Mesjid Raya District, Aceh Besar Regency, six families, 17 genera, and 47 species were found to belong to the herbivorous and coralivorous fish groups. The highest abundance of corallivorous fish was found at station 2 in October 2016, with a value of 509 ind/ha. In contrast, the lowest abundance of corallivorous fish was found at station 3 in October 2016, with a value of 182 ind/ha. The highest abundance of herbivorous fish was found at station 3 in February 2017, with a value of 498 ind/ha. In contrast, the lowest abundance of herbivorous fish was found at station 3 in October 2016, with a value of 150 ind/ha.

Table 1. Number of families of coralivorous and herbivorous fish in Krueng Raya waters

No	Family	Group	June 2016			October 2016			February 2017			Number of individuals
			St. 1	St. 2	St. 3	St. 1	St. 2	St. 3	St. 1	St. 2	St. 3	
1	Acanthuridae	Herbivorous	195	271	164	300	410	100	358	220	521	282
2	Chaetodontidae	Coralivorous	233	331	240	310	509	182	203	330	287	292
3	Kyphosidae	Herbivorous	0	0	0	373	0	0	0	0	0	41
4	Pomacanthidae	Herbivorous	205	120	400	750	153	0	400	0	400	270
5	Scaridae	Herbivorous	0	0	0	184	0	200	160	0	80	69
6	Siganidae	Herbivorous	0	0	200	300	0	0	160	0	0	73
Number of individuals			106	120	167	369	179	80	214	92	215	171

Acknowledgement. The author is very grateful to the thesis supervisor, who has helped and reminded the author of mistakes. Furthermore, the Faculty of Marine Affairs and Fisheries staff, Syiah Kuala University, have helped. In addition, Ocean Diving Club has facilitated research and research teams with help and friends who have encouraged writing. The writing of this manuscript is supported by the Matching Fund – Kedaireka funded by the Ministry of Education, Culture, Research, and Technology Indonesia awarded to the Research Center for Marine Sciences and Fisheries, Universitas Syiah Kuala.

References

1. G.D. Grimsditch, R.V. Salm. CRRRB. IUCN, Gland, Switzerland. 52pp (2006)
2. M. Adrim, Fahmi. RGMF, ORC-LIPI, Jakarta (2010)
3. M. Ulfah, S.N. Fajri, M. Nasir, K. Hamsah, and S. Purnawan. DEDIRFKR. IOP Conference Series: Earth and Environmental Science, **348** 1 (2019)
4. S. Bahri, E. Rudi, I. Dewiyanti. DEPIK, **4** 1 (2015)
5. F. Setiawan. ICFMI. WCS. Manado. Indonesia (2010)
6. G. Allen, R. Steene, P. Humman, N. Deloach. RFI Pte Ltd. Florida USA (2003)
7. J. Hill, C. Wilkinson. A Resource for Managers, ver 1. Townsville (AU): (2004)
8. H. Maddupa. ESKFD. Jakarta. IPB. (2006)
9. T. McClanahan, N. Polunin, T. Done, ESRCR. Conservation Ecology 6 (2002)
10. M. Ulfah, M. Muliari, F. Azzahara, N.M. Razi, M.R. Fazillah, M. Agustiar, N. Fadli. IOP Conference Series: Earth and Environmental Science **869** (1), (2021)
11. Suharsono. JJKI. (Jakarta: Program COREMAP LIPI. (2008)
12. M. Ulfah, C. Yolanda, S. Karina, S. Purnawan, S. Agustina. Jurnal Ilmu dan Teknologi Kelautan Tropis **10** 739-745 (2018)
13. G. Allen, R. Steene, P. Hulmann, N. Deloach. Jacksonville, Florida. (2003).
14. L.O. Anshari, B. Sadarun, Rahmadani. Sapa Laut, 5 99-105 (2019)
15. S. Frimanozi, I.J. Zakaria, Izmiarti. Jurnal Biologi Universitas Andalas, **3** 34-39 (2014)
16. C. Octavina, M. Ulfah, M.R Fazillah, M. Agustiar, N.M Razi, R. Sakinah. IOP Conference Series: Earth and Environmental Science **869** 1 (2021)
17. F. Setiawan, E. Muttaqin, S.P Tarigan, Muhidin, A. Hotmariyah, J. Trunojoyo **10** 2 (2017)
18. C. Octavina, Z.H Asri, S. Purnawan, M. Ulfah International J. of Sciences: Basic and Applied (IJSBAR) **40** 2 (2018)
19. C. Wilkinson. SCFW. AIMS: Townsville (Australia), 557 pp (2004)

Improvement on small-scale tuna fisheries data quality through the Application of e-Logbook System

Syahril Abd Raup[1,2*], *Sri* Patmiarsih[2], *Rista Devi* Juniar[2], and *Bram* Setyadji[3]

[1] Postgraduate student, Coastal and Ocean Resources Management Study Program, Faculty of Fisheries and Marine, IPB University. Jl. Raya Dramaga, Kampus IPB Dramaga, Bogor, West Java, Indonesia – 16680.
[2] Directorate of Fish Resources Management, Directorate General of Capture Fisheries, Ministry of Marine Affairs and Fisheries. Mina Bahari Building II, 14th Floor, Jl. Medan Merdeka Timur No.16, Jakarta, Indonesia – 10110.
[3] Research Institute for Tuna Fisheries, Agency for Marine and Fisheries Research and Human Resources, Ministry of Marine Affairs and Fisheries. Jl. Mertasari No. 140, Sidakarya, Denpasar, Bali, Indonesia – 80224.

Abstract. Tuna and tuna-like fisheries play a vital role in Indonesian livelihood, especially in the archipelagic waters. However, despite the importance, the concern in general data collection activities for tuna, i.e., limited, with incomplete scientific knowledge and insufficient data has hampered the assessment. The purpose of this study was to analyse on how fisheries-dependent data system could transform the data quality. For this purpose, the analysis was conducted based on e-logbook data from January to December 2020. The result showed that some robust fisheries indicators (catch, effort, fleet and species composition) could be produced from this particular dataset. In addition, e-logbook has the best attribute for improving the data quality, especially for small-scale tuna fisheries in FMA-714. Low cost and vast spatial and temporal coverage are the main characteristics on why the program should be expanded and monitored carefully in the future.

1 Introduction

Tuna and tuna-like fisheries play a vital role in Indonesian livelihood [1,2]. Among eleven Fisheries Management Area (FMA), at least nine are touted to be the migration areas and potential fishing ground for tunas [3–8]. The biggest challenge in managing the fisheries was the fact that more than 80% are determined as small-scale fisheries (incl. subsidiary and indigenous fishing) [9,10], and widely distributed in the coastal and archipelagic waters.

Two-third of the tuna catch in Indonesian EEZ (Economic Exclusive Zone) is contributed from archipelagic waters, especially from FMA-713, FMA-714, and FMA-715 [1]. Banda Sea and its adjacent waters is a trajectory of the water mass from the Pacific Ocean to the Indian Ocean [11]. The characteristics and dynamics of the waters are strongly influenced by the topography and configuration of existing islands and the monsoon wind that occurs in the region [12]. The existence of upwelling phenomena causes this region to be rich in potential fish resources, including tuna and skipjack [13]. In addition, it also acted as a nursery ground and spawning area for tunas and other large pelagic fishes [14].

Based on Ministerial Decree No. 50/KEPMEN-KP/2017, total estimated biomass (non-tuna) in FMA-713, FMA-714, and FMA-715 was about 3 million ton, whereas the average production of main tuna species (yellowfin, bigeye and skipjack tuna) from 2005-2012 in FMA-714 was around 230,000 MT. Despite of its potential, no specific management measure applied into this area, apart from Ministerial Regulati*on No. 4/PERMEN-KP/2015 related to open-close system. This drives the necessity for conducting harvest strategy as the foundation for sustainability. Hoshino et al. [15] started a pilot project in 2020 for assessing yellowfin and skipjack tuna. Although an interim harvest strategy has been adopted, but the key uncertainties i.e., abundance indices and effective sample size still need to be improved in order to get a more robust analysis.

Prior to the assessment, some reports [16,17] expressed the concern in general data collection activities for tuna, especially before 2010, which labelled as limited, with incomplete scientific knowledge and insufficient data, hence could not be generated to produce an accurate assessment. Even though the data after 2010 was available, but the scarcity i.e., lack of temporal and spatial consistencies combined with various data formats have hampered the analysis [18]. The issue was intercepted by the Directorate General of Capture Fisheries (DGCF) by promoting e-logbook system since 2018. The purpose of this study was to analyse on how e-logbook system could transform the data quality for small-scale tuna fisheries in Indonesian archipelagic waters. The assessment also covers how to use the information to generate spatial and temporal fisheries indicators. Hopefully, the result can be used as further development of e-logbook system for encouraging sustainability in the future.

* Corresponding author: chaliarrauf@yahoo.com

2 Materials and Methods

2.1 Data source

The primary data was obtained from Integrated Logbook Information System (SILOPI) restricted only within Fisheries Management Area (FMA) 714 between January to December 2020, courtesy of Directorate General of Capture Fisheries. Prior to extraction, the logbook data has been scrutinized by the verificatory officers by consider following rules, i.e., conformity of fishing ground, gear and type of species landed; suitability between total catch reported and landed; compatibility between the catch reported related to the fishing capacity of each vessel; confirmation of departure and landing port as mentioned in the license.

2.2 Data cleaning and validation

Ensuring a robust analysis, data cleaning and filtering were conducted prior to analysis. This was necessary, largely purposed to clear any potential inconsistencies and typos, which commonly found in logbook data [19]. In principle the process was commenced as follows, i.e., setting must be commenced between the departure and the arrival date; the number of day-at-sea should be the differential between the arrival and the departure date; total sets per landing should be at least 50% of the total day-at-sea; georeferenced points should not intersect with the land nor excess the boundaries from the area of interest.

As a result, a total of 6,388 set by set e-logbook data were successfully generated, which consist of 297 vessels across six tuna-related gears, i.e., lift net, hand line, pole and line, purse seine, longline and troll line (Table 1).

Table 1. Summary of dataset used for analysis.

Declared Gears	No. Boats	Total Sets
Lift net	11	452
Hand line	39	401
Pole and line	9	140
Purse seine	204	4,639
Longline	4	89
Troll line	30	667
Total	297	6,388

2.3 Data Analysis

A number of indicators will be analysed in this study, including:

1. Catch composition by gear by dominant group species. The result will be juxtaposed with Ministerial Regulation No. 86/2016 for conformity study, whether its still relevance as to date.
2. Catch composition of non-target species (by-catch)
3. Effort distribution by gear, to see any overlapping between gear operations
4. Monthly trend of catch per unit of effort (CPUE), whereas for longline declared in kg/1000 hooks [20] and the rest are in kg/day, assuming at least one set is commenced per day.

All statistical analysis were conducted using Microsoft Excel and R version 4.0.5 [21], while mapping was constructed under QGIS version 3.18 [22].

3 Results and Discussions

3.1 E-Logbook implementation

The first version of e-logbook [23–26] was introduced on August 2015 by the Marine and Fisheries Technology Research and Engineering Center (P3TKP), however it didn't fully materialized due to complication in production and installation the hardware into the fishing vessels. Therefore, a more user friendly and android-based e-logbook app (software) was launched in 2018, in parallel with Our Ocean Conference (OOC) event, held in Bali. The app supposedly working with (online) or without (offline) internet connection, thus the new platform is more convenient to use and free of charge, as more than two-third (~200 millions) of Indonesian population capable of accessing internet and smartphones [27]. It also could reach a more diverse community.

Since the introduction, the activation reaches 12,000 nationwide users by April 2021, whereas for fleets operating in the FMA (Fisheries Management Area) 714 also increasing with a total 164 boats until August, 2021 (Figure 1). The vast adoption was largely due to massive socialization throughout the years. However, since a lot of individual landing sites in eastern part of Indonesia, the implementation was slightly disturbed because e-logbook requires authorization from harbourmaster (port inspector). Hence, the activation was limited only for vessels based on official ports. As a result, only 304 boats reported the e-logbook in 2020, a considerably very low compared to the number of national activations at the same period. A possible online request for activation should be prioritized for future development to address the lingering issue.

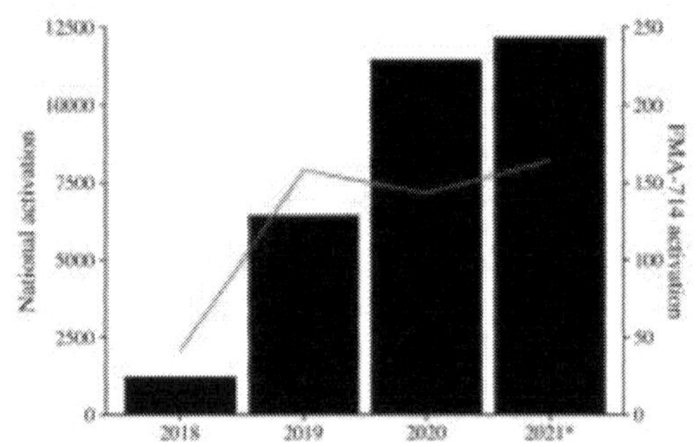

Fig. 1. Number of e-logbook activation for national (bar) and specific FMA-714 area (line) from November, 2018 to August, 2021 (Remarks: Modified from Directorate General of Capture Fisheries/DGCF)

https://doi.org/10.1051/e3sconf/202233904004

3.2 E-logbook for data improvement on small-scale fisheries

Most of the data available for determining the stock status of the fisheries in FMA-714 were from port monitoring programs [15,18,28]. However, such data requires a lot of investment, especially in human resources (enumerator), securing fund for their wage bills and other expenses. It was inevitable because manual logbook dataset was unreliable at that moment. This is where e-logbook could be a workaround, since it's specifically designed with color-coded menu which intended to be operated even with illiterate fishermen. It also compiled with some picture of dominant fish species, while fishing location, time and duration are automatically generated. Improvements on data quality can be assessed through some of these indicators as follows:

3.2.1 Fleet composition

Purse seine was the dominant fishing gear used in FMA-714. It comprised almost two-third of the total reported fleet, followed by hand line and troll line, which both contributed at least a quarter and a small portion of lift net, pole and line and longline (Figure 2). This composition aligned with previous studies in the Banda Sea for the last 10 years [29–31]. The gear mainly targeting either small pelagic fishes or tuna and tuna-like species, regardless whether it licensed for the first or later [32]. Hence, a more considerate arrangement is needed in order to encourage the sustainability in the area.

Since there was an indication of the area to be the spawning ground for tunas, a numerous regulations have been issued to control the fishing activity, particularly within FMA-714, such as: Ministry of Marine Affairs and Fisheries Regulation No. 4/PERMEN-KP/2015 on open and close system policy, No. 47/PERMEN-KP/2016 on definition of conservation area and No. 47/PERMEN-KP/2016 on moratorium of fishing permit, especially for ex-foreign vessels.

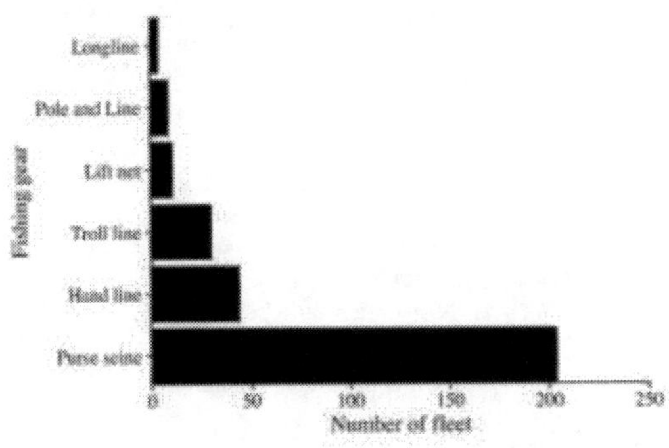

Fig. 2. Composition of fleet reporting e-logbook under FMA-714 in 2020.

3.2.2 Catch composition

The catch composition for most of the fleet operating in the FMA-714 are dominated (>75%) by tropical tuna species (i.e., yellowfin, bigeye and skipjack tuna), except for lift net and purse seine which mainly targeting small pelagic, neritic tuna and demersal fishes (Figure 3). The result on purse seine fleet was slightly different compared to previous study from Hariati [31] which the proportion of small pelagic to large pelagic was 1:3.

We consider the e-logbook data provide a better picture on the fisheries rather than the data generated from landing sites. Because most of purse seiners are known to target either small pelagic fishes (scads) [33,34] or skipjack tuna [35] in a balanced proportion. As for other gears, the result were relatively similar with previous findings [36–41]. A further prove that e-logbook data was reliable for catch composition analysis, which is a valuable information for estimating total biomass under data limited situation.

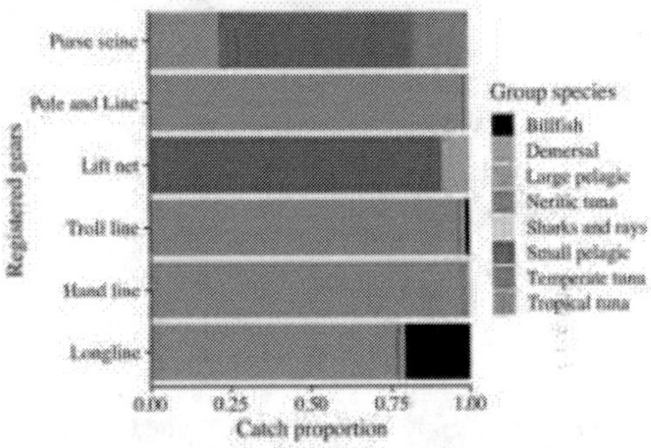

Fig. 3. Catch composition by gear operating in the FMA-714.

3.2.3 Effort distribution

The main competitive advantage of e-logbook data was the deliverance of accurate spatial information when reported accordingly. Figure 4 beautifully depicts the characteristic of fishing location for each gear. Hand line, pole and line lift net and purse seine has overlapping fishing ground, whereas longline and troll line, each has distinct preferences. Before the implementation of logbook, spatial information was mainly collected through interview [31], satellite utilization, such as: visible infrared imaging radiometer suite (VIIRS) boat detections (VBD) with vessel monitoring system (VMS) [42], and other hardware implant on the fishing vessels, e.g., spot trace [43].

However, the first approach was considerably has less robust, as fishermen are reluctant to be interviewed nor gave their fishing locations. On the other hand, deployment of VMS only mandatory for vessel above 30 GT [44] and spot trace require financial support for installing and implementation. Hence, e-logbook is a less cost solution for national spatial fisheries data collection system. Development and monitoring on this system should be the first priority for government, non-governmental organization (NGO) and other fisheries stakeholders.

3.2.4 Catch per unit of effort (CPUE)

Abundance indices, such as: catch per unit of effort (CPUE) is an important variable for relaying the current condition of stock in specific fishery [45] and very useful for running model on more detailed stock assessment [46]. Ideally, the indices are produced from fisheries-dependent data (e.g., fisheries surveys, scientific observer), however, such approach are costly, therefore most studies are exploring the port monitoring data for achieving the goal.

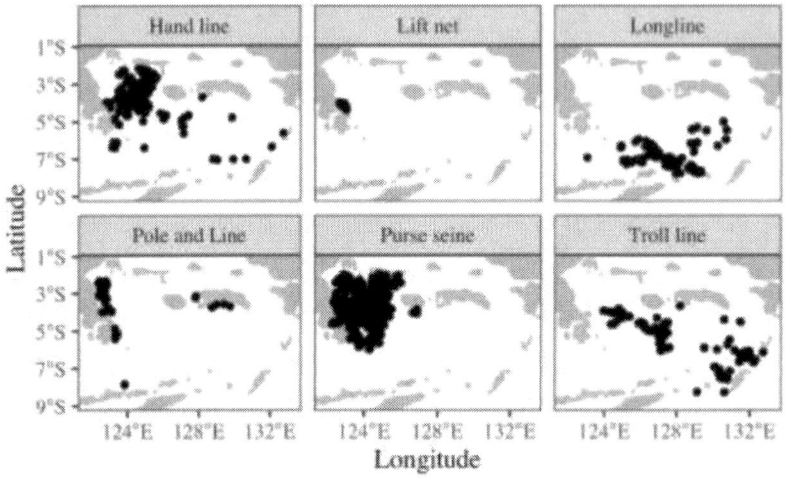

Fig. 4. Effort distribution of fishing gears operating in FMA-714 based on e-logbook information.

Since most the landing sites in FMA-714 are widespread across small islands and private jetties, enumeration can only be possibly conducted in ports or big landing sites. By utilizing e-logbook, such obstacles can be limited and the CPUE analysis can be performed either with spatial or temporal situation. Figure 5 depicts the mean monthly CPUE for main tuna species, i.e., yellowfin and skipjack tuna.

The trends showed that the fishing season for these two particular species are different, depend on the gear used. Third quarter was the peak season for handline, while catch rate from purse seine relatively stable over months, between longline and troll line performed opposite trends.

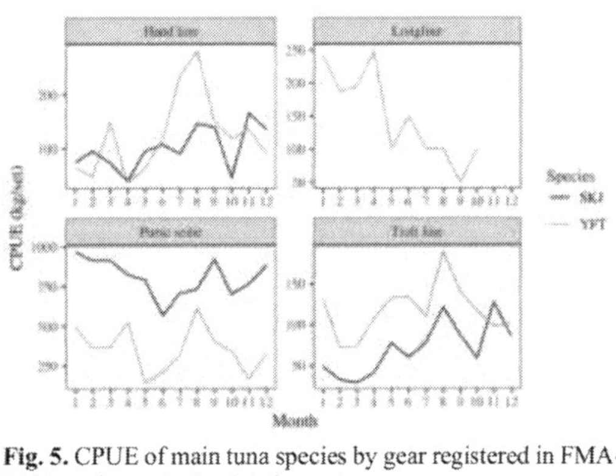

Fig. 5. CPUE of main tuna species by gear registered in FMA-714 based on e-logbook information.

Not every tuna species could be analysed due to scarcity of some data. However, if the result was cross-referenced with the available publications, they have produced similar result. For example, the fishing intensity of hand line fishery was accumulated during June to August, which was aligned with a previous studies from Enjah [41] and Nelwan et al. [47]. Moreover for purse seine and troll line fisheries which match with study from Widodo and Nugraha [29]. Interestingly, there was no publication about the monthly catch rate for longline, probably due to difficulties in collecting the data. This is where e-logbook data could be the right tools for improving the data quality by filling the missing information.

The key to this improvement lies on how the e-logbook could be expanded and maintained for years to come. Law enforcement is necessary but build an understanding to fishermen about the importance on filling and reporting the e-logbook should be the priority for the stakeholders. When proper data could be established then the analysis could be carried on, especially for harvest strategy in this particular area.

4 Conclusion

E-logbook has the best attribute for improving the data quality, especially for small-scale tuna fisheries in FMA-714. Low cost and vast spatial and temporal coverage are the main characteristics on why the program should be expanded and monitored carefully. Analysis on fisheries indicators showed a promising result, especially for filling the gap which could not be covered by research.

Acknowledgement. The authors would like extend their great appreciation to the Directorate of Fish Resource Management, Directorate General of Capture Fisheries for permission to access fisheries e-logbook data. Syahril Abd.Raup, Sri Patmiarsih, Rista Devi Juniar and Bram Setyadji are the main contributors to this scientific paper.

References

1. F. Satria, L. Sadiyah, A. A. Widodo, C. Wilcox, J. H. Ford, and B. D. Hardesty, Marine Policy **95**, 8 (2018)
2. D. E. Duggan and M. Kochen, Marine Policy **67**, 30 (2016)
3. N. Akbar, N. P. Zamani, and H. H. Madduppa, Depik **3**, 65 (2014)

4. P. D. Pertiwi, A. Sembiring, A. Mahardini, D. Cahyani, A. Wahyu, B. Nugraha, R. Kartika Sulistyaningsih, I. Jatmiko, and I. Mahardika, in (2014)
5. D. Al Malik, N. P. D. Pertiwi, A. Sembiring, N. L. A. Yusmalinda, E. Y. Ningsih, and I. A. Astarini, Biodiversitas Journal of Biological Diversity **21**, 3637 (2020)
6. I. A. Astarini, S. A. Ardiana, I. N. G. Putra, P. D. Pertiwi, A. Sembiring, A. Yusmalinda, and D. Al Malik, Musamus Fisheries and Marine Journal **3**, 107 (2021)
7. B. R. Moore, P. Lestari, S. C. Cutmore, C. Proctor, and R. J. G. Lester, ICES Journal of Marine Science **76**, 1678 (2019)
8. P. Lestari, R. J. G. Lester, and C. Proctor, Indonesian Fisheries Research Journal **23**, 23 (2017)
9. N. Ayunda, M. R. Sapota, and A. Pawelec, in *Interdisciplinary Approaches for Sustainable Development Goals*, edited by T. Zielinski, I. Sagan, and W. Surosz (Springer International Publishing, Cham, 2018), pp. 147–167
10. DJPT, *Statistik Perikanan Tangkap Indonesia Menurut Provinsi, 2016* (Direktorat Jenderal Perikanan Tangkap, Indonesia, 2017)
11. I. M. Radjawane and P. P. Hadipoetranto, Ilmu Dan Teknologi Kelautan Tropis **6**, 525 (2015)
12. W. E. Rintaka and E. Susilo, in *IOP Conference Series: Earth and Environmental Science* (IOP Publishing, Batu, Malang, Indonesia, 2018), p. 012041
13. H. A. Rachman, J. L. Gaol, F. Syamsudin, and A. As-syakur, in *IOP Conference Series: Earth and Environmental Science* (IOP Publishing, Bogor, Indonesia, 2020), p. 012015
14. Z. Nasution, K. Amri, T. Arifin, W. S. Pranowo, J. Haryadi, S. Wibowo, A. Zulham, U. Mu'awanah, S. Syamdidi, R. B. A. Nugraha, B. Sumiono, and E. Erlania, *Potensi Sumber Daya Kelautan Dan Perikanan WPPNRI 714* (Amafradpress - Badan Penelitian dan Pengembangan Kelautan dan Perikanan Kementerian Kelautan dan Perikanan, Jakarta, 2016)
15. E. Hoshino, R. Hillary, C. Davies, F. Satria, L. Sadiyah, T. Ernawati, and C. Proctor, Fisheries Research **227**, 105539 (2020)
16. R. Gillett, *West Pacific East Asia Oceanic Fisheries Management* (Western and Central Pacific Fisheries Commission (WCPFC), Pohnpei, Federated States of Micronesia, 2013)
17. S. Yuniarta, P. A. van Zwieten, R. A. Groeneveld, S. H. Wisudo, and E. C. Van Ierland, Fisheries Research **193**, 173 (2017)
18. F. Satria and L. Sadiyah, Ind. Fish. Res. J. **24**, 39 (2018)
19. D. B. Sampson, Fisheries Research **112**, **59** (2011)
20. R. Coelho, P. G. Lino, and D. Rosa, Paper Presented on 15th Working Party on Billfish, San Sebastian, Spain, 10-14 September 2017, IOTC-2017-WPB15-14 17 (2017)
21. R Core Team, *R: A Language and Environment for Statistical Computing* (R Foundation for Statistical Computing, Vienna, Austria, 2020)
22. Q. D. Team, *QGIS Geographic Information System* (Open Source Geospatial Foundation Project, 2020)
23. H. Nugroho and A. Sufyan, Jurnal Kelautan Nasional **9**, 93 (2014)
24. H. Nugroho, A. Sufyan, and R. Akhwady, Jurnal Kelautan Nasional **8**, 101 (2013)
25. H. Nugroho, A. Darmawan, and A. Sufyan, Jurnal Kelautan Nasional **11**, 53 (2016)
26. H. Nugroho, A. Sufyan, and N. N. Wiadnyana, Jurnal Kelautan Nasional **10**, 113 (2017)
27. Tim APJII, Buletin APJII **74**, 10 (2020)
28. N. R. Loneragan, B. Wiryawan, A. R. Hordyk, A. Halim, C. Proctor, F. Satria, and I. Yulianto, in *Proceedings from Workshops on Management Strategy Evaluation of Data-Limited Fisheries: Towards Sustainability – Applying the Method Evaluation and Risk Assessment Tool to Seven Indonesian Fisheries*, edited by N. R. Loneragan, B. Wiryawan, A. R. Hordyk, A. Halim, C. Proctor, F. Satria, and I. Yulianto (Murdoch University, Western Australia, and IPB University, Indonesia, 2021), pp. 11–27
29. A. A. Widodo and B. Nugraha, BAWAL Widya Riset Perikanan Tangkap **2**, 299 (2017)
30. A. A. Widodo, R. T. Mahulette, and F. Satria, Jurnal Kebijakan Perikanan Indonesia **7**, 45 (2015)
31. T. Hariati, Jurnal Penelitian Perikanan Indonesia **17**, 139 (2016)
32. U. Chodrijah and B. Setyadji, Indonesian Journal of Marine Sciences/Ilmu Kelautan **23**, 171 (2018)
33. H. Hufiadi and E. Nurdin, Jurnal Penelitian Perikanan Indonesia **19**, 39 (2016)
34. A. Mahmud and R. L. Bubun, Jurnal Teknologi Perikanan Dan Kelautan **6**, 159 (2015)
35. W. Waileruny, E. S. Wiyono, S. H. Wisudo, T. W. Nuraini, and A. Purbayanto, in *Prosiding Simposium Pengelolaan Perikanan Tuna Berkelanjutan* (WWF & KKP, Bali, Indonesia, 2015), p. 1238
36. A. Assir, M. Palo, and R. K. Sari, Jurnal IPTEKS Pemanfaatan Sumberdaya Perikanan **4**, 108 (2017)
37. B. Nugraha and S. Mardlijah, BAWAL Widya Riset Perikanan Tangkap **2**, 45 (2017)
38. E. Rahmat and M. F. Yahya, Buletin Teknik Litkayasa Sumber Daya Dan Penangkapan **13**, 119 (2016)
39. U. Chodrijah and B. Nugraha, Jurnal Penelitian Perikanan Indonesia **19**, 9 (2013)
40. T. Hidayat, U. Chodrijah, and T. Noegroho, Jurnal Penelitian Perikanan Indonesia **20**, **43** (2014)
41. E. Rahmat, Buletin Teknik Litkayasa Sumber Daya Dan Penangkapan **6**, **29** (2016)
42. F.-C. Hsu, C. D. Elvidge, K. Baugh, M. Zhizhin, T. Ghosh, D. Kroodsma, A. Susanto, W. Budy,

https://doi.org/10.1051/e3sconf/202233904004

M. Riyanto, and R. Nurzeha, Remote Sensing **11**, **995** (2019)

43. M. Doddema, G. Spaargaren, B. Wiryawan, and S. R. Bush, Fisheries Research **208**, 49 (2018)
44. M. I. Marzuki, R. Garello, R. Fablet, V. Kerbaol, and P. Gaspar, in *OCEANS 2015 - Genova* (2015), pp. 1–5
45. E. Rodríguez-Marín, H. Arrizabalaga, M. Ortiz, C. Rodríguez-Cabello, G. Moreno, and L. T. Kell, ICES Journal of Marine Science **60**, 1216 (2003)
46. M. N. Maunder and A. E. Punt, Fisheries Research **70**, 141 (2004)
47. A. F. Nelwan, M. Zainuddin, and M. Kurnia, Marine Fisheries: Journal of Marine Fisheries Technology and Management **6, 129** (2015)

https://doi.org/10.1051/e3sconf/202233905001

What contributes to micro, small, and medium enterprises' productivity in fisheries sector in Aceh, Indonesia?

M. Shabri Abd. Majid[1*], *Faisal* Faisal[2], Heru Fahlevi[3], *A* Azhari[4], and *Hijri* Juliansyah[5]

[1]Department of Economics, Faculty of Economics and Business, Universitas Syiah Kuala, Banda Aceh, Indonesia.
[2]Department of Management, Faculty of Economics and Business, Universitas Syiah Kuala, Banda Aceh, Indonesia.
[3]Department of Accounting, Faculty of Economics and Business, Universitas Syiah Kuala, Banda Aceh, Indonesia.
[4]Program Study of Master of Management, Faculty of Economics and Business, Universitas Islam Kebangsaan Indonesia.
[5]Department of Economics, Faculty of Economics and Business, Universitas Malikussaleh, Lhokseumawe, Indonesia.

Abstract. Although Micro, Small, and Medium Enterprises (MSMEs) engaged in the agricultural, livestock, forestry, and fishery sector was recorded as the second-largest contributor to the Indonesian economy in the last decade, but their productivity level was relatively low due to their small-scale business entity. To further enhance their contribution to the national economy, MSMEs needs to enhance their productivity. This study measures Total Factor Productivity (TFP) of the MSMEs in the fisheries sector across 23 districts/cities in Aceh province, Indonesia over the period from 2015 to 2019 period. The study also decomposes the TFP into its sources. Using Data Envelopment Analysis (DEA), the study documented a low level of MSMEs' TFP during the study period. The MSMEs experienced different levels of TFP. Overall, the MSMEs experienced a decline in their TFP level, contributed mainly by the deteriorations in the efficiency and technical efficiency levels. These findings indicated the urgent call for adopting advanced fisheries-related technologies and implementing good MSMEs' governance practices to further promote MSMEs' performance. Finally, the relevant government authorities are suggested to prioritize promoting less productive MSMEs in the province by providing adequate financial assistance and conducting regular professional entrepreneurial workshops, short courses, and training programs.

1 Introduction

Micro, Small, and Medium Enterprises (MSMEs) play a crucial role in the Indonesian economy. The contribution of MSMEs to the national economy has been increasing over the last decades. Apart from offering new job opportunities, MSMEs contribute to income disparity reduction and poverty alleviation. MSMEs are viewed as an engine of national economic growth as they promote foreign exchange income for the country through exports' expansion of MSMEs products to the foreign markets. Ministry of Cooperatives and SMEs reported that the existing 61.7 million units of MSMEs in 2016 have contributed 89.2% job opportunities to the labor market, 60.34% to the total national economy, 14.17% to the total exports, and 58.18% to the total investment [1]. The MSMEs' contributions to the Indonesian economy showed a growing trend. In 2020, the existing 64.2 million units of MSMEs contributed 61.07% to the national economy, 97% to the labor market, and 60.4% to the total national investment [2].

MSMEs engaged in the fisheries sector are among the leading small-scale business entities that contribute to the national economy. In 2014, MSMEs in the fisheries sector contributed 48.85% to the national GDP [3]. In the second quarter of 2021, the MSMEs in the agricultural, livestock, forestry, and fishery sector was documented as the second-largest contributor to the Indonesian economy [4]. As a maritime country, Aceh is one of the provinces in Indonesia that has great agribusiness potential in the fisheries sector [5]. Aceh is a province that has a long coastline with a water area of 295,370 km2 [6]. The geographical location of Aceh which borders the seacoast has encouraged the growth of MSMEs in the fisheries sector in Aceh Province. During the 2011-2018 period, the average yield of captured fish and fisherman cultivation in Aceh continued to show an increasing trend, respectively, by 9.60% and 12.79% [7].

The rising contribution of MSMEs to the national economy is very much related to labor productivity. Many previous studies have discovered determinants of MSMEs' productivity worldwide [8]. For instance, previous studies documented the significant role of labor, skills, and technology in the productivity of MSMEs in India [9]. Operational, products and services' innovation contributed to the MSMEs' productivity in Italy [10]. Financial supports determined the productivity of MSMEs in the Arab region [11], while

* Corresponding author: mshabri@unsyiah.ac.id

infrastructure access and competitive ability influenced MSMEs' productivity in Africa [12]. Credit availability has constrained MSMEs in Canada to improve their productivity level [13].

Many previous studies have also explored MSMEs' productivity in the ASEAN economies. For example, the improved productivity of MSMEs in Vietnam was primarily caused by their ability to secure a globally recognized certificate and to innovate products technologically [14]. The utilization of information and communication technology, tax incentives, and product innovation promoted the productivity of MSMEs in Thailand [15]. Meanwhile, the productivity of MSMEs in Semarang, Indonesia is influenced by entrepreneurial competencies, professional management, technology, innovation, and business networks [16]. Other studies documented that human resources [17], training [18], the proper mix of inputs [19], technology, financial adequacy [20], economies of scale, level of education [21] and [22], and the adoption of updated technology [23] and [24] were among the crucial determinants of Indonesian MSMEs' productivity.

To our knowledge, although the above-reviewed researches have explored the factors determining MSMEs' productivity, but we found no studies have assessed the MSMEs' productivity in the fisheries sector in Indonesia. Besides, previous studies have not evaluated the sources of MSMEs' productivity; therefore, they failed to offer comprehensive policy-designed ramifications to further enhance MSMEs' productivity. Finally, different from previous studies that focused their analyses only on specific MSMEs in a certain district in Indonesia, this study measures the relative TFP and its sources, taking the case of MSMEs in the fisheries sector across 23 districts/cities in Aceh province, Indonesia. Intended to fill the gaps in the previous researches, this study aims to measure the Total Factor Productivity (TFP) of MSMEs in the fisheries sector and identifies its TFP's sources using the Data Envelopment Analysis (DEA). Using the DEA approach enables us to measure TFP and its decompositions into efficiency change (i.e., pure efficiency and scale efficiency changes) and technical change. The findings of this study are hoped to offer the latest evidence for the SMEs and relevant government authorities in formulating policies to enhance SMEs' productivity.

2 Materials and Methods

2.1Site and time

This study empirically measures and decomposes the sources of TFP of MSMEs in the fisheries sector across 23 districts/cities in Aceh province, Indonesia over the period from 2015 to 2019. According to the Law No. 9 of 1985 and Law No. 45 of 2009 of the Republic Indonesia concerning Fisheries, MSMEs in this sector involve individual or legal entity businesses to catch or cultivate fish, including activities to store, cool, dry, or preserve fish to create added economic value for business actors (commercial business).

2.2Sampling and measurements

Of 2,178 units of MSMEs in the fisheries sector in 23 districts/cities in Aceh province, Indonesia, 138 of them were selected as the sample of study using a stratified random sampling technique. These comprise six units of MSMEs in every 23 regencies/cities. Since the study covers a period of five years, thus the study analyzes 690 observations. In measuring TFP and its decompositions of the MSMEs in the fisheries sector, this study uses three inputs, namely: number of managerial board, own capital, and external capital, and managerial board, and one output, namely: total profit. The inputs-output determination is referred to the Indonesian Co-operative Act, No. 25 (1992). According to the Act, the operationalization of the MSMEs is supported by own capital, external capital, and managerial members for the business to grasp profit. Thus, the study gathers primary data using the questionnaires' distribution to the selected sample in Aceh province, Indonesia.

2.3 Data and Analysis

In measuring the TFP of Decision-Making Units (DMUs) and identify its sources, three approaches, namely traditional, parametric, and non-parametric analyses have been commonly adopted by previous researchers [8]. A non-parametric approach of DEA has been the popular approach used to measure of TFP of the DMUs due to its strengths [25]. DEA is not only able to identify the sources of TFP, but it is also able to measure TFP through the combination of multiple inputs to produce various outputs [26] using assumptions of both Constant Return to Scale (CRS) and Variable Return to Scale (VRS). Unlike the CRS model, the VRS assumes the ratio of inputs and outputs changes are disproportionate, showing that a change in inputs by x percentage could cause changes in the output by more or less than x percentage [27].

On these grounds, thus this study measures the relative TFP of the MSMEs in the fisheries sector using a generalized output-orientation of the Malmquist index [28]. Using the DEA statistical software, the study measures the following equations:

$$M_o\left(x^t, y^t, x^{t+1}, y^{t+1}\right) = (a)x(b) \quad (1)$$

where

$$a = \frac{D_o^t(x^{t+1}, y^{t+1})}{D_o^t(x^t, y^t)}; \text{and}$$

$$b = \frac{D_o^t(x^{t+1}, y^{t+1})}{D_o^t(x^t, y^t)} \quad (2)$$

where M0 is the Malmquist Index of TFP; Do is a function of distance; x and y are inputs and output for periods t and t + 1, respectively; a is a technical change; and b is an efficiency change.

The study decomposes the TFP into two components, namely: efficiency change (EFch) and technical efficiency change (TEch). EFch is further decomposed into two sub-components, namely: pure efficiency change (PEch) and scale efficiency change (SEch). The following equations are used to decompose the sources of MSMEs' TFP in the fisheries sector in Aceh Province, Indonesia:

$$M_o\left(x^t, y^t, x^{t+1}, y^{t+1}\right) = (a)x(b) = (a)x(cxd) \quad (3)$$

where

$$a = \left[\left(\frac{D_o^{t+1}(x^t, y^t)}{D_o^t(x^t, y^t)}\right)\left(\frac{D_o^{t+1}(x^{t+1}, y^{t+1})}{D_o^t(x^{t+1}, y^{t+1})}\right)\right]^{1/2};$$

$$c = \left(\frac{D_o^t(x^t, y^t)}{D_o^{t+1}(x^{t+1}, y^{t+1})}\right); \text{and}$$

$$d = \left(\frac{D_{oc}^{t+1}(x^t, y^t)}{D_o^{t+1}(x^t, y^t)}\frac{D_o^{t+1}(x^{t+1}, y^{t+1})}{D_{oc}^{t+1}(x^{t+1}, y^{t+1})}\frac{D_{oc}^t(x^t, y^t)}{D_o^t(x^t, y^t)}\frac{D_o^t(x^{t+1}, y^{t+1})}{D_{oc}^t(x^{t+1}, y^{t+1})}\right)^{1/2}$$

a is the technical change (TEch); b is the change in efficiency; c is the change in pure efficiency (PEch); and d is the change in scale efficiency (SEch).

3 Results and discussion

Table 1 reports the relative efficiency of MSMEs in the fisheries sector across 23 districts/cities in Aceh province, Indonesia over the period from 2015 to 2019 based on the Constant Return to Scale (CSR) and Variable Return to Scale (VRS) assumptions.

As illustrated in Table 1, the MSMEs recorded the highest efficiency (62.4%) in 2015 based on the VRS and in 2016 (53.4%) based on CSR. From the year 2016 onwards, the MSMEs experienced a declining trend in their efficiency level up to the lowest level in 2019 (CRS = 44.2% and VRS = 54.6%). During the study period, on average, based on the CRS and VRS models, the MSMEs recorded efficiency levels of 49.1% and 58.3%, respectively. These findings show a low level of efficiency of the MSMEs in the fisheries sector in Aceh, Indonesia.

Table 1 also shows that, based on the CRS assumption, the study found that MSMEs in Lhokseumawe recorded the lowest efficiency level (15.04%), followed by MSMEs in the districts of Simeulue (17.06%), Pidie (18.56%), East Aceh (21.32%), and Aceh Singkil (21.36%). On the other hand, the MSMEs in the districts of Bireuen and Aceh Tamiang recorded an optimal efficient level (100%), followed by Benar Meriah (89.78%), Southwest Aceh (89.7%), Banda Aceh (71.82%), and Central Aceh (70.88%).

From the perspective of the VRS estimates, the study found that MSMEs in six districts/cities (i.e., Banda Aceh, North Aceh, Bireuen, Aceh Tamiang, Sabang, and Southwest Aceh) experienced a full efficient level (100%) during the study period. Meanwhile, the MSMEs in Lhoksuemawe recorded the worst level of efficiency (15.6%), followed by MSMEs in the districts/cities of Simeulue (19.2%), Aceh Singkil (21.9%), East Aceh (23.2%), and Pidie (23.4%). Of 138 MSMEs investigated in the study, MSMEs in Bireuen and Aceh Tamiang have been able to operate at their optimal efficiency level both based on the CRS and VRS models. Our findings showed the dissimilar level of efficiency of MSMEs across 23 districts/cities in Aceh Province.

Table 1. CRS and VRS of the SMEs in the fisheries sector, 2015-2019.

District	2015		2016		2017		2018		2019	
	CRS	VRS	CRS	VRS	CRS	VRS	CRS	VRS	CRS	VRS
Banda Aceh	0.28	1	1	1	1	1	1	1	0.31	1
Aceh Besar	0.56	0.63	0.63	0.66	0.6	0.8	0.71	0.73	0.91	0.91
Pidie	0.12	0.16	0.22	0.22	0.35	0.47	0.12	0.17	0.12	0.15
Pidie Jaya	0.79	0.99	1	1	0.41	0.56	0.32	0.45	0.33	0.41
Bireuen	1	1	1	1	1	1	1	1	1	1
Lhokseumawe	0.16	0.18	0.19	0.19	0.15	0.15	0.15	0.16	0.11	0.11
North Aceh	0.44	1	0.54	1	0.25	1	0.6	1	0.18	1
East Aceh	0.06	0.08	0.24	0.24	0.28	0.31	0.49	0.51	0.01	0.01
Langsa	0.25	0.26	0.21	0.24	0.17	0.25	0.47	0.63	0.36	0.54
Aceh Tamiang	1	1	1	1	1	1	1	1	1	1
Southeast Aceh	0.2	0.26	0.43	0.46	0.66	0.7	0.66	0.67	0.35	0.59
Gayo Lues	0.5	0.51	0.4	0.4	0.49	0.49	0.46	0.46	0.51	0.51
Central Aceh	1	1	1	1	0.54	0.66	0.3	0.42	0.7	0.83
Bener Meriah	0.89	1	1	1	1	1	0.86	0.92	0.74	0.78
Aceh Singkil	0.21	0.22	0.25	0.25	0.18	0.18	0.2	0.21	0.24	0.24
Sabang	0.66	1	0.51	1	0.28	1	0.63	1	0.3	0.3

District	2015		2016		2017		2018		2019	
	CRS	VRS	CRS	VRS	CRS	VRS	CRS	VRS	CRS	VRS
West Aceh	0.36	0.37	0.42	0.42	0.4	0.4	0.43	0.43	0.56	0.56
Nagan Raya	0.22	0.25	0.23	0.25	0.22	0.3	0.2	0.25	0.24	0.29
Southwest Aceh	0.83	1	0.94	1	0.71	1	1	1	1	1
South Aceh	1	1	0.17	0.17	0.25	0.29	0.26	0.27	0.29	0.3
Subussalam	0.29	0.33	0.23	0.23	0.26	0.35	0.23	0.3	0.3	0.41
Simeulue	0.13	0.13	0.15	0.16	0.2	0.23	0.22	0.23	0.17	0.2
Aceh Jaya	0.86	1	0.57	0.69	0.39	0.42	0.17	0.17	0.44	0.44
Mean	0.51	0.62	0.53	0.59	0.47	0.59	0.5	0.56	0.44	0.55

Source: Primary data processed using the DEA program (2021).

Table 2. TFP changes and its sources of the SMEs in the fisheries sector, 2015-2019.

District	2015-2019				
	EFch	TEch	PEch	SEch	TFPch
Banda Aceh	1.024	0.903	1	1.024	0.924
Aceh Besar	1.131	0.978	1.098	1.03	1.106
Pidie	0.991	0.887	0.982	1.009	0.879
Pidie Jaya	0.8	0.865	0.801	0.999	0.693
Bireuen	1	1.006	1	1	1.006
Lhokseumawe	0.919	0.925	0.89	1.032	0.849
North Aceh	0.8	0.915	1	0.8	0.732
East Aceh	0.697	0.919	0.646	1.079	0.64
Langsa	1.093	0.967	1.201	0.91	1.057
Aceh Tamiang	1	0.922	1	1	0.922
Southeast Aceh	1.151	0.969	1.23	0.936	1.116
Gayo Lues	1.009	0.933	1	1.009	0.941
Central Aceh	0.915	0.912	0.955	0.957	0.834
Bener Meriah	0.955	0.912	0.94	1.017	0.871
Aceh Singkil	1.032	0.93	1.013	1.019	0.96
Sabang	0.821	0.943	0.74	1.109	0.774
West Aceh	1.117	0.93	1.107	1.009	1.039
Nagan Raya	1.021	0.984	1.034	0.987	1.005
Southwest Aceh	1.047	0.992	1	1.047	1.038
South Aceh	0.732	0.834	0.741	0.988	0.61
Subussalam	1.006	0.972	1.055	0.954	0.978
Simeulue	1.075	0.93	1.103	0.974	1
Aceh Jaya	0.848	0.93	0.816	1.038	0.788
Mean	0.956	0.932	0.961	0.995	0.891

Source: Primary data processed using the DEA program (2021)

In addition, our study also indicated an unstable efficiency level of MSMEs, implying the inability of MSMEs' management to maintain and even continuously improve their efficiency level across the years. As a small scale-business entity that is not supported by adequate capital and modern management practices has contributed to the low level of efficiency of MSMEs in Aceh province, Indonesia.

Furthermore, Table 2 reports the TFP changes and its sources of the MSMEs in the fisheries sector in Aceh Province, Indonesia during the 2015-2019 period. TFP is decomposed into two main sources, namely: efficiency change (EFch) and technical efficiency change (Tech). The EFch is further decomposed into two sub-sources, namely: scale efficiency change (SEch) and pure efficiency change (PEch). The value of TFP change of greeter than 1.000 shows an improvement in the MSMEs' productivity, while the value of less than 1.00 shows a deterioration in the MSMEs' productivity.

As observed from Table 2, overall, the study found that the average TFP of the MSMEs in the fisheries sector has declined by -10.9% during 2015-2019. The declining trend of the TFP is contributed by a deterioration of efficiency change and technical change by -4.4% and -6.8%, respectively. These findings show the lack of technical innovation of the MSMEs in the fisheries sector in Aceh, Indonesia. Furthermore, the reduction in the changes in the efficiency of the MSMEs was caused by pure efficiency and scale efficiency deteriorations by -3.9% and -0.5%, respectively. The inability of the entrepreneurs in the fisheries sector to properly combine the right inputs to produce the highest level of outputs due to traditional management practices has been the main contribution to the declining in the relative efficiency and TFP of the MSMEs in Aceh Province, Indonesia.

Our findings of the low level of the MSME's efficiency and TFP levels of the fisheries sector in Aceh Province, Indonesia are in harmony with many existing empirical findings [29] and [30]. However, our findings of different MSMEs' efficiency and TFP levels contradicted the previous research that documented similar SMEs' productivity levels in 29 regions in mainland China [31].

Having different TFP levels of MSMEs across the districts/cities signifies the crucial efforts to be taken by the relevant government authorities, particularly the Offices of SMEs and Co-operatives to design proper policies to promote MSMEs TFP levels in all regions to experience a similar TFP level. The government should continuously take serious efforts to prioritize promoting MSMEs in the districts/cities that experienced a lower level of TFP by providing sufficient financial aids and improving professional entrepreneurial management skills and practices through various short courses, workshops, and training. This policy would create a just economic development across the cities/districts in the province. To further promote MSMEs' TFP, the government should suggest and provide the MSMEs with advanced fisheries-related technologies.

Finally, our empirical evidence of different MSMEs' TFP levels across the cities/districts are similar to previous studies [10], [14], [29], and [32-34]. In addition to providing financial aids, the relevant government authorities are suggested to provide regular professional entrepreneurial training, particularly for the MSMEs in the cities/districts that experienced the lower TFP levels by benchmarking the MSMEs' best practices from other districts/cities in the province.

Finally, the MSMEs should also continuously maximize the utilization of their human and financial resources for productive activities and combine properly limited inputs to produce the maximum level of outputs.

4 Conclusions

This study has measured and decomposed Total Factor Productivity (TFP) of the fisheries' MSMEs in across 23 regencies/cities in Aceh province, Indonesia during the 2015 to 2019 period. Using the Data Envelopment Analysis (DEA) analysis, the study recorded evidence of the low MSMEs' TFP level. Overall, the MSMEs have experienced a decline in their TFP level, contributed mainly by the deteriorations in efficiency and technical changes. In addition, the decline in technical efficiency has contributed most to the declining trend of the MSMEs' TFP. These empirical findings indicated the urgent call for adopting advanced fisheries-related technologies and implementing good MSMEs' governance practices to further promote MSMEs' performances. The MSMEs are also required to improve their human and financial resources for productive activities and properly combine limited inputs to produce the maximum outputs. Finally, the relevant government authorities should prioritize promoting less productive MSMEs across the districts/cities in the province by granting adequate financial supports and conducting regular professional entrepreneurial workshops, short courses, and training programs.

To provide more robust empirical evidence of the MSMEs' performance, future studies are suggested to investigate the MSMEs' TFP engaging in various economic sectors across districts/cities and provinces nationwide to produce a comprehensive picture of the MSMEs performance. Finally, future researches could adopt a mixed research method, covering both parametric and non-parametric to enrich the existing references and empirical results of the MSMEs performance.

Acknowledgement. This article is part of the research project funded by the Directorate General of Strengthening Research and Development, Ministry of Research, Technology and Higher Education, the Republic of Indonesia, under the scheme of the World Class Research No: 215/SP2H/LT/DPRM/2019.

References

1. BPS-Indonesian Statistics. Sensus Ekonomi. Jakarta: BPS (2016)
2. Ministry of Coordinator for Economic Affairs of the Republic of Indonesia. Press Release: UMKM menjadi pilar penting dalam perekonomian Indonesia [Internet]. Jakarta (2001). Available from: https://ekon.go.id/publikasi/detail/2969/umkm-menjadi-pilar-penting-dalam-perekonomian-indonesia
3. Bank Indonesia. Profil bisnis usaha mikro, kecil dan menengah (UMKM) [Internet]. Jakarta (2015). Available from: https://www.bi.go.id/id/umkm/penelitian/Documents/Profil%20Bisnis%20UMKM.pdf
4. BPS- Indonesian Statistics. Annual Report. Jakarta (2021)
5. I. Nurdin. Tantangan pengelolaan potensi perikanan Aceh. AcehTrend.Com (Internet]. Aceh (2019). Available from: https://www.acehtrend.com/2019/05/23/tantangan-pengelolaan-potensi-perikanan-aceh/
6. BPS- Indonesian Statistics. Provinsi Aceh dalam Angka 2018. Banda Aceh (2019)

7. Ministry of Agriculture of the Republic of Indonesia. Modernisasi tingkatkan produktivitas tenaga kerja sektor pertanian [Internet]. Jakarta; 2019. Available from: https://www.pertanian.go.id/home/?show=news&act=view&id=3839
8. M.S.A. Majid, H. Maulana. Gadjah Mada Int. J. Bus.**14**, 2 (2010)
9. A. Mathur, M.I. Mittal, G.S. Dangayach. Prod. Plan. Control. **23**, 10–11 (2012)
10. B.H. Hall, F. Lotti, J. Mairesse. Small Bus. Econ. **33**, 1 (2009)
11. D. Emine. J. Contemp. Manag. 1, 9 (2012)
12. P.A Igwe, A.N.A.O.M. Ogundana, O.M. Egere, J.A. Anigbo. Eur. J. Sustain. Dev. **7**, 1 (2018)
13. S. Cao, D. Leung. Econ. Model. 88 (2020).
14. E. Calza, M. Goedhuys, N. Trifković. Econ. Innov. New. Technol. **28**, 1 (2019)
15. J.W Lee, N. Suwimol. J. Transnatl. Manag. **23**, 4 (2018)
16. S.C.R. Erazo, D.A Paz, J.A. Cardona. Contribution of information technology and communications to productivity in SMEs in Santiago de Cali, in Proceeding of the 15th Iberian Conference on Information Systems and Technologies (CISTI), 1-5. IEEE (2020)
17. J.G. Vargas-Hernández. Innovation of SMEs and their effect on productivity in Jalisco, in Innovation Management and Growth in Emerging Economies. IGI Global, 155–91 (2020)
18. L. Cusmano, M. Koreen, L. Pissareva. 2018 OECD ministerial conference on SMEs: Key issues paper, in OECD SME and Entrepreneurship Paper. Paris (2018)
19. D. Kurniawati, H. Yuliando. Agric. Sci. Procedia. **3** (2015)
20. T. Suminar, A. Yusuf. People Int. J. Soc. Sci. **6**, 1 (2020)
21. J.T. Wengel, E. Rodriguez. Small Bus. Econ. **26**, 1 (2006)
22. A. Parameswara, A. Wulandari, R. Giri. Int. J. Bus. Econ. Law. **18**, 5 (2019).
23. I.R Bawono, T. Arofah, B.M.S. Ubay. J. Ilm. Mana. Ekon. Akun. **4**, 3 (2020)
24. S.U.J. Raharja, N. Kostini, H.A. Muhyi, R. Rivani. Int. J. Trade Glob. Mark. **12**, 3–4 (2019)
25. W.W. Cooper, L. Seiford. Handbook on Data Envelopment Analysis. Zhu J, editor. Boston: Boston, MA: Springer Science & Business Media (2011)
26. M.A. Omar, A.R.A. Rahman, R.M. Yusof, M.S.A. Majid, M.E.S Rasid. Asian Acad. Manag. J. Account. Financ. **2**, 2(2006)
27. T. Coelli. A guide to DEAP version 2.1 data envelopment analysis (computer) program, in CEPA Working Paper. Report No.: 96/98, Armdale (1996)
28. R.D Banker, A. Charnes, W.W. Cooper. Manage. Sci. **30**, 9 (1984)
29. S. Sulikah. Analisis efisiensi koperasi pegawai negeri Republik Indonesia Kabupaten Klaten. Thesis. Universitas Sebelas Maret (2010)
30. G. Syamni, M.S.A. Signifikan J. Ilmu Ekon. **2**, 5(2016)
31. F. Dong, A.M. Featherstone. J. Chinese Econ. Bus. Stud. **1**, 4 (2006)
32. P. Asawaruangpipop, O. Suwunnamek. Res. J. Bus. Manag. **3**, 8(2014)
33. I. Hasan, A. Azhari, M.S.A Majid. REVESCO Rev. Estud. Coop. **128** (2018)
34. M.S.A Majid, F. Faisal, H. Fahlevi, A. Azhari. Measuring and decomposing SMEs' productivity in creative economic sector in Indonesia, in IOP Conf. Ser: Earth and Envir. Sci. IOP Publishing (2012)

Social, cultural, economic and mangrove resources Gampong Lapeng, Pulo Breuh, Aceh Besar

Khansa Iklila Hanifa, *Najmi* Rinjani Ferdian, *Suha* Rakata Ferdian, *Agha* Alamulhuda Husein, *Thariq* Abdurrahman, *Maria* Ulfa*

Perkasa Alam School, Banda Aceh 23234, Indonesia

Abstract. Socio-cultural and economic a society will be able to help provide information related to the condition of the community. Mangroves are plants that are the result of cultivation activities or taken from nature. This research aims to find out the social, cultural, economic and also mangrove conditions that remain in Gampong Lapeng. In addition, to get to know the preservation of culture and customs in Gampong Lapeng. The benefit of this activity is to provide information and science. This research uses descriptive qualitative methods with a study subject of 5 people. The plot used in this investigation is as much as 1 plot covering an area of 50x50 meters. The results showed that the social conditions of lapeng community are open and friendly in accepting guests who come to visit. A very thick culture in the village of Lapeng is the advice not to go to sea on Fridays. The economic conditions of the villagers depend on the weather and the sea season. In the east wind season the income generated at sea will decrease nominally because fishermen can find fish easily. The types of mangroves in Gampong Lapeng are *Rhizophora* and *Avicennia*.

1 Introduction

Mangrove forests are plants that grow between the tidal lines, so mangrove forests are also called tidal forests. Mangroves are very rare ecosystems, because they are very small, which is about 2% of the earth's surface. Literally, this mangrove forest weld is only about 3% of almost all forest areas and 25% of all mangrove forests in the world [1].

Mangrove forest ecosystems in Indonesia are currently in critical condition, this is due to damage of almost 68% or about 5.9 million hectares of mangrove forest. This condition is very concerning, because of the magnitude of mangrove functions and benefits for marine life. To improve this condition, there needs to be a change in the landscape and the way of looking at the community related to mangroves and how to take other natural resources around mangroves. Because mangroves ecosystems have a function to maintain land from the grinding of waves and other marine biota living places [2].

Mangrove Forest area in addition to serving to keep the land from coastal abrasion, also as a provider of foodstuffs for human life. Where humans consume fish, shrimp, crabs and crabs and energy sources for life on the beach such as plankton, nectar and algae [3]. The physical function of mangrove forests is to keep the coastline stable, protect beaches and river cliffs from erosion and abrasion and resist or absorb very strong angina during the wind season. The chemical function of the mangrove ecosystem is the place where the recycling process that eliminates oxygen, as the absorption of carbon dioxide, as a processor of waste materials from pollution from industry and large ships in the ocean [4 Arief, 2003].

Geographically, Pulo Aceh District borders the Strait of Malacca and The Strait of Bengal to the North, Peukan Bada District in the South, Sabang Waters to the East, and the Indian Ocean to the West. The area of Pulo Aceh Subdistrict reaches 240.75 Km2 and consists of 3 Kemukiman, namely Nasi Island, South Breueh Island and North Breueh Island Resettled. Nasi Island consists of 5 villages, namely Alue Reuyeueng, Deudap, Lamteng, Pasi Janeng and Rabo. The resettled of South Breueh Island consists of 8 villages, namely Lampuyang, Lhoh, Paloh, Blang Situngkoh, Ulee Paya, Gugop, Seurapong and Teunom. The resettledness of North Breueh Island consists of 4 villages, namely Alue Raya, Lapeng, Meulingge and Rinon. Lapeng village located in the village of Breuh Island is one of the very remote villages, and still belongs to the remote indigenous community (KAT) of Aceh. Lapeng village is one of the village in Pulo Aceh District, Aceh Besar Regency, Aceh, Indonesia. Which the population in Lapeng village is 174 people from 46 Kartu Keluarga (Family Certificate).

Geographically the location of Lapeng village is very remote so access to the village is very difficult, as well as vehicles to the village. There are only one or two vehicles such as pick up cars that residents can use as a daily transportation. So, this greatly affects the economy of Lapeng residents. Other facilities such as health facilities and educational facilities are also very limited, making it difficult for residents. So, if the citizens want

* Corresponding author: bundaulfa1981@gmail.com

https://doi.org/10.1051/e3sconf/202233905002

to seek treatment must take a long way out of this Lapeng village. Likewise for educational facilities, there is only one elementary school and four teachers who will teach alternately. Very concerning and distressing conditions.

Based on the initial picture obtained by the Kids Survivor team, the research team together with field supervisors wanted to conduct further field studies related to the condition of mangrove natural resources in Lapeng village as well as the social conditions of the people, economy and culture of Lapeng village residents.

2 Methodology

2.1 Place and Time of Research

The research was carried out for 6 (six) days, from 27 march to 1 April 2021. The research location is in Lapeng village, Breuh Island Sub-District, Pulo Aceh, Aceh Besar District. Determining the location based on it is remote location, and also as a location for empowering remote location. Determining research locations based on their geographical location and conduct mangrove research as well as economic, social, and cultural research.

2.1.1 Type of research

The type of research we use is a type of qualitative research. According to [5], qualitative research is research used to investigate, discover, describe, and explain the qualities or privileges of social influences that cannot be explained, measured or described through quantitative approaches. Cauldron research is a type of research whose findings are not obtained through statistical procedures or other forms of calculation and seek to understand and estimate the meaning of an event of human interaction in certain situations according to the researcher's own perspective.

2.1.2 Scope of Research

The scope of research covered demographic, social, economic, cultural, and identification data on mangrove plants on the coast of Lapeng village.

2.1.3 Data Collection Techniques

Data collection techniques in the form of observations, interviews, and FGD (focus group discussion). The research team made observations by mingling directly with the village community to observe the social conditions of the community. Interviews were conducted face-to-face with village heads and local residents and informants who were mentors in the research. FGD was conducted with several villagers in the form of discussions related to the focus of research.

2.2 Data Analysis

The data obtained in the field is analysed with qualitative methods which are then presented using descriptive methods to describe mangrove's economic, social, cultural, and natural resources conditions.

3 Results

3.1 Overview and Social, Economic, Cultural Conditions of Research Locations

The research location of the Kids Survivor team is in 1 (one) village, namely Lapeng village, Pulo Aceh sub-district, Aceh Besar.

3.2 Geographical Location

Lapeng Village is one of the village located in Pulo Aceh subdistrict, Aceh Besar Regency. From bps Pulo Aceh data, Aceh Besar regency in 2020 said that Lapeng village is located on the coast and has a village area of 06.77Km2. This village includes the mukim of north Breuh island which oversees 4 (four) village, namely Alue Raya village, Rainon village, Meulingge village, and Lapeng village. In this village, there is 1 keuchik (head of a village), 1 Sekretaris Desa (secretary of a village) 2 Kepala Dusun (head of a hamlet) , and one Kepala Urusan (head of affairs). This village is very remote and located far from the capital of the district. So that this village can be called an inland area.

As a deepening area, Lapeng village has natural wealth in the form of plants. One of the potential plants is mangroves. This plant is also often called Bak Bangka in the area. Mangroves are one of the rare ecosystems that grow between tidal lines. So it needs to be a good time in preserving it. In addition to mangroves, Lapeng villages do not have many specific natural resources, this is because the villagers prefer to go to sea rather than farming. This is because the area of Lapeng village is still far away.

The village also includes a remote indigenous community empowerment location (KAT). KAT itself is a group of people who live in remote areas, where the area is in a condition of not enough both clothing and food. The area is also vulnerable to social, economic, and isolated from the outside world.

3.3 Mileage and Accessibility

The distance from the Provincial Capital (Banda Aceh) to the District Capital (Lampuyang) is reached within 1.5 - 2 hours using a fishing boat as far as the road. The cost used is also not small, because visitors have to rent a special vehicle to get to the village. This is due to the absence of public transportation to go there. Furthermore, access from the sub-district to Lapeng village can be reached by four-wheeled or two-wheeled vehicles for 90 minutes as far as approximately 5 km. This is due to the condition of the road terrain traversed across mountains and unpaved rock roads.

3.4 Population

According to Badan Pusat Statistik (BPS) (Central Bureau of Statistics) data in 2020, totaled 154 people in 56 Kartu Keluarga/KK (family certificate). 100% of the villagers are Muslim. In the Lapeng region, not each KK has each house. There are houses that have 2 or more KK in them. The population density of this village has 23 inhabitants. The condition of the village with the number of residents who are not crowded, shows the characteristics of Lapeng villagers who have a very friendly nature. This can be seen from the ethics when migrants arrive in their villages. With friendly and welcome residents warmly welcome the immigrants, whether it is guests or students and children.

3.5 Livelihoods and Economic Activities

The majority of Lapeng residents are generally gardening, trading, and sea. Plantations that are cultivated include, yam, cucumber, nutmeg, chili, and cassava. For fishermen, one of the species that are often sought is octopus which will later be traded. They also produce some fish that are sold to buy fellow villagers.

The commonly marketed crop commodity is nutmeg. Because the distance and travel between the village and the sub-district capital is inadequate, residents usually market their plantations and catches by selling with fellow residents. As for basic needs such as rice, residents do buy and sell with other villagers or by being visited by residents from sub-district villages. This is due to the lack of farming. This is due to the presence of several pests in the form of monkeys that disturb plants. Villagers also take advantage of their forests. Forest products commodities such as clean water sources, and wood sources for local purposes such as house construction and others. Forests in Lapeng village also act as NTFB (Non-Timber Forest Products).

3.6 Public Facilities and Facilities

In general, social facilities and public facilities are very minimal. Lapeng village has only one elementary school with a total of 6 classes and 2 teachers. Because of its location in the interior, the school provides a simple home to be occupied by teachers. As a deepening area, some houses in Lapeng village already have public bathing, washing, and toilet facilities. Each house has at least 1 bathroom that can be used for bathing and purifying.

For matters of worship, Lapeng village already has a mushalla that can be used for prayer and other worship. Residents also have rooms such as halls to conduct meetings, citizen associations or something like that. The hall is made of wood, which residents and children use to study. Lapeng village also has a reading garden, which children can use to read books. Many donations they get from some immigrants in the form of reading books and textbooks that will later be facilitated in the reading park.

Table 1. Facilities in Lapeng village

No.	Type of Facilities	Total	Condition			Usage		
			B	C	R	S	J	TP
1	Educational Facilities							
	• Elementary school	1	√			√		
	• Study Hall	1	√			√		
	• Reading Garden	1	√			√		
2	Religious Facilities							
	• Mushalla	1	√			√		
	• Study Hall	1	√			√		
3	Public Service Facilities							
	• Bridge	1	√			√		
4	Cemetery/Graveyard	1			√		√	
5	Transportation			√		√		

Description: Condition: B: good; C: enough; R: broken. Usage : S: frequent; J: rare; TP: never.

3.7 Social and Cultural Interaction

3.7.1 Socio-Cultural Life

This village has a unique culture, among them, local residents should not work on the day that the villagers do their studies. Also, in this village there is a middle-aged woman whose cuisine is very enjoyed by the locals. Many people buy or get some food from her. No wonder if some similar foods found in several different homes later (Table 2).

3.8 Language

The majority of villagers use Acehnese in their daily lives. Some of them can understand Indonesian but only a small number. This is because of the length of time they live in the village and almost all residents use Acehnese. Even migrants have to adapt to use the language of the region.

3.9 Mangrove

The mangrove observation site area is just behind SDN (Public Alimentary School) Lapeng. Mangrove areas are also located on the coast, where the area has become a waste area.

3.9.1 Type and number

Based on the results of mangrove analysis in the plot of 50x50 meters found 12 amounts of vegetation and consisted of 2 types. The types of mangroves located in the mangrove forests of Lapeng Village in detail will be described in Table 3.

It is known that at least the number of mangroves in the plot of 50x50 is caused by the tsunami in 2004 ago. Therefore, most of the mangroves disappeared and only a few species remained in small numbers.

Table 2. Social, economic and cultural data in Lapeng village

MENU	Forest	Garden	Settlement	River	Coast	Sea
Social & Economic	hunter	Garden Farmer	- Grocery	-	Stall	Fisherman
		Cattle	- Teacher			Fish Shelter
		Rancher	- Hall			
			- Mushalla			
			- Hall			
			- Park Read			
Culture	Prohibition of cutting down trees around settlements and water sources	Gardening ban from Thursday afternoon to Friday afternoon	Not going to sea and farming every Monday	Prohibition of cutting down trees in the river	Abeh	Gardening ban from Thursday afternoon to Friday afternoon
					Wednesday	

Table 3. Types and number of mangroves.

No	Leaf	Flower	Fruit	Rot	Type	Sum
1	Taper	Armpit leaves	Cylinder	Support	Rhizophora	5
2	Oval	Edge	Round	Breath	Avicennia	7
Total						12

3.9.2 Abundance

[5] states that relative abundance is the proportion presented by each species of an entire individual in a community. Mangrove abundance results will be detailed in Table 4.

Based on the results of the calculation of abundance in the mangrove ecosystem of Lapeng village, the overall mangrove abundance reached 0.0048.

Table 4. The abundance of mangroves

No	Plot Area (unit)	Number of species	Type	Result
1	50x50	5	Rhizophora	0,002
2	50x50	7	Avicennia	0,0028
Total			0,0048	

4 Discussion

The bottom of the waters of the mangrove area of Lapeng village is dominated by sand and a little clay. Mangrove area itself is used as a playground for village children. This is because the mangrove area located near the coast is not maintained. Even only left behind a few remaining species after the tsunami hit Aceh.

In this region, the team took data by identifying leaves, flowers, roots, and fruit from mangrove stems. This makes it easier to know the type of species being identified. The team must also know the function of mangroves in addition to their speciess. [6] mentions several functions of mangrove forests. including keeping the coastline stable, protecting the coast from erosion (abrasion), as well as dampening waves and storms, as mud retainers, flood control, treating waste materials, maintaining water quality, CO2 absorbers and O2 producers and reducing the risk to tsunami hazards. Ecologically, mangroves also function very important in playing a role as a food chain in a water, where it can ride the lives of various types of fish, shrimp and mollusks.

If treated further, mangroves can also be beneficial for social and economic. From the [6] mentioned that mangroves are useful as a search eye, including the production of various forest products, such as wood, charcoal, medicine and food. Mangroves are also useful as a source of building materials and crafts, natural attractions, educational and research objects, mooring areas, salt making grounds and plantation areas.

In mangrove identification activities, the team also taught village children about mangrove data retrieval. The reason for doing this, goes back to the title of the team activity, namely give and take. Which we take and give knowledge to. Not enough to get there, the team will also make a presentation later in front of the villagers. Not only identifying mangrove data, the team also conducted personal projects, including dumpling, cooking projects, making Thai tea, handicrafts in the form of bracelets and necklaces, teaching football techniques, and storytelling the story of the prophet. This is expected to be useful and add insight to the village children. The same goes for people who not only let us learn, but also teach. Starting from cultural, social and economic. Details per detail of the residents are ready to give us information about the village.

Around the mangrove area, there is a pier. From Wikipedia, it is said that the pier is where the ship is moored in the harbor. It is also the place where the loading and unloading activities and ups and downs of people or passengers or goods from and to the ship. The community also provides stalls on the pier. In addition, the community also buys and sells fish and other marine animals, such as octopuses. Not only in the dock, local residents also do buys and sells fellow in the village. Like nutmeg producers, which can later be sold to interested citizens. Also, there are people who work as grocery managers. Residents also buy daily necessities at the grocery such as oil, sugar, eggs, milk and so on.

Head of a village's mother also opened a meatball business in her house, which made many visitors often visit their house.

In addition to business, residents also work as hunters. It's not public works even 2/10 of working people do. Similarly, planters, who often stop working because their gardens are damaged by pests or farm animals, such as monkeys, cows, and goats. For the teaching part, residents also act as teachers in the village. Residents held studies for the village children, with one of their residents as a guide. Studies are held every afternoon between 5.00-6.00 PM. This study is not compulsory for all children to follow, but almost all children join. Unlike the school, which residents provide a teacher's home in the school yard. With that the teacher does not need to go back and forth from home to school. The study hall is also used as a facility for guests or immigrants to conduct their programs.

Lapeng village has a reading house that children use as a library and playground. Sometimes also the village is visited by immigrants or guests from outside who teach the village children many things, including reading. As did the Kids Survivor team, which taught them the story of the Prophet of choice by storytelling. To help the village, visitors can also donate some reading books that can later be read by village children.

Socio-cultural and economic a society will be able to help provide information related to the condition of the community. This research aims to find out the social, cultural, economic and also mangrove conditions that remain in Lapeng village. Mangroves are plants that are the result of cultivation activities or taken from nature. The purpose of this activity is to get to know the preservation of culture and customs in Lapeng village. In addition, to identify the types of mangroves in Lapeng village to find out the source of income of Lapeng village residents. The benefit of this activity is to provide information and science. This research uses descriptive qualitative methods with a study subject of 5 people. The plot used in this investigation is as much as 1 plot covering an area of 50x50 meters. The results showed that the social conditions of Lapeng people openly and kindly accept guests who come to add knowledge. The very thick culture in the village of Lapeng village is brotherly bond or a meal together that was done thursday night to strengthen the bonds of brotherhood. The economic conditions of the villagers depend on the weather and the sea season. In the east wind season the income generated at sea will decrease nominally because fishermen can find fish easily.

5 Conclusion

Based on the results of the analysis it be concluded that the social condition of Lapeng community has strong social ties, unpretentious with immigrants and accepts immigrants and guests in a friendly manner. Lapeng residents strongly hold fast to customs and uphold religion. The economic level of Lapeng village community has a sufficient level of economy to meet the needs of daily living, the recovery of can be from sea, farming and gardening and trading.

While the natural resources of mangrove forests have not been used as the main source of livelihood for the residents of Lapeng village. However, the residents of Lapeng village continue to be sustainable mangrove forests because children of Lapeng use mangrove forests as their playground.

Acknowledgments. Thank you to all those who have supported service activities in Lapeng village. WCS Institute, University of Muhammadiyah Aceh, HAkA which has provided both moral and material support in the implementation of this activity. The results of this research can provide information related to the condition of Lapeng village for further researchers by taking different topics or deepening the social and cultural data of Lapeng village.

References

1. C. Saparinto. Utilization of Mangrove Ecosystems. Semarang: Publisher of Dahara Prize (2007)
2. A. W. Setyawan. Biodiversity, **7**, 2 (2006)
3. M. Bismark, E. Subiandono, N. M. Heriyanto Ed. Nat. Conser., **5**, 7 (2008)
4. Arief. Hutan Mangrove Fungsi dan Manfaatnya. Yogyakarta: Knisius (2003)
5. N. A. Campbell, J.B. Reece. Biology, Eighth Edition Volume 3 (Translation: Damaring Tyas Wulandari). Jakarta: Erlangga. (2010)
6. Ujung Kulon Conservation Society. (2010). Manfaat Penanaman Mangrove. http://Ucsindonesia.Org/Index.Php?Option=Com_Content&View=Article&Id=74:ManfaatPenanamanMangrove&Catid=1:Latest&Itemid=117

Intellectual capital and islamic corporate social responsibility on the financial performance of sharia commercial banks in Indonesia

Rahayu Sri[1*], *Fauzi* Mahdi[1], *Julkarnain*[1], *Ningsih* Heny Triastuti Kurnia[1], *Azwansyah* Habibie[2]

[1]Universitas Islam Sumatera Utara, 20217, Indonesia
[2]Universitas Harapan, Medan

Abstract. This study aims to examine Intellectual Capital, Islamic Corporate Social Responsibility on the Financial Performance of Islamic Commercial Banks in Indonesia based on the Islamicity Performance Index. The dependent variable is the Financial Performance of Islamic Commercial Banks in Indonesia based on the Islamicity Performance Index. The independent variables used in this study are Intellectual Capital and Islamic Corporate Social Responsibility. The population in this study is 14 Islamic banking companies. The sampling technique used purposive sampling, as many as 8 Islamic banking 2016-2019. The data were analyzed using Partial Least Square (PLS) tools. General Sharia based on Islamic City Performance Index with indicators of Equitable Distribution Ratio and Profit Sharing Ratio. Islamic Corporate Social Responsibility has no effect on the Islamic City Performance Index with indicators of Equitable Distribution Ratio and Profit Sharing Ratio.

1 Introduction

Since their inception in the mid-1970s, Islamic banks have expanded into the majority of the world's financial markets. Based on the theoretical point of view of Islamic banks, based on the principle of profit and loss sharing, instead of interest-bearing deposits or loans found in conventional banks, two conflicting legal views emerge in modern Islamic banks. According to conventional financial practice, the Islamic banking system should change the old contract by strictly incorporating Islamic business practices and social responsibility into banks [1]. The objective of Islamic banks is to promote and develop the application of Islamic principles in financial, banking and related business transactions.

Management of Islamic banks eliminates cruelty and injustice in the financial system due to usury and ensures economic development by introducing halal businesses [2]. Islamic banks are better in deposit growth business development, profitability, liquidity and solvency, asset quality, capitalization solvency of financial stability, risk, cash, cash deposits ratio and financial stability [3]; [4].

Competing in today's business world is becoming increasingly difficult. This condition necessitates that every company consider its competitors when developing company strategies and carrying out activities, so that company goals can be met, which can ultimately reflect good company performance. The achievement of peak performance is unquestionably desired by the company's management as well as its shareholders. One performance that we can see is the financial performance, financial performance is a measure of the success of management in carrying out activities or operational company. It is also inseparable from the banking sector which has a significant impact on the economic development of a country. In any country, banking plays an important role in regulating the economy in its movement and development process. This sector is considered the main factor for the development and success of various industrial and development projects of the country [5].

At present, the growth of the bank sharia in Indonesia has increased rapidly after the enactment of Law No. 21 of 2008 concerning Islamic Banking. This development can be seen from the number of banks and the number of offices of both Sharia Commercial Banks (BUS), Sharia Business Units (UUS), and Sharia People's Financing Banks (BPRS) [6]. Judging from the survey listed on the website of the Financial Services Authority (OJK), the development of Indonesian banking is good and shows quite significant progress. Because within 28 years there have been around 198 sharia banks registered on the OJK site. If in the future the development of Islamic banking continues to show progress, it can be ascertained that Islamic banking will be even bigger as and as big as conventional banks. Based on statistical data from the Financial Services Authority (OJK) the number of Islamic Commercial Banks in Indonesia continues to increase, in January 2015 the number of Islamic Commercial Banks was 12 banks with total assets of 213,423 billion then in December 2019 increased to 14 banks with total assets of 350,364 billion. Conventional banks, the number of banks in 2014 was 118 banks with total assets of 6,095,908 billion, until 2019 the number of banks was 110 banks with total assets of 8,562,974 billion [7].

The current growth of Islamic banking in Indonesia has not shown positive indicators. The growth rate can be assessed based on the number of banks, the amount of accumulated funds and the value of assets owned, which shows a minimal increase and is significantly different from conventional banking operations. When

*Corresponding author: sri.rahayu@fe.uisu.ac.id

compared to Malaysia, 20% dominates the Islamic banking market share of 20%, while Indonesia is around 5% [8]. Various institutions identified the reasons for the slowdown in the development of the Islamic banking sector, one of which was the OJK which stated that the delay was triggered by several things, such as low government support, the value of capital owned by Islamic banks, the structure of Islamic banking funding, products and services that were not varied, the number of quality of human resources and inadequate use of technology, knowledge and public awareness of sharia banking as well as regulation and supervision that is not optimal for sharia banking [9]. Seeing the backwardness of Islamic banking, further analysis is needed to overcome these problems and make Islamic banking a wheel that has the opportunity to boost economic conditions in Indonesia.

Along with changes in the science-based economy and with the application of knowledge management, the prosperity of a company will depend on an achievement of transformation and capitalization of the knowledge possessed by the company. The use of science and technology at this time is known as Intellectual Capital or intellectual capital. Intellectual Capital is the main source of value creation, business performance, and competitive advantage. Knowledge resources in the form of employees, customers, processes or technology which the industry can use in the process of creating value for the company [10]. Intellectual Capital consists of 3 components, namely, human capital, structural capital, and customer capital. The concept of intellectual capital has received much attention in various fields, both in the fields of management, information technology and accounting. This paper discusses the concept of intellectual capital, how to measure intellectual capital, intellectual capital management and how intellectual capital should be reported in financial statement [11].

Several studies related to improving company performance have been carried out previously, including 12, 13], states that intellectual capital has a positive effect. That Intellectual Capital (IC) has an effect on the company's financial performance. The higher the value of intellectual capital owned by a company, the higher the company's ability to generate profits for the company. So that the higher the Intellectual Capital (IC) value of a company, the ROA is also "higher." Here are some studies on the influence of IC on financial performance as reported by, the variable Intellectual Capital (IC) and the supervisory board sharia (sharia supervisory board) has a positive and significant influence on financial performance". Meanwhile, research conducted by [14] concluded that Intellectual Capital has no effect on the financial performance of companies proxied by the Islam city Financial Performance Index.

The next researcher stated that Islamic Corporate Social Responsibility had an effect on company performance based on the Islam city performance index [15]. Islamic Corporate Social Responsibility is a corporate social responsibility towards all stakeholders based on the values Islamic [16]". Then [17] "found significant positive results related to the relationship between ICSR and company reputation and performance. Research conducted [18] concluded that Islamic Corporate Social Responsibility has no effect on company performance.

The Islam city performance index is a tool for evaluating a company's financial performance based on the materialistic and spiritual values found in the BUS annual report [19]. According to the Islamic finance institution, ICSR may be viewed as an entity that the Muslim community can rely on to channel their funds. [20] so as increase Islamic financial performance.

This study will look at the impact of intellectual capital and Islamic Corporate Social Responsibility on financial performance at Islamic Commercial Banks in Indonesia from 2016 to 2019.

1.1 Literature review and Hypotheses.

1.1.1 Islam city Performance Index (IPI)

Islamic Banking Performance Evaluation Islam city Performance Index is a performance measurement tool that can reveal the materialistic and spiritual values found in Islamic banks. In a performance measurement method for Islamic banks, financial ratios used by [21, 22].

a. Islamic Income vs. Non-Islamic Income Ratio (IsIR). The measurement scale in the IsIR value uses the ratio scale of Islamic Income divided by Total Income.
b. Equitable Distribution Ratio (EDR). EDR = (Qard + Employees + Net Profit)/ (Revenue-(Zakat + Tax))
c. Zakat Performance Ratio (ZPR). ZPR=Zakat divided by Net Asset
d. Profit Sharing Ratio (PSR). PSR= (Mudharabah Financing + Musyarakah Financing)/Total Financing

1.1.2 Intellectual Capital (IC)

Intellectual Capital (IC) is data and information that is applied within the organization to make the organization superior. The IPI scale measurement indicator in this study uses a ratio that has been studied by.

a. Value Added (VA).

$$VA = OUT - IN \quad (1)$$

Information:
VA = Value Added
OUT = Output (total revenue)
IN = Input (total expenses other than employee expenses).

b. Value Added Capital Employed (VACA).

It is the capacity of the organization or association in supervising the assets of capital resources.

VACA= Value Added (VA) divided by Capital Employed (CE) (2)

c. Human Capital (VAHU).

VAHU= Value Added (VA) divided by Human Capital. (3)

d. Structural Capital (STVA).

 https://doi.org/10.1051/e3sconf/202233905003

It is the capacity of the organization to include data frameworks, authoritative cultural systems, and foundations to assist workers' organizations.

STAVA= Structural Capital (SC)/Value Added (VA) (4)

e. Value Added Intellectual Coefficient (VAIC TM)..

VAIC is the sum of its components, namely VACA, VAHU, and STVA.

IB-VAIC = IB-VACA + IB-VAHU + IB-STVA (5)

1.1.3 Islamic Corporate Social Responsibility (ICSR)

Islamic "Corporate Social Responsibility (ICSR) is the social, natural and religious responsibility of a company based on Islamic norms in carrying out it. [23] provides an explanation of the indicator research ICSR quality using a 0-3 scale for each indicator information disclosed, namely:

1) Value 0: "if not disclosed in the annual report

2) Value 1: if it discloses, but only provides a statement about the indicator itself without any other explanation.

3) Value 2: if it discloses and there is a qualitative explanation.

4) Value 3: if it discloses and there is a qualitative and quantitative explanation.

ICSR indicators include:

a. Islamic economic responsibility

b. Legal responsibility I slam

c. Islamic ethical responsibilities

d. Islamic Philanthropic Responsibilities

Furthermore, to get the ICSR value, the total ICSR disclosure quality score of each Islamic bank is compared with the total maximum disclosure score. The formula is:

$$\text{QICSR i} = ICSR = \frac{\text{SQICSR}}{SQMAX} \quad (6)$$

Where: QICSRit : Value of ICSR quality for Islamic commercial banks; iSQICSRi : ICSR quality score of Islamic commercial banks; iSQMAX : Maximum score of ICSR quality

In this study, the researcher determined that there were 2 problem formulations, with three variables, namely 2 independent construct variables (X) and 1 dependent construct variable (Y). To make it easier to understand the flow of data testing, the researchers first designed a structural model. The following is a structural model of the research construct variables as follows:

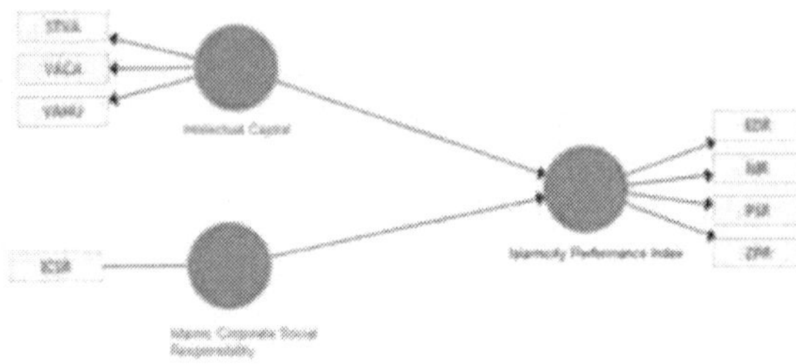

Fig.1.Structural Model with Partial Least

The hypotheses of this research are:

H1 : Intellectual Capital has an effect on financial performance based on the Islam city Performance Index of Islamic commercial banks.

H2 : Islamic Corporate Social Responsibility has an effect on financial performance based on the Islam city Performance Index of Islamic commercial banks.

2 Research Methods

2.1 Population and Sample

In the study, the population is recorded in the Islamic banking Financial Services Authority (FSA) from 2016 to 2019. This investigation's population consists of 14 Islamic financial organizations. Purposive testing is the examination strategy in this study, and there are 8 organizations that meet the size (providing a total exploration factor and fulfilling the year of investigation) that can be used as examples in this study.

2.2 Data analysis

In " this investigation data analysis using Partial Least Square (PLS) and analysis of research data is supported by previous studies that " research [21] using the method of Partial Least Square (PLS).

3 Results

3.1 Outer Model Hypothesis 1

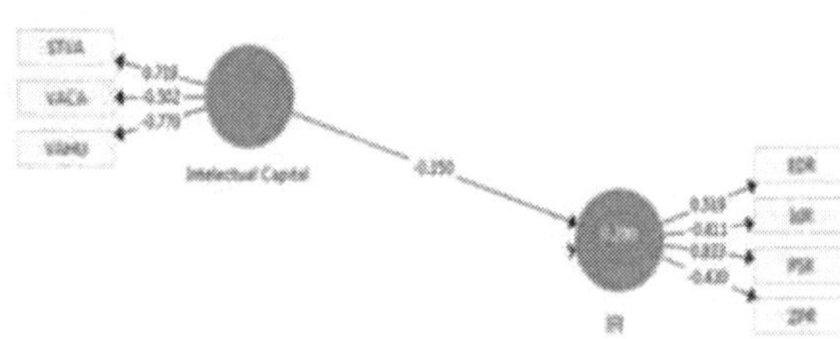

Fig. 2. Structural Model with Partial Least Square H1

From the picture above, it can be seen that the loading factor value of each indicator of the two variables, namely Intellectual Capital and Financial Performance of Syrian Commercial Banks (Islamicity

Performance Index). There are indicators that have a low and insignificant loading factor value, namely the Value Added Capital Employed (VACA), Value Added Human Capital (VAHU), Islamic Income Vs Non Islamic Income (IsIR) and Zakat Performance Ratio (ZPR) indicators, it is necessary to retest by eliminating insignificant indicators and only involving significant indicators. Shows the consequences of retesting after the markers Value Added Capital Employed (VACA), Value Added Human Capital (VAHU), Islamic Income versus Non-Islamic Income (IsIR) and Zakat Performance Ratio (ZPR) are removed.

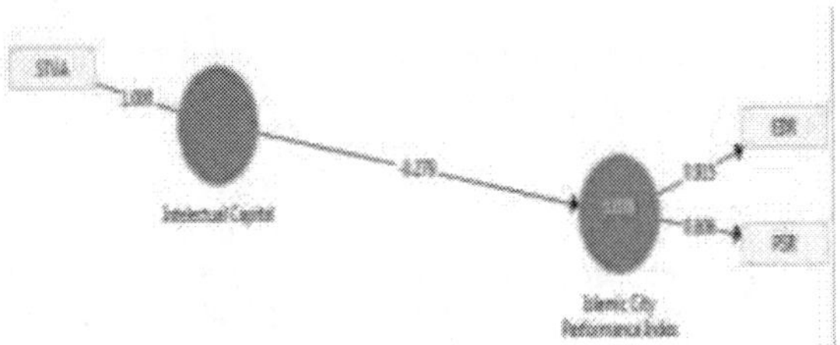

Fig. 3. Structural Model with Partial Least Square H1 (Recalculate)

After eliminating indicators that are not significant and only involve significant indicators, it can be seen that the indicator that has a loading factor above 0.5 is the Structural Capital Value Added (STVA) indicator in the Intellectual Capital variable. Meanwhile, in the Syrian Commercial Bank Financial Performance (Islam city Performance Index) variable, only the Profit Sharing Ratio (PSR) indicator, and the Equitable Distribution Ratio (EDR) have a loading factor value above 0.50.

3.2 Discriminant Validity

Table 1. Value of Discriminant Validity H1
Intelectual Capital Islamicity Performance Index

	Intelectual Capital	Islamicity Performance Index
EDR	-0.271	0.915
PSR	-0.138	0.606
STVA	1.000	-0.279

From the table above, it can be seen that each factor related to pointers is higher than the relationship variables with different markers. This shows that dormant factors, particularly Intellectual Capital (IC) and Islam city Performance Index (IPI), anticipate markers in their squares to be higher than other squared markers. For example, in the square of the Intellectual Capital (IC) variable, the value of the pointer in this square is higher than the different factors.

3.3 Composite Reliability

Table 2. Composite Reliability H1

Variable	Average Extracted Variance (AVE)
Intellectual capital	0.446
Financial Performance	0.566

This is indicated by the value of Intellectual Capital Variable composite reliability under 0.50 while the Financial Performance of Commercial Bank sharia (Islam city Performance Index) above 0.5 as the recommended criteria.

3.4 Outer Test of Hypothesis Model 2

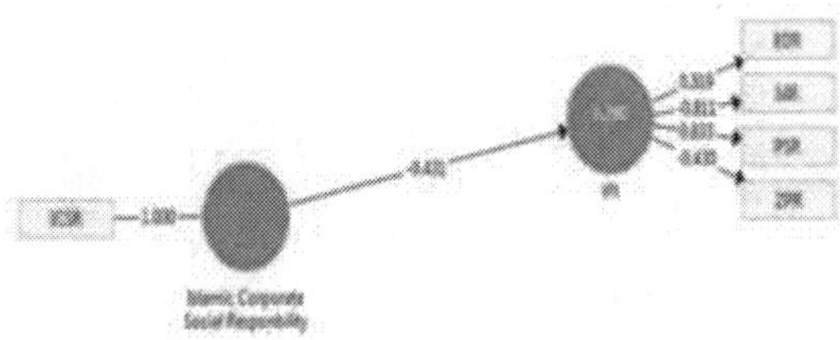

Fig. 4. Structural Model with Partial Least Square H2

From the figure above, it can be seen that the loading factor value of each marker of the two variables is Islamic Corporate Social Responsibility (ICSR) Financial Performance of Universal sharia Bank (Islam city Performance Index). There are markers that have a low and insignificant loading factor value, namely the indicator, Islamic Income Vs. Non Islamic Income (IsIR) and Zakat Performance Ratio (ZPR) so it needs to be retested by eliminating indicators that are not significant and only linking indicators that are not significant. Significantly and only relates significant indicators.

Table 3. Average variance Extracted (AVE).

Variable	Average Variance Extracted (AVE)
Islamic Corporate Social Responsibility	1.000
Financial Performance	0.566

It is very clear that the marker that has a loading factor above 0.5 is the Islamic Corporate Social Responsibility (ICSR) pointer on the Islamic Corporate Social Responsibility (ICSR) variable. Meanwhile, on the Islam city Performance Index variable, only pointers from Profit Sharing Ratio (PSR) and Equitable Distribution Ratio (EDR) have a loading factor esteem above 0.50.

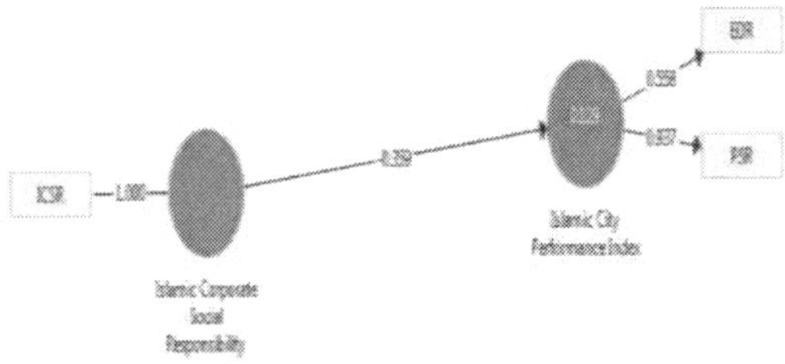

Fig.5. Structural Model with Partial Least Square H2 (Recalculate)

3.5 Discriminant Validity

From the table above, it can be seen that each relationship factor with markers is higher than connection variables with different pointers. This shows that the dormant variables, namely Islamic Corporate Social Responsibility (ICSR) and the Financial Performance of Syrian Commercial Banks (Islam city Performance Index), predict that the marker in its square is superior to other squared markers.

Table 4. Value of Discriminant H2

	Islamicity Performance Index	Islamic Corporate Social Responsibility
EDR	0.558	-0.150
ICSR	-0.359	1.000
PSR	0.937	-0.358

3.6 Composite Reliability

Table 5. Composite Reliability H2

Variable	Average Variance Extracted (AVE)
Islamic Corporate Social Responsibility	1.000
Financial Performance	0.566

Based on the calculation above it can be concluded that

The variable has a quality that is not "deterred. This is indicated by the immovable quality of the composite

Variabel	Sampel Asli (O)	Rata-rata Sampel (M)	Standar Deviasi (STDEV)	T Statistik (|O/STDEV|)	P Values
Intellectual capital -> Commercial Banks Financial Performance	-0.279	-0.335	0.102	2.748	0.007

which is valued above 0.50 as the recommended size.

3.7 Reliability Test with Composite Reliability

Determining the composite reliability test, if the composite reliability value is > 0.8 it can be said that the construct has high reliability or is reliable and > 0.6 is said to be quite reliable.

Tabel 6. Cronbanch alpha

Variable	Composite Reliability	Information
Intellectual capital	1.000	Very Reliable
Islamic Corporate Social Responsibility	1.000	Very Reliable
Financial Performance	0.760	Reliable

From the measurement results above, two latent variables have a composite reliability value > 0.8, and one variable > 0.61 means that all independent latent variables Intellectual capital (IC) and Islamic Corporate Social Responsibility (ICSR) are appropriate and feasible to be tested variables to determine their effect. On the dependent latent variable, namely the Financial Performance of Syrian Commercial Banks.

3.8 Designing the Inner Model

Table 7. R Square

Variabel	R Square	Adjusted R Square
Financial Performance	0.208	0.154

From the R Square table above, the estimated R Square is 0.208. The results of this R Square can be clarified that the influence of the variable Intellectual Capital (X1) and Islamic Corporate Social Responsibility (X2) provides an estimate of 0.208 which can be explained that " by the independent latent variable is 20.8%, while 80.2% is explained by other variables outside of the study. this . From these numbers, it can be categorized that the dependent variable can be explained by an independent variable with a moderate scale.

3.9 Hypothesis Testing

3.9.1 Hypothesis Testing 1

The influence of "variable Intellectual capital against variable Financial Performance of Commercial Bank " Sharia (Islamicity Performance Index).

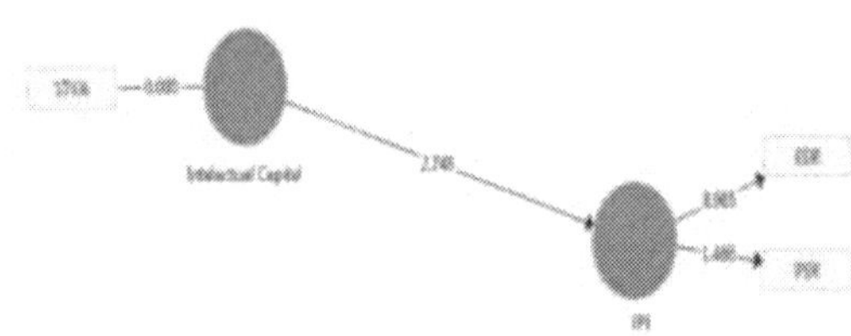

Fig. 6. Bootstrapping H1

The t-table value is 2.042 and the t-statistic is 2.748. The measurement results show that t-statistic > t-table so that Intellectual capital (IC) has a positive and significant influence on Dependent Variable Y Financial Performance of Islamic Commercial Banks (Islam city Performance Index).

3.9.2 Hypothesis Testing 2

The Influence of "Islamic Corporate Social Responsibility Variables on the Financial Performance of Commercial Banks" Sharia (Islam city Performance Index).

T-table 2.042 and t-statistics 1.782. The measurement results showed that the t-statistic < t-table so that Islamic Corporate Social Responsibility is not impact and not significant against Dependent Variable Y Financial Islamic Banks (Islam city Performance Index).

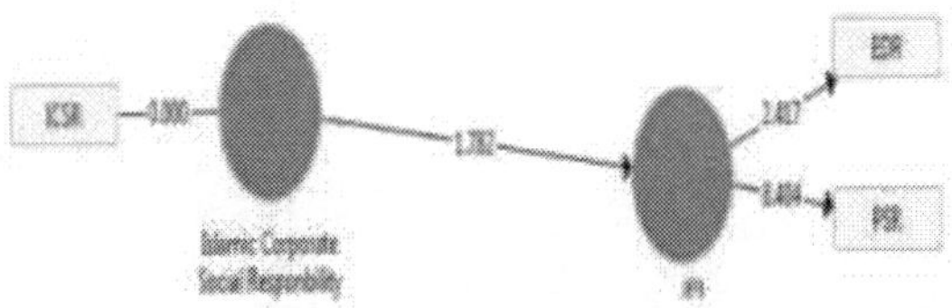

Fig. 7. Bootstrapping H2

Table 8. Bootstrapping Hypothesis 2

Variabel	Original Sample	Sample Average	Standard Deviasi	T Statistics	P Values
Islamic Corporate Social Responsibility -> Financial Performance of Islamic Commercial Banks	-0.359	-0.375	0.202	1.782	0.078

4 Conclusion

Based on the test results, the following conclusions can be drawn:

1. Variable independent (X1) Intellectual capital indicators Structural Capital Value Added (STVA) has a positive effect on draft significantly on Financial Performance Islamic Banks (Islam city Performance Index) indicator Equitable Distribution Ratio (EDR) and Profit Sharing Ratio (PSR).

2. The independent variable (X2) Islamic Corporate Social Responsibility (ICSR) tubs unaffected and significantly to Financial Performance Islamic Banks (Islam city Performance Index) indicator Equitable Distribution Ratio (EDR) and Profit Sharing Ratio (PSR).

Acknowledgments. This research was aided by fellow lecturers and students, so we sincerely thank you for publishing a journal that will benefit Islamic banking and future researchers.

References

1. H. Farag, C. Mallin, K. Ow-yong, J. Econ. Behav. Organ, (2014).
2. A. A. Jalil, R. A. Rahman, Institutional investors and earnings management : Malaysian evidence, **8**(2), (2010)
3. T. Beck, A. Demirgüç-kunt, O. Merrouche, J. Bank Finan., **37,** 2 (2013).
4. S. Ali, M. Ullah, Int. J. Econ. Manag, **05**, 03 (2016).
5. W. Djuanda, A. R. Tanjung, Kamaliah, J. Ekon., **27,** 3 (2019).
6. P. RI, Undang-undanG Republik Indonesia Tentang Perbankan Syariah 1998, (2008)
7. OJK., ShariA Banking Statistics, (2020).
8. D. Supriyadi, Strategi MEningkatkan Kinerja Perbankan Syariah Melalui Value Added Creation Intelectual, (2019).
9. D. Aditiasari, https://finance.detik.com/moneter/d-3076959/7-hambatan-yang-buat-bank-syariah-lambat-berkembang-di-ri, (2015).
10. S. Suroso, T. Widyastuti, M. N. Salim, I. Setyawati, International Journal of Economics and Financial Issues, *7*, 4, (2017).
11. A. D. Cahyati, JRAK, **3,** 1 (2012).
12. D. R. Wijayani, Kompartemen: Jurnal Ilmiah Akuntansi, **17,** 1 (2017).
13. W. Kholilah, *Pengaruh Intellectual Capital Dan Islamic Corporate Governance Terhadap Kinerja Perbankan Syariah Indonesia*, (2019).
14. D. Dinaroe, I. Mulya, E.Mutia, JPED, **5**, 1 (2019).
15. H. N. Husna, *Pengaruh Islamic Corporate SOCIAL Responsibility Terhadap Reputasi* (2020).
16. Gustani. J. Chem. Inf. Model (2017).
17. I. Sidik, Pengaruh zakat dan Islamic Corporate Social responbility (ICSR) terhadap reputasi dan kinerja perusahaan. 1–82 (2016).
18. Khairiyani, Islamic Corporate Social Responsibility terhadap Kinerja Keuangan dan Nilai Perusahaan, **8,** 2, (2020)
19. R. S. Hanafi, Dampak Islamic Corporate Governance , Islamic Social Reporting Pada Kinerja Keuangan Bank Syariah di Indonesia, **8**, 2 (2019).
20. A. A. Afandi, N. W. N. Supaijo, JIMAT, **7**, (2019).
21. N. Hidayanti, Pengaruh intelectul capital dan penerapan islamic corporate governance terhadap kinerja keuangan bank umum syariah di indonesia berdasarkan islamicity performance index. **6** (2017).

22. S. Hameed, A. Wirman, B. Alrazi, M. Nazli, S. Pramono, Alternative Disclosure and Performance Measures for Islamic Banks. *Second Conference on Administrative Sciences: Meeting the Challenges of the Globalization Age, King Fahd University of Petroleum & Minerals, Dhahran, Saudi Arabia*, 19–21 (2004).
23. Anggraini, D. Yuni R. Kartika, J. Akun. Keuangan Islam, **3,** 1 (2019).

Social, economic and institutional study of purse seine fisheries based on the ecosystem approach in the Fishing Port of Kutaraja, Banda Aceh

Salmarika Salmarika[1*], *Am* Azbas Taurusman[2*], *Sugeng* Hari Wisudo[2] and *Muhammad* Irham[3,4]

[1]Department of Fisheries Resources Utilization, Faculty of Marine and Fisheries, Universitas Syiah Kuala, 23111 Banda Aceh, Indonesia
[2]Department of Fisheries Resources Utilization, Faculty of Fisheries and Marine Science, IPB University, 16680 Bogor, Indonesia
[3] Geographical Information Systems Laboratory, Faculty of Marine and Fisheries, Universitas Syiah Kuala, 23111 Banda Aceh, Indonesia
[4]Department of Marine Sciences, Faculty of Marine and Fisheries, Universitas Syiah Kuala, 23111 Banda Aceh, Indonesia.

Abstract. Purse seine is the most dominant fishing gear used by fishermen at the Fishing Port of Kutaraja. It affects increasing the number of fishing fleets and fishing efforts per year. The increase is considered to have a negative impact on fishers because it can trigger competition and conflict between fishers, violate the law, and even affect fishermen's income. This study aims to assess the status and formulate fisheries management actions regarding social, economic, and institutional aspects of fishing based on purse seines in the Fishing Port of Kutaraja. This research was conducted in 2018, and data were collected through a survey method by conducting in-depth interviews with several respondents and analyzed using the Flag Modelling technique. The results showed the composite value of the social domain was 91.66% in a 'very good' category, the economic domain was 33.33% in 'poor' condition, and the institutional domain was 67.66% in a 'good' shape. These results indicate that economic management actions need to be better pursued so that purse seine fisheries remain sustainable in accordance with social, economic, and institutional conditions by; (1) implement a profit-sharing system in accordance with applicable regulations; (2) strengthening good cooperation between stakeholders; (3) open dissemination of the importance of Panglima Laot customary rules; (4) improve supervision and law enforcement; (5) increase the synergy of regulations, institutions and stakeholder capacities.

1 Introduction

Aceh Province has one of the largest fishing ports as a fishery center, namely the Fishing Port of Kutaraja. Fishers conduct fishing activities in the Indian Ocean waters. The dominant fishing gear based on Kutaraja Ocean Fishing Port is purse seine with large pelagic fish and small pelagic fish [1]. Meanwhile, other fishing gears used by local fishermen are handlines and longlines. The number of purse seine fleets from 2013 to 2018 has increased in line with the increase in fishing effort per year. In 2014 the purse seine fishing fleet was 238 units and continued to increase until 2018 reaching 264 units [2].

The high number of purse seine fleets in the Fishing Port of Kutaraja causes decreasing the proportion of catch per vessel. This triggers fishers to increase their fishing intensity. As a result, there is competition among fishers to increase the catch as much as possible. Competition between fishers generally caused conflict and increases the number of lawlessness related to fishing activities. On the other hand, an increased fishing intensity impacts the abundance of catches in the Fishing Port of Kutaraja, and it will also affect the price of fish. This is supported by [3] that fish prices are influenced by the number of catches and other important factors in fish marketing. Based on these conditions, if the number of fish stocks is not proportional to the number of buyers, then the price of fish becomes unstable. The price of fish is an important factor in supporting the increase in fishermen's income.

The application of economic and social aspects of purse seine fishers is related to each other, so efforts are needed to keep these two aspects in balance. The balancing of economic and social's fishers cannot be separated from the role of stakeholders as managers. Furthermore, the success of fisheries management efforts is highly dependent on stakeholder's implication to participate and work actively towards the goals to be achieved [4]. Fisheries management is an effort to maintain the sustainability of fishery resources by considering aspects to related in an integrated, such as the ecosystem approach. The ecosystem approach aims to manage fisheries sustainability by balancing ecosystem well-being, human well-being, and harmonious governance. Based on the problem focus that exists in the Fishing Port of Kutaraja, the researchers are interested to research the management of

*Corresponding author: azbas@apps.ipb.ac.id

purse seine fisheries based on the EAFM in terms of economic, social, and institutional aspects.

2 Method

2.1 Site and Location

This research was conducted in the Fishing Port of Kutaraja in Aceh Province from September to October 2018. The selection of Kutaraja Ocean Fishing Port as the research location was based on the consideration that the Fishing Port of Kutaraja is the largest fishing port and fish landing center in Aceh Province.

2.2 Data Collection Methods

Research data includes both primary and secondary data. The Primary data were obtained from the results of interviews with the distribution of questionnaires. Secondary data was collected from annual report data and documents related to the amount of Province minimum wage in Banda Aceh City, the percentage of loan interest rates, the number of fishermen conflicts, the customary rules documents, rules, or policies that apply in the community related to fisheries management.

Respondents were determined to use the purposive sampling technique. The sample of fishers was selected 25% of the total population [5]. Based on the references, the number of fishermen respondents were 65 respondents. The basic considerations for selecting a sample were the presence and availability as a respondent, reputation, position, and credibility as an expert, experiences, and knowledges related to the problems of the study [6]. Then, each one expert respondent represents a stakeholder consisting of the Aceh Provincial Marine and Fisheries Service, the Marine and Fishery Resources Supervision Unit, the Marine Police, Customary Institutions, and Academics.

2.3 Data Analysis

The data were analyzed descriptively and quantitatively to evaluate the existing condition of each indicator in the social, economic, and institutional domains. Data analysis consists of 2 stages, namely identifying and assessing each indicator.

2.3.1 Indicator identification

The fisherman's household income indicator was measured by the total income of fishermen during fishing activities and compared to the Provincial Minimum Wage from Aceh Province. The fisherman's savings ratio indicator serves to determine the potential of fishermen's households in saving their income as an indicator of the level of fishermen's welfare. The formula for saving ratio according to [7] is as follows:

$$\textit{Saving Ratio} = \frac{(Income-Outcome)}{Income} \; x \; 100\%$$

Stakeholder participation aims to determine the level of involvement in the management of the purse seine fishery. The failure of fisheries management can be seen in how often conflicts occur in waters. The more often fishers conflict occurs, the more it shows an imbalance in fisheries management so that conflict indicators become important. The indicator of local knowledge is one of the benchmarks for the level of effectiveness of fisheries management. The more local knowledge is applied in the community, the greater the chance of successful fisheries management. The local knowledge in question includes traditional ecological knowledge such as applicable customary laws.

Indicators of compliance level with the principles of responsible fisheries were measured by the frequency of violations by fishers, both formal and informal. The completeness of the rules was reviewed from the availability of policies related to fishing business licensing and fishing operations. The decision-making mechanism for fisheries management was seen by the decision-making procedures related to the rules for fisheries management. Furthermore, the fishery management plan was reviewed from its availability related to purse seine fisheries management. The synergy and capacity of stakeholders were measured by the number of policies and institutions that contribute.

2.3.2 Indicator assessment

The assessment of EAFM domains and indicators was conducted by Multi Criteria Analysis (MCA) approach with the development of a composite index using a Likers scale score (ordinal based 1, 2, and 3). Indicators in the ecosystem approach are used to assess fisheries management performance in a particular area [8]. The stages of indicator assessment are (1) identifying the criteria for each indicator; (2) assigning a score for each indicator (ordinal based 1,2,3); (3) determine the weight of each indicator; and (4) develop a composite index for each indicator. According to [7], the equation that can be used to calculate the index value of the indicator is as follows:

$$\text{Index value} = \text{score x 100 x quality}$$

The next step is to determine the composite score for each EAFM domain, which can be calculated using the following equation:

$$\textit{Composite Score}_i = \frac{C_i}{C_{max}} \; x \; 100$$

The composite value is visualized by the flag model in Table 2. The model describes the condition of fish resource management in the categories of 'bad', 'poor', 'moderate', 'good', and 'excellent' [7].

Tabel 1. Domain and aggregate value score limits

Composite Score	Flag Model	Category
1 – 20		Bad
21 – 40		Poor
41 – 60		Moderate
61 – 80		Good
81 – 100		Excellent

3 Results and Discussion

3.1 Economic Domain

The status of purse seine fisheries management from the economic domain is viewed from the fishermen's household income and saving ratio indicators. The composite value of the economic domain, which was 33.33, means that it is included in the poor category (Table 2). This economic condition is feared to be a problem that fisheries management cannot be optimal if the welfare of the actors has not been realized because one of the goals of fisheries management is to ensure human well-being.

Tabel 2. Economic domain composite assessment results

Indikator	Skor	Quality	Index value
Household income of fishermen	1	60	6000
Saving ratio	1	40	4000
Total		100	10000
Domain maximal value			30000
Domain composite value			33.33
Category			Poor

Income is the result obtained from business activities in the form of money and goods that can meet the needs of life [9]. Fishermen's household income is one of the determining factors for the success of the fishing business. A source of fishermen's income was drawn from the sale of their catch. Information of fish price is very important, the price of fish gives an income overview that fishers get from their catches [10]. The higher the fish price, the higher the proportion of fishermen's income, but if the price of fish decreases, the proportion of fishermen's income will also decrease.

The difference in the profit-sharing system for purse seines affects the proportion of the income of each fisher. The division is adjusted to the job duties. The average income of purse seine fishers at the Cempae Soreang Fish Landing Base in Parepare City was around IDR 671,875 per month, while at the Ocean Bungus Fishery Port, Padang City around IDR 1,870,200 per trip [11, 12]. Based on the results of the business analysis, the average household income of fishers was IDR 1,003,822 per month. When compared with the Aceh Province Minimum Wage stipulated in Aceh Governor Regulation No.67/2017 concerning the 2018 Aceh Province Minimum Wage Determination, which was IDR 2,700,000 it can be concluded that the household income of the fishermen was very low. The low income of fishermen's households can impact the ability of fishers to meet their daily household needs.

The savings ratio of purse seine fishermen based in Kutaraja fishing port was -1.73. If the savings ratio value is positive, the fisherman can increase their assets such as household appliances, electronics, but if it is negative, the crew tend not to be able to increase their total assets [13]. Based on the ratio of savings owned by purse seine fishermen, it shows that income as a fisherman has not been able to increase the ability to increase the number of household assets. Furthermore, the value of the fisherman's savings ratio was also very much smaller when compared to the prevailing loan interest rate at Bank Indonesia (10.81%), so this further indicates that the income earned by fishermen has not been able to guarantee the welfare of the fishermen themselves. Meanwhile, human welfare can be realized from the sustainability of its ecological, social and economic systems [14].

3.2 Social Domain

The status of the social domain can be seen from stakeholder participation, fishermen's conflicts, and local knowledge related to fisheries management indicators. The composite value of the social domain obtained is 91.66, which is interpreted that the social management of fisheries is included in the 'very good' category (Table 3). The research results by [15] in Banda Aceh also showed that the social domain was categorized in 'good' condition. This condition is very supportive to realize fisheries management can be run properly.

Stakeholder participation has been 100% conducted for the purse seine fishery management at Kutaraja Ocean Fishing Port. This percentage was seen by the participation of stakeholders in activities in the surrounding area related to fisheries management. Stakeholder participation is an important indicator as a reflection that management efforts can be carried out together by involving many parties. This type of management is often equated with joint fisheries management or co-management. Co-management is defined as joint fisheries management by involving stakeholders to formulate policies for the common interest [16]. Co-management is considered the main approach in implementing fisheries with an ecosystem approach [17]. Efforts to increase stakeholder participation for fisheries management must be carried out effectively to realize synergies between institutions and policies properly.

Conflicts between fishers based in the Kutaraja fishing port were less than two times per year. Based on data obtained from the traditional Panglima Laot institution, fishers conflicts occurred in 2015 due to FAD ownership claims and in 2017 there was a ship collision at the port dock. Conflicts are usually resolved with Panglima Laot by deliberation and are given sanctions according to applicable customary rules in the form of compensation, confiscation of catch, or prohibition from going to sea within a mutually determined period.

On the other hand, the principle of fisheries management is not only limited to scientific information but also obtained by utilizing the best available information. Information can be obtained from a combination of qualitative and quantitative data by utilizing various sources of knowledge, which is traditional knowledge or local wisdom [17]. Positive local wisdom values are recognized and maintained by the community, and can also play a role in environmental protection and management [18]. In Indonesia, there are several practices of customary rules, namely Awig-Awig in West Lombok, Mane'e in North Sulawesi, Sasi Laut in Maluku [19]. Aceh province also has a privilege that other regions do not have, namely the Panglima Laot institution. This institution has customary rules that have been passed down from generation to generation and are used by the local community as a guide for social life with others according to customs, customs, and religion. So far Panglima laot has carried out its role to manage fisheries well, especially in Banda Aceh City [20]. However, the implementation of these customary rules is not effective yet because some fishermen do not fully understand about customary law. The same thing also happened to fishermen in Sumbawa in grouper and snapper fisheries, they have not utilized local knowledge for fish resource management efforts [21]. As a result, it can reduce the essence of local knowledge itself.

Tabel 3. Social domain composite assessment results

Indicators	Skor	Quality	Index value
Stakeholder participation	3	40	12000
Fisherman conflict	3	35	10500
Local knowledge of fisheries management	2	25	5000
Total			27500
Domain maximal value			30000
Domain composite value			91.66
Category			Excellent

3.3 Institutional Domain

The institutional domain consists of 6 indicators, namely the compliance level with the principles of responsible fisheries, the completeness of the rules, decision-making mechanisms, Fisheries Management Plans (FMP), the level of synergy between policies and institutions of fisheries management, and the capacity of stakeholders. The composite value for the institutional domain is 67.77, which is included in the 'good' category (Table 4).

The level of compliance with the principles of responsible fisheries is related to legal norms in the Code of Conduct for Responsible Fisheries (CCRF). The frequency of formal violations committed by fishermen was 62 violations and 4 informal violations. The high level of violations indicates that the principles of good fishing business are difficult to implement, even though the government has implemented the rules well so far. The rules that have been applied at the Kutaraja fishing port include rules for the use of fishing gear, permits, as well as fishing areas and routes that are enforced following standards and sanctions for each violation. However, in practice, it has not been optimally implemented due to limited human resources. On the other hand, the available rules of the game have not been able to be used as a reference in making fisheries management decisions. Because they often conflict with the habits that have been passed down from generation to generation by fishermen. So far, the government has legalized the fisheries management plan, especially for economically important fish such as Tuna, Skipjack, Bullet Tuna stated in KEPMEN KP No. 107/2015. The FMP is related to purse seine fisheries because Tuna, Skipjack, Bullet Tuna are the targets of purse seines.

Tabel 4. Institutional domain composite assessment results

Indicators	Skor	Quality	Index value
The compliance level with the principles of responsible fisheries	1.5	26	3900
The completeness of the rules	2	25	5000
Decision-making mechanisms	2	18	3600
Fisheries management plans	2	15	3000
The level of synergy between policies and institutions of fisheries management	3	11	3300
The capacity of stakeholders	3	5	1500
Total			20300
Domain maximal value			30000
Domain composite value			67.66
Category			Good

The synergy level between policies and institutions can be described through good cooperation. Based on interviews, the synergies built between stakeholders and policies do not overlap and support each other. The policies cover the level of customary institutions, districts, provinces, to the central government. This is because the harmonizing of governance and management systems is an important point that must be possessed for catalyzing fisheries management action successfully in various conditions [22]. In addition, efforts to increase the capacity of stakeholders are also continuously pursued by participating in activities that support the ability to implement fisheries management. This is a great opportunity for good fisheries management.

3.4 Management Recommendations

The assessment of the EAFM indicator in the economic domain shows an unfavorable condition (Table 2). Based on this value, it makes the economic domain a priority in the management of purse seines. The management action that needs to be taken is to implement a profit-sharing system by applicable regulations so that fishermen's income will ensure the welfare of fishers. A non-standard profit-sharing system can cause income inequality for fishers [23]. The ideal fishery product sharing should be implemented as stated in Law No. 16 of 1964 article 3, namely 60:40%. Opinion by [24] that the ideal profit-sharing system for fishermen is 52:48%, which are 52% is for boat owners and 48% for the crew. This comparison hopefully can improve the living standard of the fishermen. Another solution is to socialize business management, provide capital and create alternative livelihoods for fishers [25, 26].

On the other hand, the management of the purse seine fishery from a social and institutional point of view has worked very well (Table 3 and 4). However, in its implementation, it still requires management actions to maintain this situation. Actions that need to be taken are strengthening good cooperation between stakeholders to increase synergies in managing fisheries together with co-management. Co-management can increase participation, communication, human resources, trust, and information between stakeholders and fishers [27]. Teamwork between stakeholders will also reflect efforts to prevent conflicts between fishers because conflicts are prone to occur between parties with different views. [28] also argues that conflict can start because of the inequality of one party with another. Cooperation is also helpful in improving law enforcement and supervision. So, it is essential to continue to increase cooperation.

Another management action is socialization regarding the importance of open Panglima Laot customary rules so that local knowledge can be useful for management, especially the purse seine fishery at the Kutaraja Ocean Fishing Port. Socialization can be done by re-notifying the points of the customary rules along with the sanctions and the benefits of implementing the rules themselves. Socialization can also be done through information media placed around the port so that its implementation can be more effective than holding a meeting because it takes into account the availability of fishermen's time.

4 Conclusion

This study concludes that the assessment of purse seine fishery through economic domain indicators was in the 'poor' category while the social and institutional domains were in excellent and good category. The actions of purse seine fisheries management that need to be focused on are implementing a profit-sharing system following applicable regulations, strengthening good cooperation between stakeholders, and publicizing the importance of open Panglima Laot customary rules.

References

1. Salmarika, A.A Taurusman, S.H Wisudo, Jurnal Penelitian Perikanan Indonesia **24**, 263-272 (2018)
2. Syahbandar Perikanan Lampulo Fishing Port Data jumlah armada penangkapan di PPS Lampulo dari tahun 2013 sampai 2018 (Lampulo: Syahbandar Perikanan PPS Lampulo) (2018)
3. N. Zulbainarni Teori dan Praktik Permodelan Bioekonomi dalam Pengelolaan Perikanan Tangkap (Bogor: IPB Press) p 310 (2012)
4. I Muliawan, A. Fahrudin, A. Fauzi, M. Boer, J. Sosek KP **9**, 233-246 (2014)
5. Food and Agriculture Organization *Handbook for fisheries socioeconomic sample survey* (Rome, FAO (Food and Agriculture Organization). (2017)
6. Edwarsyah and Gazali, Jurnal Perikanan Tropis **2**, 37-43 (2015)
7. National Working Group on Ecosystem Approach to Fisheries Management Modul indikator pengelolaan perikanan dengan menggunakan pendekatan EAFM (Ecosystem Approach to Fisheries Management) (Jakarta, Direktorat Sumber Daya Ikan Kementerian Kelautan dan Perikanan Republik Indonesia) p 163. (2014)
8. M.R.S Damanik, M.R.K. Lubis, A.J.D. Astusi, Jurnal Geografi **8**, 165-176 (2016)
9. Manggabarani and Ishak, Jurnal Agrovital **1,** 27-33 (2016)
10. L. Richmond, D. Kotowicz, J. Hospital, Ocean and Coastal Management **106**, 87-96 (2015)
11. F. Indahyani and Khairuddin, Jurnal Galung Tropika **5**, 63-70 (2016)
12. W. Alan, Hendrik, F. Nugroho, *JOM FAPERIKA* **2**, 1-11 (2015)
13. D. J Tarigan Pengelolaan daerah penangkapan gurita (*Octopus cyanea*) secara berkelanjutan di Kabupaten Banggai Laut, provinsi Sulawesi Tengah [tesis] (Bogor, Institut Pertanian Bogor) (2018)
14. S. E. Rees, M. Ashley, L. Evans, S. Mangi, E. V. Sheehan, T. Mullier, A. Rees, M. J. Attrill, Fisheries Research **235**, 1-10 (2021)
15. E Miswar, D. Natasya, M. Irham, R. M Aprilla, S. Agustina, J. M Affan, *Analysis of social and economic aspects of management of marine and coastal resources based on EAFM (Ecosystem Approach to Fisheries Management) method in Banda Aceh City*, in Proceedings of the International Conference on Fisheries, Aquatic and Environmental Sciences, ICFAES, 25-26 September 2018, Banda Aceh, Indonesia (2018)
16. R Indrawasih, J. Bijak dan Riset Sosek KP **3,** 175-198 (2008)
17. Food and Agriculture Organization *Fisheries management: the ecosystem approach to fisheries, the human dimensions of the ecosystem approach to fisheries* (Rome, FAO (Food and Agriculture Organization) (2009)
18. D. Oktaviani, E. Prianto, R. Puspasari, J. Kebijak. Perikan. Ind. **8**, 1-12 (2016)
19. S. Oktaviyani, Oseana **40**, 49-58 (2015)
20. A. Rahmah. S. Salmarika, E. Miswar, *The role of panglima laot towards fisheries management based on ecosystem approach in Banda Aceh City,* in Proceedings of the International and National Symposium on Aquatic Environment and Fisheries, INSAEF, 24 September 2020, Banda Aceh, Indonesia (2021)
21. S. Hilyana, S. Gigentika, M. T. Rafandi, Hernawati, *Assesment of grouper and snapper fisheries with EAFM approach and sustainable strategy management in Sumbawa Indonesia*, in Proceeding of the 3rd International Conference on Bioscience

and Biotechnology, 12-14 October 2020, West Nusa Tenggara, Indonesia (2020)

22. R. S. Pomeroy, L. R. Garces, M. D. Pido, J. E. Parks, G. Silvestre, Marine Policy **106**, 1-10 (2019)
23. L. D Rejeki, S. Kasimin, T Fauzi, Jurnal Ilmiah Mahasiswa Pertanian Unsyiah **1**, 520-531 (2016)
24. R. H. S. Tawari, D. Simbolon, A. Purbayanto, A. M. Taurusman, Buletin PSP **21**, 237-245 (2013)
25. Ihsan, E. S. Wiyono, S. H. Wisudo, J. Haluan, J. Kebijak. Perikan. Ind. **7**,25-36 (2015).
26. R. Widihastuti and L. Rosyidah, J. Kebijakan Sosek KP **8**, 63-75 (2018)
27. M. Cucuzza, J.S. Stoll, H. M Leslie, Marine Policy **126**, 1-10 (2021)
28. M.F.A Sondita 2010 *Manajemen sumber daya perikanan* (Jakarta, Universitas Terbuka).

Fiqh review on zakah of marine resources

Hafiz Mubarraq Haridhi[1] and *Fikriah* Fikriah[2*]

[1] Department of Islamic Economic, Faculty of Economic and Business, Universitas Syiah Kuala, Banda Aceh 23111, Indonesia.
[2] Department of Economic Development, Faculty of Economic and Business, Universitas Syiah Kuala, Banda Aceh 23111, Indonesia.

Abstract. The obligation of zakah for Muslims is imposed on valuable assets and it reaches the nishab level. The objectives of the shari'ah of zakah are to create an even distribution of income, avoid social inequality and raise the standard of living of the poor. Every asset that is obtained by a human being in the form of findings, results of efforts, work, gifts, and others that is fully owned, by achieving the nishab, must be paid its zakah. In classical fiqh, the classification of zakah maal is still limited, so that scholars continue to develop its classification with various legal istinbath methods. Zakah on marine resources also raises interpretations among scholars so that it needs to be studied from a fiqh perspective. Marine resources, both valuable goods and fisheries in large quantities which are used as sources of income and reach the nishab, need to be reviewed for its zakah regulation. This study is a literature review that reviews classical to modern zakah fiqh. This normative research found that marine resources that reach the nishab must be taken out as zakah for agriculture, commerce, mining, and others. Differences of opinion among scholars in arguing whether or not zakah on the sea resources is obligatory are based on the presence or absence of sources of livelihood from marine products at that time, so that differences in times led to differences of opinion. However, marine resources today have become a big source of livelihood for certain people. So that all forms of income that reach the nishab must be issued zakah. It can be concluded, that the zakah level of the traded marine products can be verified to commercial zakah or gold and silver zakah, which is 2.5%, or zakah on findings, which is between 2.5% and 20%, it depending on the type of resources obtained.

1 Introduction

The word of Maal comes from arabic 'mala' which means leaning or turning from the middle to one side, and al-maal is defined as everything that pleases humans and they maintain, both in the form of material and in the form of benefits [1]. The attitude towards wealth should not be excessive because humans are only a temporary given amanah. Allah tests people with wealth and gives pleasure from it, one of which is also to achieve falah and hayatan tayyibah. Both of these can be realized through the fulfillment of the objectives of the Shari'a. Imam Ghazali includes cases that can protect religion, soul, mind, lineage and wealthy into maqasid [2]. Among the forms of protection of property is to clean it through the payment of zakat, because on every wealth that reaches the nisab, there are few other people's rights. So that zakat is also a maqashid of wealth protection.

The scholars of Fiqh differ on the ruling on items that are explored from marine. Marine products also apply to fish caught. The results of the fish are very large and generate large amounts of money since being worked on by large companies with modern equipment. So it is not natural at all if fish are not subject to the obligation of zakat, which in this case can be analogous to mining goods, agricultural products, and others [3].

Zakat on assets is obligatory zakat on assets owned after reaching haul and nisab. Zakat is obligatory for every individual who has met the requirements to be obliged to issue zakat. The obligation to issue zakat is part of the pillars of Islam as the foundation of Islam whose obligations are always juxtaposed with prayer. Even in the Qur'an, the editor's call to pay zakat is often juxtaposed with the obligation to perform prayer, which shows that prayer and zakat are forbidden to be abandoned under any circumstances. In contrast to the fasting of Ramadan and Hajj in which there is an option not to do it for certain people.

The scholars of the school of fiqh have different editors in interpreting zakat. The Hanafi school defines it as the ownership of a certain part of a certain property for a certain party who has been determined by syari (Allah swt.) to expect His pleasure. The Malikiyah school defines Zakat as issuing a special portion of the property that has reached the nisab for those who are entitled to receive it, if it is perfect and reaches haul apart from mining, plants, and rikaz. The Syafi'iyah School stated that zakat is a name for something that is issued from property and body in a certain way. While the Hambali school considers zakat as an obligatory right in certain assets for certain groups [4].

* Corresponding author: fikriah.mimi@gmail.com

1.1 Urgency and nature of zakat distribution

Classical scholars and contemporary scholars have agreed that zakat must be carried out by every Muslim who has property and has reached the nisab and haul [3]. The wisdom of zakat is fortifying wealth from the reach of criminals, helping the poor/people in need purify themselves from miserly and miserly diseases, and a form of gratitude for the blessings of wealth [4]. Zakat plays a role in the problem of the rich-poor difference, the problem of begging, the problem of envy and broken relationships with others, the problem of disaster, the problem of celibacy, and the problem of refugees [3]. Zakat is the most important means in overcoming the gap between rich and poor, and realizing solidarity or social security in Islam [4]. Qur'an said:

"And give his due (zakat) when reaping the results" [4:141]

In this verse, zakat is called a right because the portion / level that must be issued zakat from every asset that reaches the nisab is the right of another person who has been determined by Allah from someone who can afford it. So, the reluctance to pay zakat, indicates equality against taking the rights of others. Therefore, in the Qur'an it is also commanded to set aside a portion of wealth for zakat in order to purify and clean one's property [9:103].

The source of assets for Zakat includes mandatory Zakat for those who have completed nishab, complete property, and proper transportation (one year), rather than debt yield, which exceeds basic needs. The collected Zakat funds are then distributed to mustahik, the people who are entitled to receive it under Sharia law, including: Fakir, Poor, Amil, Mualaf, Riqab, Gharim, Sabilillah and IbnSabil [5].

There are some key notes that can be used as input to assess the progress of zakah institutions to increase the role and contribution of zakah in the national economy. First of all, the zakaah system is still voluntary, which is clearly visible in Article 12, paragraph 1, UU No. 38 1999. The zakaah system should be placed in a mandatory position so that zakah can play the role of an economic tool. Second, the Zakat system and mechanism are still under the jurisdiction of the Ministry of Religion. Therefore, Zakat should be used as a financial institution designated by the Minister of Finance or the government in the economic authority of the government. This can make Zakat an economic tool, so when philanthropy becomes a tool of economic policy, its effect will be more obvious [6].

2 Methods

2.1 Types of research

This research is in the form of literature research. The research was purposing to analyze the marine resources zakat on the perspective fiqh. The approach used is a qualitative normative approach in the form of detailed elaboration on marine zakat.

2.2 Targets/subjects of research

The research indicates the marine zakat as a target of research. With this, the study will focus on matters relating to fiqh analyze.

2.3 Data analysis technique

Data analysis in this research based on literature review that was conducted in an inductive descriptive manner.

3 Result and discussion

3.1 Zakat of marine resources

Economic growth in this modern era has led to the diversity of assets that can be obtained and has a very large productivity value. These potential assets do not mention the obligation of zakat in the Qur'an and hadith, but it does not mean that these assets are not an obligation to be issued zakat. The development of the times followed by the development of technology and science must also be accompanied by the development of thought in religious science/fiqh [7].

There are differences of opinion among scholars. However, the scholars of the four schools of thought agree that there are five types of assets that are obligatory to pay zakat, namely livestock (camels, cows, buffalo, goats/sheep), gold and silver (including paper money), trade, mining and found treasures, and agriculture. wheat, dates, grapes) [8].

The scholars of Fiqh differ on the ruling on items that are explored from the sea such as pearls, and Marjan and perfumes such as Amber which one piece can weigh up to 1000 mitsqal. According to Abu Hanifah and his friends and Hasan bin Salih and the Shia Zaidiyah school, nothing was imposed. Agreeing with that, Ibn Abbas narrated by Ibn Syaibah and others, said that Amber is not a treasure but is something produced by the sea and therefore is not subject to anything [3].

However, Ibn Abbas also once argued that if there is something in the amber, the zakat is 20%. The mujtahid imams also differed in opinion, as narrated through Hasan bin Imara from Ibn Abbas from Umar Bin Khattab that amber and sea pearls are obligatory for zakat of 20%. Another hadith narrated from Ibn Abbas also states that Ya'la bin Mina once wrote a letter to Umar about the law of images found in the coastal area, which Umar then asked his friends about, suggested that 20% of the zakat be withdrawn, which Umar replied to. the letter and ordered amber and all kinds of decorations taken from the sea to be issued zakat 20%. But Umar is also reported to have another opinion about that which he replied to the letter to take zakat from Ambar and sea decorations of 10%. The narrations originating from Umar above actually have not reached the level of authenticity. Some say the zakat is 20% like treasure 10% like agricultural products and 2.5% like dirhams and dinars [3].

Hasan bin Imara from Ibn Abbas from Umar bin Khattab said that amber and sea pearls must be paid zakat. And as has been narrated by the tabiin such as, and Hasan Basri and from Ibn Shihab Zuhri Abd Razak and Ibn Syaibah also narrated from Umar bin Abdul Aziz that he had collected zakat from amber, and added by the opinion of Abu Yusuf who stated that amber and any form of decoration that is removed from the sea, zakat is obligatory [3].

From the *qiyas*/analogy, goods released from the sea have the same position as ghanimah, which has already been legalized, and if it cannot be accepted, then at least it has the same position as mining goods that are removed from the earth because they are of the same nature. In this case, goods released from the sea are more reasonable if they are not exempt from the obligation of zakat and the amount of zakat is left to ijtihad and the decisions of experts. Marine products also apply to fish caught. The results of the fish are very large and generate large amounts of money since being worked on by large companies with modern equipment. So it is not natural at all if fish are not subject to the obligation of zakat, which in this case can be analogous to mining goods, agricultural products, and others. Abu Ubaid narrated from Yunus Bin Ubaid, "Umar once sent a letter to his officer in Oman so that he did not collect anything from fish that was less than 200 dirhams in value. If it is worth 200 dirhams, which is a large amount of money, then zakat must be collected." It was also narrated from the source of Ahmad. According to the Imamiyah school, the amount of zakat on fish is 20% because they view it as the same as ghanimah [3].

Imam Ahmad is of the opinion that goods produced from the sea such as fish, pearls, etc. are subject to zakat if the total price is the same as the price of the produce of the earth. This opinion was reinforced by Abu Yusuf from the Hanafi school, especially regarding rocks [9]. Basically, every asset that is obtained by a person, which has the potential to reach the nisab on the amount, must be paid zakat. The fulfillment of the obligation of zakat on every asset that reaches the nisab can be seen from the generality of the verses of the Qur'an in talking about infaq/zakat of assets from 'business results'. The results of operations (mimma kasabtum) are an indication or indication of all types and sources of income which are the result of business. So that the classification of zakat commodity goods contained in the hadith is basically an explanation of examples of the main sources of income that produced large incomes at the time of the Prophet, so that from the generality of this verse, it can be applied to other types of assets that have high potential for payment. zakat and the achievement of nisab\ Qur'an says:

"O you who believe, spend (in the way of Allah) some of your good 'results' and some of what We bring out of the earth for you," [2:267]

Among the provisions of zakat is to clean one's property from the rights/parts of others that have been determined by Allah. Thus, it is inappropriate for a Muslim not to issue zakat from the 'business results' obtained which reach the nisab only because it is not classified in the zakat commodity at the time of the Prophet. Additionally, the verse also indicates zakat from the produce of the earth, which in this case can mean all goods and sustenance given by God that come from the earth, including ocean products. Qur'an said:

"Take zakat from some of their wealth, with that zakat you cleanse and purify them," [9:103]

Zakat can cleanse a person of stinginess and excessive love for material possessions and cultivate good qualities in the heart. The verse above is seen as very appropriate if we apply it to zakat on marine products, especially considering marine products now which have become commodities with high income value. Therefore, it further emphasizes that the law of issuing zakat on marine products is obligatory.

In a study, it was stated that zakat on marine products is the domain of ijtihadi, therefore an analysis is needed that can prove that marine products have provided sufficient profits to pay zakat. In this study with a sample in an area in Indonesia, proving that the fishing profession can also be subject to compulsory zakat if income from the sea reaches the nishab of zakat. In this case, of course, a large-scale exploration of marine products by a company gives more indication of its zakat obligation [10].

In addition to being recognized through the indications above, zakat on marine products is also required based on *qiyas* (analogy) which links something that does not have texts yet with something that already has texts because of an illat. In formulating the provisions of zakat on assets that are not mentioned in the hadith of the Prophet, it is done by applying *qiyas*. 'Illat which can be equated to zakat on marine products is the result/goods which is the result of the earth, which is found based on a search. In this case, it can be analogous to zakat on mining goods and rikaz (ancient relics), and it can also be analogous to zakat on plants which are also crops.

Zakat on marine products can also be said to be obligatory because the acquisition is no longer solely for consumption, but has been developed nowadays for very large productivity, both local trade and exports to various countries, whether carried out by individuals or controlled by companies. So that the obligation to pay zakat on marine products can also be applied through zakat money, gold and silver, or trade zakat.

Taking into account the number of 1.6 million fishermen, it is very important to extract zakaah from seafood, but in the past 10 years, as of February 3, 2017, the number of small-scale fishermen has decreased to 800,000 (Susi Pujiastuti, 2017). In 2016, it reached 125 trillion or 6.83 million tons, the highest output value in 5 years. Marine resources can exceed the income of farmers. Former Minister of Ocean Affairs and Fisheries Sharif Sisip Sutarjo revealed that the annual income of fishermen reached 52 million, which was higher than the income of farmers, which reached 20 million per year, and the income of fishermen was 26 million per year. year. In 2009, the fisheries sector grew by approximately 6%, while the agricultural sector grew by 3.5% (Sharif Cicip Stutardjo, 2014). All types of zakah that have never been mentioned are zakah from marine resources. Note that the vast area of the Republic of Indonesia includes a land area of 1,910,931.32 square kilometers, and a sea area = 3,544,743.9 square kilometers, two-thirds of which are marine resources. The collection of zakaah from fishermen is part of the legal investigation conducted by scholars, as a reference for decision-making, and is stipulated by the fishermen's zakah obligations [5].

3.2 Terms and levels of spending zakat on marine products

The condition for paying zakat is full ownership. Ownership is an innate nature that humans have from birth. In the Islamic perspective, obtaining it must be in a lawful way, not in a prohibited or unlawful way [11]. Wealth in the true sense belongs to Allah, and humans are only creatures who are given the mandate to use them in His way as vicegerents and representatives of Allah on earth and will be held accountable in the hereafter [12]. As a result of the fact that humans are only temporary owners of assets, these assets must be managed and used in accordance with the rules and limitations that have been outlined by Allah SWT. So that someone who has full possession of the property, is obliged to spend in the way of Allah, including through zakat.

Nisab according to jumhur is a condition that must be met in the issuance of zakat. The wealth that is subject to the obligation of zakat must reach the sensab agreed upon by the scholars, except for agricultural products, fruits, and precious metals. Abu Hanifa is of the opinion that a lot or a little, the results that grow from the soil must be paid zakat of ten percent. Likewise, the opinion of Ibn Abbas, Umar bin Abdul Aziz and others, that in ten bunches of vegetables that grow from the ground, one bunch of alms is obligatory. But Jumhur Ulama are of the opinion that nisablah is a provision that requires zakat on all wealth, whether it is wealth in the form of land or not [3].

The next requirement in the issuance of zakat is zakat is an excess of needs (in this case the intended need is a permanent need as a defense of life, so that the part of the property that will be issued zakat is not part of the property for that need), free from debt, and reaching haul (except in zakat on plants and minerals).

While the level / size of zakat expenditure, Ibn Taimiyah supports a varied ratio of zakat expenditure, which is due to the difference in the number of efforts from various jobs to produce. Theoretically, there are 4 kinds of percentages of zakat expenditure, namely 2.5% of income which is equated with gold and silver, 10% or 5% of land plants, livestock which is adjusted to the details of the amount, and 20% of assets found from the earth [13].

There are three opinions of scholars who oblige zakat on what is produced from the sea, as on the basis of *qiyas* from other zakat, including [14]:

a. The zakat of 1/5 (20%) is analogized (*qiyas*ed) to ghanimah and minerals produced from the bowels of the earth, in this case some jurists are of the opinion that there is no nisab in it and there are those who think that it must reach the nisab by *qiyas* it on zakat. gold [3].

b. The zakat of 1/10 (10%) is analogous to agricultural zakat. For the expenditure of zakat, the yield from plants and fruits, the nisab is 653 kg [4].

c. The 2.5% zakat is analogous to trade zakat. In this case, trade zakat is equated with zakat on gold and silver. The nisab for gold and silver is 20 dinars (gold) and 200 dirhams (silver). Based on the opinion of Yusuf Qardawi, 20 dinars is equivalent to 85 grams of gold. And 200 dirhams are equivalent to 595 grams of silver. When this number has been reached, the zakat that must be issued is 2.5% of the nisab. There are also differences in the level of zakat on mining goods, some are of the opinion 20% and some are of the opinion 2.5% [3].

The amount of zakat will increase if the level of difficulty decreases. On the other hand, the amount of zakat will decrease if the effort and costs increase, as in the zakat of agricultural products irrigated with rain water is different from that irrigated with irrigation [3]. By considering the determination of zakat on marine products as a source of zakat by the government, the community (especially coastal communities) who are not financially capable can be helped from the existence of marine zakat sources which are known that marine resources with very large amounts have the potential to be issued zakat [7]. Strategic solutions that are prioritized in zakat associations are on the System, External, and Internal aspects. Thus, fighting for zakat through the legislative and executive channels becomes of high urgency [15].

4 Conclusion

There is also a need for seafood zakah based on qiyas (analogs), which connects things that have not yet been text with things that already have text due to illat. When formulating zakah clauses for assets not mentioned in the hadith of the prophet, it was done through the application of qiyas. The obligation to pay for seafood Zakat can also be fulfilled through Zakat currency, gold and silver or trade Zakat. The collection of zakaah from fishermen is part of the legal investigation conducted by scholars, as a reference for decision-making, and is stipulated by the fishermen's zakah obligation.

Acknowledgment: Sincere thanks are due to the Research Center for Marince Sciences and Fisheries, Universitas Syiah Kuala for their effort in organizing the 10th ICMR in conjunction with the 2nd INSAEF. The writing of this manuscript is supported by the Matching Fund – Kedaireka funded by the Ministry of Education, Culture, Research, and Technology Indonesia awarded to the Research Center for Marine Sciences and Fisheries, Universitas Syiah Kuala.

References

1. R. Ritonga, Figh Muamalah (Edaran Kalam, Kuala Lumpur, 1999)
2. M. U. Capra, Islam Dan Tantangan Ekonomi (Gema Insani, Jakarta, 2000)
3. Y. Qardawi, Hukum Zakat (Pusataka Litera Antar Nusa Bogor Baru, Bogor, 1988)
4. W. Az-Zuhaili, Fiqih Islam Wa Adillatuhu (Gema Insani, Jakarta, 2011)
5. A. Muhasim, H. Hirsanuddin, and H. Ul Haq, J. Lib. Amd Int. Aff. **5**, (2019)

https://doi.org/10.1051/e3sconf/202233905005

6. M. E. Nasution, in Potential Socio Islam. Financ. Institutions Islam. Financ. Syst. (2007), p. 5
7. D. S. Riwanto, Int. J. Zakat **2**, 29 (2017)
8. A. Asnaini, Zakat Produktif (Pustaka Pelajar, Yogyakarta, 2008)
9. Z. dan W. Proyek Peningkatan Sarana Keagamaan Islam, Pedoman Zakat 9 Seri (Proyek Peningkatan Sarana Keagamaan Islam, Zakat dan Wakaf, Jakarta, 1998)
10. R. Firdausiyah, Analysis of Fishermen's Zakat from Sea Catch (Case Study of Tambak Village, Tambak District, Bawean) (State Islamic University of Sunan Ampel, Surabaya, 2020)
11. M. Alim, Asas-Asas Negara Hukum Modern Dalam Islam; Kajian Komprehensif Islam Dan Ketatanegaraan (LkiS, Yogyakarta, 2010)
12. R. Rozalinda, Ekonomi Islam (Rajawali Pers, Jakarta, 2014)
13. M. S. Ibrahim, Int. J. Business, Econ. Law **5**, 49 (2014)
14. M. A. Hasan, Zakat Dan Infak: Salah Satu Solusi Mengatasi Problema Sosial Di Indonesia (Kencana, Jakarta, 2008)
15. A. Ascarya and D. Yumanita, Work. Pap. **9**, (2018)

Regulation of tourist dress ethics in the realization of maritime-based halal tourism in Sabang city

Naili Sumaiya[1] and *Muhammad Adli* Abdullah[2*]

[1]Research Center for Marine Sciences and Fisheries, Universitas Syiah Kuala, Banda Aceh 23111, Indonesia.
[2]Department of Custom Law, Faculty of Law, Universitas Syiah Kuala, Banda Aceh 23111, Indonesia.

Abstract. The indications of halal values in the local regulation need to be looked at further on the overall coverage of halal aspects included dress ethics that need to be applied to support halal tourism with local identity and global competitiveness. The Aceh government which has special autonomy under the leadership of the mayor of Sabang, has issued special regulations to support the implementation of 'good' and 'halal' tourism. This regulation is stated in Qanun Number 9 of 2019 concerning the Master Plan for Tourism Development of Sabang City for 2019-2027. The stipulation of these regulations clearly indicates the spirit of halal, including the ethics of dress which is a local identity in the management of marine tourism in the city of Sabang which can be seen in many poins of mentioned qanun. Normative research by looking at the scope of government regulations that is presented qualitatively shows the need for the scope of rules or derivative rules from the side of non-muslim tourists in terms of the obligation to comply with local halal norms and ethics, such as etiquette for beachfront dress and supervision. This is because the embodiment of halal tourism in products, governance, facilities, and supervision must support social, ethical, customary and local communities. The preference for halal tourism that has been declared by the government must not prioritize the quantity of tourism but prioritize the quality of halal, wisdom and social culture of the local community.

1 Introduction

Sharia law in every field, both in the field of muamalat and in the field of worship, aims to bring people to the achievement of a falah life/happiness in the world and the hereafter. The field of muamalah is the dominant aspect of activity carried out in human life. Focused attention in this area is needed in order to create the goals of the Shari'a.

The term of Halal is the norms and rules of Muslim life that are inherent and should be maintained in every practice of life, including tourism. Aceh, which is an area that holds the norms and rules of life of Islamic law, is certainly more stressed to practice halal principles in tourism development.

Certainly, the archipelagic area is the main natural marine tourism location which has a great attraction for tourists to visit the area/city. Aceh has many natural marine tourism sites, even in the form of special islands. Although many islands in Aceh are still isolated from visits by foreign tourists, the city of Sabang has attracted their arrival because it has several islands as attractive destinations. Of course, in improving regional tourism, the Aceh government, which has special autonomy under the leadership of the mayor of Sabang, has issued special regulations to support the implementation of good and 'halal' tourism as a characteristic of Aceh. The Qanun that regulates Sabang tourism is Qanun Number 9 of 2019 concerning the Master Plan for Sabang City Tourism Development for 2019-2027 [1]. The spirit embodied in the qanun has briefly shown the purpose of realizing halal tourism. However, the practice of tourism in the field still shows that it is not halal or contrary to local ethics which is shaded by the labeling of the area as 'veranda of Mecca'. Ethics is explained as differentiating between correct and incorrect, it is imbibed by the individual, group in society and business organization [2]. In this case, especially the practice of how to dress tourists when traveling to the city of Sabang which is very contradictory. This gap in practice shows that there is a gap in regulation or even if it is not regulated.

1.1 Halal tourism as Islamic tourism

Fundamentally, halal tourism and Islamic tourism are the same terminology because one is part of the other. However, on the other hand, they both have different meanings. Islamic tourism is a term commonly used in the world to refer to tourism areas that are centers of Muslim visits. Meanwhile, halal tourism is part of Islamic tourism which can cover a wider range than that, which in this case does not only include Muslim tourists but also non-Muslim tourists in maintaining halal norms. Tourism is generally defined as:

"Tourism is a complex phenomenon with sociological, behavioral, economical, political, cultural and environmental dimensions influencing every aspect

* Corresponding author: bawarith@unsyiah.ac.id

https://doi.org/10.1051/e3sconf/202233905006

of life in modern societies and scientists have difficulties in drawing its borders. Much has been achieved in tourism studies in the 1990s. A number of books and academic journals have been published in tourism, hospitality and related areas. However, it is hard to make the same conclusion for studies related to tourism and its relationship with the religion of Islam and Islamic consumption [3]. Leisure and mass tourism usually related to negative impacts to the local residents especially for social and cultural impact [4]."

Meanwhile, Islamic tourism includes activities (eg, Muslims), locations (eg, Islamic destinations), products (eg, accommodations, food and beverage), dimensions (economic-cultural-religious) and management of the service production process (eg, marketing and ethics) [3]. In Islam, the halal measure is used to indicate whether something is allowed to be consumed or used. In other words, halal terminology is used to indicate the consumption and use of something that is not prohibited by the Qur'an or Sunnah. The relationship between tourism and religion is:

"competition, mutual influence, being complimentary and co-habitual. An Islamic tourism concept also brewed with economic concept, cultural concept and religion conservative concept. However, the relationship of tourism and religion can reduce the risk of negative encounters between tourist and the host [4]."

2 Methods

2.1 Types of research

This research is in the form of literature research. The research was purposing to analyze the local regulation and the opportunity to realize the real of halal tourism. The approach used is a qualitative normative approach in the form of detailed elaboration on halal tourism and its connectivity to tourist dress ethics.

2.2 Targets/subjects of research

This research indicates the regional regulation on tourism as a target of research. With this, the study will focus on matters relating to how deep-covered the local rule to dress ethics.

2.3 Data analysis technique

Data analysis in this research based on literature review that was conducted in an inductive descriptive manner.

3 Result and discussion

Some of the general parameters used in measuring halal tourism are the quality of the attitude of the population in applying Islamic law, the city life system/regulation in placing Islamic law as a guide to life, and obedience to Allah [5]. The guide to the implementation of halal tourism is a complex guide that focuses on the implementation of halal tourism for Muslim tourists [6]. In summary, the Indonesian Minister of Tourism provides parameters for halal tourism as the provision of halal facilities and infrastructure which includes amenities, attraction, and accessibility [5].

3.1 General practice of halal tourism

Various countries have practiced halal tourism to attract tourists around the world, especially Muslims. From a study in Malaysia, it is stated that Islamic tourism can have a great influence on the Malaysian tourism industry from a social, environmental and political economic perspective [4].

"Islamic tourism can give high impacts on Malaysia's tourism industry from the perspectives of economic, social, environmental and political. This is because Islamic tourism provides new peculiar and universal needs which will influence to the development of tourism industry in Malaysia [4]."

In various applications of halal tourism in various regions of the world, Muslim tourists are the main target for tourists. This can be seen as various studies showing the application of halal tourism to increase Muslim tourist visits such as in the Andaman Gulf of Thailand [7] and the Malaysian tourism industry which is also developing halal tourism to attract Muslim visits in the world, even by establishing an Islamic Tourism center (ITC) by Ministry of Tourism. ITC defines Islamic tourism as "Sphere of Interest or activity that is related to travel to explore Islamic history, art, culture and Heritage and/or to experience the Islamic way of life, in conformity with the Islamic faith [4]." From this it can be seen that the main target in the practice of halal tourism should be aimed at Muslim tourists, without excluding non-Muslim tourists with the obligation to comply with halal norms in traveling.

3.2 Halal tourism regulation and supervision of tourist dress ethics in Sabang city

Tourist attitudes in decision making can be classified in five forms, namely need recognition, information seeking and information acquisition, evaluation of alternatives on choices and preferences, purchase or consumption decisions, and post-purchase behavior. This study categorizes tourist needs into three, namely hotels, restaurants and public areas [7].

In general, the focus of the development of halal tourism in various regions that is currently happening is the development of the application of halal from the tourism provider side, so that it is not offended by supervision on the tourist side. Tourist alludes are only focused on Muslim tourists, but not with non-Muslim tourists. So, this will cause problems when there are foreign tourists who travel to Islamic regions/cities whose people live in the norms and customs of Islamic law, especially in the ethics of dressing for non-Muslim tourists. Marine tourism destinations are tourist locations that are most in demand by foreign tourists who are non-Muslims, and become a "special" place for them to dress. Meanwhile, Islamic cities have strict norms and ethics in dressing in a closed manner. So that this opposite is the problem to focus on in terms of halal tourism.

Based on a study that the Aceh Ulama Consultative Assembly has carried out socialization in collaboration with the Aceh Islamic Shari'a Service, the Tourism Office, the wilayatul hisbah in order to implement, enforce and develop halal tourism. This party has also given a will for the government in the development of halal tourism such as in terms of dressing on the beach, where it is illegal to travel to the beach with minimal clothing [8].

The government has issued regulations governing Sabang city tourism as outlined in Sabang City Qanun number 9 of 2019 concerning the Sabang City Tourism Development Master Plan For 2019-2027, which is the latest qanun for the Sabang city tourism development plan. However, this qanun also does not explain in detail about the implementation of Halal tourism from various sides, both tourism providers and tourist visitors. So, this is what provides an opening for non-Muslim tourists to disobey halal norms, especially in dressing. Although this Qanun has alluded to in general the description of management based on halal values and local wisdom values as the identity of Sabang City Tourism. In article 10 point a it is stated that the tourism destination development policies include:

"Development of marine tourism products based on international standards of history and geology with management based on halal values and local wisdom values as the identity of Sabang City Tourism, and developing other community-based tourist attractions to strengthen the identity and competitiveness of tourism destinations [1]."

In the qanun mentioned above, Article 6 Letter b states that the principles of regional tourism development include "Tourism development with ethical principles and principles of halal values (tourism development that implements, respects, and upholds socio-cultural norms and halal values, as well as applies the code of conduct of world tourism ethics [1]."

Article 15 paragraph 1 letters c and d states that the strategy in realizing the development of a tourism marketing system based on information technology includes:

"Setting the tourist market from Singapore, Malaysia, Thailand, and Australia as the main target market for foreign tourists, as well as the tourist market from the Middle East and the foreign tourist market with high interest in marine tourism activities and halal tourism as a potential target market for foreign tourists [1]."

And,

"Determining residents and tourists visiting all areas of Aceh Province, North Sumatra, DKI Jakarta, provinces on the islands of Sumatra and Java, as the main target market for domestic tourists in the city of Sabang, as well as the tourist market from marine tourism destinations and halal tourism in Indonesia as the target market. potential Indonesian tourists [1]."

The spirit of halal tourism development in the qanun above is only limited to the application by tourism managers as is done by halal tourism managers in the world. But in general, this is done because tourism managers focus and target Muslim tourists to visit these tourist sites. In relation to the city of Sabang as marine tourism, this city's tourism planning targets various countries in the world from various religious circles, because the mention is only on the classification of foreign tourists and domestic tourists. This is certainly very well done to be open to every tourist visit, provided that there are special regulations that clearly regulate norms and ethics in halal tourism visits that must be obeyed in detail, including in dressing well, and being socialized and supervised openly to every visitors.

The elaboration of the Ripparkot Qanun above, showed the spirit of realizing halal tourism has been proclaimed by firmly adhering to halal values and social norms. Certainly, this indicates that halal norms and values (a quality of tourism) are more important than promoting the number of visits to non-Muslim tourists who do not maintain social norms and halal values (a quantity of tourism). Basically, the quality and quantity of halal tourism can be realized at the same time by maximizing a complex and globally competitive development plan, by targeting visits by Muslim tourists and non-Muslim tourists who maintain the ethics and norms of halal attire.

3.3 Opportunities for the development of kaffah halal tourism in the city of Sabang

The tourism development of the city of Sabang must be committed to the vision and mission of its development. One of its visions as a sustainable maritime-based global halal tourism destination, is pursued with several missions, including as stated in the Ripparkot Qanun in article 8 letter a, namely

"Building an integrated sustainable world marine tourism destination, and based on halal values and local wisdom as the identity of Sabang City Tourism [1]."

In the article, it is stated about the values of local wisdom, which the City of Sabang is one of the cities in Aceh which also applies Islamic law and the local wisdom of its residents has been attached to an Islamic standard of living. In this case, the realization of halal tourism in the maritime-based city of Sabang does not only hold halal values but also local wisdom values, which means that in this case there are two values that must be maintained in the mission of halal tourism development in Sabang City.

The political strategy for the development of the city of Sabang is carried out with a mixed approach between top-down and bottom-up. This development policy was initiated by the formulation by the upper class, namely the central government through the Ministry of Tourism of the Republic of Indonesia to formulate policies regarding halal tourism which were then passed down to several provinces that have the potential to develop halal tourism in the regions, including Aceh. Furthermore, the halal tourism regulations were also handed down to the district/city government, in this case the City of Sabang. In the development of halal tourism, bottom-up actors are also involved, namely the tourism awareness group which is a community initiative group towards the importance of tourism awareness, and is fostered by the Tourism and Culture Office of Sabang City directly [8].

The development of Sabang city halal tourism by the above actors is carried out in accordance with the Sabang City Tourism Development Master Plan (Ripparkot) as stated in Qanun Number 9 of 2019, by also lowering complex rules on halal tourism. In article 2 points d, f, and h, it is stated that Ripparkot (master plan for urban tourism development) aims to:

"Creating a tourism industry with international competitiveness and high credibility that is responsible for the natural, social and cultural environment through management that is based on halal values [1]".

And,

"Realizing the integration of tourism marketing with trade and investment, with provincial and national tourism marketing, as well as with maritime and halal tourism destinations in the world to build an image as a maritime-based world halal tourism destination [1]."

As well as,

"Realizing the integrated governance of Sabang City Tourism across sectors in the regions, between the central government and city governments, industry, society, academia, and the media in building halal tourism destinations that are internationally competitive and sustainable [1]".

Basically, the city of Sabang has a very big opportunity in the development of halal tourism because this city has been capitalized by the application of Islamic law in Aceh. However, the absence of special regulations regarding halal tourism causes the development to be not optimal.

There are several actors who play a role in the development of halal tourism, namely the Ministry of Tourism of the Republic of Indonesia, the Indonesian Ulema Council/Institute for Food, Medicine and Cosmetics, the Indonesian Ulema Council, the Aceh Culture and Tourism Office, the Aceh Ulema Consultative Assembly, the Halal Tourism Acceleration Team with the branding "The Light of Aceh", the mayor/deputy mayor of Sabang, the Tourism and Culture office of Sabang City, the Islamic law service/Satpol PP WH of Sabang City, the Sabang City People's Representative Council, the Sabang area concession agency, the Sabang City Ulemas Consultative Assembly, the Aceh traditional assembly, the association Indonesian HPI guides for the City of Sabang, the green community forum FKH Teurawah which is under the guidance of the Ministry of Public Works and Public Housing, pokdarwis/tourism awareness groups in the City of Sabang consisting of Gampung youth and funded by the APBG, travel agents, and lodging managers. In this case, the actors from the national, local to the individual and private levels of Sabang City as the driving force of development must synergize with each other and strive as much as possible in order to realize the direction and development goals to be achieved, namely Sabang City halal tourism [1]. According to the Fatwa of the Ulama Consultative Assembly of Aceh (MPU Aceh) number 7 of 2014 concerning tourism in Islam, that:

"Tourism is everything related to tourism, including the exploitation of tourist objects and attractions as well as businesses related to these fields. Tourism which contains elements of disobedience is haraam, and tourism which contains beneficial values is legally permissible. The Fatwa of the Aceh MPU has also called on the Aceh government to prioritize Islamic sharia values in tourism development in Aceh, socialize Sharia tourism to tourism managers and the community, prepare human resources for professional tour guides who understand sharia local wisdom, prioritize the promotion of Sharia tourism outside the region and country. Muslim countries prepare adequate worship facilities at tourist sites, place WH wilayatul hisbah personnel and other relevant officers at tourist sites, provide sanctions for tourism managers and tourists who violate Islamic sharia values. And the people of Aceh are also encouraged to participate in supervising tourism activities [9]."

The MPU fatwa above can be used as a guide for the government in implementing halal tourism in a kaffah manner and providing special derivatives for the Sabang city halal tourism regulations. This is because one of the challenges in the development of halal tourism in Sabang City is the absence of regulations/legal coverage that specifically regulate the concept of derivative halal tourism which includes tourism management and regulations for halal tourism visitors. The absence of detailed rules on halal tourism compliance for tourists will have an impact on tourists' non-compliance, including clothing. This can be seen clearly from the non-compliance of non-Muslim tourists in minimally dressed, which is still difficult to organize. So it is necessary to have supervision under a clear law to organize tourists to dress in halal tourist areas. By maximizing the application of halal identity in tourism through strengthening regulations and supervision, it will require every tourist including non-Muslims to comply with halal values and local wisdom. This can be seen as one of the great mosque tours in Malaysia, a form of respect by every religion by giving a cloth covering. The island of Sabang is likened to a valuable house for the Muslim community with a qanun (Islam-based law) in it, so that tourists who visit still have to meet the elements of halal and local ethics.

In addition, maximizing the development of halal tourism in Sabang City must also be carried out through the construction of facilities that are globally competitive to attract more tourists. The provision of facilities and development that is internationally competitive is a very important step, by making Muslim tourists the main target and non-Muslim tourists who can accept the rules and norms of halal as the second target market. The level of visits to tourist areas by domestic and foreign tourists is very influential from modernity, facilities, convenience and beauty available.

Basically, this Qanun regarding Ripparkot Sabang is sufficient to regulate tourism in terms of development planning, management, and control. However, the application needs to be done more and more elite and modern tourism development to attract tourists in the world. On the other hand, in terms of managing the spirit of halal tourism, it is necessary to bring down special rules that regulate halal tourism in the city of Sabang to regulate all halal coffers that are kaffah. This is needed as a guide for tourists to visit, including socialization and supervision carried out on non-Muslim tourists who plan to have a maritime tour in Sabang City.

4 Conclusion

Marine tourism destinations are tourist locations that are most in demand by foreign tourists who are non-Muslims, and become a "special" place for them to dress. Meanwhile, Islamic cities have strict norms and ethics in dressing in a closed manner. So that this opposite is the problem to focus on in terms of halal tourism. The challenges in the development of halal tourism in Sabang City is the absence of regulations/legal coverage that specifically regulate the concept of derivative halal tourism which includes tourism management and regulations for halal tourism visitors.

The absence of detailed rules on halal tourism compliance for tourists will have an impact on tourists' non-compliance, including clothing. This can be seen clearly from the non-compliance of non-Muslim tourists in minimally dressed, which is still difficult to organize. So, it is necessary to have supervision under a clear law to organize tourists to dress in halal tourist areas. By maximizing the application of halal identity in tourism through strengthening regulations and supervision, it will require every tourist including non-Muslims to comply with halal values and local wisdom.

Dissemination of information and travel rules to tourists who arrive must also be massive by involving government officials, village officials, and also the community. The halal identity that remains inherent in every aspect of tourism control can even give a different prestige in the eyes of foreign tourists. Therefore, the government needs to consider the derivatives of these regulations in order to create universality in regulations and supervision of travel to foreign tourists.

Acknowledgment: Sincere thanks are due to the Research Center for Marince Sciences and Fisheries, Universitas Syiah Kuala for their effort in organizing the 10th ICMR in conjunction with the 2nd INSAEF. This research is funded by the Ministry of Education, Culture, Research, and Technology through Universitas Syiah Kuala, under the Research Institutions and Community Service with contract number: 174/UN11/SPK/PNBP/2021. The writing of this manuscript is supported by the Matching Fund – Kedaireka funded by the Ministry of Education, Culture, Research, and Technology Indonesia awarded to the Research Center for Marine Sciences and Fisheries, Universitas Syiah Kuala.

References

1. P. K. Sabang, Rencana Induk Pembangunan Kepariwisataan Kota Sabang Tahun 2019-2027 (2019), p. Qanun of Sabang City No. 9
2. S. Khan, J. Hosp. Tour. Syst. 7, (2014)
3. T. Duman, in World Islam. Tour. Forum (2011), p. 720
4. L. M. Kamarudin and H. Nizam, in Proc. Int. Conf. Tour. Dev. (2013), p. 403
5. R. Hakim and Muslikhati, J. Ilm. Ekon. Islam 5, 170 (2019)
6. K. Pariwisata, Panduan Penyelenggaraan Pariwisata Halal (2019)
7. P. Sripraset, O. Chainin, and A. Rahman, Underst. Behav. Needs Halal Tour. Andaman Gulf Thail. a Case Asian Muslim 2, 218 (2014)
8. M. Ulfa, H. Kusmanto, and Warijo, J. Adm. Publik 9, 78 (2019)
9. M. P. U. (MPU) Aceh, Fatwa of Islamic View on Tourism (2014), p. Number 7

The impact of the spread of covid-19 on marine tourism-based small businesses and suggested recovery plans

Syafruddin Chan[1*], and *Kurnia* Asni[2]

[1]Management Department, Syiah Kuala University, Banda Aceh, 23111, Indonesia
[2]Master in Management Department, Abulyatama University, Aceh Besar, 23272, Indonesia

Abstract. The purpose of this study is to find out the real impact of the COVID-19 epidemic on tourism-based small businesses in the island, as well as what recovery strategies that should be undertaken from the standpoint of company owners, including potential for leveraging digital business technology. This research was conducted using a quantitative descriptive method. Data collected using online questionnaires. 50 questionnaires distributed using non-probability sampling. The distribution of Covid-19 had a detrimental effect on the small companies who participated in this study. The greatest consequence is a decrease in revenues due to a decrease in demand for its products and services. As a result, firms downsized by reducing their business scale and laying off staff. Companies are beginning to employ technology to enter digital industries, such as joining online apps and selling more aggressively in e-markets. Virtual reality tour business opportunities can be tried to generate alternative income.

1 Introduction

Throughout 2020, the Covid-19 pandemic has caused an economic crisis throughout the world. Not only experienced by developing countries and countries categorized as poor, developed countries also feel the negative impact of the outbreak of Covid-19. Most developed countries as well as developing countries are even caught in a deep economic recession. Only a handful of countries in the world are still able to survive and grow positively to face the impact of the Covid-19 pandemic in 2020, namely China (2.3 percent), Vietnam (2.9 percent), and Taiwan (2.98 percent) [1].

Outside of these three countries, the Covid-19 pandemic has ravaged the economies of countries in the world, including developed countries, such as the United States (US), Germany, Britain, and Japan. The United States (US) economy was recorded to have shrunk to minus 3.5 percent in 2020. Germany grew by negative 5 percent. Britain contracted 9.9 percent. Japan shrank 4.8 percent. Meanwhile, France and Italy closed 2020 with contractions of 8.2 percent and 8.9 percent, respectively. In Southeast Asia, only Vietnam grew positively by 2.91 percent in 2020. Meanwhile, other countries in the ASEAN region grew negatively. Singapore contracted 5.4 percent, Malaysia minus 5.6 percent, and Thailand minus 6.1 percent. Meanwhile, the Philippines was the worst because its economy contracted by 9.5 percent in 2020 [1].

A number of international institutions predict that global economic growth in 2020 is predicted to fall deeper than in previous years. In January 2021, the IMF estimates global economic growth at minus 3.5 percent. Meanwhile, the World Bank in January 2021 and the OECD in December 2020 each predicted a deeper decline in global economic growth to minus 5.2 percent and minus 4.2 percent. The sluggish global economy cannot be separated from the effects of the Covid-19 pandemic which has spread to world economic and financial problems. His sudden arrival put great pressure on the supply and demand sides. The world's production chain is not only disrupted, it is even cut off, because many countries choose regional quarantine (lockdown) to contain the spread of Covid-19 [2].

The impact of the COVID-19 pandemic has also spread to Indonesia. The World Bank projects that Indonesia's gross domestic product (GDP) growth throughout 2021 will reach 4.4 percent after the previous year's contraction of 2.1 percent. This economic recovery depends on the effectiveness of the Covid-19 control program and the government's ability to take advantage of the momentum of the global market revival. This has a very serious impact because the trade sector tax revenue has a major contribution in boosting state revenues, precisely which is in the second largest [3]. This pandemic has also greatly affected market projections. Investors may tend not to invest due to changing market assumptions and unclear supply chains [4]. The impact of COVID-19 on the tourism sector is also not without threats. Data compiled from the Central Statistics Agency explains that in 2019 foreign tourists from China who came to Indonesia touched 2.07 million travellers or 12.8% of the total number of foreign tourists throughout 2019. The COVID-19 pandemic resulted in a decline in tourists coming to Indonesia. Tourism supporting sectors, namely restaurants, hotels and retail entrepreneurs are also affected by the COVID-

* Corresponding authors: syafruddin.chan@unsyiah.ac.id

19 pandemic. Hotel profits have decreased by up to 40% so that it has an impact on hotel operations and threatens the continuity of its business. The decline in foreign visitors also affects the income of restaurants or restaurants whose customers are more dominantly visitors from abroad [3].

The COVID-19 pandemic has also affected the micro, small and medium business sector in the tourism sector. This is because foreign visitors who come to a destination will usually buy souvenirs to take home [5]. Other micro businesses in the tourism industry that have been affected by the COVID-19 pandemic include diving and snorkeling businesses, culinary businesses, lodging accommodations, laundry services, communication services, retail and many others. If foreign visitors visit decrease, it is certain that the income for micro, small and medium enterprises will also decrease [6]. Bank Indonesia has released data in 2016 regarding the micro, small and medium enterprises sector which states that the enterprises are the most vulnerable business during pandemic

For business entities, the loss of income is due to no sales, but expenses still occur, although not completely. Real losses will vary depending on what type of expenditure is being made. Among the expenses that are relatively fixed are the rent of the place (or depreciation of the building if it is self-owned) along with the routine costs that accompany it, salaries of staff who cannot be laid off for various reasons, security, payments to suppliers that can no longer be postponed and so on. Another impact is the emergence of fines/penalties due to inaccuracy in delivery times [7]. Damage to goods if stuck in the warehouse or on the road, or payment of additional premiums when applying insurance. Another impact is the emergence of severance costs if you are laid off. Fines/interests arise when using bailout funds or debts, or if they are unable to pay in accordance with the provisions of the amount and time. Losses if the company is forced to sell assets at low prices. A priceless loss if the crisis causes the company's capital to be drained so that the scale of the business shrinks or even goes bankrupt. However, the company may have income from previous sales invoices, which can still reduce the loss or budget deficit in the current month. With a successful billing record, the client is able to pay according to the provisions [7]. A priceless loss if the crisis causes the company's capital to be drained so that the scale of the business shrinks or even goes bankrupt. However, the company may have income from previous sales invoices, which can still reduce the loss or budget deficit in the current month. With a successful billing record, the client is able to pay according to the provisions [7-8]. A priceless loss if the crisis causes the company's capital to be drained so that the scale of the business shrinks or even goes bankrupt. However, the company may have income from previous sales invoices, which can still reduce the loss or budget deficit in the current month. With a successful billing record, the client is able to pay according to the provisions [9].

The purpose of this study is to find out the real impact of the COVID-19 epidemic on tourism-based small businesses in the weh island, as well as what recovery strategies that should be undertaken from the standpoint of company owners, including potential for leveraging digital business technology.

2 The Government's Response to Help MSMEs against the Pandemic

When the Covid-19 outbreak spread throughout the world, a number of countries struggled to contain the spread and contain the outbreak. Governments around the world are seeking to implement extraordinary policy measures to save lives and protect the livelihoods of people hit hard by the Covid-19 pandemic. In terms of handling the pandemic, the governments of countries generally carry out a combination of 3M (maintaining distance, wearing masks, washing hands) and 3T (testing, tracing, treatment) as well as social distancing policies to regional closures. In terms of handling economic impacts, various policies are directed at three priorities, namely providing funds for health care, distributing social safety net assistance, and providing economic stimulus/support to the business world [10-11].

Governments of countries in the world generally provide stimulus either through monetary policy or fiscal policy. In the monetary sector, the policy of lowering interest rates is carried out by almost all over the world. This was done with the aim of encouraging economic activity, both on the consumption and investment side, by reducing bank lending rates. The policy is expected to continue in 2021. The central bank will tend to lower or maintain interest rates due to the ongoing economic recovery period [12].

Just like other countries, Indonesia has also taken several steps to reduce the effects of the COVID-19 pandemic, including reducing the BI 7-Day Reverse Repo Rate by 25 bps to 4.75%, the Deposit Facility interest rate by 25 bps to 4.00% and the interest rate on the Deposit Facility by 25 bps to 4.00%. Lending Facility interest rate of 25 bps to 5.50%. This measure was implemented to stimulate domestic economic growth amidst the delayed prospects of global economic recovery due to the COVID-19 pandemic. In addition, to keep inflation and external stability under control and to strengthen the momentum of economic growth, Bank Indonesia must be able to observe global and domestic economic developments [2] and [7].

Overall, the response made by the Indonesian government during this pandemic, especially against MSMEs, including those engaged in the tourism sector, can be broken down into five schemes, namely (1) Social support; (2) SMES tax incentives; (3) Credit restructuring and relaxation (4) Working capital support stimulus; and (5) Collaboration with State-owned enterprise as the off-taker for SMES products. On the other hand, there are some SMES business strategies in adjusting amidst the Covid-19 pandemic in Indonesia. Those strategies are (1) Products innovation; (2) Customer and stakeholder relationship management (3) Digital marketing and distribution; and (4) Strategic Agile [12].

For small and medium enterprises which are categorized as poor and vulnerable to being affected by

Covid-19. Efforts made to ensure that business actors in this scheme become recipients of social assistance from the Government. According to Minister of Social Affairs Number 1/2018, Government must ensure that they are included as part of the recipients of social support namely PKH (Family Hope Program), food packages, cash assistance, direct cash support, as well as exemption from reducing electricity rates and pre-employment cards. The second scheme is a scheme regarding tax incentives that applies to SMES actors with a turnover revenue of below Rp. 4.8 billion per year (12). The Government has lowered the final income tax rate (the final Income Tax (tax income) or PPh rate for SMEs from 0. 5 to 0 per cent over six months starting from April to September 2020 this policy written in Minister of Finance Regulation Number 44/PMK.03/2020. The third scheme is relaxation and SMES credit restructuring. This scheme includes the postponement of installations and interest subsidies for recipients of KUR (People's Business Credit), UMi (Ultramicro Credit), PNM Mekaar (Fostering Prosperous Families), LPDB (revolving fund management institutions), and recipients of capital support from several ministries. The Government will also impose expansion financing support for SMEs, in the form of working capital assistance stimulus. The Government prepares emergency working capital assistance designed explicitly for SMES entrepreneurs who are severely impacted by Covid-19. The data shows that currently, there is 41 million SMES business have already connected with financial institutions and banks. Meanwhile, there are still 23 million SMEs who have never received financing from financial institutions or the banking system. Therefore, for SMEs that are bankable, the financial support distribution will go through the expansion of the KUR program and at the same time encourage financial inclusion. On the other hand, for those who are not bankable, the distribution can be done through UMi, Mekaar, or other program schemes [12]. For SMEs that are bankable, the financial support distribution will go through the expansion of the People's Business Credit (KUR) program and at the same time encourage financial inclusion. On the other hand, for those who are not bankable, the distribution can be done through Ultra Micro Financing (UMi), Fostering the Economy of Underprivileged Families (Mekaar), or other program schemes [12].

3 Opportunities in the Pandemic Era for MSMEs

The spread of the Covid-19 outbreak in Indonesia has caused the Government to urge the public to limit physical contact. Daily activities such as work, study, and worship are recommended to be done at home. This limitation of physical contact, aka physical distancing, has a significant impact on many parties, one of which is business people. Not a few business actors, ranging from the SME level to large scale, have revealed a decline in income over the past few weeks. However, my friend as a business actor, of course, should not give up. In the midst of the widespread spread of Covid-19, it turns out that there are still promising business alternatives [13].

The Covid-19 pandemic is indeed quite influential on business continuity and is considered to bring changes in business going forward, especially in consumer behavior and consumption. However, working and doing business from home (Work Form Home - WFH) apart from being a limitation, it also brings opportunities. This period of social distancing can be an opportunity to optimize business strategies. This encourages the importance of identifying consumption and consumer behavior in producing goods needed by the community. In addition, it is also a momentum for the development of the digital and creative economy.

As we all know, the tourism sector is the sector most severely affected by the COVID-19 pandemic. The existence of new business opportunities in the pandemic era can be used as a momentum to rise and take advantage of existing opportunities to increase income. For SMEs who are engaged in the culinary field, which previously served mostly tourists, now they can join delivery businesses such as Gojek and Go Food to serve the domestic community. The types of food sold can also be expanded by serving other frozen food, homemade side dishes packaged in bottles, pastries and drinks such as milk coffee, and others.

4 Methodology

In contrast to the previous section, which was conducted through a literature review, this section uses primary data. The population of this study are small businesses engaged in the tourism industry, especially marine tourism on Pulau Weh, Aceh, Indonesia. Data were collected using an online questionnaire. The respondents were MSMEs with 50 (fifty) questionnaires distributed using non-probability sampling. Non-Probability Sampling is a technique of taking data or samples so that all data selected as samples are not the same size. Based on the researcher's policy in determining the sample elements to be used

5 Entrepreneur Response for Business Recovery

The spread of the Covid-19 outbreak in Indonesia has caused the government to urge the public to limit physical contact. Daily activities such as work, study, and worship are recommended to be done at home. This limitation of physical contact, aka physical distancing, has a significant impact on many parties, one of which is business people. Not a few business actors, ranging from the MSME level to the enterprise level, have revealed a decline in sales over the past few weeks. In line with changes in people's behavior, the Online Shop is excellent. Before the emergence of the Covid-19 case in Indonesia, online business had indeed become one of the solutions because it was easy for customers to shop without being limited by distance and time. Despite having to limit store opening hours due to physical distancing [4].

 https://doi.org/10.1051/e3sconf/202233905007

One of the most significant determinants of the success of SMEs during this pandemic is their flexibility in paying attention to customer needs and expectations. When there are drastic changes due to this pandemic, many business models or businesses cannot survive, but new opportunities will be created. Under these conditions, it is understandable that SMEs are very difficult to focus on identifying and exploring new opportunities by losing profits, customers and even key employees. However, with the survival of SMEs at stake, owners must exercise resilience and adopt a mind-set to recognize and identify opportunities to bounce back and move forward. Thinking positively in the midst of difficult times is not always easy. However, adaptation and flexibility are the main keys that must be held by anyone who is in the hospitality & tourism industry. After all, change is the only thing that keeps happening in the world. So, changing the strategy to be able to survive during the pandemic is a realistic step that tourism activists can take.

Just like other sectors, SME business actors in the tourism sector must also be creative to innovate their offerings, so that they can still exist in today's difficult times. However, the role of the government or other parties is still expected to help these SMEs get out of the crisis due to the impact of the COVID-19 pandemic.

When asked if you have ever received assistance, to help your company avoid business difficulties? So, 46% answered yes and 54% answered no.

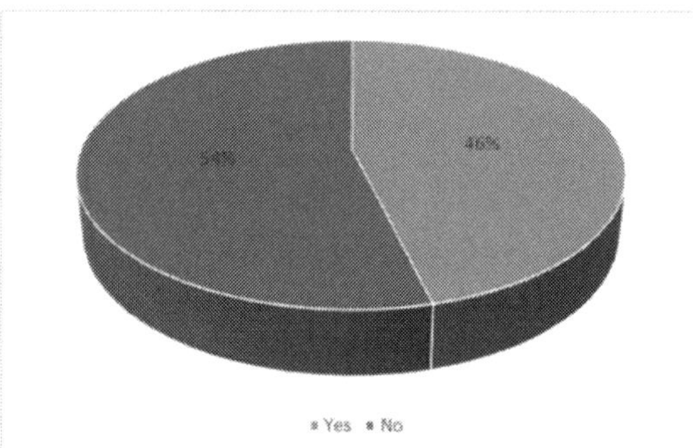

Fig. 1. Is there any Receiving Help?

To those who answered yes, they were asked what kind of assistance they had received. So, 17% said they had the opportunity to have a CHSE (Cleanliness, Health, and Safety & Environment Sustainability) certificate which is a certificate of recognition for the implementation of Cleanliness, Health, and Safety & Environment Sustainability in their business units. Cleanliness, Health, Safety & Environment Sustainability (CHSE) certification is given to owners/managers of tourism businesses and tourism destinations from all over Indonesia for free by the Ministry of Tourism and Creative Industries (Kemenparekraf).

All stages of the certification process are financed by the Ministry of Tourism and Creative Economy, so that the costs are not charged to the destination managers and tourism businesses. The use of CHSE is a guarantee to tourists and the public that the products and services provided have complied with the hygiene, health, safety, and environmental sustainability protocols. Because the key to the successful recovery of the tourism sector and the creative economy is the application of standard health protocols in the tourism sector. The certificate, known as Indo Care, which stands for Indonesia Care, provides certification for hospitality businesses, restaurants, tourist attractions, white water rafting, home stays, diving, tourism transportation businesses, and sports recreation such as golf clubs, Spa business and other Creative Economy Business Because the key to the successful recovery of the tourism sector and the creative economy is the application of standard health protocols in the tourism sector. The certificate, known as Indo Care, which stands for Indonesia Care, provides certification for hospitality businesses, restaurants, tourist attractions, white water rafting, home stays, diving, tourism transportation businesses, and sports recreation such as golf clubs, Spa business and other Creative Economy Business Because the key to the successful recovery of the tourism sector and the creative economy is the application of standard health protocols in the tourism sector. The certificate, known as Indo Care, which stands for Indonesia Care, provides certification for hospitality businesses, restaurants, tourist attractions, white water rafting, home stays, diving, tourism transportation businesses, and sports recreation such as golf clubs, Spa business and other Creative Economy Business.

Other assistance they received was staple food (26%), to ease the burden of the cost of meeting basic needs for tourism practitioners during the pandemic. Then through the Provincial Tourism Office.

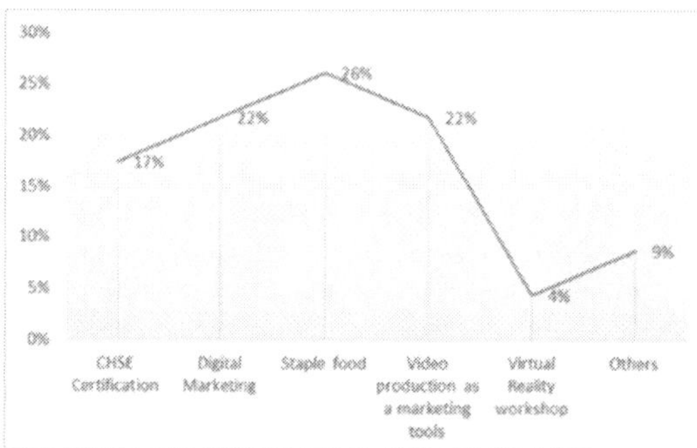

Fig. 2. What assistance have you received?

Aceh, tourism actors receive digital marketing training (22%), especially for those who are engaged in the souvenir business, then training on making video marketing (22%) as a tool in marketing their business units, as well as virtual workshop training (4%) aimed at enabling tourists to enjoy tourist destinations in a fun way without having to come to the physical location of the destination.

It has been more than a year that the pandemic has hampered the running of business in all sectors. So far there is no sign of abating. The number of people exposed to and infected with the Corona virus continues to grow. Most tourism actors view pessimism that this suffering will end in the near future. They really hope that there will be further assistance provided by the government so that the difficulties they face can be

https://doi.org/10.1051/e3sconf/202233905007

helped a little. When asked what form of assistance they expect from the government, the answer can be seen in the following figure:

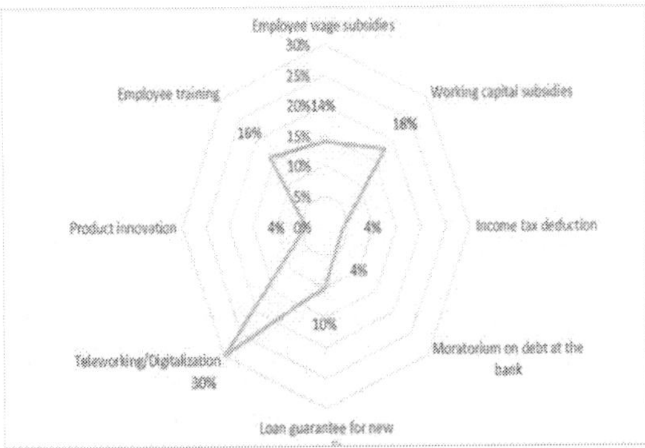

Fig.3. what help do you expect from the government?

Most of them (30%) expect assistance in digitizing their business to take advantage of increasingly promising online business opportunities. Meanwhile, other respondents expect the subsidies for their employees' wages which are still maintained at the current company (18%). There are also those who ask for the scheduling of bank loan repayments for the loans they have made (10%) as well as a reduction in income tax if possible (4%).

The next question is what he has done in the past year to keep his business afloat. They stated that they had started to take advantage of the online business by opening an online shop to sell their products, especially souvenir products (30%). Another respondent stated that what they were doing was cost efficiency to keep the company's cash flow liquid (20%). There are also those who have tried to obtain additional loans as additional working capital (20%). Meanwhile, other respondents have diversified their business outside the tourism sector, such as the fishing business (16%) and 10% of the respondents have laid off their employees, and hope to re-employ them when the situation improves.

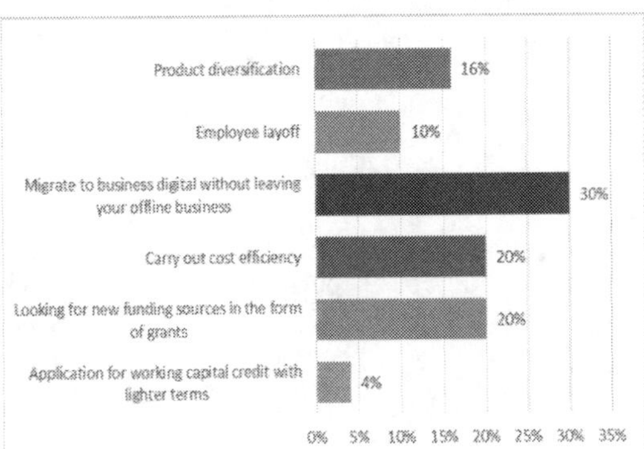

Fig.4. What has been done to get out of the Pandemic

When further investigated regarding their statement to switch to a digital business, the majority answered that they had used social media such as the Instagram shop or Facebook business to market their products (80%). Meanwhile, the other efforts they have made in utilizing the advantages of technology for business benefits are still lacking. Even though they currently have a desire to build their own company website (20%), in addition to utilizing online stores by joining the e-market place they have been running so far (20%)

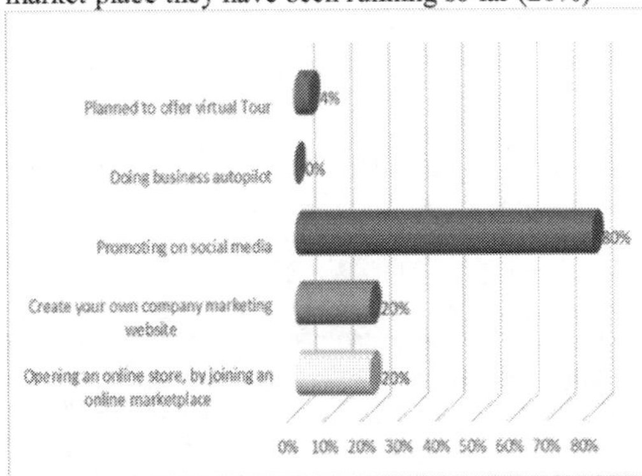

Fig. 5. Strategy Online

From the respondents' answers above, we can conclude that all have tried to keep their businesses running in these difficult times, including lowering selling prices and offset by steps to cut operational costs so that the reduced profit can still cover the existing cost needs. During the interview session, it was also revealed that it is very difficult to expect business from new customers. Therefore, efforts to re-establish relationships with old customers continue to be carried out while hoping that they will still make visits when the situation is getting better. In addition, the use of digital technology for companies engaged in the tourism industry is also very crucial. New opportunities are wide open when all activities take place online like now. Maximizing digital marketing is the best solution to maintain the company's image in the tourism world for both domestic and foreign guests. Because there are already several companies that are pioneering in creating and having their own websites, it is better to use the web and social media to display interesting, useful, as well as educative content. For example, browse YouTube channels with hotel reviews, web blog pages with light articles about destinations and travel tips, to microblogs about the history of tourist attractions on Instagram.

The application of adaptation to new habits (new normal) clearly changes the behavior of tourists when traveling. Likewise with the trend of Indonesian tourist destinations. Along with the easing of lockdown rules in various regions, slowly the hospitality & tourism industry is expected to start throbbing again. Although not fully recovered, experts predict the following tourist destinations will attract tourist visits. Positioning selling tourist destinations in remote areas far from crowds such as hiking, climbing, or hidden beaches will be a tourist wish list [14]. The open nature allows them to maintain physical distance and avoid crowds of people, while also being able to breathe fresh air and not worry about being exposed to the corona virus. Along with the development of a new trend, namely vacationing at a hotel near home (Staycation) it can be used as a marketing policy suggestion by targeting millennials who are bored at home and want to change the atmosphere as the reason for those who choose a staycation as a tourist activity. As long as hotel managers are able to implement strict health protocols according to WHO standards, this option certainly opens the hopes of many parties [15].

Another trend that can be used as business diversification in the tourism industry is virtual tours utilizing VR (virtual reality) technology. Traveling to various world tourist destinations can now be done through virtual tours. If you can't take advantage of VR technology, virtual tours can be carried out through live social media or conference meetings. Top museums offer an online exploration of the building's special corners and collections. In fact, there are virtual tour operators complete with tour guides who are ready to accompany and share exciting stories throughout the session. This idea can be tried to add diversification of opinion to cover the revenue deficit due to the lack of guests who visit physically [16].

6 Conclusion

SME business owners who are engaged in tourism have tried to make various efforts to keep their business throbbing at the current time. From migrating and starting to take advantage of online businesses by opening an online shop to sell their products, making cost efficiencies to keep the company's cash flow liquid, trying to get additional loans as additional working capital, diversifying businesses outside the tourism business such as the fishery business and to do employee layoffs, although they will be rehired when conditions improve. However, all of these efforts have not been good enough, so they still hope for further assistance from the government in the form of support for business digitization, subsidizing employee wages,

Further efforts that can be made by tourism entrepreneurs are to maximize the use of virtual reality (VR) technology, so that tourists who do not yet have the ability to physically visit their pleasure tourism destinations can still enjoy them through virtual reality tours. Promoting the use of VR can reduce the possibility of spreading the virus, increase the sustainability of the destination, and provide other options for a wider audience of visitors to see how prospective and viable the site is. This company also generates alternative income. Innkeepers can take advantage of the staycation trend that is booming among millennial teenagers to vacation in hotels near their homes, while still implementing health protocols.

References

1. Litbang Kompas. Ekonomi-dunia-di-masa-pandemi-covid-19-dari-dampak-hingga-proyeksi-pertumbuhan-2021-2022 [Internet]. 2020. Available from: https://kompaspedia.kompas.id/baca/paparan-topik/ekonomi-dunia-di-masa-pandemi-covid-19-dari-dampak-hingga-proyeksi-pertumbuhan-2021-2022
2. D.W. Nugraha. Proyeksi bank dunia ekonomi indonesia tumbuh 44 persen tahun ini. KOMPAS [Internet]. 2021; Available from: https://www.kompas.id/baca/ekonomi/2021/03/26/proyeksi-bank-dunia-ekonomi-indonesia-tumbuh-44-persen-tahun-ini/
3. D. A. D. Nasution, E. Erlina, I. Muda. J Benefita. 5, 212–24 (2020).
4. S. A. Ni Kadek , S. D. Ni Nyoman. Kondisi ekonomi Indonesi -dimasa pandemi covid-19 [Internet]. 2021. a8452d541df489606ccd2/kondisi-ekonomi-indonesia-dimasa-pandemi-covid-19
5. M. Irham. Virus corona: UMKM diterjang pandemi Covid-19 sampai "kembang kempis" [Internet]. 2021. Available from: https://www.bbc.com/indonesia/indonesia-51946817
6. L. Iswari, M. Muharir, J Ilm Mhs Ekon Syariah. 1, 13–20 (2021).
7. W. Hadiwardoyo W. Baskara J Bus Entrep. 2, 83–92 (2020).
8. N. F. Fabeil, K. H. Pazim, J. Langgat. J Econ Bus. 3, (2020).
9. K. Sandy. 72,6 Persen UMKM Turun Omzet akibat Pandemi Covid-19. (2021).
10. M. Albert. Australia's Economic Measures And Employment Laws Changes In Response To Covid-19 [Internet]. Covid-19 Pandemic: Labor and Employment Law measures around the world. Editora Conhecimento Livre; 2020. Available from: http://dx.doi.org/10.37423/200501001
11. LiveCareer Staff Writer. Government-policy [Internet]. 2020. Available from: https://www.livecareer.com/resources/careers/planning/government-policy
12. R. Hidayati, N. M. Rachman. Niagawan. 10, 1–9.
13. L. Susan, A. Madgavkar, J. Manyika, S. Smit, K. Ellingrud and OR. whats next for consumers-workers and companies i -the post covid 19 recovery [Internet]. 2020. Available from: https://www.mckinsey.com/featured-insights/future-of-work/whats-next-for-consumers-workers-and-companies-in-the-post-covid-19-recovery
14. J.J. Laird. Valuing the quality of strategic ferry services to remote communities. Res Transp Bus Manag [Internet]. 2012 [cited 2016 Jun 28];4:97–103. Available from: http://www.sciencedirect.com/science/article/pii/S2210539512000223
15. T.O. Priasni, D. Absharina, M. Rizal, Pd. Arnesia. Pros Sentik. 4, 49–56 (2020).
16. L.E. Potter, L. Carter, A. Coghlan. Virtual reality and nature based tourism [Internet]. Proceedings of the 28th Australian Conference on Computer-Human Interaction - OzCHI '16. ACM Press; 2016.

The impact of economic growth on technological developments, emoneys and fluctuations interest rates and exchange rates in Indonesia

Milla Naeruz[1*], *Syaad* Afiffudin[2], *Dede* Ruslan[3], and *Muhammad* Syafii[4]

[1]Universitas Sumatera Utara 1 (Medan, Sumatera Utara, Indonesia)
[2] Universitas Sumatera Utara 2 (Medan, Sumatera Utara, Indonesia).
[3] Universitas Negeri Medan 3 (Medan, Sumatera Utara, Indonesia).
[4] Universitas Sumatera Utara 4 (Medan, Sumatera Utara, Indonesia)

Abstract. Technological developments have an impact on the payment system, namely Electronic Money moved very fast in 2018 and 2019, in 2018 it was 50.3% of the money in circulation and economic growth increased by 5.4% even though the interest rate in that year was at 6%. This means that some Indonesians have started to make changes to the payment system. Changes in digitalization in the financial sector, especially with new fundamental changes in the behavior of people's lives from the social and economic fields. The concept of Financial Technology is very good in the formation of digital financial infrastructure based on sustainable technological innovations that are considered effective in financial markets, including for small and medium-sized companies, this article focuses on three factors that affect economic growth, namely capital, labor, and technological developments. This study uses secondary time series data for the 2004-2019 quarter using Multiple Regression (OLS/One Least Square) and processed using the eviews 10 application. This study aims to determine the impact of technology on economic growth. And the results show that Emoney has a negative and significant effect on economic growth, interest rates and exchange rates have a negative effect on economic growth, and technology has a positive effect on economic growth.

1 Introduction

Technology is growing faster in the future, judging by the system of its use. Technology is a means or system that functions to provide comfort and convenience for humans. It is very clear that technology has changed the economic system, especially in terms of transactions and payments, which used to be done manually, now they are done online. It is highly probable that this change will affect the financial system of Bank and Non-Bank Financial Institutions as well as society in general. Technology makes it easier for people to send money and other transactions online using their mobile phones. This tool seems more convenient than physically transferring cash (Mccaffrey & Schiff, 2017). Advanced technology can significantly affect lifestyle changes, especially in rural areas (Wilson & Edwards, 2008). Keynessian Roy F. Harrod and Evsey D. Domar focus on economic growth functions such as capital, demand, and investment, and are the most important factors for economic improvement in a country. The New Economic growth theory was developed by Robert Lucas and Paul Romer. Where this theory focuses its cycle on human resources which are the main capital for increasing production and the national economy. According to Lucas and Romer, a workforce with broad knowledge, higher education, and professional training can accelerate the development of industry and technology. As a result, national production activities can be increased more quickly.

Table 1. Indonesian Community Activities in Using the Internet.

Year	User Internet (Juta)	Online Transaction (Rp. Triliun)
2017	143	30.942
2018	154	77.766
2019	196.7	182.02

Source: APJII.co.id

From the data above, according to the Association of Indonesian Internet Service Providers, from 2017 to 2019 Indonesia is one of the countries with the most internet users and has experienced a significant increase from year to year. Online transactions obtained from 2017 to 2019 continue to increase, meaning that the Indonesian nation is a developing country that is participating in the digitization process. Because the Indonesian people have started to make changes to transactions from offline to online and the internet is considered a very important thing. Technological developments have an influence on the growth of e-

* Corresponding author: milla.naeruz@gmail.com

commerce, especially in 3 important parts of the Indonesian economy, employment, public spending levels, and regional economic growth which will be discussed further in the discussion below based on McKinsey & Company (2018, 2020).

Table 2. Macro Indicator.

Year	Economic Growth (%)	Internet User (Juta jiwa)	*EMoney Growth* (Rp. Triliun)
2015	4.8	88	535.6
2016	5.0	132.7	683.1
2017	5.1	143	943.3
2018	5.4	171.6	2900
2019	5.02	196.7	2700

Source: BPS

(McKinsey & Company, 2020) stated that technological developments will affect economic growth, employment and spending levels and based on macro data from the Central Statistics Agency, it shows that technological developments will affect Indonesia's macro economy, including economic growth, Electronic Money growth. in 2018 economic growth was 5.4% down 5.02% but internet users rose to 196.7 million people from 171.6 million people while emoney growth fell from 2700 trillion rupiah in 2019 from 2900 trillion rupiah

Table 3. Changes in Interest Rates and Exchange Rates from.

Item	2017	2018	2019
Economic Growth	5.1%	5.4%	5.02%
Interest rate	4.25%	6%	5%
Exchanges rate (Rp-US$)	Rp. 13.548	Rp. 14.481	Rp. 13.901

Source : BPS

based on the data above, economic growth fluctuated from 2017 to 2019, where in 2018 the highest economic growth was 5.4%. while the interest rate was also the highest in 2018 at 6%, the exchange rate was also the same at 14,481 rupiah against the dollar. This means that based on 2018 if economic growth increases, interest rates and exchange rates will also increase, or in other words two indicators that are very influential on economic growth in Indonesia are interest rates and exchange rates

The development of Financial Technology contributes to economic growth through the financial sector, namely increasing e-commerce and financing the real sector as well as creating profitable loans for small and medium enterprises (Sadigov, Vasilyeva, & Rubanov , 2020). If the rupiah exchange rate decreases, the government will implement a policy to reduce the amount of rupiah in circulation, namely by increasing the value of the rupiah through raising interest rates, so that the law of supply and demand balance can be achieved (Mankiw, 2006). Mobile technology, the Internet and all digital currencies have a strong influence on the demand for money. So it can be concluded that financial technology and monetary policy have a high role in the economy of a nation (Mumtaz & Smith, 2020)

1.1 Problem Statement

Technological developments have triggered an increase in the number of internet users and along with that, digital payment users have also increased. This change caused economic growth to fluctuate from 4.80% in 2015 to 5.02% in 2019. Not only economic growth, interest rates and exchange rates also had an impact on this change. Harrod Domar focuses on the function of economic growth, namely demand and investment capital, while Romer focuses on economic growth on capital, namely human resources who have knowledge and insight. the findings of mumtaz Financial technology and monetary policy have a significant effect on the economy of a nation, especially in controlling the financial system on the stock market, sadigov and friends state that the development of financial technology contributes to economic growth in the financial sector, especially the increase in e-commerce turnover and real sector financing..

1.2 Formulation Problem

Based on the introduction above, it can be obtained a problem statement, namely Do Electronic Money, Interest Rates, Exchange Rates and Technology (internet users) affect Economic Growth?

1.3 Research Purposes

To analyze, test, develop and modify the theory or previous research on the effect of Electronic Money, Interest Rates and Exchange Rates and Technology (internet users), on Economic Growth whether partially or completely accepted either in changes or in the development of modifications.

2 Literatue Review

2.1 Economic Growth

Robert M Solow argued that developing or underdeveloped countries in improving the economy by increasing the accumulation of physical capital, labor and human resources as well as efficiency in their use. The role of technology has not been seen as a driving force for economic growth. so that if a failure is found in the development process, it can be resolved by placing an efficient investment allocation. because the sollow theory states that to increase savings and investment it will directly increase resources this will affect the output obtained. and will continuously affect the improvement of the economy of a nation (Mankiw: 2000). $Y = A K\alpha L\beta$
Where
Y = Aggregate output
L = Number of workers
K = Total capital

A = Multifactor productivity or total factor productivity
= Elasticity of capital output
= Elasticity of labor output

Roy Harrod (1948) and Evsey Domar (1957 Harrod and Domar use Keynes's thinking that the development process in the long run will affect the level of savings and the level of investment through economic growth. Harrod-Domar theory concludes that economic growth is determined by high savings and investment. If savings and investment are low, the economic growth of the community or country will be low. or in other words Harrod focused on saving and investment as an important factor in economic growth.

2.1 Technology

Technology comes from the Greek Technologia which means systematic treatment or handling something systematically, while techne as the basis of the word technology means skill or expertise, skills and knowledge. The word technology is often understood by ordinary people as something in the form of a machine or things related to machinery.

Technology according to Roger is an action design tool that reduces uncertainty with a causal relationship in achieving a desired result. While the opinion of Jacques Ellul defines technology as a whole method that rationally leads and has efficient characteristics in every human activity.

Gary J Anglin argues that technology is the application of the behavioral and natural sciences and other knowledge in a systematic and systematic way to solve problems. Meanwhile, according to Vaza technology is a process carried out in an effort to realize something rationally.

Technology is at the center of changing payment arrangements, with better solutions and capabilities influencing consumer behavior and expectations, also driving considerable industry transformation (Broom, 2015). The most common approach used to measure the impact of the digital economy is the technology that is included in the regulation of the production function Van Reenen, Draca, and Sadun (2007).

2.3 Emoney

Electronic Money has an insignificant positive effect on economic growth (Pitriani Ritonga 2018). In the 21st century, non-cash payments are considered a very practical and convenient means of payment and are considered as alternative payments from cash. Plastic money is considered more practical because it is more convenient and safe in everyday life in its use (Japan-METI, 2017). The increase in people's use of non-cash transactions is driven by the ease of transactions which can encourage a decrease in transaction costs and in time will stimulate economic growth. In addition to making transactions easier It is hoped that in the future non-cash transactions can reduce the circulation of money from the central bank, where this will affect the central bank's task in controlling monetary policy (Costa Storti & De Grauwe, 2001)

Financial Technology has changed the payment system in society and has helped many companies to reduce their high capital and operating costs. FinTech is also able to replace the role of formal financial institutions such as banks in terms of payment systems, among others; Provide a market for business actors Become a means of payment Settlement clearing Assist in the implementation of more efficient investments Minimize risk from conventional payment systems Assist parties in need to save, borrow funds and equity participation

2.4 Interest Rate

The interest rate according to classical theory is divided into two, namely the nominal interest rate and the real interest rate. Where the nominal interest rate has a very important role in financial developments because the nominal interest rate is a determinant in increasing real interest rates. If the government does not set a ceiling on the nominal interest rate, the nominal interest rate will tend to adjust to the movement of inflation. However, with a nominal interest rate ceiling, the nominal interest rate can be less than inflation, so creating a small real interest rate reduces the amount of savings in the economy. A small interest rate will increase investment, where investment will affect the increase in income.

Demand The amount of goods depends on the interest rate, where the interest rate is an indicator of investment financing. Healthy investment will increase the amount of production of goods and services so that it will increase future income. Mobile phones, Internet technology, Litecoin, and Ethereum have a very strong relationship to Financial Technology and real interest rates, GDP, inflation, financial development indices, and stock market indices (Mumtaz, 2020)

2.5 Exchange Rate

The nominal exchange rate is a rate at which a person can trade the currency of one country for another country's currency. If a currency appreciates, it is said that it is strengthening, because it can buy more foreign currency. Similarly, when a currency depreciates, it is said that the currency is weakening. While the real exchange rate is a value where a person can trade goods and services from one country with goods and services from other countries (Mankiw Gregory, 2000

3 Equations and mathematics

3.1 Conceptual Framework

This study uses a quantitative approach, where the approach has been determined previously using secondary data and data in numeric form. Connecting variables in problems and hypotheses. Quantitative data, namely data in the form of numbers or qualitative data that is numbered (Sugiyono, 2014). Secondary data with the type of time series (time series) in the period 2004 – 2019 quarterly. Data collection methods in this study are secondary data collection obtained from Bank

Indonesia, the internet, BPS (Central Statistics Agency), Data Words, Infokom and library techniques, are by collecting data from books, journals, and research results as well as reading sources or writing materials. relevant to this article.

Data processing using eviews 10. The data analysis method carried out in this study will use regression equation using the least squares regression method or Ordinary Least Square (OLS)

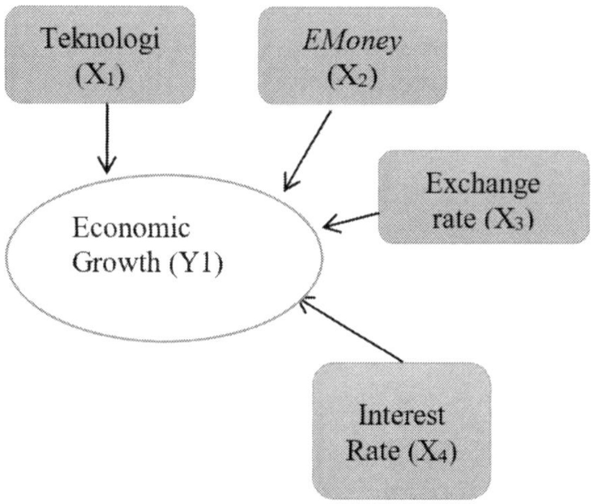

Figure 1 Conceptual Framework

3.2 . Growth Economy Equation

$$Y = \alpha_1 + \beta_1 X_1 + \beta_2 X_2 + \beta_3 X_3 + \beta_4 X_4 + \varepsilon_1$$

Y_1	**= Economic Growth**
X_1	**= Technology**
X_2	**= Emoney**
X_3	**= Interest Rate**
X_4	**= Exchange rate**
$\alpha\beta$	**= Koefisien**
ε	**= Error Term**

4. Results

Table 4. Results

Variabel	Koefisien	t-statistic	Prob
TEK	1.505866	4.245711	0.0001
EMoney	-0.335027	-4.476063	0.0000
IR	-0.050739	-1.218920	0.2277
ER	-2.105060	-4.815125	0.0000
R-Squared	0.414369		
Prob>F	0.000000		

Y = 7.80 - 0.335*EMONEY + 1.505*TECHNOLOGY - 0.05*IR - 2.1*ER (4.4)

The estimation results find that simultaneously or together have an effect on Economic Growth, this can be seen from Prob F which is smaller than 5%. The goodness of fit test (R2) Based on the estimation results, the inflation equation model obtained a coefficient of determination (R2) of 41.4%, meaning that variations in money, exchange rate interest rates, inflation and economic growth were 41.4%, while the remaining 58.6% was influenced by various factors. other. This shows that Economic Growth is not only influenced by Technology, Money, Interest Rates, Exchange Rates, but there are other factors not discussed in this study that affect Economic Growth.

5. Discussion

5.1 The effect of money on economic growth

Emoney's research results have a negative and significant effect on Economic Growth. Does not fit the hypothesis. This means that if Emoney transactions increase, economic growth will decrease. Emoney is Electronic Money used in transactions through electronic devices and also the internet network. It is very easy to use and can save costs. Increasing Electronic Money will reduce the workforce and what is left is only an educated and skilled workforce in following the development of the technology. Because all transactions use the system through technology. As a result, a reduced workforce will reduce economic growth. The use of online purchase transactions grew in 2019 by 96%, previously in 2018 by 53.4% and 2017 by 32.19%. The rise of digital money issuance, has not been directly proportional to the current digital innovation players. Because there are still many Indonesian people who do not understand emoney. So that this digital payment is widely used by millennial children who are generally students and still do not have a steady income. And innovation actors who do not understand the use of the E-Wallet and generally Indonesian people still like to use cash. Factors that cause economic growth that is not immobile are human resources that are not optimal, limited natural resources, lack of knowledge and technology, lack of capital and bad public attitudes

Schumpeter's Theory of Economic Growth (1934) the economic progress of a society can only be applied to entrepreneurs who have innovation and an entrepreneurial spirit and says that technological progress is largely determined by those who have an entrepreneurial spirit who are able to see opportunities and take risks.

Pitriani Ritonga (2018) Emoney has an insignificant positive effect on economic growth

Nizar, Sholeh (2021) show that the digital economy has business resilience and is significantly able to make a positive contribution to economic growth.

5.2 The Effect of Technology on Economic Growth

The results of the study found that technology has a positive and significant effect on economic growth. according to the hypothesis. This means that the more people use technology, the more economic growth will grow. This is because Indonesia has shown rapid development in the last five years. Not only by households but also by industry. And Indonesia has started to become the largest digital economy growth and development magnet in Southeast Asia and at the global level

To overcome these technological developments, the Government has prepared policies that are able to optimize the impact and results of implementing technological advances in Indonesia. namely, improving the quality of digital services to increase the efficiency and competitiveness of the sector driving Indonesia's economic growth and improving the quality of human resources (HR) so that they can adapt to future employment needs

Sollow's theory (1956) states that the higher a person's education, it is expected that the stock of human capital will be higher, because humans have a positive relationship with economic growth. Sollow theory states that Encouraging technological progress will increase income per labor market so that the opportunity to innovate in the private sector will have a major impact on economic growth. And according to Sollow, technology is an exogenous variable, while human capital is a factor that influences technology. Because capital per worker will only make a small contribution to output.

Romer's (1997) economic growth states that technological change is endogenous growth, namely technology is a process of economic growth. Meanwhile, human capital, investment, savings and physical capital also determine economic growth. According to Romer, technology is considered an endogenous factor that can have a large effect or impact on human capital who has broad insight, higher education and who has professional training so that it will create a digital financial system industry that affects economic growth aimed at the welfare of the Indonesian people.

William Straus and Neil Howe (2012) Generation Y prefers non-cash payments and non-cash payments will replace cash and tend to use technology in their work.

Raef Bahrini, et al (2019) in the Middle East and Africa technologies such as cell phones, internet users and broadband are the main drivers of economic growth and increasing human resource allocation.

Edna Meyen et al (2020) in Africa the use of technology and information stimulates the economy and creates jobs

Rinaldo Evangelista, et al (2014) in Europe digitalization can boost job productivity and have a positive effect on economic growth.

Garces & Daim (2012) in America Technological and HR innovation has a positive effect on economic growth.

5.3 The Effect of Interest Rate and exchange rate on Economic Growth

The results of this study found that interest rates have a negative and insignificant effect on economic growth. In accordance with the theory and previous research. This happens because high interest rates will cause banks to withdraw money and this will cause inflation. High inflation will cause a decline in economic growth. And high loan interest rates will reduce investors to invest because economic activity will weaken and cause economic growth to decline

The results of the study found that the exchange rate was negatively and significantly related to economic growth. In accordance with theory and research. This means that if the exchange rate of the rupiah against the dollar increases, economic growth will decrease. This is due to many factors that made the rupiah weaken because the dollar was one of the first by the American central bank to raise the benchmark interest rate as a result the dollar interest rate rose. Both demand for the dollar rose. Third, the maturity of foreign debt payments. Fourth, consumption imports increase and so on

Keynes's theory concludes that what increases GDP is an increase in consumption, investment, government spending and net exports. These four factors are again influenced by various factors, namely income, price level, interest rate, inflation rate, money supply, exchange rate.

Susanto's research (2017) finds that interest rates and exchange rates have a negative and significant effect on economic growth.

Heri Sasono's research (2020) which found that interest rates and exchange rates have a negative and insignificant effect on economic growth

6. Conclusion

Emoney has a negative and significant effect on economic growth, interest rates have a negative and insignificant effect on economic growth, the exchange rate has a negative and significant effect on economic growth, and technology has a positive and significant effect on economic growth

References

1. Ariff, M., Chung, T. F., & M., S, (2012). *Money supply, interest rate, liquidity and share prices*: A test of their linkage. GFJ, 10 (2012)
2. Ferguson, K. K., Soutter, L., & Neubert, M, (2019). *Digital payments in Africa - how demand, technology, and regulation disrupt digital payment systems. International Journal of Teaching and Case Studies,* IJTS, (2019)
3. Heri. Sasono, (2019). Analisa Pengaruh Tingkat Suku Bunga, Nilai Tukar, Inflasi, Harga Minyak Dunia, Indeks Harga Saham Gabungan Dan Produk Domestik Bruto Terhadap Pertumbuhan Ekonomi, Trijunal, (2019)
4. Khiewngamdee, C., & Yan, H. D, (2019). *The role of Financial Technologie e-payment on APEC economic development.*, In Conference Series, JOP, (2019)
5. McKinsey & Company. (2020). McKinsey on Climate Change. *McKinsey on Climate Change*

6. Mumtaz, M. Z., & Smith, Z. A. (2020), *Empirical examination of the role of Financial Technologie in monetary policy*. PER, (2020)
7. Nampewo, D., & Opolot, J. (2016), *Financial Innovations and Money Velocity in Uganda,* ADR, (2016)
8. Ningsih, S., & Kristiyatnti, L. (2016), Analisis Pengaruh Jumlah Uang Beredar, Suku Bunga, Dan Nilai Tukar Terhadap Inflasi Di Inodnesia Periode 2014-2016, JEMSD, (2016).
9. Nizar, N. I., & Sholeh, A. N. (2021), Peran Ekonomi Digital Terhadap Ketahanan dan Pertumbuhan Ekonomi Selama Pandemi COVID-19, JMIPTH, (2021)
10. Rodoni, A., Mufraini, A., & Ansori, R. (2010). Analisis Pengaruh Tingkat Inflasi, Sbi, Jumlah Uang Beredar, Dan Tingkat Pendapatan Terhadap Nilai Tukar Rupiah Terhadap Dollar Amerika.
11. Sadigov, S., Vasilyeva, T., & Rubanov, P. (2020), *Financial technologie in economic growth: cross-country analysis*. In Prceeding 55 Th International Scientific Conference on Economic and Social Development-Baku, (2020)
12. Susanto (2017), Pengaruh inflasi, tingkat suku bunga, dan nilai tukar terhadap pertumbuhan ekonomi indonesia. JSTIB, (2017)
13. Vovchenko, N. G., Galazova, S. S., Sopchenko, A. A., & Dzhu, O. S, (2019). *Financial Technologie ecosystem as an instrument of sustainable development provision*. IJOEBA, (2019)
14. Zhang, L., & Ma, Q. (2011), *The Effect of New Electronic Money on Money Supply*. IEEE AIMSEC, (2011)

The function of the attitude of Islamic banks in coastal societies in mediating the effect of emotional attachment on the continuance intention of Islamic banks

Nasir Nasir[1], *Syafruddin* Chan[1*], and *Cut* Aprilia[1],

[1]Management Department, Faculty of Economics and Business, Syiah Kuala University, Indonesia

Abstract. This study aims to analyse the factors that cause the reluctance of Islamic bank customers in coastal areas to be loyal to the particular banks. The population in this study are conventional bank customers who live in Aceh, North Sumatra and Riau. The sample size in this study was 350 respondents. Sampling was carried out using a stratified random sampling method with the following proportions: Aceh 25%, North Sumatra 50% and Riau 25%. In this study, the sampling method used the census technique. Data analysis used SEM (Structural Equation Modelling) to verify the path of the relationship between variables. The results showed that of the 3 direct hypotheses tested, the effect of Emotional Attachment on Attitude. The effect of emotional attachment on continuance intention, the effect of Attitude on continuance intention, all of which show a significant effect. The significance that occurs in this direct test indicates that there is a match between the research model developed and the facts in which this research was conducted. On the indirect effect tested, although it showed significant results, the role of attitude in mediating the effect of emotional attachment on continuance intention through attitude is partial.

1 Introduction

Indonesia, as a Muslim majority country that is dominated by middle class population is a prospective market for Islamic banking [1]. However, the road to success for Islamic banking in Indonesia is not a smooth sailing. According to the Financial Services Authority of Indonesia or "OJK" in the beginning of the establishment, Islamic banking did not attract the attention of Indonesian.

The current phenomenon is the low penetration of coastal communities, which are dominated by the fishery sector, to the bank services, both for savings products, micro-credit, transfer/payment facilities, and the use of EDC machines. Although many efforts have been made to improve access to banking services in these coastal communities, these efforts have not been fully successful. Penetration to Islamic banks among coastal communities only occurred when "the qanun" for Islamic financial institutions was enacted in Aceh which made all banking services in Aceh based on sharia. Even if there is a switch, there is no guarantee that they will stay there and not switch back to conventional banks. From previous research conducted by [2] revealed that there are still material and infrastructure aspects that are the reluctance of conventional bank customers to switch to Islamic banks. Many of them prioritize rational aspects such as more facilities, more advanced technology, better benefits and cheaper transaction costs than Islamic banks [3]. The question is whether it is the mere rationality aspect that is considered by consumers to continue using Islamic bank products. Other researchers [4] are revealing the aspect of religiosity asone of the differentiators. The higher people's level of religiosity, the higher the probability of switching to an Islamic bank.especially those related to the aspect of unsury which is forbidden in Islam [5]. However, according to [1] there are still doubts among customers, especially Muslims, regarding the implementation of the business, is it really sharia and compliance with Islamic law? If they are not sure about this, it will certainly raise doubts about continuing to use (continuance intention) Islamic banking products even among sharia customers themselves [6].

Continuance intention is the study that measures customer intention to continue using/purchasing a product/service [7]. Numerous scholars stated that the sustainability of business depends on the continuity of product/service consumption. Therefore, to further increasing the performance of Islamic banking, it is important to measure the predictors of continuance intention of Islamic banking customers.

Consumer decision to continue or discontinue purchasing a product/ service depends on their evaluation of the performance of the product and their state of emotion. According to [8] customers emotional state affect their decision. While, through service quality, consumer can measure the overall performance of the service offered [9].

Many previous studies have discussed the determinants of the continuance intention factors for Islamic banking products, as was done by [10] that

* Corresponding author: syafruddin.chan@unsyiah.ac.id

prioritizes aspects of technology, product features and other benefits. However, as far as the author's knowledge, discussing aspects of emotional attachment and attitude as an antecedent of continuance intention has not been done much.

This study aims to measure the role of emotional attachment and attitude in predicting consumers' continuance intention to use Islamic banking products/services for financial transactions.

2. Literature Review

2.1. Emotional attachment

Originated from the study of psychology, emotional attachment is considered as an emerging construct in consumer behaviour study [11]. The scholars of marketing science believed that the attachment that human feel on each other also applied to objects like product and place [12].Therefore according to [13] elucidated emotional attachment as a reflection of consumers feeling or mental state that is associated with a brand.

Building an emotional attachment between consumer and brand is important for a company [13]. stated that an emotional attachment consumers have a strong relationship with the brand and commitment to keep the relationship going by giving a positive behaviour. The study done by [11] found that brand attachment has a significant effect on continuance intention to use mobile apps. [14] also found that emotional attachment leads consumers to support the brand, purchase the brand and have an intention to continue purchasing and consuming the brand. Hence, this study hypotheses that:

H1: The effect of emotional attachement on attitude

H2: Emotional attachment influence continuance intention

2.2 Attitude

In the context of marketing, the attitude formed is related to buying behaviour, actual experience, and product trials [15]. Attitudes are formulated from the media and from advertising messages or other promotional tools, such as direct marketing or the internet, and through words of mouth received from others who have tried the product [16]. There is a direct relationship between attitude and behaviour. Attitude can lead to or result in behaviour, but these terms are not interchangeable or synonymous. Behaviour reflects positive or negative evaluations and actions towards the object of attitude [17]. Attitudes have motivational factors, as they may drive consumers towards certain actions or may hinder certain behaviours [18].

Attitude consists of three main elements: cognitive, affective, and conative [19]. This is the tri-component attitude model adopted in this study. Cognitive elements relate to the types of information, data, knowledge, and perceptions collected by consumers [20]. The source of the data and knowledge is direct experience with the attitude object itself. This information and knowledge is formulated in the form of trust. Consumers believe that attitude objects have certain benefits and features or that certain behaviours or actions will lead to certain outcomes and preferred outcomes [19]. The affective element is related to the feelings and emotions that consumers have towards the attitude object. Consumer researchers, consider respondents' emotions as a factor that helps consumers in evaluating attitude objects [15]. The emotional element is the main factor involved in judging attitude objects, because consumers tend to classify objects as favourable or unfavourable based on their emotions [19]. The conative element is the action taken by the individual in relation to the attitude object. The action can be in the form of a consumer's tendency to buy their actual intention to buy an attitude object. In measuring the conative element, marketers assess buyer intentions on a certain scale to measure the probability of purchasing decisions or behaviour [19].

H3: Attitude influence continuance intention

2.3 Continuance Intention

Continuance intention is part of behavioural intention that describes a consumer willingness to continue consuming a service in the stage of post consumption [21]. It is an important construct to measure due to its ability to influence actual consumption [22]. According to [21] consumer continuance intention is the combination of intentional, rational, and emotional response. Therefore, consumers' perception on the bank service performance and their attachment to the bank contributes on their decision to continue or discontinue to consume the bank's service.

H4: Emotional attachment influence continuance intention through attitude

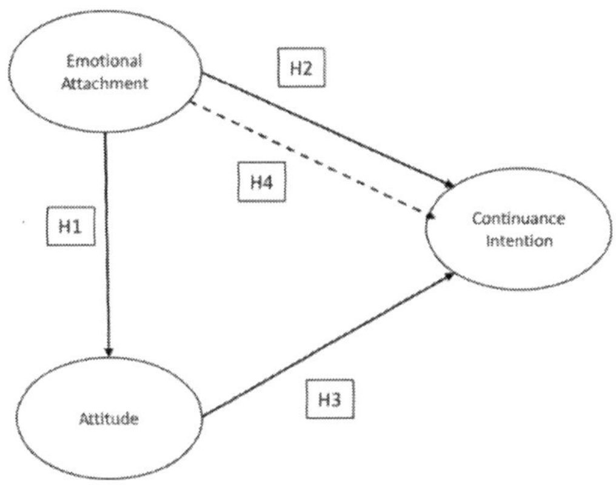

Fig. 1. Conceptual Framework

3. Research methods

3.1 Sample Design and Data Collection

The population in this study are conventional bank customers who live in coastal area in Aceh, North Sumatra and Riau. The size of the sample in this study was 350 respondents [23]. Sampling was carried out using a cluster random sampling method with the following proportions: Aceh 25%, North Sumatra 50% and Riau 25%. The proportion of North Sumatra is larger because the number of bank customers in North Sumatra is also larger than Aceh and Riau

3.2 Data analysis

The data analysis technique in this research is descriptive and verification. Descriptive analysis was conducted to assess the demographic profile of the respondents and the internal consistency of construction. While the verification analysis uses SEM (Structural Equation Modelling) to verify the path of the relationship between the variables in this study. In addition, the SEM analysis software is IBM SPSS-AMOS version 22.

4. Results and Discussion

4.1 Characteristics of Respondents

In terms of age, the most respondents came from the age group 30-40 years, namely 199 (56%). regarding the gender, there are more men than women, namely 250 (71.4%). Then from the educational background the majority are those with undergraduate education, namely 175 (50%). Most of their jobs are in the fishery sectors 188 (53.7%), the rest 162 (46.2%) work outside the fisheries sector such as trade, culinary and other tourism services.

4.2 Validity with Measurement Model

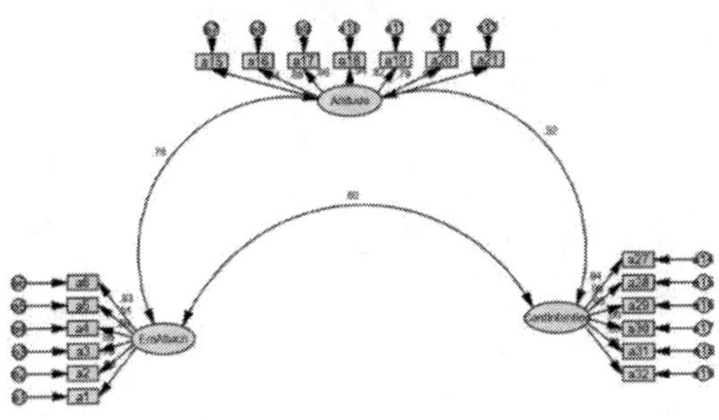

Fig. 2. Initial Measurement Model

Item validity is indicated by the existence of a correlation or support for the total item (total score), the calculation is done by correlating the item score with the total item score. If we use more than one factor, it means testing the validity of the item by correlating the item score with the factor score, then continuing to correlate the item score with the total factor score (the sum of several factors). An indicator is said to be valid if it has a standardized loading factor value > 0.50. For indicators that do not meet the requirements will be eliminated from the model. From the results of calculations that have been carried out, it turns out that all indicators contained in this model have a loading factor value of > 0.50, so it can be said that all of them are valid and nothing should be eliminated. The loading factor can be seen in the Table 1.

Table 1. New Loading Factor

			Estimate
a1	<---	EmAttach	.683
a2	<---	EmAttach	.792
a3	<---	EmAttach	.849
a4	<---	EmAttach	.859
a5	<---	EmAttach	.813
a6	<---	EmAttach	.829
a15	<---	Attitude	.935
a16	<---	Attitude	.878
a17	<---	Attitude	.956
a18	<---	Attitude	.937
a19	<---	Attitude	.916
a20	<---	Attitude	.787
a21	<---	Attitude	.850
a27	<---	ContIntention	.944
a28	<---	ContIntention	.959
a29	<---	ContIntention	.951
a30	<---	ContIntention	.957
a31	<---	ContIntention	.853
a32	<---	ContIntention	.815

Based on the table above, it can be explained that all the variable items used in this study are all valid, because they have a loading factor number > 0.50.

4.3 Fit Model

Table 2. Modification Indices

ITEMS		ITEMS	MI
e12	<-->	e13	33,142
e19	<-->	EmAttach	31,373
e6	<-->	Attitude	31,183
e6	<-->	e16	28.893
el	<-->	e2	25,374
el	<-->	e3	25,033
e9	<-->	EmAttach	22.02
e17	<-->	Attitude	19,809
e2	<-->	e3	19.262
e10	<-->	e13	16,577
e18	<-->	e19	15,708
e8	<-->	e11	13,677

Model fit at the measurement stage is one of the requirements in SEM to see if there is a feasibility of a model that has been developed through a literature review process with data collected from the field in the form of primary data. A model is said to be fit with the existing data if it meets several criteria tested at the goodness of fit test stage. The criteria used include CMIN/DF, GFI, TLI, IFI and RMSEA. At the time of initial testing, it turned out that there were several fit criteria that were not met, therefore it was necessary to modify the model referring to the modification indices table (Table 2).

Based on the modification indices table above, modifications were made to the model so that the results of the model modification, after adding some covariance lines as suggested by the MI table, are shown in the Figure 3.

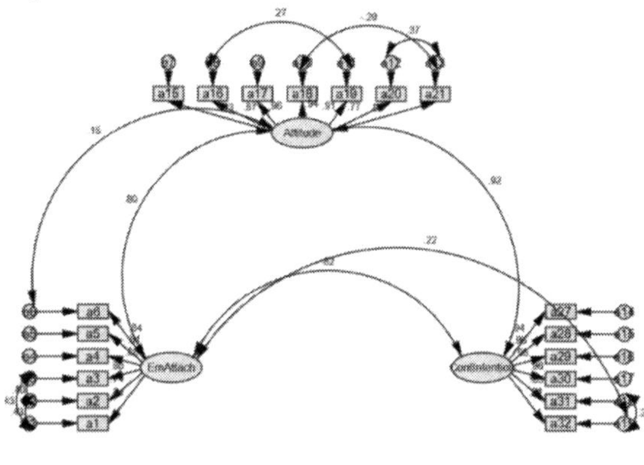

Fig. 3. Measurement Model after Modification

After modifying the model, it turns out that some criteria that were not fit before, such as CMIN/DF and RMSEA, are now better and have become fit. The cut off values for each criterion and test results before and after model modification can be seen in the Table 4.

Table 4. Goodness of Fit

No	Criteria	Cut off value	Value Before	Value After	Fit/Not Fit
1	CMN/DF	<2.00	3,602	1.902	Fit
2	GFI	>0.90	0.787	0.901	Fit
3	TLI	>0.90	0.921	0.962	Fit
4	IFI	>0.90	0.932	0.969	Fit
5	RMSEA	<0.08	0.107	0.065	Fit

4.4 Reliability Test

The reliability test intended in this study is to determine the extent to which the measurement results remain statistically consistent, namely by calculating the magnitude of the composite reliability of the data based on the estimated output obtained using Cronbach alpha. The results are as described in Table 4.3 which shows that the instrument in this study is reliable because its reliability coefficient value is greater than 0.60 [24].

Table 3 Reliability Using Cronbach Alpha (CA)

No	Variable	CA	Information
1	Emotional attachment	.918	Reliable
2	Attitude	.966	Reliable
3	Continuance intention	.967	Reliable

Based on the reliability analysis, it can be seen that the alpha for each respondent's perception variable can be seen from several variables Emotional attachment is 91.8%, Attitude is 96.6% and Continuance intention is 96.7%. Thus the reliability measurement of the research variables shows that the reliability measurement meets the requirements of Cronbach Alpha (CA) where the CA coefficient value is at least or greater than 60 percent.

4.4 Hypothesis Testing

Hypothesis test verification in this study was conducted to test and analyze the effect of Emotional Attachment, Attitude and Continuance Intention to use Islamic banking products and services. The verification hypothesis testing consists of testing the direct influence hypothesis and testing the indirect effect hypothesis. The results of testing the direct influence hypothesis can be seen in the following figure and table.

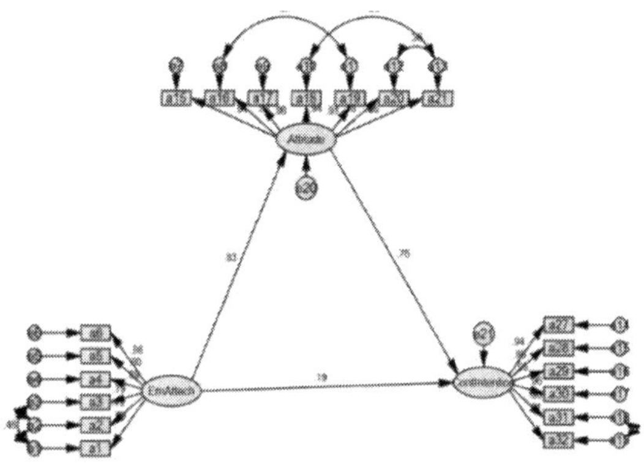

Fig. 4. Structural Model

Table 5. Hypothesis Testing

			CR	P	Beta
Attitude	<---	EA	9.073	***	.825
CI	<---	Attitude	12.356	***	.762
CI	<---	EA	3.116	.002	.191

EA: Emotional attachment
CI: Continuance intention

H1: The effect of Emotional Attachment on Attitude

Testing the effect of Emotional Attachment on Attitude shows a CR value of 9.073 and with a probability value of ***. The two values obtained have met the requirements for the acceptance of Ha, namely the CR value should be greater than 1.96 and the probability value is less than 0.05. The magnitude of the coefficient of the effect of Emotional Attachment on Attitude is 0.825

H2: The effect of Emotional Attachment on Continuance Intention

Testing The effect of Emotional Attachment on Continuance Intention shows a CR value of 12,356 and with a probability value of *** Thus it can be stated that the effect of Emotional Attachment on Continuance Intention is significant. The magnitude of the coefficient of the effect of Emotional Attachment on Continuance Intention is 0.762

H3: The effect of Attitude on Continuance Intention

Testing the effect of Attitude on Continuance Intention shows a CR value of 3.116 and with a probability of .002. The two values obtained have met the requirements for acceptance of Ha. The magnitude of the coefficient of the effect of Attitude on Continuance Intention is 0.191.

H4: The Effect of Emotional Attachment on Continuance Intention through Attitude

Testing the Effect of Emotional Attachment on Continuance Intention through Attitude shows a probability of .001. The p value obtained has met the requirements for acceptance. Ha. The coefficient of the Effect of Emotional Attachment on Continuance Intention through Attitude is 0.629 or 62.9%.

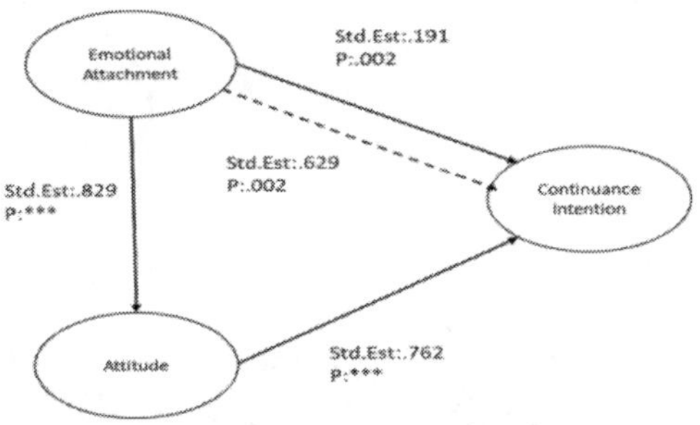

Fig. 5. The Effect of Emotional Attachment on Continuance Intention through Attitude

Because the direct effect of emotional attachment on continuance intention is significant and the indirect effect of the effect of emotional attachment on continuance intention through attitude is also significant, we can say that the role of attitude in mediating these two variables is a Partial Mediating.

5. Discussion

Of the 3 direct hypotheses tested, the effect of Emotional Attachment on Attitude, The effect of Emotional Attachment on Continuance Intention, The effect of Attitude on continuance intention, all of which show a significant effect. The significance that occurs in this direct test indicates that there is a match between the research model developed and the facts in which this research was conducted.

Based on the results or findings presented in the previous section, it means that Emotional Attachment and Attitude are indeed determinant variables -in improving the intention of continuance bank's customer in using Islamic banking. The management can use these two variables as tools to improve the intention of the bank's customer to continue using Islamic banking. This result is in line with research conducted by [12].

To support the strongest increase in the intention to use Islamic banking services, it can be done through the increasing of attitude, because this variable has the largest magnitude number compared to the other variables, namely 82.5% compared to emotional attachment which is only 19.1% . On the indirect effect tested, although it showed significant results, the role of attitude in mediating the effect of emotional attachment on continuance intention through attitude is quite good. The path of the effect of emotional attachment on continuance intention through attitude has a coefficient of 62.5%

6. Conclusion

From the research findings that have been discussed in the previous section, we can see that the main factor that can increase coastal societies interest in continuing to use Islamic banking services is to increase the customer's positive attitude towards Islamic bank products. Because this factor has the highest coefficient compared to other variables. This attitude improvement must of course be accompanied by an integrated marketing program, including generating emotional feelings about the importance of using Islamic banks in conducting financial transactions, especially for customers who are Muslim, because this banking product is free from usury.

On the other hand, the Islamic banking business practices carried out must also follow sharia business practices in a *"kaffah"* manner, so that there is no impression, Islamic banks are only the casing in an effort to attract the sympathy of Muslims to make Islamic banks an institution where they entrust financial transactions, saving, borrowing and lending.

Further suggestions that need to be conveyed to sharia business actors, in addition to the sharia financial regulations that have been initiated in Aceh, are also necessary to issue regulations for the conversion of non-bank financial institutions such as sharia-based health and loss insurance, and management of labour social security funds with sharia schemes.

References

1. D. Suhartanto, J Islam Mark **10**, 4 (2019)
2. S. Ayyub, Xuhui W, Asif M, Ayyub RM, Int J Islam Middle East Finance Manag **13**, 1 (2019)
3. I. Muslichah, Sanusi S, Asian J Islam Manag **1**, 2 (2019)
4. H. Basri, Tabrani M, Management and Financial Transparency of Islamic Religious Organizations: The Case Study of Modern Islamic Boarding School in Contemporary Indonesia, Glob J Humanit Soc Sci [Internet]. 2015 [cited 2016 Jun 28]
5. H. Usman, Projo NWK, Chairy C, Haque MG. The exploration role of Sharia compliance in technology acceptance model for e-banking (case: Islamic bank in Indonesia), J Islam Mark, (2021)
6. N. Tara, Irshad M, Khan MR, Yamin M, Rizwan M, J Public Adm Gov **4**, 3 (2014)
7. H.L. Song, Customer adoption of internet banking: An integration of TAM with trust, perceived risk,

https://doi.org/10.1051/e3sconf/202233905009

and quality. In: Proceedings - 2010 2nd International Conference on Multimedia Information Networking and Security, MINES (2010)

8. J. Gountas, Gountas S, J Bus Res, (2007)
9. H. Chahal, Manag Serv Qual An Int J [Internet]. (2010)
10. H. Herjanto, M. Amin, Int J Bank Mark, **38**, 6 (2020)
11. C. Y. Li, Y. H. Fang, Predicting continuance intention toward mobile branded apps through satisfaction and attachment, Telemat Informatics (2019)
12. P. Cheng, Z. OuYang, Y. Liu, Transp Res part A policy Practice **124** (2019)
13. S. Levy, H. Hino, Int J Bank Mark, (2016)
14. A. S. Prastyaningsih, Suyadi I, Yulianto E, J Business Adm **16,** 1 (2014)
15. M. M. H. Abou-Youssef, W. Kortam, E. Abou-Aish, N. El-Bassiouny, Int J Bank Mark **33**, 6 (2015)
16. N. Souiden, M. Rani, Int J Bank Mark **33,** 2 (2015)
17. J. Bananuka, M. Kasera, G.M. Najjemba, D. Musimenta, B. Ssekiziyivu, S. N. L. Kimuli, J Islam Mark **11**, 1 (2019)
18. M. Janssen, M. Mntymäki, J. Hidders , B. Klievink, W. Lamersdorf, B. van Loenen, Open and Big Data Management and Innovation: 14th IFIP WG 6.11 Conference on e-Business, e-Services, and e-Society, I3E 2015 Delft, The Netherlands, October 13–15, 2015 Proceedings. Lect Notes Comput Sci (2015)
19. L. Schiffman, L. L. Kanuk. Consumer Behavior (seventh edition) consumer behavior. Jakarta PT Index Qual Res net" Syst Versus Interpret Anal with two CAQDAS Packag Nvivo MAXQDA. (2008)
20. A. Kaakeh, M.K. Hassan, S.F. Van Hemmen Almazor, Int J Emerg Mark **14**, 4 (2019)
21. J. Hepola, M. Leppäniemi, H. Karjaluoto, J Retail Consum Serv **57** (2020)
22. A.S. Oertzen, G. Odekerken-Schröder, Int J Bank Mark, (2019)
23. J.F. Hair, WC Black, BJ Babin, RE Anderson, RL Tatham, Prentice hall Upper Saddle River, NJ **5** (1998)
24. L.R. Kahle, N.K. Malhotra, J Mark Res [Internet] **31**, 1 (1994)

The role of customer satisfaction in mediating the effect of the evaluation of the bank's compliance with Islamic law on the continuous intention to use Islamic banking among coastal communities

Syafruddin Chan[1*], *Cut* Aprilia[1], and *Jalaluddin* Jalaluddin[2]

[1]Management Department, Faculty Economic and Business, Universitas Syiah Kuala, Indonesia
[2]Accounting Department, Faculty Economic and Business, Universitas Syiah Kuala, Indonesia

Abstract. This study aims to analyze the factors that cause low intention of Islamic bank customers to continue using banks' services, especially among those who live in coastal areas. The population in this study is the coastal community who are customers of Islamic banks and domiciled in the coastal areas of Aceh and North Sumatra. The sample size in this study was 200 respondents (Hair, 2009). Sampling was done by stratified random sampling method with the following proportions: Aceh 40%, North Sumatra 60%. In this study, the sampling method used the census technique. Data analysis used SEM (Structural Equation Modelling) to verify the path of the relationship between variables in this study using IBM SPSS-AMOS software version 22. The results showed that Of the 3 direct hypotheses tested, the effect of Evaluation of bank's compliance with Islamic law on Customer satisfaction, the effect of Evaluation of bank's compliance with Islamic law on Continuous Intention to Use Islamic Banking and The effect of Customer Satisfaction on Continuous Intention to Use Islamic Banking, all of which show a significant effect. On the indirect effect tested, the effect of Evaluation of bank's compliance with Islamic law on Continuous Intention to Use Islamic Banking mediated by Customer Satisfaction has also a significant effect. In contrast to previous studies, which mostly analyzed the reluctance of customers to switch to sharia banks because of the profit motive, this study tries to look at the phenomenon of the reluctance of coastal communities to stay in Islamic banks, especially related to the aspect of compliance with Islamic law which is carried out by Islamic bank practitioners as seen from customer perceptions.

1. Introduction

1.1 Research Background

The growth of the Islamic banking industry in Indonesia is not as expected. The slowdown in the growth of Islamic banks can at least be seen from several indicators including Return on Assets (ROA), profitability and the ratio of financing to funding (finance to deposit ratio / FDR). The monthly ROA (%) growth from August 2019 compared to August 2021 showed a downward trend from 1.64% to 1.38%. Likewise the company's ability to generate profits by comparing them in the same period. Profitability in September 2019 was 1.83 while in September 2021 it also showed a decline to 1.36. Another indicator, namely the financing to deposit ratio (FDR) in the same period also decreased from 80.8 to 79, 56 (Statistics of Islamic Banking – OJK, 2021). This phenomenon was also recognized by the Financial Services Authority (OJK) through the Deputy Commissioner for Banking Supervision of the Financial Services Authority (OJK) who stated that there was a slowdown in the growth of Islamic banks before and after the Corona virus (COVID-19) pandemic (finance.detik.com, 8 November 2021).

The slowing growth in the performance of Islamic banks, both Islamic Commercial Banks (BUS) and Islamic Rural Banks (BPRS) cannot be separated from active the attractiveness of Islamic banking compared to conventional banking The customers are more logical, in determining the choice, which bank to trust as the bank where they invest and conducts their financial transactions. Research conducted by [1] reveals that religious motivation is not the dominant factor considered for choosing a sharia bank, but the bigger motivation is how much profit he gets from the activities he does. the investment [1]. This means that as long as their return on investment in conventional banks is higher than Islamic banks, then they will choose a conventional bank as a place for them to make transactions and financial investments. Another identified reason as the cause of people's reluctance to switch to Islamic banks is the indication that Islamic banks are still carrying out the concept of bank interest even though they are wrapped in different terms [2].

The public's reluctances to access Islamic banking products and services are evidenced by the emergence

* Corresponding author: syafruddin.chan@unsyiah.ac.id

https://doi.org/10.1051/e3sconf/202233905010

of understanding and expressions from several groups such as: Islamic banks are no different from conventional banks, borrowing from Islamic banks or Baitul Mal wat Tamwil (BMT) is more expensive than conventional banks or financial institutions. Another thing identified as the cause of people's reluctance to switch to Islamic banks is the indication that Islamic banks are still carrying out the concept of bank interest even though they are wrapped in different terms [2] . The public's reluctance to access Islamic banking products and services is evidenced by the emergence of understanding and expressions by several groups, such as: Islamic banks are no different from conventional banks, borrowing from Islamic banks or Baitul Mal wat Tamwil (BMT) is more expensive than conventional banks or financial institutions. Another thing identified as the cause of people's reluctance to switch to Islamic banks is the indication that Islamic banks are still carrying out the concept of bank interest even though they are wrapped in different terms [2]. The public's reluctance to access Islamic banking products and services is evidenced by the emergence of understanding and expressions by several groups, such as: Islamic banks are no different from conventional banks, borrowing from Islamic banks or Baitul Mal wat Tamwil (BMT) is more expensive than conventional banks or financial institutions [2].

The purpose of this study is to analyze customer evaluations of the application of the concept of sharia in sharia banking (evaluation of the bank's compliance with Islamic law) which is the expectation of many bank customers in an effort to encourage sustainable intentions in using sharia banking services for their customers to be the main issue raised in the study.

Many previous studies have discussed the determinant factors of repurchase intention towards Islamic bank products, as was done by [4] who found aspects of religiosity and bank reputation as the main determining factors to make customers of Sariah banks not switch to other banks. Then [5] who found the main factor for customers to stay in a bank because of perceived security concerns. Other researchers, namely [6] who reported the ease of use and speed of transaction variables as the dominant variables to increase customer interest in staying in a bank.

In contrast to previous research, this study tries to look at the phenomenon of community reluctance, especially coastal communities, to survive in Islamic banks, especially related to the aspect of compliance with Islamic law which is carried out by Islamic bank practitioners from the customer's perspective. This is important to do because, as previously indicated by previous researchers, there are customers who consider Islamic banking practices no different from conventional banks [2], or borrowing from Islamic banks is more expensive than from Islamic banks [3]. Coastal communities are defined as groups of people who live in coastal areas and whose sources of economic life depend directly on the utilization of marine and coastal resources. They consist of owner fishermen, fishing workers, cultivators of fish and other marine organisms, fish traders, fish processors, suppliers of fishery production facilities. In the non-fishery sector, coastal communities can consist of sellers of tourism services, sellers of transportation services, and other community groups who use marine and coastal non-biological resources to support their lives [7].

The results of this study will be a reference for banking practitioners to redesign their marketing strategy not only to retain existing customers but also to acquire more and more conventional customers switching to sharia. In addition, the output of this research also adds to the repertoire of references and contributes to the enrichment of marketing concepts, especially those related to customer retention in the banking industry. Thus, the market share of Islamic banks in Indonesia, which is only 5% [3], nationally can continue to increase in the future.

1.2 Formulation of the problem

There is still a reluctance of people in Indonesia, including Aceh, to switch to Islamic banks and remain customers of Islamic banks, even for the Muslim community. This is evidenced by the still sluggish growth of Islamic banks as evidenced by the sloping growth curve of return on assets (ROA), finance to deposit ratio (FDR) and Profitability for the period August 2019 to August 2021 (Islamic Banking Statistics, 2021). Several factors suspected to be the cause of the low public interest in switching to Islamic banks such as the bank's compliance with Islamic law that has been identified for further exploration in this study.

1.3 Research Objectives and Types of Outcomes

The purpose of this study is to introduce a conceptual framework that can facilitate the exploration of the causes of people's reluctance to become customers of Islamic banks, both from the aspect of applying Islamic law and ease of transacting using multichannel moderated by perceived compatibility and customer satisfaction.

2. Literature review

2.1. Equations should be centred and should be numbered with the number on the right-hand side.

The discussion on Evaluation of the bank's compliance with Islamic law, cannot be separated from the variable Religious attitude towards Islamic Bank. With regard to retail banking services, previous research has shown that a number of factors influence repurchase intention and customer loyalty [8]. A study conducted by [9] examines these determinants, including the reasons customers intend to do business with Islamic banks in a sustainable manner. [10] also examined customer attitudes towards sharia banking compliance practices. Other researchers examine the evaluation of their compliance in their business practices with Islamic law, because customers consider the main reason to keep doing business is their religiosity motive [11].

Many studies related to the influence of Religious attitude towards Islamic Bank or Evaluation of bank's compliance with Islamic law on Continuous Intention to Use Banking (Repurchase Intention) in Islamic banks have been carried out, including by [10], [8] and [12]. Meanwhile, research related to the influence of Religious attitude towards Islamic Bank or Evaluation of bank's compliance with Islamic law on customer satisfaction has been conducted by [13].

H1: There is an effect of Evaluation of bank's compliance with Islamic law on customer satisfaction

H2: There is an effect of Evaluation of bank's compliance with Islamic law on Continuous Intention to Use Islamic Banking

2.2 Customer satisfaction

The concept of customer satisfaction and repurchase intention conceptually has been widely discussed and is one of the most popular theories in the field of marketing management. [14] suggests that satisfaction is mainly based on emotion rather than reasoning. He defines satisfaction as the fulfilment response of the consumer'. His statement is based on consumers' evaluations of the after-sales services of service providers and whether they meet the level of customer satisfaction or not. [15] on the other hand measure service and service satisfaction through seven components such as; interest, pleasure, surprise, anger, wise choice, and doing what is right. Because customer satisfaction is primarily based on emotion as previously pointed out by [14] it is natural that each person experiences emotions in different ways and therefore also assesses customer satisfaction in different ways. Research related to the effect of customer satisfaction on repurchase intention has been carried out by [16] and [10]

H3: There is an effect of customer satisfaction on Continuous Intention to Use Islamic Banking

2.3 Continuous Intention to Use Islamic Banking

According to [17] repurchase is described as a customer's real action in buying or reusing the product. Once, a customer buys an item, it has the potential to repeat the purchase [16]. This means that customers repeatedly consume similar services or products from similar sellers. While repurchase is an actual act, repurchase intention indicates the customer's decision to engage in future activities with retailers or suppliers [18]. Customer repurchase intention has been conceptualized in several recent studies based on two aspects namely repurchase intention and repurchase intention to engage in promotions and positive recommendations (referrals) [16].

H4: There is an effect of Evaluation of bank's compliance with Islamic law on Continuous Intention to Use Islamic Banking through customer satisfaction

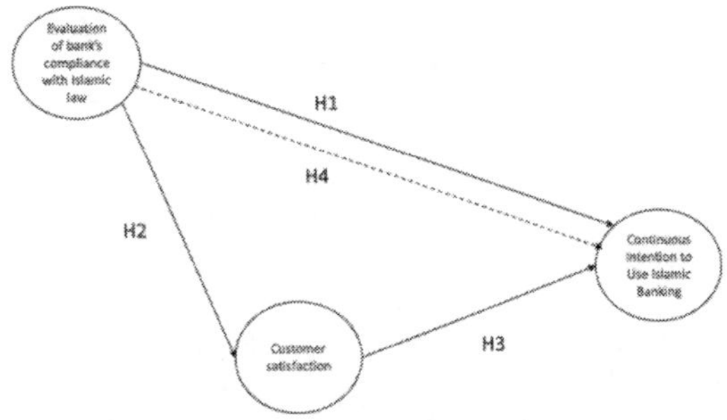

Fig. 1. Research Model

3. Research methods

3.1 Research Instruments

All measurement items were taken from previous studies to ensure validity; however, slight changes to the statement were made to suit the current analysis. The eight-item of Evaluation of bank's compliance with Islamic law was adapted from [9]. Four items adopted from [19] were used to measure customer satisfaction. Then five items are used to measure Continuous Intention to Use Islamic Banking was adopted from [16]. A questionnaire with a 5-point Likert scale was used to collect data. In addition, this study uses in-depth interviews with several sources (informants) who represent the elements of providers and customers.

3.2 Sample Design and Data Collection

The population in this study is the coastal community who are customers of Islamic banks and domiciled in the coastal areas of Aceh and North Sumatra. The sample size in this study was 200 respondents [20]. Sampling was done by stratified random sampling method with the following proportions: Aceh 40%, North Sumatra 60%. The proportion of North Sumatra is greater because the number of coastal communities who are customers of Islamic banks in North Sumatra is also greater than Aceh

3.3 Data analysis

The data analysis technique in this research is descriptive and verification. Descriptive analysis was conducted to assess the demographic profile of the respondents and the internal consistency of construction. While the verification analysis uses SEM (Structural Equation Modelling) to verify the path of the relationship between the variables in this study. In addition, the SEM analysis software is IBM SPSS-AMOS version 22.

4. Results and Discussion

4.1 Characteristics of Respondents

In terms of age, the most respondents came from the age group 31-35 years, namely 42 (34.7%). From gender, there are more men than women, namely 83 (68.6%).

Then from the education group the most are those with undergraduate education, namely 60 (49.6%). The majority of their jobs are in the fishery sector (60%), the rest (40%) work outside the fisheries sector such as trade, culinary and other tourism services.

4.2 Validity with Measurement Model

An indicator is said to be valid if it has a standardized loading factor value > 0.50. For indicators that do not meet the requirements will be eliminated from the model. From the calculation results, it turns out that there are 5 indicators that do not meet the requirements because they have a loading factor number below the required one, namely 0.5. So indicators a1, a2, a16 and a23 must be eliminated.

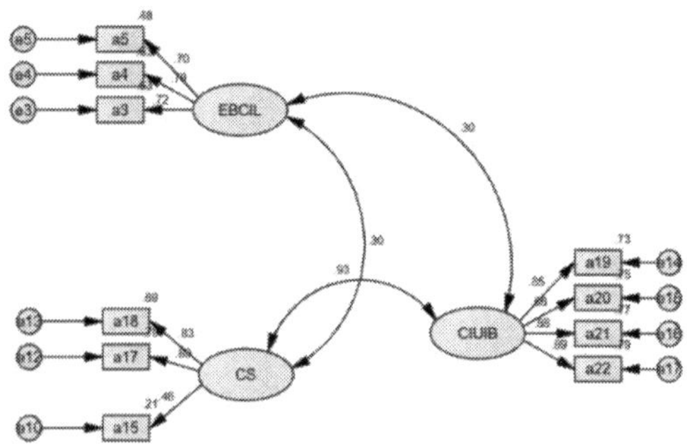

Fig. 2. Measurement Model

The new loading factor after elimination can be seen in the table below.

Table 1. New Loading Factor

Item		Var	Estimate
a3	<---	EBCIL	.725
a4	<---	EBCIL	.793
a5	<---	EBCIL	.695
a15	<---	CS	.463
a17	<---	CS	.887
a18	<---	CS	.832
a19	<---	CIUIB	.853
a20	<---	CIUIB	.865
a21	<---	CIUIB	.878
a22	<---	CIUIB	.891

Based on the table above, it can be explained that all the variable items used in this study are all valid, because they have a loading factor number > 0.50.

4.3 Reliability Test

The reliability test intended in this study is to determine the extent to which the measurement results remain statistically consistent, namely by calculating the magnitude of the composite reliability of the data based on the estimated output obtained using Cronbach alpha. The results are as described in Table 4.3 which shows that the instrument in this study is reliable because its reliability coefficient value is greater than 0.60 [21].

Based on the reliability analysis, it can be seen that the alpha for each respondent's perception variable can be seen from several variables the bank's compliance with Islamic law is 74.5%, Customer satisfaction is 79.8%, and Continuous Intention is 88.3%. Thus the reliability measurement of the research variables shows that the reliability measurement meets the requirements of Cronbach Alpha (CA) where the CA coefficient value is at least or greater than 60 percent.

Table 2. Reliability Using Cronbach Alpha (CA)

No	Variable	CA	Items	Desc.
1	Evaluation of the bank's compliance with Islamic law	.745	5	Reliable
4	Customer satisfaction	.798	4	Reliable
5	Continuous Intention to Use Islamic Banking	.883	5	Reliable

4.4 Hypothesis Testing

Hypothesis test verification in this study was conducted to test and analyze the effect of Evaluation of the bank's compliance with Islamic law, Customer satisfaction, and Continuous Intention to Use Islamic Banking. The verification hypothesis testing consists of testing the direct influence hypothesis and testing the indirect effect hypothesis. The results of testing the direct influence hypothesis can be seen in the following figure and table.

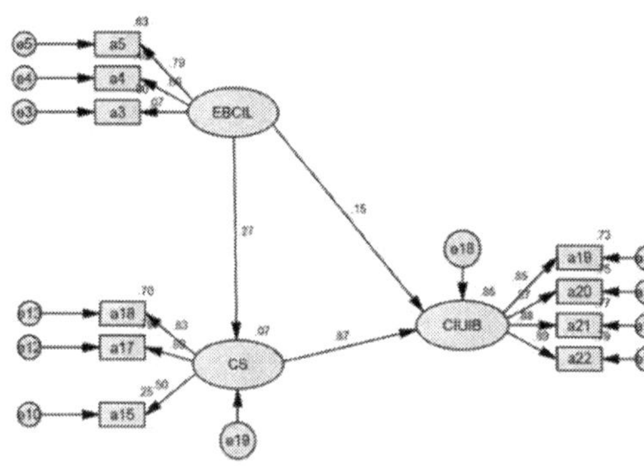

Fig. 3. Structural Model

Table 2. Hypothesis Testing

Variables			CR	P	Beta
CS	<---	EBCIL	2.406	.016	.269
CIUIB	<---	EBCIL	8,620	***	.153
CIUIB	<---	CS	8,629	***	.871

EBCIL: Evaluation of the bank's compliance with Islamic law
CS: Customer satisfaction
CIUIB: Continuous Intention to Use Islamic Banking
H1: The effect of Evaluation of bank's compliance with Islamic law on Customer satisfaction

Testing the effect of Evaluation of bank's compliance with Islamic law on Customer satisfaction shows a CR value of 2.406 and with a probability of .016. The two values obtained have met the requirements for the acceptance of Ha, namely the CR value greater than 1.96 and the probability less than 0.05. The magnitude of the coefficient of the effect of Evaluation of bank's compliance with Islamic law on Customer satisfaction is 0.269

H2: The effect of Evaluation of bank's compliance with Islamic law on Continuous Intention to Use Islamic Banking

Testing The effect of Evaluation of bank's compliance with Islamic law on Continuous Intention to Use Islamic Banking shows a CR value of 8,620 and with a probability of *** Thus it can be stated that the effect of Evaluation of bank's compliance with Islamic law on Continuous Intention to Use Islamic Banking is significant. The magnitude of the coefficient of the effect of Evaluation of bank's compliance with Islamic law on Continuous Intention to Use Islamic Banking is 0.153

H3: The effect of Customer Satisfaction on Continuous Intention to Use Islamic Banking

Testing The effect of Customer Satisfaction on Continuous Intention to Use Islamic Banking shows a CR value of 8,629 and with a probability of ***. The two values obtained have met the requirements for acceptance of Ha. The magnitude of the coefficient of the effect of Customer Satisfaction on Continuous Intention to Use Islamic Banking is 0.871

H4: The effect of Evaluation of bank's compliance with Islamic law on Continuous Intention to Use Islamic Banking through Customer Satisfaction

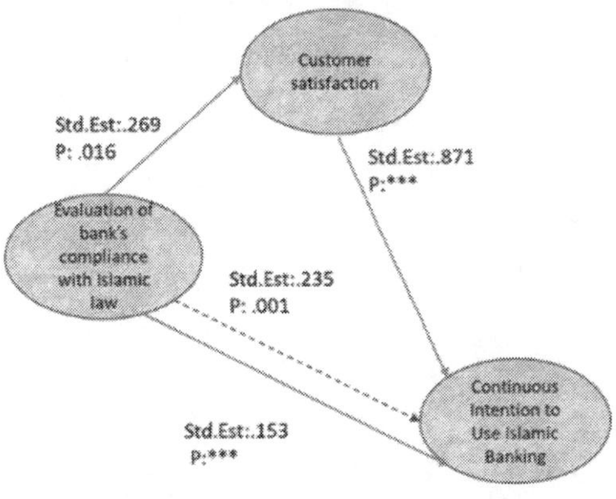

Fig. 5. The effect of Evaluation of bank's compliance with Islamic law on Continuous Intention to Use Islamic Banking through Customer Satisfaction

Testing the influence of Evaluation of bank's compliance with Islamic law on Continuous Intention to Use Islamic Banking through Customer Satisfaction shows a probability of .002. The p value obtained has met the requirements for acceptance. Ha. The coefficient of the effect of Evaluation of bank's compliance with Islamic law on Continuous Intention to Use Islamic Banking through Customer Satisfaction is 0.235 or 23.5%.

Because the effect of Evaluation of bank's compliance with Islamic law on Continuous Intention to Use Islamic Banking is significant and the effect of Evaluation of bank's compliance with Islamic law on Continuous Intention to Use Islamic Banking through Customer Satisfaction is significant, we can say that the role of Customer Satisfaction in mediating these two variables is Partial Mediating.

5. Discussion

Of the 3 direct hypotheses tested, the effect of Evaluation of bank's compliance with Islamic law on Customer satisfaction, the effect of Evaluation of bank's compliance with Islamic law on Continuous Intention to Use Islamic Banking and The effect of Customer Satisfaction on Continuous Intention to Use Islamic Banking, all of which show a significant effect. The significance that occurs in this direct test indicates that there is a match between the research model developed and the facts in which this research was conducted.

Based on the results or findings presented in the previous section, it means that Evaluation of bank's compliance with Islamic law and Customer satisfaction are indeed a determinant variable of improving the intention of bank's customer to continue using Islamic banking. So that the management can use these two variables as tools to improve the intention of the bank's customer to continue using Islamic banking. This is in line with research conducted by [22]. The phenomenon of the tendency of customers to switch to other banks should be addressed because of the low level of satisfaction. Low satisfaction means that customers' expectations of banking services are not fully obtained. From the research conducted by [3] it was revealed that there was dissatisfaction of Islamic bank customers with the level of profit they got after investing in Islamic banking products. On the other hand, the costs incurred when using transaction facilities provided by Islamic banks are also relatively high compared to conventional banks. Not to mention the cost of borrowing funds that they have to pay, the cost of funds is also relatively higher than conventional banks. The various facts that they mostly feel are below their expectations so that it has an impact on their dissatisfaction with the offerings made by Islamic banks. Of course this will have an impact on their desire to switch to competitors.

In order to support the strongest increase in the intention to use Islamic banking services, it can be done through the increasing of customer satisfaction, because this variable has the largest magnitude number compared to the other variables, namely 87.1% compared to the bank's compliance with Islamic law which is only 26.9% On the indirect effect tested, although it showed significant results, the role of customer satisfaction in mediating the effect of Evaluation of bank's compliance with Islamic law on Continuous Intention to Use Islamic Banking is quite

https://doi.org/10.1051/e3sconf/202233905010

good. The path of the effect of Evaluation of bank's compliance with Islamic law on Continuous Intention to Use Islamic Banking mediated by Customer Satisfaction has a coefficient of 23.5%.

6. Conclusion

From the discussion that has been carried out in the previous section, research findings related to hypothesis testing were carried out either directly or indirectly. In principle, the model built is based on a literature review and preliminary research based on references related to the field under study, in this case improving the intention of Islamic bank customer to continue using Islamic bank services. . Based on the literature review, a research model was built as presented in the previous section in this research paper. And in fact, after going through a series of tests, it turned out that all the hypotheses that were built involving the Bank's compliance with Islamic law, Customer satisfaction and Continuous Intention to Use Islamic Banking variables proved significant and had a positive effect.

For further research development, variable synthesis can also be added to create new variables from the results of the synthesis of two or more variables in this study. The synthesis of these variables must of course be carried out based on existing field facts combined with theoretical concepts related to the variables to be synthesized, which can influence customers to use goods or services from the Islamic banking.

References

1. C. Erol, El-Bdour R, Int J Bank Mark. (1989)
2. A. Al-Mawarid, J Islam Law **10** (2003)
3. S. Bahri, Iqtishaduna, J Ilm Ekon Kita. **9**, 1 (2020)
4. N. Tara, M. Irshad, M.R. Khan, M. Yamin, M. Rizwan, J Public Adm Gov **4**, 3 (2014)
5. H. S. Yoon, L. Occeña, J Comput Inf Syst **54**, 3 (2014)
6. A. L. N. Chima, V. K. Chikasanda, The Impact of Internet Banking on Service Quality Provided by Commercial Banks [Internet]. Lecture Notes of the Institute for Computer Sciences, Social Informatics and Telecommunications Engineering. (Springer International Publishing, 2014)
7. C. T. Karubaba, G. B. Dietriech, V. P. Nikijuluw, Indones J Coast Mar Resour **3**, 3 (2001)
8. N. Alam, M. S. Rajjaque, Shariah-Compliant Equities: Empirical Evaluation of Performance in the European Market during Credit Crunch. Islamic Finance. (Springer International Publishing, 2016)
9. H. Kamiyama, K. Kashiwagi. J Financ Serv Mark **24**, 3-4 (2019)
10. H.M. Awan, K. Shahzad Bukhari. J Islam Mark **2**, 1 (2011)
11. N. Souiden, M. Rani, Int J Bank Mark **33**, 2 (2015)
12. S.N. Maswadeh, Procedia Econ Financ **23** (2015)
13. K. Naser, A. Jamal , K. Al-Khatib, Int J Bank Mark **17**, 3 (1999)
14. R. L. Oliver, Customer Satisfaction with Service [Internet]. Handbook of Services Marketing & Management. SAGE Publications, Inc.; p. 247–54.
15. J. J. Cronin, M.K. Brady, G. T. M. Hult, J Retail **76**, 2 (2000)
16. N.K. Mu'azu Saidu Badara, N. Mat, A.M. Mujtaba, A.N. Al-Refai, A.M. Badara, F.M. Abubakar, Management **3**, 1 (2013)
17. H. Herjanto, M. Amin, Int J Bank Mark **38**, 6 (2020)
18. J. Kandampully, D. Suhartanto, Int J. (2000)
19. R.L. Oliver, J Mark [Internet] **63** (1999)
20. J.F. Hair, W.C. Black, B.J. Babin, R.E. Anderson, R.L. Tatham, Multivariate data analysis 6th Edition (New Jersey, Prentice Hall, 2006)
21. L.R. Kahle, N.K. Malhotra, J Mark Res **31**, 1 (1994)
22. M. Brady, J. Cronin, R. Brand, J Bus Res, (2002)

Alternative kits and worksheets on light and optics as science experimental devices at home for middle school student

Nana Mardiana*, and *Sheila* Fitriana

Physics Education Department, Islamic University of North Sumatera, 20143 Medan, Indonesia

Abstract. The aims of this research concerned to develop an alternative kits and worksheet on light and optics as science experimental devices at home for middle school student. The method used is a research and development. Alternative kits and worksheet have been tested for feasibility through empirical testing and validation. The research data were analyzed by calculated the index of each indicator instrument experiment. The result showed that science experimental devices is appropriate for home learning activity according to evaluation by a material expert and peer reviewers as well as according to result of middle school students try outs with respective high category.

1 Introduction

Teaching science in schools is divided into two major parts, namely science as a product and science as a process. The context of science as a product is in the teaching of facts, theories, principles and natural laws; while science as a process is the development of students' abilities in scientific methods and problem solving in science. For science as a process, teaching through laboratory practicum is an activity of applying the scientific method by students. There are many claims that laboratory practicum activities can improve students' critical attitude, science process skills, or scientific attitude [1]. Besides that, science learning is also expected to provide psychomotor skills, scientific attitude skills (affective), understanding, habits and appreciation in finding answers to a problem [2].

The most suitable learning method for developing process skills is the experimental method. The experimental method is a method in which students experience and prove for themselves something they have learned [3]. The description above shows the importance of conducting experiments for students, but in general, the implementation of experiments that take place in school laboratories is rarely carried out by teachers. Based on the results of Jeperis' research, the reasons why teachers rarely do practicum are as follows: 1) unavailability of laboratories in schools, 2) unavailability of science practicum tools, 3) teachers do not have a science educational background, 4) teachers rarely attend training on the use of science laboratories, 5) the implementation of training on the use of laboratories is rarely carried out, 6) teachers lack understanding of the use of labor, and 7) there are no laboratory assistants. In addition, based on Sumintono's research to carry out practicum in the laboratory in supporting the learning process, there are several difficulties, including; Insufficient tools and materials, lack of time, limited space, and the absence of a laboratory [1].

Practicum become less effective, due to inadequate lab space, incomplete laboratory equipment, less effective practice time given the absence or lack of equipment preparation (no laboratory assistant), irregular practice due to the absence of a practice schedule. However, sometimes problems occur due to insufficient time with limited materials and equipment, which makes it difficult to apply the experimental method in schools, especially at the middle school level. In order to conduct the practicum, experimental devices are needed which consist of tools and materials used for experiments as well as guidelines. For this reason, experimental devices are needed which covered in the material that student learned during their study without being constrained by non-existent laboratories and unavailability of tools. One solution is to utilize a set of tools sourced from the environment that can be used as learning media. This set of tools is then referred to as an alternative kit.

The alternative kit in question is a set of experimental tools that are practical and can be taken home so that they can overcome the problem of limited time and space. The development of this alternative light and optical kit is also accompanied by the development of effective and efficient Student Worksheets (LKS), so that teachers can provide understanding to students. Light and optics are chosen because light and optics are one of the physical concepts that have many applications in nature and therefore a device is needed to make observations and provide an explanation of the concepts of light and optics. The objectives of this research are 1) assembling tools and materials and packaging them in an Alternative Kit for the topic of light and optics, 2) testing the validity of the Alternative Light and Optics

* Corresponding author: mdna.nana1@gmail.com

Kit, 3) testing the validity of the LKS Alternative Kit for Light and Optics.

2 Research Methods

The form of research is used research and development method, because it aimed to find and develop a new or existing prototype in order to improve and develop so that more productive, effective and efficient results are obtained.

Research and development is a method used to develop and test a product. Research and development is a research approach to produce new products or improve existing products. There are three steps of research and development; a preliminary study to examines theory and observes existing products or activities; developing new products or activities; testing and validating new products or activities [4]. The research flow can be seen in Figure 1.

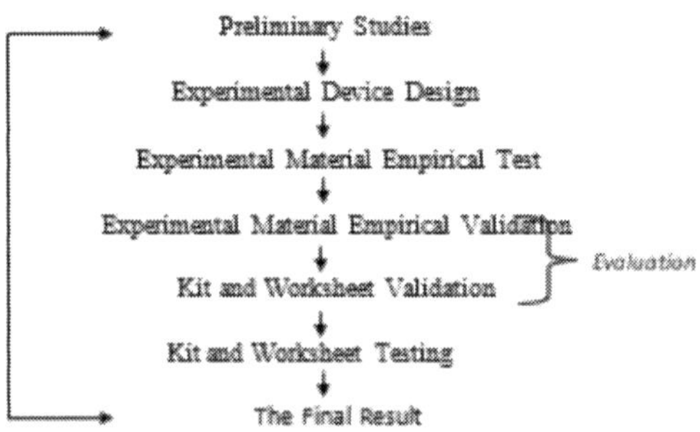

Fig. 1. Research Flow of Light and Optical Alternative Kit Development

The beginning of this research is start from analyze the constraints on the needs of students in the use of standard practicum kits used in schools as well as information related to the problems to be studied. Based on information from experts and literature studies, Experimental Device Design, is done by determining the experimental topics to be developed. The design at this stage, based on the analysis of concepts, is relevant to the characteristics of the light experimental device.

Development of Experimental Equipment in the form of preliminary preparation of experimental devices and light worksheets by selecting tools and materials to be used as alternative media for these devices, as well as designing the appearance of kits and worksheets in accordance with the development objectives. Experimental material empirical test, in this stage the alternative device material is tested and the best is selected in showing the expected physical symptoms so that it can be used to carry out light experiments. To validate the experimental devices, we asked four senior lecturer with a list of questionnaires in the assessment list. On the assessment sheet there are indicators that will be assessed by the validator along with a suggestion column. Repairs and improvements to the device are carried out based on the suggestions given.

Testing of Experimental Equipment, at this stage the implementation of testing of experimental devices is carried out through the implementation of limited practicum trials by 10 students of middle school level. The selection of research subjects was based on consideration of the diversity of academic abilities. The practicality test process was continued with the activity of filling out the practicality test assessment sheet for the light experiment.

Evaluation, in this study, a formative evaluation took place at each of the four stages above because it aimed at the need for revision or improvement of the experimental device. The final result is an experimental device consisting of an alternative kit and light and optical worksheets in sizes that can be brought home by individual students who have been tested for feasibility.

The data analysis technique used in this research is descriptive analysis, namely by calculating the index of each LKS validation indicator and testing the practicality of the experimental equipment. The validity of the LKS is determined by the score of the validation results by expert lecturers. While the practicality test of the experiment is determined by the assessment score given by the students after doing the practicum.

Analysis of the validation data using the following steps:

1) Summing up each indicator of the validation questionnaire.
2) Questionnaire assessment categories by validators and practicality tests by students using the guidelines as shown in Table 1.
3) Find the average of each indicator of the validation questionnaire.
4) Find the average of the whole validation questionnaire.
5) Determine the average category of indicators based on the table using a Likert scale.

Table 1. Questionnaire Assessment Categories

No	Category	Skor
1	Strongly agree	4
2	Agree	3
3	Disagree	2
4	Do not agree	1

Table 2. Categories of Validation and Practicality by Practitioners

No	Average Score	Category
1	$> 3{,}25 - 4$	Very high
2	$> 2{,}5 - \leq 3{,}25$	High
3	$< 1{,}75 - \leq 2{,}5$	Low
4	$1 - \leq 1{,}75$	Very low

To determine the average per indicator used the formula:

Average per indicator = (Number of Scores Per Indicator)/(Number of Validators) (1)

3 Results and Discussion

3.1 Result

Preliminary studies. Based on the information and literature study, the researcher obtained information that laboratory practicum activities can improve students' critical attitude, science process skills, or scientific attitude. Besides that, science learning is also expected to provide psychomotor skills, scientific attitude skills (affective), understanding, habits and appreciation in finding answers to a problem. The most suitable learning method for developing process skills is the experimental method. The experimental method is a way of presenting lessons in which the subject learns to experiment by experiencing and proving himself something learned.

Experimental Device Design, based on the results of the preliminary study, the learning device developed in this research is the Light and Optical Alternative Kit and Student Worksheet which is effective and efficient, which is able to solve the problems described in the preliminary study. Development of experimental devices, at this stage the development of worksheets is carried out which is adapted to experiments or activities designed by researchers so that students can easily observe light symptoms in a simple way but still can master and understand the concept of light. During this stage, several revisions were made as needed.

The results of the assessment of alternative kits and worksheets can be seen in the following description. From the assessment of the Alternative Light Kit by the Validator, the results of the data analysis are obtained in Table 3.

Table 3. Validation Results of Alternative Light and Optical Kits

No.	Indikator							
	1	2	3	4	5	6	7	8
Average	3,47	3,5	3,72	3,78	3,09	3,97	4	3,35
Conc	VH	VH	VH	VH	VH	VH	VH	VH

Based on the analysis table above, it can be seen that the alternative kit developed in this study can be declared valid for experiments of light material. The validation, the results are summarized in Table 3. Based on the recapitulation table for assessing the validity of the worksheets above, it can be seen that the worksheets is valid to be used by middle or junior high students and equivalent to show the symptoms contained in the light material. Testing of experimental devices is carried out by students by filling out an experimental practicality test assessment sheet. Based on the recapitulation table of the worksheets practicality assessment above, it can be said that the alternative kit and light worksheets have a very high level of practicality, and high for indicator 2.

Based on the analysis table above, it can be seen that the alternative kit developed in this study can be declared valid for experiments of light material. The validation, the results are summarized in Table 3. Based on the recapitulation table for assessing the validity of the worksheets above, it can be seen that the worksheets is valid to be used by middle or junior high students and equivalent to show the symptoms contained in the light material. Testing of experimental devices is carried out by students by filling out an experimental practicality test assessment sheet. Based on the recapitulation table of the worksheets practicality assessment above, it can be said that the alternative kit and light worksheets have a very high level of practicality, and high for indicator 2.

Fig. 2. Alternative light and optical kit

3.2 Discussion

After evaluating the alternative kits and light worksheets, both validators and practitioners generally gave a positive assessment, so that no significant problems were found. This is shown by the assessment scores ranging from high to very high, as well as enthusiasm in providing feedback and suggestions on various aspects of developing alternative kits and light worksheets.

The alternative kits and light worksheets developed are expected to overcome the problems described in the preliminary study. Some of the advantages possessed by this device are; The kit is developed by using tools and materials that are easily obtained, the existing experiments are experiments designed to be carried out by individuals so that they do not require many people/groups, the device is designed in a small size so that it can be brought home as a learning medium that can increase interest and knowledge students, with kits that can be taken home, can save time for practicum in schools that spend a lot of study time, the device is accompanied by worksheets that help students develop their process skills. As an alternative media if the school does not have a national standard light kit. Helping teachers in learning.

In the implementation of this research, from the results of the practicality assessment by the practitioner, it was obtained that in several aspects, the alternative kits and worksheets developed could only show symptoms or inculcate basic concepts regarding light and optics but have not been able to provide a deeper understanding of light material at the junior high school level. /MTS equivalent. Some experiments that are considered weak in this alternative light kit are light propagation, light reflection on a curved mirror, light refraction on the lens. Propagation of Light, experiments can only show the nature of light propagates in a straight line and penetrates a clear object. Reflection of light in a curved mirror, experiments have not been able to provide an overview of the real or virtual properties that exist at the time of image formation by a curved mirror (spherical).

Refraction of Light in Lenses, the experiment of light refraction on the lens only shows the nature of refracting rays but does not clearly show the nature of the image formed because an image of the direction of the light is not obtained which determines the nature of the image.

4 Conclusion and Suggestions

The experimental device developed in this study consisted of alternative kits and Student Worksheets (LKS) on the subject matter of light which were packaged in small sizes so that they could be brought home as learning media. Based on the results of data collection and analysis conducted in this study, the following conclusions can be drawn; The experimental equipment developed in this study has gone through a feasibility test (empirical test of experimental device material) and has been declared feasible as an experimental device that can show symptoms in light material. This experimental device has been validated by four experts and is declared suitable for experiments that show the symptoms contained in light matter. The worksheet which was developed as an experimental guide has been validated by four experts and declared eligible with very high criteria, but for display indicators it is quite attractive with high criteria. The experimental set developed in this study had very high practicality of experiments by practitioners for all experiments, and with high scores on the worksheets for indicators of the language used being easy to understand.

The experimental device developed was appropriate as a learning medium, however, because this device was only tested in a limited group of students, it is recommended that this device can be tested again in schools during the learning process. So that all students can use it and get hands-on learning experience and proper concept understanding from this device.

References

1. Praktikum Laboratorium: Perspektif dari guru-guru Sains SMPN di kota Cimahi, *Jurnal Pengajaran MIPA*, Fakultas Pendidikan Universitas Teknologi Malaysia, Johor bahru Malaysia. (2010)
2. Trianto, *Model Pembelajaran Terpadu*, Bumi Aksara, Jakarta. (2010)
3. Irianti, M., *Pengembangan Program Pengajaran Fisika*, Cendekia Insani, Pekanbaru. (2009)
4. Sukmadinata, N., *Metode Penelitian Pendidikan*, PT Remaja Rosdakarya, Bandung. (2005)
5. Arsyad, A., *Media Pembelajaran*, Rajagrafindo Persada, Jakarta. (1996)
6. Bahri, S., *Stratregi belajar Mengajar,* Rineka Cipta, Jakarta. (2006)
7. BSNP, Standar Isi Untuk Satuan Pendidikan Dasar dan Menengah, BSNP, Jakarta. (2006,)
8. Depdiknas, *Standar Kompetensi Mata Pelajaran Fisika*, Balai Pustaka, Jakarta. (2006)
9. Djamarah, S. dan Aswan Z., *Strategi Belajar Mengajar*, Rineka Cipta, Jakarta. (2006)
10. Giancoli, D.C., *Fisika Jilid 1*, Erlangga, Jakarta. (2001)
11. Irianti, M., *Pengembangan Program Pengajaran Fisika*, Cendekia Insani, Pekanbaru. (2009)
12. Irianti, M., 2006, *Dasar-Dasar Pendidikan MIPA*, Cendekia Insani, Pekanbaru. (2009)
13. Jeperis, Studi kompetensi guru IPA Fisika SMP Negeri Kota Pontianak Dalam melakukan Praktikum di Laboratorium. (2009) http://jeperis.wordpress.com/.(Februari,2012)
14. Muslich, M., *KTSP Dasar Pemahaman dan Pengembangan*, Bumi Aksara, Jakarta. (2007)
15. Mahmudin, *Pendekatan Keterampilan Proses dalam Pembelajaran IPA* (2009) [online].http://mahmuddin.wordpress.com/2009/11/05/pendekatan-keterampilan-proses-dalam-pembelajaran-ipa/
16. Prastowo, A., *Panduan Kreatif Membuat Bahan Ajar Inovatif*, Diva Press, Yogyakarta. (2011)
17. Roestiyah, *Strategi Belajar Mengajar,* Rineka Cipta, Jakarta. (2008)
18. Sanjaya, W., Strategi Pembelajaran Berorientasi Standar Proses Pendidikan, Kencana, Jakarta. (2007)
19. Silvinia, *Pendidikan IPA*, Departeman Pendidikan Nasional, Jakarta. (2005)
20. Sudjana, N., *Media Pengajaran*, Sinar Baru Algesindo, Bandung. (2003)
21. Sudrajad, A., *Pendekatan, Strategi, Metode, Teknik, dan Model Pembelajaran* (2008), http://akhmadsudrajat.wordpress.com
22. Susilana, R. dan Cepi R., *Media Pembelajaran*, CV Wacana Prima, Bandung. (2007)
23. Sutrisno, Laboratorium Fisika Sekolah, (2006) http://file.upi.edu/Direktori/FPMIPA/JUR._PEND._FISIKA/195801071986031SUTRISNO/Perkuliahan/Bahan_ajar/Laboratorium_Fisika_Sekolah_I/LABORATORIUM_FISIKA_SEKOLAH.pdf
24. Villa, *Pengertian dan Fungsi Pendekatan*, (2011) http://www.vilila.com/2011/11/pengertian-dan-fungsi-pendekatan.html#ixzz1ybGcXjLI.
25. Wena, M., Strategi Pembelajaran Inovatif Kontemporer, Bumi Aksara, Jakarta. (2009)

https://doi.org/10.1051/e3sconf/202233906002

Learning management strategy new normal era

Ratna Soraya[1], *Nana* Mardiana[2*], *Nani* Mardiani[3], *Rika* Kartika[1], and *Wildawani* Siregar[4]

[1]Indonesian Language Department, [2*]Physics Education Department, [4]Chemistry Education Department, Islamic University of North Sumatera, 20143 Medan, Indonesia
[3]Physics Education Department, SMAN 3 Pujud, 28983 Rokan Hilir, Riau, Indonesia

Abstract. The Covid-19 pandemic has an impact on various sectors of life, including in learning activities. The government as a decision maker makes policies to carry out online learning activities in order to slow the spread of Covid-19. Principals as leaders in learning activities require the right strategy in implementing online learning that supports government policies. The strategy used in school management in the era of new habits due to the Covid-19 pandemic is expected to be able to maintain the quality of learning activities as in the pre-pandemic period in the schools they lead. The purpose of this study was to determine the principal's leadership strategy in dealing with learning in the era of new habits. The method used was a qualitative method. From the results of the study, it can be concluded that the principal's leadership strategies in dealing with learning in the new era of habits is; 1) to make online learning policies in schools by involving relevant teachers and school officials; 2) to improve the facilities and quality of ICT infrastructure in learning activities in schools; 3) to improve the quality of human resources in integrating IT for learning activities; 4) to communicate with parents or guardians of students; 5) to provide understanding to students about online learning activities; 6) to be ready and open in dealing with the changes in learning activities in the future.

1 Introduction

The pandemic of Covid-19 has an impact on various sectors of life, including in learning activities. The government makes policies to carry out online or online learning activities in order to slow down the spread of Covid-19. The implementation of learning from home (Learn from home) or online issued by the Government [1]. An Emergency Period for the Spread of Corona Virus Disease (Covid-19) through a circular by the Ministry of Education and Culture Number 15 of 2020 concerning Guidelines for Organizing Learning from Home. The emergency spread of Corona Virus Disease (COVID-19), so that it can fulfill the rights of students to obtain educational services during the implementation of learning is carried out from home.

Principals as leaders in learning activities need the right strategy in carrying out the online learning that supports government policies. The strategies were used in managing schools in the new era of habits due to the Covid-19 pandemic was expected to be able to maintain the quality of learning activities as they were before the pandemic in the schools they lead. The responsible of principal for the implementation of learning activities in the midst of a pandemic crisis situation so that learning takes place online or online. This sense of security and comfort must be felt by teachers, students, and parents. This includes security and comfort during the response period of Covid-19 emergency [2]. The threat or crisis currently being faced is not entirely bad for schools. On the contrary, it can actually motivate schools to be better in the future [3].

Based on research, the best of three leadership practices for navigating unexpected adaptive challenges such as the corona virus pandemic [4, 5]. First, by adopting kind of servant leadership, which can emphasize empowerment, collaboration, engagement, and, academic leaders with emotional intelligence and stability should put the interests of others above their own. Second, academic leaders should distribute the leadership of responsibilities to a network of teams and improve the quality of decisions made in crisis resolution. Third, the teaching and learning process should be designed by the principal as the highest leader in the school, started from preparation, implementation, monitoring, and evaluation in other hand the teachers and students can be service users could follow the teaching and learning process well. The school management which includes, independence, responsibility, accountability, transparency, and justice affect the loyalty and satisfaction of school personnel [6]. Various managerial and operational activities were aimed at supporting the achievement of the vision, mission, and strategies that were prepared so the education will be able to take place effectively and efficiently in school management [7].

Based on the explanation above, this study is considered important to be carried out with the intent and purpose of this study, namely to determine as the leadership of principal strategy deal with learning in the era of new habits.

* Corresponding author: mdna.nana1@gmail.com

2 Methodology

This study used a descriptive qualitative approach, which is accurately describe existing phenomena found in the field, then describe them in a narrative manner used qualitative methods through critical analysis.

The research method was used descriptive qualitative research method. the research was carried out in natural conditions *settings*; also called the ethnographic method, can affect qualitative research methods were often called as naturalistic research methods, because at first this method was used more widely for research in the field of cultural anthropology; called the qualitative method, because the data were collected and analyzed more qualitative in nature [8]

According to Denzin and Lincoln (1994) stated that qualitative research was a research that used the intention of interpreting the phenomena which occurred and carried out by involving some of the existing methods. Erickson (1968) stated that qualitative research looks for and describe in a narrative way the activities could impact of the actions taken on their lives.

3 Results and Discussion

The results of the research data analysis, it was found that several strategies were implemented by the principal in carrying out online learning activities during the new habit era due to the Covid-19 pandemic, including; 1) to make online learning policies in school by involving relevant teachers and school officials; 2) to increase the facilities and qualities of ICT infrastructure in learning activities in schools; 3) to make the higher quality's human resources in integrating IT for learning activities; 4) to communicate with parents or guardians of students; 5) to provide understanding to students about online learning activities; 6) to be ready and open with the changes in learning activities in the future.

3.1 Online Learning Policy in Schools

Through a circular by the Ministry of Education and Culture, Number 15 of 2020 concerning "Guidelines for Organizing Learning from Home in an Emergency Period for the Spread of Corona Virus Disease (Covid-19)". In other hand to fulfill the right of students could obtain educational services during Corona Virus Disease (COVID-19), learning is started from home. Schools are given the freedom to manage teaching and learning activities through online facilities [9].

The principals as leaders in learning activities need the right strategy in carrying out the online learning that supports government policies. The strategies used in managing school in the new era of habits due to the Covid-19 are expected to maintain the quality of learning activity as they were before the pandemic. The principal has responsibility for the implementation of learning activity in the midst on a pandemic crisis situation so that learning takes place online [10]. The leader must have criteria as someone who can lead in educational organizations.

3.2 Online Learning Policy in Schools

Increasing the procurement of supporting facilities and infrastructure such as computer laboratory, internet network, learning resource, teaching aid, and ICT-based learning media, is an effort that can be made by school principals increasing the facility and quality of infrastructure used in school.

Principal as a manager as well as a school leader has the authority to develop ICT in schools [11]. The leaders who follow technological developments can influence, encourage, guide, and direct others in implementing and developing education so that professional principals will understand the needs of the school.[12].

3.3 Schools Improvement of Human Resources in Integrating IT

Training and educational programs can be used to increase teacher competence's in ICT learning. The role of leadership to advance schools is to provide ICT training for teachers to know how to use it [13].

3.4 Building Communication with Parents and/or Guardians of Students at Home

Communication leader must be good at communicating well so that the information obtained is conveyed properly to members of the organization and the public correctly so as not to cause misunderstandings. The fourth collaboration of the leadership ability is collaborating well.

3.5 Providing Understanding of Online Learning to Students

The principal with the help of teachers and other school officials provide understanding and motivation to students are able to have online learning activity at home well with parents and or guardians during the pandemic Covid-19. Research [14] stated that the important elements that determine digital leaders are organizational goals, evaluating task and work results together with team members, making an open atmosphere with the effect of a collaborative atmosphere to handle conflict situations, communication and innovation.

3.6 Readiness and Openness in Facing Future Learning

The next strategy is the adaptability of school principals to face the development to deal with thing that are going to happen in the future, especially during the crisis due to the current pandemic to improve the quality of education (Y. Wulandari et al., 2018).

Openness by filtering information obtained in the outside world to improve the world of education. Before receiving information, it is necessary to build awareness and maturity in oneself by responding to current developments when information flows without clarity and truth (Haris, 2019). Being a good leader is certainly

 https://doi.org/10.1051/e3sconf/202233906002

critical in dealing with rapid changes (Peramesti & Kusmana, 2018).

4 Conclusion

It can be concluded that the leadership of principal strategy deal with learning in new era of habits are;

1. To make online learning policies in schools by involving relevant teachers and school officials;
2. To increase the facilities and quality of ICT infrastructure in learning activities in schools;
3. To increase the quality of human resources in integrating IT for learning activities;
4. To communicate with parents / guardians of students at home;
5. To provide understanding to students about online learning activities;
6. To be ready and open in facing the changes in learning activities in the future.

The suggestions and inputs for this study are to carry out the further research in-depth and measurable analysis of the principal's strategies and policies in online learning activities so that the effectiveness of the strategies used and the advantages and disadvantages of implementing the strategies taken by the principal can be known.

References

1. Karasan, A., Erdogan, M. Prioritization of Influence Factors for Selecting E–Learning Systems.Vol. 1197 AISC.Advances in Intelligent Systems and Computing (pp. 550-556) (2021)
2. Firmansyah, Y., & Kardina, F. The Influence of New Normal Amid the Covid-19 Pandemic on School Management and Students. BUANA ILMU, 4(2), 99-112 (2020).
3. Lo Moro, G., Sinigaglia, T., Bert, F., Savatteri, A., Gualano, MR, & Siliquini, R. Reopening schools during the COVID-19 pandemic: Overview and rapid systematic review of guidelines and recommendations on preventive measures and the management of cases. [Review]. International Journal of Environmental Research and Public Health, 17(23), 1-21 (2020).
4. D'Angelo, D., Sinopoli, A., Napoletano, A., Gianola, S., Castellini, G., del
5. Fernandez, AA, & Shaw, GP Academic Leadership in a Time of Crisis: The Coronavirus and COVID-19. Journal of Leadership Studies, 14(1), 39-45 (2020b).
6. Kusmayadi, D., Rusdiana, D., & Badruzaman, J. Modeling Good School Governance after the International Standard School Pilot (RSBI) in Tasikmalaya City. Sustainable Competitive Advantage (SCA), 5(1) (2015).
7. Khairuddin, K. Principal Leadership Amid the Covid-19 Pandemic Education, 8(2), 171-183 (2020).
8. Sugiyono, Quantitative Approach Educational Research Methods, Qualitative R&D, Bandung: Alfabeta, Cet, XXIII, 2016.
9. Lo Moro, G., Sinigaglia, T., Bert, F., Savatteri, A., Gualano, MR, & Siliquini, R. Reopening schools during the COVID-19 pandemic: Overview and rapid systematic review of guidelines and recommendations on preventive measures and the management of cases. [Review]. International Journal of Environmental Research and Public Health, 17(23), 1-21 (2020).
10. Novitasari, D., & Asbari, M. The Role of Readiness to Change on Teacher Performance in the Covid-19 Pandemic. Journal of Industrial Engineering & Management Research, 1(2), 219-237 (2020).
11. Apandi, I. Innovative and Visionary Principals in the Industrial Revolution Era 4.0.8 December (2018). Retrieved from https://www.kompasiana.com/idrisapandi/5c0b431bbde57540ec66d978/principal-inovatif-dan-visioner-di-erarevolution-industri-4-0
12. Hope, E. The Principal's Vision as a Driver of Education Quality. Journal of Management, Leadership, and Educational Supervision, 1(2), 133–145 (2016). (2016).
13. Alenezi, ATechnology leadership in Saudi schools. Education and Information Technologies, 22(3), 1121–1132 (2017) https://doi.org/10.1007/s10639016-9477-x
14. Oberer, B., & Erkollar, A. Leadership 4.0: Digital Leaders in the Age of Industry 4.0. International Journal of Organizational Leadership, 7(4), 404–412 (2018). https://doi.org/10.33844/ijol.2018.60332

Determination of superior agriculture commodities in North Sumatra Province

Desi Novita[1*], *Teja* Rinanda[2], and *Mhd.* Ilham Riyadh[1]

[1]Universitas Islam Sumatera Utara, Agribusiness Department, 20223, Medan, North Sumatra, Indonesia.
[2]STIE Graha Kirana, Management Department, 20223, Medan, North Sumatra, Indonesia.

Abstract. This study aims to determine the contribution of the agriculture sector in the economy in North Sumatra Province and to find out the superior agriculture commodities. The data used in this study is secondary data, namely data on the production of various agricultural commodities in North Sumatra Province and Indonesia for the period 2019-2020. Quantitative description analysis in the form of contribution analysis and growth analysis is used to determine the contribution of the agricultural sector in the economy in North Sumatra Province in 2016-2020. Static Location Quotient (SLQ) and Dynamic Location Quotient (DLQ) analyses are used to analyze superior agricultural commodities. The results show that the contribution of the agricultural sector has decreased to the economy in North Sumatra Province in the 2016-2020 period. The results of the SLQ and DLQ analyses show that the superior agricultural commodities that are cross-sub-sectoral are Red Chili, Cabbage, Tomato, Eggplant, Durian, Salak, Rubber, and Fisheries. The superior agricultural commodities come from the Horticulture Sub-sector, Plantation Sub-sector, and Fisheries Sub-sector. Meanwhile, commodities originating from the Food crop Sub-sector and Livestock Sub-sector did not contribute to the creation of superior agriculture commodities in North Sumatra Province in 2019-2020.

1 Introduction

National development goals can be achieved through achieving development goals in each region [1]. Regional development cannot be separated from the ability of a region to manage its resources to be able to create jobs, alleviate poverty and be able to increase economic activities in the future. One of them is the achievement of agricultural development. This is because the agricultural sector is still a sector that contributes greatly to national development [2].

North Sumatra Province is known as one of the provinces that a strategic position in the development of the agricultural sector in Indonesia. The agricultural sector is the dominant sector in the economy in North Sumatra. This is because the agricultural sector is the main contributor to the formation of Gross Regional Domestic Product (GDP). In 2017, the GRDP of the agricultural sector reached 121,300.05 billion rupiahs, or 24.88 percent of the total GRDP [3]. In addition, another important position is the high export value of agricultural commodities (18.58%), the value of industrial exports, which are mostly agriculture-based industries, and the agricultural sector is still where most of the people work (38%).

"Organizing the City and Building the Village" is the words of wisdom currently being carried by the North Sumatra Provincial government. Building a village in question describes the development of the agricultural sector to realize the foundations of a people's economy that relies on the agricultural sector as its leading sector One of the development priorities contained in the Regional Medium-Term Development Plan (RPJMD) for 2018-2023 is increasing competitiveness through the agrarian sector.

This increase in competitiveness, one of which is directed at the development of international standard agricultural areas with natural potential and agricultural cultivation with location specifications. Some of the problems faced in the development of the agricultural sector include:

1. Still spreading agricultural production centers
2. The farmers' cropping pattern is still high, which is adjusted to the assistance program provided by the government.
3. Farmers' Exchange Rate (NTP) is still low (99.14 in 2017)
4. . Population pressure on food security is getting bigger
5. Inequality of rural development.

The above problems are caused by agricultural development planning which has not been fully developed based on superior commodities produced in each region. Development planning must be based on the local potential that can encourage production economic activities [4].

One of the keys to achieving successful agricultural development is through planning, coordination, and integration between sectors and decision-makers. With this, the efficiency and effectiveness of regional

* Corresponding author: desi.novita@uisu.ac.id

independence will be achieved [5]. Sectoral development must be adjusted to the potentials possessed by each region.

Optimization of agricultural development can be achieved through the development of superior commodities in an area. Superior commodities are commodities that have a strategic position, both in terms of land suitability, social, economic, and institutional [6]. In addition, superior commodities are commodities that in proportion are sufficient to meet the needs of their regions and also the needs of other regions.

The development of agricultural potential in North Sumatra Province can be done by determining the superior commodities owned. Determination of superior commodities in an area is important to support the development of an efficient agricultural sector. Regional development is based on superior commodities in increasing people's income and welfare [7]. Therefore, the purpose of this study is to determine the role of the agricultural sector in the economy in North Sumatra Province, as well as to find out which commodities in the agricultural sector are the leading commodities in North Sumatra Province.

2 Materials and Methods

This research is a descriptive research with a quantitative approach. Research location is in North Sumatra Province. The type of data used in this research is secondary data which are the GRDP data for the Province of North Sumatra in 2016-2020 and production data for 51 agricultural commodities at the provincial and Indonesian levels in 2019-2020. Data were obtained from the Central Statistics Agency of Indonesia, the Central Statistics Agency of North Sumatra Province, the Ministry of Agriculture of the Republic of Indonesia, and other related reference sources.

Data analysis method used is quantitative descriptive analysis to find out the role of the agricultural sector on the economy in North Sumatra Province. SLQ (Static Location Quotient) and DLQ (Dynamic Location Quotient) methods are to determine leading commodities in North Sumatra Province.

2.1 Static Location Quotient (SLQ) Analysis

SLQ is a data analysis method used to identify basic and non-basic commodities in a region. SLQ is usually abbreviated as LQ or Location Quotient. The SLQ formula is as follows:

$$SLQ = \frac{Rip/Rtp}{Nin/Ntn} \qquad (1)$$

Information :

Rip = Production of commodity i at the provincial level of North Sumatra

Rtp = Total production of all commodities at the provincial level of North Sumatra

Nin = Production of commodity i at the Indonesian level

Ntn = Total production of all commodities at the Indonesian level

If SLQ > 1, then the commodity is a base commodity, and if SLQ < 1, then the commodity is a non-basic commodity. The structure of the SLQ formulation provides several values as follows:

- SLQ > 1, this means the growth rate of commodity i at the provincial level of North Sumatra is greater than the growth rate of the same commodity at the Indonesian level. Thus, commodity i is a basic commodity to be further developed at the provincial level of North Sumatra
- SLQ < 1, this means that the growth rate of commodity i at the level of North Sumatra Province is smaller than the growth rate of the same commodity at the Indonesian level. Thus, commodity i is not a base sector
- SLQ = 1, this means the growth rate of commodity i at the level of North Sumatra Province is the same as the growth rate of the same commodity at the Indonesian level

2.2 Dynamic Location Quotient (DLQ) Analysis

DLQ is an LQ analysis conducted in the form of a time series/trend. In this case, the development of LQ can be seen for a particular sector at different times, whether it has decreased or increased [8]. DLQ is a modification of SLQ by accommodating the large production value of horticultural commodities from time to time. DLQ is calculated using the formula:

$$DLQ = \left[\frac{(1+gik)/(1+gk)}{(1+gtp)/(1+gp)}\right]t \qquad (2)$$

Information :

gik = average production growth of commodity i at the provincial level of North Sumatra

gk = average growth in total commodity production at the level of North Sumatra Province

gip = average production growth of commodity i in Indonesia

gp = average growth of total commodity production in Indonesia.

T = difference between end year and beginning year

The concept of DLQ analysis is:

- DLQ > 1 = potential development of commodity i faster than the same sector
- DLQ < 1 = potential development of commodity i is lower than the same sector

The next step is to combine the SLQ and DLQ values into four commodity criteria, namely leading, mainstay, prospective and underdeveloped commodities. The matrix of the four criteria can be seen in Table 1.

Table 1. Classification of commodity criteria based on the combined value of SLQ and DLQ

criteria	SLQ > 1	SLQ < 1
DLQ > 1	Superior Commodity	Mainstay Commodity
DLQ < 1	Prospective Commodity	Underdeveloped Commodity

Source : Kuncoro, Mudrajad (2012)

3 Results and Discussion

3.1 The Contribution of the Agricultural Sector in the Economy of North Sumatra Province

Table 2. Distribution of Gross Regional Domestic Product (GRDP) on the basis of current prices by business field in North Sumatra Province in 2016-2020

No	Business Fields	Years				
		2016	2017	2018	2019	2020
1	Agriculture	21.5	21.3	20.9	20.5	21.3
2	Mining	1.35	1.30	1.29	1.27	1.28
3	Industry	20.1	20.3	20.0	19.1	19.3
4	Electricity	0.11	0.12	0.11	0.11	0.11
5	Water Supply	0.10	0.11	0.10	0.10	0.10
6	Constrcution	13.5	13.7	13.9	14.2	13.6
7	Retail	17.8	17.5	18.1	18.8	18.9
8	Transportation	5.1	5.02	4.99	5.07	4.48
9	Provision of Accomodation	2.39	2.37	2.38	2.42	2.18
10	Information and Communication	1.95	2.01	2.04	2.14	2.28
11	Financial and Insurance	3.31	3.17	3.05	2.92	2.90
12	Real Estate	4.75	4.97	5.03	5.12	5.27
13	Company Service	1.00	1.03	1.03	1.08	1.07
14	Government Administration	3.67	3.71	3.68	3.69	3.73
15	Education	1.88	1.82	1.82	1.85	1.90
16	Health Service	0.95	0.96	0.98	1.02	1.03
17	Other Services	0.56	0.58	0.57	0.58	0.57

The agricultural sector is the mainstay sector as a contributor to GRDP and also as the sector that absorbs the most labor compared to other sectors. Based on Central Statistics Agency's data, the agricultural sector is still the highest sector contributor to GRDP in North Sumatra, although it has decreased. The agriculture sector contributes to GRDP around 20 – 22% over the last 5 years.

Meanwhile. the growth rate of the agricultural sector in the 2016-2020 period is still lower than the general GDP growth rate. The growth rate of the agricultural sector shows an increase every year. Especially when the Covid 19 Pandemic occurred. it was seen that the agricultural sector in North Sumatra Province grew by 1.95%. while the level of the economy in general decreased by -1.07%. This condition shows that the agricultural sector in North Sumatra Province is a priority in regional economic development.

Table 3. GRDP Growth Rate of Agriculture Subsector in North Sumatra Province in 2016-2020 (%)

Subsector	Years					Average
	2016	2017	2018	2019	2020	
1. Food Crops	7.57	5.58	2.15	4.20	-0.26	3.63
2. Horticultural Crops	2.03	5.69	0.16	-0.37	1.54	1.81
3. Plantations	4.47	5.52	6.59	7.47	3.68	5.55
4. Ranc	6.78	6.64	6.78	6.28	-1.96	4.90
5. Agriculture Services	1.98	5.02	-0.45	4.37	0.90	2.36
6. Forestry	-3.54	0.27	6.86	2.54	2.07	1.64
7. Fishing						
Agriculture Sector	**4.65**	**5.31**	**4.87**	**5.13**	**1.95**	**4.38**
GRDP of SUMUT	**5.18**	**5.12**	**5.18**	**5.22**	**-1.07**	**3.93**

3.2 Determination of Leading Commodities in the Agricultural Sector in North Sumatra Province

3.2.1 Stage 1. Static LQ (SLQ) Analysis

Based on the results of the SLQ calculation. it is known that of the 51 existing agricultural commodities. there are 20 agricultural commodities that have a SLQ value >1. This means that these 20 agricultural commodities have a comparative advantage compared to other agricultural commodities in North Sumatra Province.

Agricultural sector commodities classified as basic commodities have a comparative advantage because they have an average value of SLQ>1. This value indicates that the production value of the commodity is high so that it has the ability to meet the needs of other regions. As for the 20 basic commodities, the five highest include salak, pork, fisheries, capture fisheries, and cabbage.

Table 4. Results of SLQ Analysis of Agricultural Sector Commodities in North Sumatra Province (2019-2020)

No	Commodity	SLQ Average
1	Paddy	0.49
2	Corn	1.09
3	Soybean	0.23
4	Peanuts	0.16
5	Mung Beans	0.07
6	Cassava	0.94
7	Sweet Potato	0.74
8	Red Onion	0.18
9	Red Chili	1.78
10	Cayenne Paper	0.48
11	Potato	1.24
12	Cabbage	1.95
13	Tomato	1.74
14	Garlic	0.18
15	Petsay/Mustard	1.52
16	Carrot	1.92
17	Eggplant	1.53
18	Spring Onion	0.30
19	Durian	1.04
20	Siamese Orange	1.71
21	Mango	0.15

22	Pineapple	1.27
23	Papaya	0.40
24	Snakefruit	3.27
25	Jackfruit	0.22
26	Ginger	0.37
27	Turmeric	0.14
28	Cardamon	0.01
29	Galangal	0.16
30	Aromatic Ginger	0.05
31	Banana	0.18
32	Palm Oil	1.55
33	Rubber	1.49
34	Tea	0.84
35	Cocoa	0.63
36	Coconut	0.46
37	Sugarcane	0.09
38	Tobaaco	0.08
39	Coffeee	1.29
40	Clove	0.10
41	Pepper	0.01
42	Beef	0.36
43	Buffalo Meat	0.91
44	Lamb	0.21
45	Pork	2.43
46	Broiler	0.56
47	Eggs	1.39
48	Fishing in The Sea	2.24
49	Public Water Fisheries	1.70
50	Catch Fishing	2.13
51	Aquaculture	0.33

3.2.2 Stage 2. LQ Dynamic Analysis (DLQ)

The dynamic LQ calculation is a development of the SLQ calculation. DLQ analysis is developed in the form of a time series/trend. DLQ analysis in determining the priority commodities in the amount of production for each commodity. Based on the DLQ calculation, there are eight agricultural sector commodities that fall into the basic commodity category. These commodities include red chili (DLQ 4.98), tomatoes (DLQ 4.71), cabbage (DLQ 4.16), durian (DLQ 3.95), Eggplant (DLQ 3.88), fisheries (DLQ 3.55), rubber (DLQ 1.14), and salak (DLQ 1.13). The eight commodities have a DLQ value > 1 which indicates that the potential for growth of these commodities is faster than other commodities in North Sumatra Province.

3.2.3 Stage 3. Combined Static LQ and Dynamic LQ Analisis Analysis Results

Based on the calculation results of Static LQ and Dynamic LQ analysis. the combined results of these two methods are able to classify commodities into 4 categories. The 4 categories are superior commodities, mainstay commodities, prospective commodities, and underdeveloped commodities. The results of the calculation of the SLQ and DLQ analysis showed that there were 8 agricultural commodities which became the leading commodities in North Sumatra Province. These commodities are red chili, cabbage, tomatoes, eggplant, durian, salak, rubber, and fisheries. Meanwhile, there are 10 agricultural commodities that are categorized as prospective commodities including corn, potatoes, Chinese cabbage, carrots, Siamese oranges, pineapple, palm oil, coffee, pork, eggs, and chicken eggs. There are 11 agricultural commodities which are mainstay commodities and 8 commodities which are underdeveloped commodities. In more detail, the division of the agricultural commodities into 4 categories can be seen in Table 5.

The superior commodities are commodities that have DLQ and SLQ values > 1 (Category 1). This commodity is said to have a good comparative advantage at the moment and has the potential to remain a leading commodity in the future. The second category, mainstay commodities are agricultural commodities as indicated by the value of DLQ > 1 and SLQ <1. This mainstay commodity means that this commodity has not become a commodity that has advantages at this time. but has the opportunity to have advantages in the future.

Meanwhile in the third category, prospective commodities are commodities that have a SLQ value >1 and DLQ <1 which means that this commodity has a comparative advantage at the moment but is not expected to have an advantage in the future. The fourth category is the category that is categorized as underdeveloped commodity with DLQ and SLQ values < 1. This lagging commodity means that the commodity does not have a comparative advantage at this time and also has no potential to develop in the future.

Table 5. Matrix of combined analysis of SLQ and DLQ

	SLQ > 1	**SLQ < 1**
DLQ > 1	**SUPERIOR COMMODITIES :** Red Chili Cabbage Tomato Eggplant Durian Snakefruit Rubber Fishing in the sea	**MAINSTAY COMMODITIES:** Mung Beans Sweet Potato Cassava Red Onion Ginger Turmeric Coconut Beef Bufallo Meat
DLQ < 1	**PROSPECTIVE COMMODITIES :** Corn Potato Petsai/Mustard Carrot Siamese Orange Pineapple Palm Oil Coffe Pork Eggs	**UNDERDEVELOPED COMMODITIES :** Garlic Papaya Cocoa Tea Broiler Clove Pepper Aquaculture

4 Conclusion

The agricultural sector provides the largest contribution to the formation of the GRDP structure in the economy of North Sumatra Province with a declining contribution every year. The agricultural sector was able to survive and grow amid the Covid-19 pandemic that occurred.

Commodities in the agricultural sector which are the leading commodities in the province of North

Sumatra include 8 commodities namely Red Chili, Cabbage. Tomato, Eggplant, Durian, Salak, Rubber and Fisheries. These commodities can meet needs within the province and also able to meet needs originating from outside the province so that it has a comparative advantage compared to other commodities in the agricultural sector.

Proper planning is needed to increase regional income through the 8 leading commodities in the agricultural sector. Increasing regional income based on superior commodities is the focus of development so that the direction and policies of agricultural development in North Sumatra Province can be right on target.

Acknowledgments. The authors would like to thank the Ministry of Education, Culture, Research. and Technology of Republic Indonesia for funding this research through PDUPT Schema in 2021.

References

1. Ayubi. A.A. *Economic Potential Analysis of Banyuwangi Regency*. Journal of Development Economics. 12(1): 1-15(2014)
2. Rasyid, A. *Potential Analysis of Agricultural Potential Sector in Kediri Regency 2010-2014*. Journal of Development Economics, 14(02): 100-111 (2016)
3. Central Bureau of Statistics of North Sumatra Province. *Sumatra in Figures 2018*. Medan. North Sumatra (2019)
4. Saragih. J. R.. Siburian. A.. Harmain. U.. and Purba. T. *Leading Commodities and Potential Agriculture Sector Simalungun Regency. North Sumatra Province*. Agro Bali: Agricultural Journal. 4(1). 51-62 (2021)
5. Gafur. Safri. M. dan Hodijah. S. *Analysis of Leading Sector/Sub-Sector in Bungo Regency*. Journal of Regional Development and Financing Perspectives. 3(3): 2338-4603 (2016)
6. Yulianti M.*Determination of Priority Commodities of Fruits in North Minahasa Regency, North Sulawesi Province: Application of LQ Analysis and Attractiveness-Competitiveness*. Journal of Rural Agribusiness.1 (3) : 206-218 (2011)
7. Kurniawan, K. *Analysis of the Leading Economic Sector of Kerinci Regency, Jambi Province*. el-JIZYA Islamic Economics Journa, 4(1): 1-26 (2016)
8. Tarigan. R. *Regional Economy*. Theory and Application.PT. Bumi Aksara. Jakarta (2009)
9. *Kuncoro,Mudrajad. Regional Planning: How to Build Local Economy, Cities and Regions?*.Jakarta Selatan: Salemba Empat (2012)

Analysis of rice production trends and the effect of harvest failure on insured land area in Aceh Province

Anwar Deli[1,2*], *Muhammad R.* Pahlevi[1], *Rahmaddiansyah* Rahmaddiansyah[1], *Ali* Afwanudin[2], *Muhammad* Authar ND[3], *Edy* Miswar[4], and *Muhammad* Irham [4,5]

[1]Department of Agribusiness, Faculty of Agriculture, Universitas Syiah Kuala, Banda Aceh, Indonesia
[2]Center for Environmental and Natural Resources Research (PPLH-SDA), Universitas Syiah Kuala, Banda Aceh, Indonesia
[3]Program Studi Agribisnis, Fakultas Pertanian, Universitas Malikussaleh, Lhokseumawe, Indonesia
[4]Center for Marine and Fisheries Research (PRKP), Universitas Syiah Kuala, Banda Aceh, Indonesia
[5]Laboratorium Sistem Informasi Geografis, Fakultas Kelautan dan Perikanan, Universitas Syiah Kuala

Abstract. Agricultural insurance is a risk transfer that can provide compensation due to farming losses so that the sustainability of farming can be guaranteed. Rice farming insurance (AUTP) is a pilot program or pilot project that has been implemented in Aceh Province since 2015. However, the realization of the insurance program is still very low from the target set. The central government coordinates with local governments to implement the AUTP program by issuing a policy of providing premium subsidies of 80 percent or IDR.144,000/Ha/MT of the total premium of IDR.180,000/Ha/MT. AUTP participants only pay a premium of 20% per hectare of land insured with an insured value of IDR.6,000,000/Ha with conditions damaged or failed in the event attack of the pest and natural disasters at least 75%. The purpose of this study is to see the effect of a simple regression analysis of the area of failed rice harvest on the area of land insured in Aceh Province. This study uses secondary data from 2010-2019. The results of this study via a simple regression analysis indicate that the effect of the area of rice harvest failure on the area of land insured in Aceh Province is not significantly affected.

1 Introduction

Aceh Province is one of the provinces in Indonesia which has extensive agricultural land for rice plants and is a center for rice production. Aceh province is targeted to become self-sufficient in rice and become a national rice barn. Aceh Province itself has 3 (three) main rice-producing centers, namely North Aceh, Aceh Besar and Bireuen Regencies. The data from the Central Statistics Agency (BPS) of Aceh Province in 2020[1] shows the development of harvested area and rice production in Aceh Province during 2015 to 2019 (Table 1).

Table 1. Harvested area and rice production in Aceh Province 2015-2019.

Year	Harvested Area (Ha)	Rice Prod. (Ton)
2015	461.060	2.331.046
2016	429.486	2.205.056
2017	470.351	2.494.613
2018	329.516	1.861.567
2019	310.012	1.714.438

Source: Aceh Central Statistics Agency, 2020

The area of agricultur†al land tenure is something that is very important in the production process of agricultural business [2]. The total area of agricultural land in Aceh Province has decreased almost every year in the amount of rice production [3]. The decrease in the amount of land is due to an increase in the number and activities of the population and development [4]. Rice farming is very vulnerable to climate change [3], [5]–[8]. This activity has a high level of dependence on climatic and weather conditions[9]-[10]. Therefore, rice farming will always be faced with quite high risks such as a reduction in harvested area and rice production [11].

According to [12] farmers find it difficult to predict when a disaster will occur. Thus the inability of farmers to handle the impact of the risk personally. Therefore the option to transfer or share the risks faced by farmers to fellow farmers or companies is the most rational [13]. Agricultural insurance shows partiality to anticipate the risk of farming losses. This is in line with the research in Nigeria [14] which states that agricultural insurance helps farmers reduce risk.

To minimize farmer losses, the Indonesian government through the Ministry of Agriculture issued the Minister of Agriculture Regulation No. 40 of 2015 concerning the facilitation of agricultural insurance which is technically rice farming insurance (AUTP).

*Corresponding author : anwar_deli@unsyiah.ac.id

Agricultural insurance is a risk transfer that can provide compensation due to farming losses so that the sustainability of farming can be guaranteed [13]. With the existence of government policies in the agricultural sector related to agricultural insurance, at least farmers get a guarantee of certainty of the reimbursement of production costs in the event of crop failure and natural disasters that befall the farmers. Agricultural insurance is one of the important methods to protect farmers from big losses and ensures that farmers will have sufficient working capital to finance their rice farming in the following season from January to April 2021[3].

2 Research Method

2.1 Place and time of research

This research was conducted in Aceh Province using secondary data in the form of time series data in 2010-2019. This research was conducted from April to August 2020.

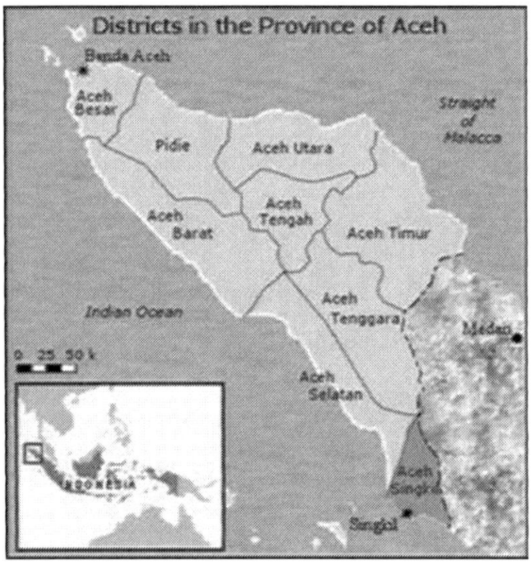

Fig. 1. Aceh Province, Indonesia

2.2 Object and Scope of Research

The object of this research is data on rice production in 2010-2019 in districts that are familiar with rice farming insurance (AUTP) in Aceh Province and data on rice harvest failure in 2015-2019. The scope of this research is limited to rice production trends and projections in districts that are familiar with rice farming insurance (AUTP) in Aceh Province, and to analyze the effect of crop failure on the area of land insured in Aceh Province.

2.3 Data Types and Sources

The types of data and their sources are from the Central Statistics Agency (BPS))[1] and related agencies. The type of data used in this study is secondary data in the form of a time series from 2010 to 2019.

2.4 Analysis Method

The data analysis methods used in this research are trend analysis and simple regression analysis. The first analysis is trend analysis to make an estimate or forecast in the future. Linear trend is the trend of data where the change based on time is constant (constant). The linear trend has the following model [15]:

$$Yt = \beta o + \beta 1T \quad (1)$$

Where; Yt is Rice production value in year t; βo is Constant, which shows the value of the data in the initial year; β1 is The amount of data change from one period to another; T is Year

The second analysis is a simple regression analysis to determine the effect of the area of crop failure (x) with the variable the amount of land area insured (y) where the variable is causal (influence) positive or negative. The simple regression formula is as follows[16]:

$$Y = a + bX + e \quad (2)$$

Where, Y is Total area of land insured; X is area of crop failures; e is Error term; a is Constant (Y value if X is 0); b is Number of regression coefficient direction (value increase or decrease).

The use of sections to divide the text of the paper is optional and left as a decision for the author. Where the author wishes to divide the paper into sections the formatting shown in Table 2 should be used.

2.4.1 Normality test

The normality test aims that the data that is good and suitable for use in research is data that has a normal distribution. The normality test Kolmogorov Smirnov is used to test whether the distribution is normal or otherwise, the results of this analysis are then compared with the critical value. Normality test can be done by looking at the Kolmogorov Smirnov quantity with the following test criteria: Significant number (Sig) > = 0.05, the data is normally distributed; and Significant number (Sig) < = 0.05, the data is not normally distributed.

2.4.2 Classic assumption test

The classical assumption test used in this study is the heteroscedasticity test. Heteroscedasticity test aims to test whether the variance of the residual variance from one observation to another observation occurs in the regression model. The glejser test is carried out to predict the presence or absence of heteroscedasticity in a model by performing absolute regress of residuals from the estimated model on the independent variables. The data is said to be free from heteroscedasticity if the significance value obtained is greater than the confidence level (value) which is 0.05.

2.4.3 Hypothesis testing

a. T Statistic Test (Partial)

The t-statistical test shows how far the influence of one variable area of crop failure individually in explaining the variation of the variable area of insured land. To perform the t test, namely is by looking at the probability value and the degree of confidence determined in the study. If the significance value < the level of confidence (value) is determined then an independent variable individually affects the dependent variable

b. Coefficient of Determination (R2)

The coefficient of determination (R2) essentially measures how far the independent variable's ability to explain the variation of the dependent variable. The value of R2 is used to indicate the magnitude of the regression that is able to explain the dependent variable. The value of the coefficient of determination is between zero and one. A small value of R2 indicates the ability of the independent variables to explain the variation of the dependent variable with a very limited extent and a value close to one means that the independent variables provides almost all the information needed to predict the variation of the dependent variable.

3 Result and Discussion

Analysis of Rice Production Trends and Projections in Regencies That Know Rice Farming Insurance (AUTP) Analysis of trends and projections of rice production is a trend movement (tendency) up or down in the long term in a straight line. The average change can increase or decrease. If the trend has an upward trend then the trend is positive. On the other hand, if the trend has a downward trend, the trend is negative[17][18]. The following is an analysis of trends and projections of rice production in districts that are familiar with rice farming insurance (AUTP) in Aceh Province. Districts that have recognized rice farming insurance (AUTP) consist of South Aceh District, Southeast Aceh District, East Aceh District, West Aceh District, Aceh Besar District, Pidie District, Bireuen District, North Aceh District, Southwest Aceh District, Aceh Tamiang District and Aceh Jaya District.

The results of the analysis of trends and projections of rice production in districts that have recognized rice farming insurance (AUTP). There are 7 districts experiencing trends and projections of increased rice production which are Southeast Aceh District, West Aceh District, Aceh Besar District, Pidie District, Bireuen District, North Aceh District, and Aceh Jaya District. However, there are 4 districts whose rice production trends and projections have decreased, namely South Aceh District, East Aceh District, Southwest Aceh District, and Aceh Tamiang District.

3.1 Issues Related to the Lower Level of Rice Production Trends in Several Districts in Aceh Province

3.1.1 South Aceh District

Geographically, the position of the South Aceh District includes the West-South coast of Aceh Province, where most of the residential areas are directly adjacent to the South West coast of Aceh. According to BPS data from 2010-2019 the results of the amount of rice production in South Aceh District tend to fluctuate, wherby in the past few years there had been a decline in harvested area and yields of rice production. It was recorded in 2019 that the rice harvested area in South Aceh District was 8,607.52 Ha, with a total rice production of 43,980 tons. Quoted from bpbaacehprov.go.id in 2019 there was a flood in South Aceh District which affected 9 sub-districts and also caused landslides in several areas. The impact of the floods that hit a number of areas in South Aceh District caused hundreds of hectares of rice crops owned by farmers in several sub-districts to experience crop failure.

Quoted by [9] in 2018 at least a total of 68.5 hectares of rice fields were flooded, caused by heavy rain that hit several submerged sub-districts namely Tapaktuan, Kota Bahagia, and Bakongan in South Aceh District. In 2017, quoted from Antaranews.com in North Kluet District, tens of hectares of rice plants experienced crop failure caused by the attack of sparrows[21]. This is because farmers in North Kluet District have not implemented a simultaneous rice cropping pattern system, which makes them very vulnerable to the attack by sparrows.

3.1.2 East Aceh District

Quoted from [20] in 2018, East Aceh District experienced drought that hit a number of sub-districts in East Aceh District. Rice fields that experienced drought on average were rainfed rice fields where water sources only existed during the rainy season, other factors the land experienced is the unavailability of irrigation channels in the area. The drought in the area is due to being in a non-technical irrigation area. In 2019, quoted from Baranewsaceh.com, rat pests that attacked rice plants in the Serbajadi Sub-district also greatly disturbed the residents, especially the rice farmers in the villages of Loot, Sekualan, Jering Sunti and Umah Tarin. Rat attacks begin when the plants are 60 days old or when the rice plant begins to contain.

Quoted from [21], thousands of hectares of rice fields experienced drought that hit a number of sub-districts in East Aceh District. Rice fields that are threatened with crop failure include the sub-districts of Nurussalam, Darul Falah, Julok, Darul Aman and Idi Rayeuk. On average, rice fields that experience drought are rainfed rice fields. During the drought, farmers only used river water, which is now running dry. The absence of irrigation or large reserves of water stored in the area during a drought has an impact and the threat of crop failure.

3.1.3 Southwest Aceh District

In early 2018 the Southwest Aceh Agriculture and Food Service (Distanpan) distributed M400 variety seeds to farmers in Southwest Aceh Regency. The M400 seed variety is one of the superior rice seeds developed by the central government, considering the very good yields due to having larger stems with full grains of rice on each stalk, M400 rice varieties are also resistant to pests and weather anomalies. However, many of the seeds of the M400 variety that had been sown by the farmers did

https://doi.org/10.1051/e3sconf/202233906004

not grow, and some farmers experienced a decrease in the number of harvests, which usually reached 7.5 tons/hectare. Now the average harvest is only about 4.5 tons/hectare caused by planthoppers and rats, and also water drought.

In 2018 in Tangan-Tangan District hundreds of hectares of rice plants owned by farmers were threatened with crop failure caused by the high attack of the pest walang sangit (Leptocorisa oratorius). Hundreds of hectares of rice plants that were attacked by the pest walang sangit (Leptocorisa oratorius) in the Tangan-Tangan sub-district are now producing panicles. Almost every year the pests attack rice plants in the sub-district, even though the attacked rice plants are already thriving and will soon enter the harvesting period. Walang sangit (Leptocorisa oratorius) is a pest that generally attacks rice plants during the panicle releasing phase to ripening by sucking the liquid rice grains. In 2018 in Kuala Bate District, about 10 hectares of the rice plants experienced crop failure caused by planthoppers and rats, and also water drought.

3.1.4 Aceh Tamiang District

Quoted from [20] in 2018 Aceh Tamiang Regency experienced a number of problems in rice farming which resulted in a decrease in the amount of rice production. The problems that occurred in Aceh Tamiang Regency during 2019 included water shortages and uneven water flow to irrigate rice fields, drought in the dry season which resulted in cracked land in the rice fields, no construction of irrigation networks and land conversion of rice fields into palm plantation. Quoted from analysisdaily.com in 2019 in Manyak Payed District and the Bendahara District of about 450 hectares of rice plants were attacked by planthoppers and clovers. The average age of rice plants attacked by the pests is 80 days to three months before harvest. The impact of the planthopper epidemic made the grains of rice blackened like scorched fire which resulted in the rice production of farmers in Aceh Tamiang Regency falling sharply during the harvest season.

The dry season that hit parts of Aceh Tamiang Regency was also the cause of the decline in rice production. At the beginning of 2019 in Manyak Payed District, farmers in the area experienced two time harvest failures due to lack of water and drought in the dry season which caused the rice fields to crack. This is the impact of minimal rainfall, because farmers in the area only rely on rainwater for their rice fields due to the unavailability of irrigation channels in the area.

The decline in the amount of rice production in Aceh Tamiang Regency is related to the increasing conversion of agricultural land to non-agricultural uses. The problems caused are the conversion of rice fields to oil palm plantations will have an impact on disrupting food availability and decreasing the amount of Padi production. The impact of the loss of agricultural land results in permanent loss of agricultural yields, so that if conditions are not controlled, it is certain that there will be a decrease in rice production and the Aceh Tamiang District area will ultimately experiences instability in food security. Data from BPS Aceh Tamiang District shows that the area of oil palm plantations in Aceh Tamiang District has increased from 19,702 hectares in 2018 to 20,429 hectares in 2019.

3.2 Analysis of the Effect of Harvest Failed Areas on the Insured Land Area

3.2.1 Normality test

The normality test aims to test whether the regression model has a normally distributed residual value or otherwise. A good regression model has a residual value that is normally distributed or close to normal[22]. The normality test in this study used the Kolmogrov-Smirnov non-parametric statistical test. In the Kolmogrov-Smirnov test, if the significance value is below 0.05, it means that the data is not normally distributed and if the significance value is above 0.05, the data is normally distributed.

Table 2. Kolmogorv-Smirnov Normality Test

	Unstandardized Residual
N	5
Asymp. Sig. (2-tailed)	0,753

Based on the results of the normality test regression using the One Sample Kolmogorov Smirnov method via the SPSS application in Aceh Province, the Asymp.Sig value 0.753 was obtained which is greater than the significance level of 0.05 (0.753> 0.05), It can be concluded that the data is normally distributed. This can be interpreted that the variable data on the area of crop failure is normally distributed.

3.2.2 Classic assumption test

Classical assumption test is a requirement in performing a simple regression analysis test. Heteroscedasticity test aims to detect whether in the tested regression model there is an inequality of variance from the residuals of one observation to another observation. If the variance of the observation residuals remains with each other, it is called homoscedasticity and if the variance is different then it is called heteroscedasticity. A good regression model is one that does not occur heteroscedasticity

Table 3. Hypothesis Test Results

Model	t	Sig
Constant	.360	.743
Rice Harvest Failed Area	.326	.766

Where, H0 is The Variable Area of Rice Harvest Failure has no effect on the Insured Land Area variable, and Ha is The Variable Area of Rice Harvest Failure affects the Insured Land Area variable.

From the results of the SPSS output of linear regression analysis, there is a t-count value of the variable area of crop failure of 0.326 which is greater than the t-table of 3.182. So, it can be concluded that H0 is accepted, meaning that the variable area of crop failure has no effect on the variable area of the insured land.

3.2.3 Coefficient of Determination Test (R2)

The coefficient of determination (R2) is a coefficient to measure how far the ability of a model to explain the variation of the dependent variable. To see more details or to find out how far the variable of rice harvested area to the area of land is insured, it can be seen in the following table.

Table 6. Coefficient of Determination Test Results (R2)

Model	R	R2	Adj R2	Std Error of The Estimate
1	0.185	0.034	-.288	1129.35137

Based on the regression results that have been obtained, it is known that the R square value is 0.034 = 3.4%. This means that the independent variable of rice harvested area affects the dependent variable of insured land area by 3.4% and the rest is influenced by other variables (not included in this study). The results of this study strengthen the findings of previous researchers which state that research on rice farming insurance (AUTP) is still in the pilot project stage.

4 Conclusion

Trends and projections of rice production in districts that have recognized rice farming insurance (AUTP) in Aceh Province, are increasing. However, there are 4 districts with declining trends and projections for rice production, namely South Aceh District, East Aceh District, Southwest Aceh District, and Aceh Tamiang District in 2010-2024. There is no significant effect between the area of rice harvest failure and the area of land insured. This shows that the rice farming insurance program (AUTP) has not been the main strategy for farmers in Aceh Province to prevent financial risks. This research proves that the rice farming insurance program (AUTP) is just a pilot project.

Acknowledgment. We acknowledge the support received from the Department of Agribusiness, Faculty of Agriculture, Universitas Syiah Kuala and Center for Environmental and Natural Resources Research (PPLH-SDA), Universitas Syiah Kuala who have supported and helped to provide data, information and facilities for this research.

References

1. Badan Pusat Statistik, Provinsi Aceh dalam Angka, 2020. Banda Aceh.
2. D. Clarke, N. Das, F. De Nicola, and R. Hill, The value of (customized) insurance for farmers in rural Bangladesh, in Proceeding Res. Conf. Micro Insur. Twente, August (2012)
3. S. M. Pasaribu, J. AASPRO **1** (2010)
4. A. Haris, L. B. Subagio, F. Santoso, and N. Wahyuningtyas, JPG **5**. 2 (2019)
5. K. Y. Clement, W. J. Wouter Botzen, R. Brouwer, and J. C. J. H. Aerts, J.IJDRR **28** (2018)
6. H. T. Abebe and A. Bogale, J. Energy Environ., **1**, 2 (2014)
7. N. K. Budhathoki, J. A. Lassa, S. Pun, and K. K. Zander, J. Landusepol **85** (2019)
8. M. Mustika, A. Fariyanti, and N. Tinaprilla, FAGB **9**, 2 (2019)
9. R. Zhang and D. Fan, J. Agric. Sci., **8**, 3 (2016)
10. C. Mina, C. Reyes, R. Gloria, and S. Mercado, Discuss. Pap., (2015)
11. M. Algazali, M. H. Jamil, and N. Lanuhu, Persepsi Petani Terhadap Asuransi Usahatani Padi (AUTP) Di Kecamatan Paleteang Kabupaten Pinrang,
12. P. C. Ajieh, J. Agric. Sci., **1**,1 (2010)
13. Insyafiah and I. Wardhani, Kajian Persiapan Implementasi Asuransi Pertanian Secara Nasional, Kementeri. Keuang. Badan Kebijkan Fiskal, Asuransi pertanian (2014)
14. F. M. Epetimehin, SSRN Electron. J., (2012)
15. A. B. Santoso, J. Penelit. Pertan. Tanam. Pangan, **35**, 1 (2016)
16. N. Utika, . M., and H. Yozza, J. Mat. UNAND, **2**, 2 (2013)
17. A. M. Razmy and A. N. Ahmed, J. Manage., **HI**, (2005)
18. A. Arifin and M. A. Sadat, J. Agribis, **13**, 1 (2021)
19. TDMRC, Buku Studi Kelayakan Penanganan Banjir Trumon Aceh Selatan Kerjasama dengan Perguruan Tinggi (Swakelola) (2019) [Online].
20. R. Fachlevi, Agussabti, and Ismayani, J. Bisnis Tani, **4**, 2 (2018)
21. R. Afrian and Z. R. Islami, J. Pendidik. Geogr., **24**, 2 (2019)
22. A. Nasrum, Uji Normalitas Data Untuk Penelitian, Jayapangus Press Books (2018)
23. N. N. Suindah, D. P. Darmawan, and I. K. Suamba, J. Sos. Ekon. Pertan.,**4**,1 (2020)

The effectiveness of TOEFL app for final-year students of Marine and Fisheries Faculty Universitas Syiah Kuala, Indonesia

Riyan Maulana[1], *Irwan* Irwan[2*], *Cut* Maila Hanum[3], *Mujiburrahman* Mujiburrahman[1] and *Lidiana* Lidiana[1]

[1]STMIK Indonesia Banda Aceh, Information System Department, Aceh-Indonesia
[2]Department of Marine Sciences, Faculty of Marine and Fisheries, Universitas Syiah Kuala, Aceh- Indonesia
[3]STIK Pante Kulu, Forestry Science Department, Aceh-Indonesia

Abstract. The paper reports the final-year students' needs of English and its effectiveness in joining TOEFL Class conducted by Academic of FKP using TOEF FKP application in learning process. There were about 45 final-year students in the academic year 2020-2021 as the primary objects of this research. The researcher is qualitative research and administered questionnaires and interviews to collect data and constructed as. The results found that the students need English skill to face challenges in the future for their career. It was also found that there was a high demand for TOEFL Certificate as a proof as English capacity, as a graduation requirement in the faculty and as a tool in order to work in international maritime industries in Aceh. Therefore, most of the students placed English skill as their highest priority. This application can facilitate the process of learning English in particular because it utilizes technological developments in this case the android application where they have been introduced since childhood to them at home. It concluded that the results of this investigation should be taken into account in developing teaching, learning materials, improving the TOEFL FKP app and its regulation for marine students.

1 Introduction

Education is a human attempt to develop and increase physical and spiritual capability based on the society and culture values especially for those who are studying in a higher university. Each student needs to perform well and gets good learning results especially in English. Moreover, to make it happens there are some variables such as motivation, teacher, etc to accomplish all that things and in this case in pandemic era learning tool based on android system is as one of the factors affecting students achievement in English skill that could assist the effectiveness of that process [1]. In this pandemic era a media for learning process especially TOEFL Class such as application should be available to conduct good academic process [2]. The Law of the Republic of Indonesia Number 20 of 2003 mentions that it is related to distance education where students are separated from educators and learning uses various learning sources including online media. [3].

As we know that TOEFL mastery is a must for someone who wants to complete their study in Syiah Kuala University as a requirement. Without a good TOEFL score, it is quite impossible for students to conduct final examination. Because English is as a requirement for final year students and now is pandemic era so it is would be better if there will be learning media such as an application to support TOEFL learning process either online or offline [4]. It means that the aim of learning media such as TOEFL FKP is to let the final year students enjoy the last learning process before completing thei²r study in Syiah Kuala university. Clark said that at present time Internet use through mobile devices increasingly dominant, so e-learning will easily become mobile learning without special changes in content [5].

Therefore, it is deemed necessary to develop interactive multimedia as a learning medium [6]. Delphi argues that merging between mobile-based computing and e-learning will broaden access to resources, robust search capabilities, enrich interactions and support unrestricted access to space and time [7], [8]. In addition, Multimedia has an important role in learning, therefore it is deemed necessary to develop interactive multimedia as a learning medium [9].

Some previous research about using mobile app in learning already conducted by some researcher for example a research conducted by the Acehnese researcher from STMIK Indonesia Banda Aceh entitle An Android-Based Vocabulary Model for Autism showed that mobile app can improve students outcome after learning and many more advantages [10]. Based on the description of the problems above, the writer wants to see the quality of learning of students in mastering English through the use of mobile learning application to support teaching and learning activities, especially TOEFL for students as a support for the independent learning process during pandemic era.

*Corresponding author : irwan@unsyiah.ac.id

1.1 Research Question

Based on the research background above. This study specifically aims to find answers about the following questions:

1. What are students' perceptions toward TOEFL FKP application in Marine and Fisheries Faculty, USK?
2. How are students' motivations toward TOEFL FKP application at Marine and Fisheries Faculty, USK?

1.2 Research Significant

Based on the research statement, this particular study aimed at finding out:

1. To know students' perceptions toward TOEFL FKP application in Marine and Fisheries Faculty, USK
2. To know students' motivations toward TOEFL FKP application at Marine and Fisheries Faculty, USK.

2 Methodology

2.1 Research Design

The method of this research is qualitative research. It is based on the research, there are focus; to exploring students perceptions toward the use of TOEFL FKP application by distributing questionnaire and interview to some final year students. This research focuses on descriptive qualitative method. Descriptive method has not purpose to test the certain hypothesis, but to describe some variable and condition naturally [11]. This research has purpose to look at the effectiveness of TOEFL FKP App for Final-Year Students of Marine Science Faculty Syiah Kuala University, Indonesia.

To prove the influence of the students' perception on TOEFL FKP and their motivation to learn, the researcher used the chi-square test and obtained the following results:

The calculation used to find out the percentage level of answer scores from each respondent is used a formula:

$$P = \frac{f}{N} \times 100\% \quad (1)$$

Where, P is Percentage of anwers; f is Frequency of anwers; N is Total frequency.

Based on the results of the percentage value above, the effectiveness between students' perceptions suggested that the management of TOEFL FKP section would increase and ultimately affect learning performance because good responses from the questioners.

2.2 Data Collection

In this research, the researcher uses questionnaires in collecting data. According to Yusuf, questionnaire is an investigation carried out by sending a questionnaire to respondents who have been determined and after being filled out the questionnaire is returned to the researcher [12]. In addition, Deddy said that the questionnaire is data collection instruments that the most widely used in social research, whether qualitative or quantitative [13]. The purpose of questionnaire is to find out their mind, what they think or how they feel about something [14].

In this research, the researcher uses closed-ended questionnaire. The questionnaire consisted of ten questions that focused on students' perceptions and motivation of TOEFL FKP application. Moreover, to make it easier for students to fill, the questionnaire is prepared by using Indonesian. To distribute the data, the researcher made an electronic questionnaire in the form of a Google Form [15].

2.3 Data Analysis

In analysing data the researcher uses descriptive qualitative analysis based on Miles and Huberman Interactive models theory [16]. There are three main components, they are: reduction the data, presentation/display the data, verification or drawing conclusion.

2.3.1 Population and Sample

This part talks about the sample that is going to be chosen by the researcher. Research sample is a part of a subject that is going to be observed or analysed. The sample is partial or representative of the subject. The population of this study is 45 final-year students in the academic year 2020-2021. The researcher used purposive sampling technique in choosing sample for this study because purposive sampling is an effective way as a sampling technique for qualitative research.

According to Sugiyono purposive sampling technique is a type of non-probability sampling that is most effective when one needs to study a certain cultural domain with knowledgeable experts within and purposive sampling may also be used with both qualitative and quantitative re- search techniques.

3 Results and Discussion

Through a number of features and menu options in the application, the user can carry out a series of activities that are in accordance with the purpose of application development. Here are some menus and features provided in the application:

3.1 Main Display

The main display is the media interface for the menu to be selected, looks like the following picture 1. In Figure 1 above it can be explained that users can easily learn TOEFL based on the menus that have been arranged in such a way. They only need to click on one of the menu in the Home application view.

Fig. 1. Main Display

Table 1. Questionnaire Perceptions toward TOEFL FKP App.

No.	Item	Option	Frequency	Percentage (%)
1.	I like to use the TOEFL FKP application to study	Always Often Sometimes Seldom Never	0 4 5 32 4	0 8,9 11,1 71,1 8,9
2.	Using the online application like TOEFL FKP is my style	Always Often Sometimes Seldom Never	2 10 14 11 8	4,4 22,2 31,1 24,4 17,8
3.	The TOEFL FKP application provides information/materials/questions that are easy to understand	Always Often Sometimes Seldom Never	5 4 17 17 2	11,1 8,9 37,8 37,8 4,4
4.	The TOEFL FKP application allows me to study whenever and wherever I am	Always Often Sometimes Seldom Never	5 4 17 14 5	11,1 8,9 37,8 31,1 11,1
5.	The TOEFL FKP application facilitates the learning needs of me	Always Often Sometimes Seldom Never	3 11 20 9 2	6,7 24,4 44,4 20,1 4,4
6.	I intend to use the TOEFL FKP application for future	Always Often Sometimes Seldom Never	4 12 12 10 7	8,9 26,7 26,7 22,2 15,6
7.	The TOEFL FKP application is easy to use	Always Often Sometimes Seldom Never	2 10 26 5 2	4,4 22,2 57,8 11,1 4,4
8.	I use the TOEFL FKP app because it's fun	Always Often Sometimes Seldom Never	6 13 8 9 9	13,3 28,9 17,8 20 20
9.	I believe that using the TOEFL FKP application will improve my learning outcomes.	Always Often Sometimes Seldom Never	6 7 9 12 11	13,3 15,6 20 26,7 24,4
10	I am used to using new technology like the TOEFL FKP app	Always Often Sometimes Seldom Never	6 7 9 12 11	13,3 15,6 20 26,7 24,4

4 Conclusion

Briefly, this application has been designed to be used specifically for students especially for TOEFL subject and has been tested using descriptive statistical methods with different calculation models and has run well, and can then be installed on an Android-based mobile phone.

This application can facilitate the process of learning English in particular because it utilizes technological developments in this case the android application where they have been introduced since childhood to them at home. So hopefully the language learning process for them can attract attention by displaying features and menus in the application.

From the results and the previous discussion, it can be concluded several things related to the application of the TOEFL FKP as follows;

a. The application of TOEFL FKP received a positive response in several ways, namely the practicality of sharing material, independent learning, and saving time and costs in the learning process.

b. Students still cannot have strong motivation to learn if they have to study alone without face to face with the lecturer.

c. Students cannot understand easily if they have to do self-study of the material provided.

d. Some students have not been able to enjoy and accept TOEFL FKP, since they used to the face-to-face model that has been done so far.

Finally, this application should be taken into account in developing teaching, learning materials, improving the TOEFL FKP app and its regulation for marine student.

References

1. M, Deddy. *Qualitative Research Methods.* Bandung. PT Teen Rosdakarya. (2003)
2. K, Nyiri. *Towards a philosophy of learning, Wireless and Mobile Technologies in Education,* IEEE International Workshop, 121,124. (2002)
3. Kaur, M. *Blended learning - its challenges and future.* Procedia - Social and Behavioral Sciences, 612-617. (2003)
4. J, Chen. *The Effectiveness of Peer Assessment in EFL Blended Learning Environments.* Paper presented at the 2021 2nd International Conference on Computers, Information Processing and Advanced Education. (2021).
5. J. W, Creswell,.*Research design: qualitative, quantitative, and mixed approaches.* Yogjakarta: PT PustakaPelajar. (2010).
6. A, Fergina, Prancisca, S., Ismiyani, N., Suparjan, S., Barella, Y., &Linarsih, A. *The*

Training of Digital Media Utilization in English Online Learning for Students and Alumni in Pontianak. JPKM (JurnalPengabdianKepadaMasyarakat) UNTAN, 4(1), 9-15. (2021).

7. Arifin, Z. (2012). *Penelitian Pendidikan.* Bandung: PT. RemajaRosdakarya.
8. S. E, Smaldino, Lowther, D. L., & Mims, C. *Instructional Technology and Media for Learning.* (2019).
9. D, Bandi. *Learning for children with special needs (in inclusive education settings) Prints I.* Bandung: RafikaAditama. Combined. Jakarta :prenadamedia group. (2006).
10. R Maulana, Bahruni. *An Android-Based Vocabulary Model For Autism.* International Journal of Information Systems & Technology. (2019)
11. Sugiyono. *Quantitative, Qualitative and R&D Research Methods.* Bandung: Afabeta. (2011).
12. A. Muri Yusuf..*Quantitative, Qualitative & Research Methods.* (2014)
13. B. M, Miles and M Huberman. *Qualitative Data Analysis Source Book About New Methods.* Jakarta: UIP. (1992)
14. C Quinn., *M-Learning: Mobile, Wireless, In Your-Pocket Learning*, Retrieved October 28, 2021, from http://linezone.com: http://linezone.com. (2014).
15. M. M. E. I, Bali &Holilah, N. *The Role of Foster Caregivers in the Effectiveness of Online Learning in Pesantren.* EdukasiIslami: JurnalPendidikan Islam, 10(01), 339-362. (2021).
16. Alam, M. M., Ahmad, N., Naveed, Q. N., Patel, A., Abohashrh, M., &Khaleel, M. A. *E-learning services to achieve sustainable learning and academic performance: An empirical study. Sustainability*, 13(5), 2653. (2021)

 https://doi.org/10.1051/e3sconf/202233906006

Profile of rational thinking ability skills and student learning motivation in physics learning

Sheila Fitriana*and *Nana* Mardiana

Universitas Islam Sumatera Utara, Department of Physics Education, Medan 20217, Indonesia

Abstract. Based on observations in the field, the learning process of Physics is still teacher-centered and does not provide opportunities for students to build their own knowledge through rational thinking skills. The research aims to describe the profile of rational thinking skills and student motivation. The research method used descriptive research with a single case design for the 32 students of SMPS AL WASHLIYAH 8, who are studying temperature and heat. The technique sampling used accidental sampling or convenience sampling. The results showed that students' rational thinking skills were still low, aspects of students' remembering, classifying and evaluating skills were at the highest achievement with percentages of 94%, 95%, and 95% respectively, aspects of generalizing skills were in the medium category with a percentage of 70 %, while the aspect of comparing and analyzing skills is in the very low category with percentages of 30% and 48%, respectively. The student's learning motivation profile includes the variables of persevering in facing tasks, being tenacious in facing difficulties, showing interest, and independent learning showing results in the medium category. Based on the results of the study, it is necessary to conduct an in-depth experimental study to see the success in improving the rational thinking skills of junior high school students in learning physics.

1 Introduction

Physics is one of the fields of Natural Sciences (IPA). Science is related to how to find out about nature systematically, so that science does not only have a collection of knowledge in the form of facts, concepts, or principles but is also a process of discovery [1]. Science has a goal that is closely related to learning with the environment and everyday life. This is in line with the various science learning objectives in the 2013 curriculum. Science education emphasizes providing direct experience to develop competencies so that students are able to explore and understand the natural surroundings scientifically [2].

Based on this context, a student must use scientific methods, namely exploring knowledge through investigation or research, communicating his knowledge with others, using thinking skills, using scientific attitudes and values. One of the abilities that students must have is the ability to think rationally. Rational thinking belongs to the stage 1 thinking hierarchy, because aspects of basic thinking skills are prerequisites for complex thinking [3]

Thinking activities are carried out using basic thinking skills and complex thinking skills. Basic thinking skills include qualifications, classifications, variable relationships, transformations, and causal relationships [4]. Complex thinking skills include problem solving, decision making, critical thinking, and creative thinking. Rational thinking skills are the basis of complex thinking skills that can be trained on students.

In rational thinking students are required to use data, principles, and logic, to determine cause and effect and draw conclusions. Students are required to use certain appropriate cognitive strategies to test the reliability of problem-solving ideas. The emergence of the idea to develop process skills and the development of CBSA is an operational manifestation of the emphasis on thinking skills in the teaching and learning process. Development in the field has not been carried out properly. The ability to think is still not developed, the teacher conditions and motivates students to learn to think through various subjects. Physics is one of the subjects that can train rational thinking skills. According to [4] we are not teaching students to think (teaching of thinking) but through learning students are invited to think (teaching for thinking). The teacher provides more opportunities for students to think, through planned activities. Rational thinking activities are: remembering, imagining, classifying, generalizing, comparing, evaluating, analyzing, synthesizing, deducing, and inferring [5].

In detail, rational thinking activities are described as follows: (1) remembering does not mean remembering "dead" knowledge but remembering those that use reason/mind; (2) imagining is creating works that can be in the form of writing, images from imagination; (3) to classify is to classify objects based on certain criteria; (4) generalizing is looking for an orderly pattern from some observed objects; (5) comparing is looking for

* Corresponding author: sheila.fitriana01@gmail.com

similarities and differences of existing objects based on existing criteria; (6) evaluating is writing something by stating relevant reasons; (7) analyzing is looking for a pattern of regularity through classifying, comparing or generalizing; (8) synthesizing is through the aspects of classifying, generalizing, comparing, and evaluating to find a new regularity pattern; (9) deduction is connecting between concepts and facts that occur to find a solution; (10) conclude is the integration of rational thinking activities.

The ability to think rationally is needed in analyzing a phenomenon or event scientifically. Thinking rationally students become more able to think sensibly and criticize all events based on scientific. Science is also one of the characteristics or attitudes needed in independent and modern science learning [6].

Based on the results of interviews with teachers and initial observations at one of the schools at SMPS Medan City, it shows that there are several problems in learning physics, including: there are still many students who talk to their friends during teaching and learning activities, do not pay attention to teacher explanations, are reluctant to ask, very a lot while the time allocation for teaching and learning activities is limited, and students' reading interest is lacking. This lack of interest in reading is proven by the lack of students reading physics textbooks that are already available in the library. So that students do not understand the material presented by the teacher. Another student's difficulty is understanding questions that are in the form of calculations, as well as questions that are in the form of grouping several statements at the time of practice at the end of learning.

In line with the results of the latest Study Program for International Students Assessment (PISA), in 2018 that Indonesian students were ranked 75th out of 80 countries in the field of science. The assessment was published by The Organization for Economic Cooperation and Development (OECD). The average math score of Indonesian children is 375, the average reading score is 396, and the average score for science is 382. In fact, the average OECD scores are 494, 496, and 501 respectively. The scientific literacy ability of Indonesian students is still classified as low and experiencing a decline, according to the results of the Program for International Student Assessment (PISA). In the field of scientific literacy, the results of the 2015 PISA assessment showed that Indonesian students only achieved a score of 403 from the average score of 493 OECD countries [7]. While the results of the 2018 PISA assessment, Indonesian students experienced a decrease in their score to 396 from the average score of OECD countries, which was 489 [8]. Physics learning that has been designed has not been realized optimally.

Based on the description above, the authors are interested in revealing the ability of rational thinking skills and students' motivation to learn physics in junior high school. The purpose of this study is to describe the profile of students' rational thinking skills and describe the profile of students' motivation to learn physics.

2 Methodology

The research method used in this research is descriptive research with a single case design research design. Descriptive research is research that examines the status of the state of something in accordance with current conditions or past conditions. Research like this only describes the achievements, attitudes, behaviors or other characteristics of a group of subjects. Descriptive research reports as it is and does not involve manipulation of independent variables [9]. The single case research design is case study research that places a case as the focus of the research.

The population in this study were all seventh-grade students who took or followed the temperature and heat material at SMPS AL WASHLIYAH 8. The sample in this study was class VIIA, totaling 32 students who were studying temperature and heat material. Sample selection was done by accidental sampling or convenience sampling.

The research instruments used in this study are:
The research instruments used in this study are:

2.1 Learning Observation Sheet

This observation sheet was made with the aim of knowing the extent to which the implementation of learning activities carried out by the teacher. The observation sheet consists of an observation sheet for teachers and an observation sheet for students. The observation sheet that is made contains the model/method of learning carried out by the teacher and the observation sheet of student activities in learning activities.

2.2 Rational Thinking Skills Test

The test used to measure rational thinking skills is in the form of essay questions. This question is based on indicators of rational thinking skills which include remembering, classifying, comparing, evaluating, generalizing and analyzing. In this study, the test instrument used is an instrument that has been tested for validity and reliability so that it remains to be used in research.

2.3 Student Learning Motivation Questionnaire

Students' learning motivation in this study focused on four variables, namely, persevering in facing tasks, being tenacious in facing difficulties, showing interest, and independent learning consisting of each indicator (found in the appendix). The variable instrument of student learning motivation used was 20 statement items which were divided into positive and negative statements. The statement is equipped with 4 answer options, namely SS (Strongly Agree), S (Agree), TS (Disagree), STS (Strongly Disagree), Each proposed answer choice has a score between 1 to 4. Variable scores can be obtained by adding up all item scores.

In this study the variable instrument of student learning motivation used is an instrument that has been tested for validity and reliability so that it remains to be used in research.

2.4 Teacher and Student Interview Sheet

The teacher interview sheet was made with the aim of exploring the learning system carried out by the teacher starting from the opening, core and closing activities. The aspects that were asked in the interview were adjusted to the learning aspects contained in the questionnaire including the clarity of the subject matter, the existence of real experiences in learning (demonstrations and experiments), student involvement in learning, practice questions, evaluations and student paradigms about physics subjects.

The data obtained from questionnaires and tests in this study are quantitative data that will be analyzed descriptively by percentage with the steps as follows [10]:

1. Calculating the value of the respondent and each aspect or sub variable.
2. Recap values.
3. Calculate the average value.
4. Calculate the percentage with the formula:

$$DP = n/N \times 100\% \quad (1)$$

Information:
DP = Descriptive Percentage (%)
n = empirical score (score obtained)
N = Ideal Score for each question item

To determine the type of descriptive percentage obtained by each indicator in the variable, and the descriptive percentage calculation is then interpreted into sentences [11].

Table 1. Descriptive Analysis Criteria Percentage

No.	Percentage	Criteria
1.	90 ≤N≤ 100	Very High
2.	80≤ N < 90	Tall
3.	65≤ N < 80	Medium
4.	55≤ N < 65	Low
5.	0 ≤ N < 55	Very Low

3 Results and Discussion

3.1 Observation Results

3.1.1 The results of the observation of the learning model/method carried out by the teacher

During the case study, the researcher also observed the teacher in teaching. In general, the teacher gives lessons with the lecture method interspersed with several questions and discussion of questions. The teacher performs apperception and motivation, in learning the teacher sometimes demonstrates first and then forms a group, the teacher does not assess aspects of student performance and the teacher only assesses reports on the results of the practicum carried out by students, the teacher does not ask students to convey the results in front of the class, the teacher only going around while confirming student answers in the form of calculation results, the teacher does not openly conclude in front of the class on the results of the practicum but only confirms to each group. The results of other observations of researchers are:

1. The teacher uses the lecture method from beginning to end which is interspersed with questions and answers and discussion of sample questions
2. Students are passive, just sitting quietly watching, but there are some who talk to each other
3. Students answer the teacher's questions on the sidelines of learning, some are biased to answer but some are silent without answering the teacher's questions
4. Only a few students dared to try in front of the class to answer the exercises given by the teacher, while others did not dare to even do the practice questions given by the teacher.
5. At the end of the lesson the teacher concludes the subject matter that has been studied and sometimes gives assignments at the end of the lesson.

3.1.2 The results of observing student activities

All students pay attention to apperception and motivation by the teacher. There are some students in the back who don't listen to the teacher's explanation carefully and only a few students in the front pay attention to the demonstration made by the teacher. Students are quite active in asking and answering questions posed by the teacher. Students want to work together in groups and exchange opinions.

3.1.3 Rational Thinking Skills Test Results

Based on the results of the recapitulation of the rational thinking skills test, the following data were obtained:

Table 2. Rational Thinking Skills Test Results

No	Aspect	Question Number	Percentage (%)	Criteria
1	Remembering	1	94	Very High
2	Classifying	2	95	Very High
3	Comparing	3	30	Very Low
4	Evaluating	4	95	Very High
5	Generalizing	5	70	Medium
6	Analyzing	6	48	Very Low

The next diagram the average percentage ratio of rational thinking skill tests as a whole are presented in the following diagram:

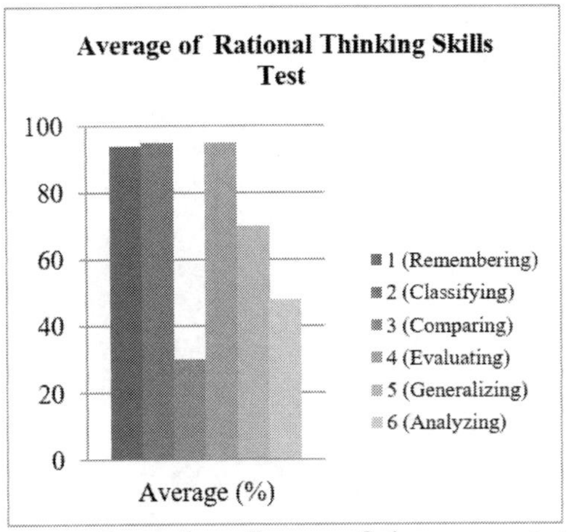

Fig. 1. Comparison diagram of the average test of rational thinking skills

3.1.4 The results of the questionnaire on students' motivation towards learning physics

Based on the results of the student motivation questionnaire recapitulation, the following data were obtained:

Table 3. Results of Student Motivation Questionnaire Recapitulation

Variabels	Question Number	Average	Criteria
Persevere to face Task	1,4,5,13,14	75	Medium
Persevere in the face of adversity	2,3,7,15,16	68	Medium
Show interest	8,10,12,17, 18	69	Medium
Learn to be independent	6,9,11,19,20	70	Medium

Next is a diagram that shows the students' overall learning motivation:

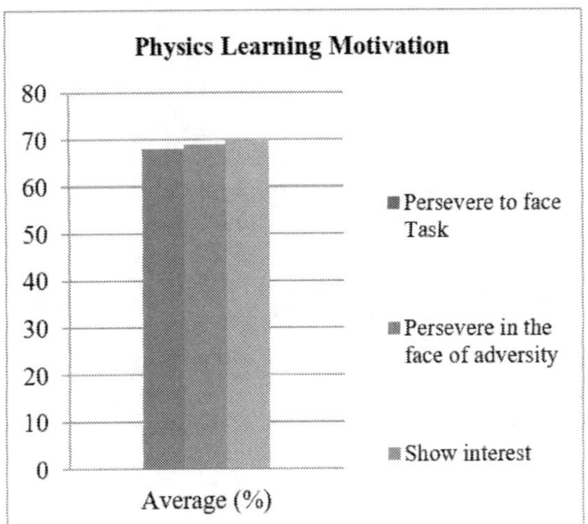

Fig. 2. Diagram of students' motivation to learn physics

3.2 Discussion

Based on the test results of the rational thinking skills test, it can be seen that the students' rational thinking skills can be described as follows:

1. Aspects of students' remembering, classifying and evaluating skills are at the highest achievement with percentages of 94%, 95%, and 95% respectively
2. Aspects of generalizing skills are in the medium category with a percentage of 70% for questions related to explaining the effect of heat on the increase in object temperature.
3. The aspect of comparing and analyzing skills is in the very low category with percentages of 30% and 48%, respectively. This is indicated by the results of student answers relating to the ability of students to sort the temperature from the highest to the lowest temperature in a simple experiment and using the equation $Q = m.c.\ \Delta T$ in events in everyday life in the questions most of the students are unable to answer correctly.

The results of the student's rational thinking skills test have not been maximized, allegedly caused by several things as follows:

1. Students are less accustomed to answering teacher questions, expressing their ideas, and responding to other students' ideas, reporting and comparing group discussion results so it is recommended for teachers to guide, motivate, and facilitate students to practice students' rational thinking skills.
2. Students are not accustomed to working on test questions that measure rational thinking skills, because usually students only get questions that only touch aspects of memory or understanding. Therefore, it is recommended that teachers facilitate more students to practice questions that are able to construct their rational thinking skills.

In addition to students' rational thinking skills, data on the motivation to learn physics for junior high school students was also obtained. Students' motivation to learn physics is still in the medium category. This is shown from the results of interviews and questionnaires that have been given to students. Based on the motivational questionnaire conducted by students, it is known that most students feel less interested and enthusiastic about physics lessons. In general, students are less motivated to study physics, this is caused by several factors, including:

1. Students feel bored with physics lessons because the teacher always teaches by the lecture method
2. Too many formulas to understand and memorize
3. Students feel incomplete in understanding the lesson, this is because the teacher does not fully explain and limited time.

After exploring what students want when studying physics, the results are as follows:

1. Students want teachers to apply innovative and creative learning models so they are not boring
2. Students want to do physics experiments

3. Learning is interspersed with games related to physical phenomena.

In general, the teacher gives lessons with the lecture method interspersed with several questions and discussion of questions. The following are the results of the researchers' observations:

1. The teacher uses the lecture method from beginning to end which is interspersed with questions and answers and discussion of sample questions
2. Students are passive, just sitting quietly watching, but there are some who talk to each other
3. Students answer the teacher's questions on the sidelines of learning, some are biased to answer but some are silent without answering the teacher's questions
4. Only about 4 students dared to try in front of the class to answer the exercises given by the teacher, while the others did not dare to even do the practice questions given by the teacher.
5. At the end of the lesson the teacher gives homework.

4 Conclusions

1. Rational thinking skills are still relatively low, aspects of students' remembering, classifying and evaluating skills are at the highest achievement with percentages of 94%, 95%, and 95% respectively, aspects of generalizing skills are in the medium category with a percentage of 70%, while the aspect of comparing and analyzing skills is in the very low category with percentages of 30% and 48%, respectively.
2. Most student learning motivation towards learning of physics is medium.

References

1. Depdiknas, Pengembangan Bahan Ujian dan Analisis Hasil Ujian: Materi Presentasi Sosialisasi KTSP Jakarta: Departemen Pendidikan Nasional (2006)
2. Rahayu. P, Mulyani. S., et al, Jurnal Pendidikan IPA Indonesia., **1(1)**, 63–70 (2012)
3. Farida. Ida, Jurnal Pendidikan MIPA, **1**,8 (2006)
4. Costa. Arthur., Developing Minds A Resource Book for Teaching Thinking. Virginia: Association for Supervision and Curriculum Development, 44-45 & 20-21 (1985)
5. Novak. J.D, Gowin. D. B., Learning How to Learn , 255 (1984)
6. Nurachma. D. E., Irawan. E., Integrative Science Education and Teaching Activity Journal., **1(1)**, 34–44 (2020)
7. OECD, PISA Result in Focus (2015b)
8. OECD, Programme for International Student Assessment (PISA), 1–10. (2018)
9. McMillan. J.H, Schumacher. S., Research in Education. New York: Longman, Inc, 397 (2001)
10. Riduwan, Akdon, Rumus dan Data dalam Analisis Statistika, 71-95 (2010)
11. Arikunto, S. Prosedur Penelitian Suatu pendekatan praktik. Jakarta: PT. Rineka Cipta (2011)

Community based tourism in Nagari Lawang, West Sumatera: Participation Approach Analysis

Wempie Yuliane[1*], *Vivi* Ukhwatul KM[2], and *Moh.* Sholeh[3,4]

[1]Faculty of Economics, University of Andalas Padang
[2]Research and Development Agency of West Sumatera
[3]Faculty of Fisheries and Marine Science, Brawijaya University
[4]Departement of Aquatic Resource Management, Trunojoyo University Malang, Indonesia

Abstract. Community-Based Tourism (CBT) is one of the emerging tourism models in the last few decades as an alternative to mass tourism in all over the world. This can be categorized as sustainable tourism because it includes local people participation, and conservation. Lawang Village (known as Nagari Lawang) is one of the village in the West Sumatera Indonesia that has been trying CBT since 2012 with rich potential tourisms: culture, nature, adventure, traditional culinary, and handycraft. However, this is not as easy for CBT conducted sucessfully benefiting stakeholders because of two major dominated private-owned tourism objects. Initally the CBT was promoted and initiated by one of the owner of private tourism, now it is just has lost its power and continuity and faced problems such as lack of human resources, loss global promotion, and finally the collaps of many homestays that have been designed before. However some local economic activities like traditional beverage production called "tebu lawang" and traditional gift shop are still running well. The purposes of this study are to find the community participation level in CBT Lawang and to determine the proper policy to recover CBT's profitable business. To achieve the study purposes this research used qualitative research design through Focus Group Disucssion and SWOT Analysis. Primary data were collected through observation, interview, and focus group discussion (FGD). The study results shows that based on Arnstein Participation Theory, CBT Lawang lies on the second step of participation where local people support tourism but has no access to resolve problems. Based on the SWOT analysis, the local government rules needed to regulate private tourism object in order to evade competition and to strengthen cooperation with CBT.

1 Introduction

Tourism sector in West Sumatera Province based on Central Bureau Statistics became the second highest growth among the other sectors in West Sumatera. The average growth of tourism activity which is represented by accomodation and food service activities, the last five year (2015-2019) is in average, 8.31 percents while regional economics growth is in average 5.31 percents. Nevertheless the contribution to regional economy is still small around 1 percent, but the growth and the linkage with other sector such transportation, communication, industry and agriculture, make tourism sector in West Sumatera priority. Communication sector depicted the highest economic growth, 8.91 percents and transport has 1,68 percents higher economic growth than regional growth.

Community Based Tourism (CBT) is kind of tourism that link natural, social and culture resources through local communities as a complementer activities which is based parimarily on agriculture and livestock farming [1]. As a results, CBT gives benefits to economy, social, culture and environment [2]. But it needs some pre condition, what we call, collaboration process between different public administrations: Non-Governmental Organisations (NGO), private institutions and local community. Additionally, local communities have some limitations like financial resources, infrastucture or know how, limitations of cultural kind and conflicts between different public administrations [1].

Furthermore, the need of collaboration between CBT-private enterprise-NGO and university-local government is important [1, 3, 4]. CBT can not stand alone on their own feet because most CBT still has lack of capacity [4], insufficient institutional capacity [3,5], and inadequate government intervention as well as private enterprise support [4].

Tourism office of West Sumatera Province has been refocusing on Community Based Tourism (CBT) since 2017. Nevertheless, it has been developing by community and local champion people long time before 2017. Many tourist object based on community arised in the last ten year. The first CBT in West Sumatera is Forum Agrowisata Madani Matur or we can call CBT Lawang as part of it, Agam District, was held in 2012. Secondly, CBT LA Adventure Nyarai, Padang Pariaman District, was held in 2013. Followed by CBT Rantih, Sawahlunto City, in 2015, CBT Kubu Gadang, Padang Panjang City, in 2017, CBT Tungkal Selatan, Pariaman City, in 2018, CBT Kapalo Banda, 50 Kota District, in

* Corresponding author: wempie_y@yahoo.com

2018, CBT Simarosok, Agam District, in 2018, and CBT Silokek, Sijunjung District, in 2018. Data from the tourism promotion departement office, majority above new comer tourist destinations still exist until now except CBT Lawang.

As the first CBT in West Sumatera, we would like to have lesson learn why CBT Lawang became less success and is there any prerequisite to improve it. This study limit the investigation into how social CBT through participation of local people in CBT Lawang, regardless other two important factors: economic and environment. Then how to improve the way local people, local government and private enterprises interact with CBT Lawang.

2 Methodology

This research method is a qualitative method with descriptive approach and quantitave data collected as data from primary and secondary data source. According to Creswell [6], qualitative research method deal with data collection, analysis, interpretation and report writing. Research activities like purposive sampling, collection of open-ended data, analysis of text or pictures, representation of information in figures and table, and personal interpretation of the findings, are all part of qualitative methodology research. This research was aimed to produce descriptive analysis to determine level of participation of local people in CBT Lawang using participation hierarchy model by Anrstein and then Okazaki specific to community-based tourism model. This research was conducted for six months starting from April-September 2018. Types of data used is primary and secondary data. Primary data collection techniques were conducted by in-depth interviews, focus group discussion, questioner and direct observation. Secondary data sources originated from village government, village potential data (pondes) and Tourism Office of Agam District.

In-dept interviews was conducted to answer the questions related to CBT Matur and CBT Lawang as part of it related to organization, human resources, administrator, financial capacity, and problems arose for continuity of CBT in the long term. Some key person that we had in interview such owner of one private own tourist object that is also local champion of CBT Matur, another administrator of CBT Matur, village government, administrator of CBT Lawang, souvenir entertprise, religious leader, traditional public figure and youth figure.

Questioner and FGD was held at the same time to get information about people perception on tourism and what kind of strategy to have in follow to increase the role of tourism especially generated by local people.

Furthermore, secondary data collection is an information in documents, reports, scientific papers and secondary data sources relevant to this research. Specific data from office district was collected to confirm some information that we already had in the first step primary data collection like tourism district office, sub district office and regional development planning agency and village potential data. Mostly secondary data that we had in this step was to get more information of district government policy and financial support from APBD to tourism of Lawang as private own tourism site, government own tourism site and CBT Lawang. Based on all of primary and secondary data, we used SWOT analysis to have strategy to rectify CBT Lawang performance in the future.

3 Results and Discussion

3.1 CBT Matur and Lawang History

CBT Lawang held in 2012 as a part CBT Matur or we call Forum Agrowisata Madani in Matur Sub District [7]. As a village (nagari) in Matur Sub District, Nagari Lawang was projected to became a center of nature tourism for its well known tourism object Lawang Park and Puncak Lawang. The existense of these two objects created some businesses related to tourism such as homestay for accomodation, traditional food and home industry souvenir, and traditional baverage, water cane drink, called "tebu lawang". According to the local champion man who has pioneered CBT Matur it has produced higher value added for cane.

Lawang Park and Puncak Lawang are the main tourist object in Lawang owned by private besides other tourist object in Nagari (Village) Lawang such as Tigo Baleh Nan Basa, Green View, Street View which owned by village government and managed by local people and lastly Ambun Tanai owned by district government under district tourism office. Most visitors from Malaysia and domestic, visit Lawang because of scenic of Lake Maninjau, called "negeri di awan". It lies at 2.000 metres above sea level and has cold and windy climate.

The support and participation of the community in the management of Lawang tourism are good enough, that is, there are channels to convey their aspirations even though they cannot solve the existing problems. The Conflict between natural tourism object management (private) with the community is relatively non-existent. However, the level of community participation in CBT has not been optimal so that it has not been able to have a large impact on improving the welfare of the people of Nagari Lawang. This is due to the lack of facilitation from the government to increase human resource capacity, financial and equipment support, regulations and the provision of adequate infrastructure. Issues that have still not been resolved to date are the maintenance of tourism roads, the need for sugar cane refining machine tools, and regulations on permits for managing tourist objects by the private sector.

The synergy between CBT Lawang and the private sector in the past few years has been intensively established when the new CBT Matur was established as the mother of CBT Lawang. However, the current condition of this synergy only occurs at certain moments, where there is some collaboration made between the private sector and CBT Lawang but is still very sporadic by following the request of visitors visiting the Lawang Park attraction. The role of the private sector as investors, promoters and facilitators of capacity building for community human resources around tourism objects has not been seen in the development of CBT Lawang.

CBT Lawang organization as a driving force in community-based tourism management, still cannot be said to be active. The majority of management was no longer active due to various reasons. CBT Lawang organization has not been able to provide incentives for management to be involved both full and partime. The absence of capacity building carried out routinely both by the regional government and through partnerships with the private sector or by universities has made the management cadre process not run well. The existence of a driving force in CBT Lawang becomes a very important issue and determines CBT Lawang journey. The driving force for CBT Lawang is West Sumatera tourism activists and also tourism object owner of Lawang Park, private tourism object in Lawang, which is now no longer active in CBT Lawang in particular as well as CBT Matur in general.

Based on Focus Group Discussion arose the role of the government is felt to be still not optimal in the development of CBT Lawang. There are still many problems such as the maintenance of tourism roads whose responsibility is unclear, the completion of the permit management of the Lawang Park tourism object and the lack of facilitation and support both morally and materially for Lawang tourism, especially CBT Lawang from the local government. Furthermore, the development model for CBT Lawang is aimed at increasing the determinants of CBT, starting from community participation, CBT synergy with the private sector, institutions and government support.

3.2 CBT Lawang Participation

Phillips and Roberts [8], community participation determined by sustainable interest among local people and institutional (government and local) support to CBT. Nurhudayati [5] explain three elemements for CBT: economy, social and culture. Each element of CBT in Batu Malang has different result: employment, social capital, local organization that determine CBT sustainability.

Based on Arnstein participation, define community partisipation level into three ladders: non participation, degrees of tokenism dan degrees of citizen power. Each level has sub level, manipulation and therapy as part of non participation. Informing, consultation and placation (satisfaction/reconciliation) as sub level of degrees of tokenism. Finally the degrees of citizen power can be categorized into partnership, delegated power and citizen control.

Social capital in CBT is the ultimate goal to develop CBT. Okazaki [9] categorized CBT into four quadrants: (1) Quadrant I, good social capital relation, good government support that link the social capital. Quadrant 1 can be categorized as the best situation in CBT development because social and economic aspects are interconnected; (2) Quadrant II, bad social capital relation but has good government support or can be defned as latent conflict or exlusion. The effort needed to improve this situation is to redistribute resources to exluded groups; (3) Quadrant III, bad social relation and bad government support. This is the worst situation in CBT development process while government needed to participate and facilitate CBT at the same time social partnership needed to be energized; and the last, Quadrant IV, good social capital relation but lack of government support. This situation is named by coping or something that can be substituted by other parties to replace government role like NGO involvement.

CBT Lawang based on category can be determined through some description in the table below:

Table 1. CBT Lawang Descriptive Explanation

No.	Conditions	Criteria
1.	High level of participation: social relation is good as well as social institution. Relatively low horizontal / latent conflict social instituion. Relatively low horizontal / latent **conflict**	Social institution like Custom Body (Kerapadatan Adat Nagari) appointed as the rule model in West Sumatera
		Participation in tourism activities high but not in CBT development and management (planning, organizing, monitoring and evaluating)
2.	Society have a job related to tourism and relatively have more income after tourism development exist both as a worker in private enterprise and as a micro enterprise	Economically, community's involve in tourism as a worker for private tourism enterprise and as a micro related tourism enterprise for example as a fruit sellers, food traders, souvenir traders and producers, sugarcane water sellers
		There use to be homestay project but now left only one homestay.
		CBT organization does not run well anymore
3.	Lack government asssistance, aid and support	Local government budget allocation for tourism site in 2018 was given to other relatively new tourism site (Banto Royo dan Pantai Mutiara)
		No assistance for homestay and CBT Lawang from technical side (tourism office) and organizational management (cooperative trade and industryoffice)
		In 2018, cooperative trade and industry office allocate sugar cane processing equipment not for Nagari Lawang as a main cane producer but for neighborhood village

3.3 SWOT Analysis

Considering 3 aspects of development CBT: social, economy and environment, based on survey to 30 people as representative of community in Lawang such as religious leaders, public figure, youth figure and traditional figure, we can have conclusion that Lawang people do not reject the tourism activity both, by private as well as by community. From social approach, we have a question whether tourism disturb local people live. Two of three does not agree with that statement. Means, they think tourism does not harm local people life. From point of view of local identitiy, value and culture, 60 % agree that tourism does not make local people lost their identity, value and culture. At the same time majority respondent think that tourism makes local lifestyle being preserved (87 %). From interaction between local people and visitor, more than 80 % feels that visitors respect them and vice versa. Seven of eight people approved tourism create new idea and cooperation among them. And openness to outsider to get involved in their problem is something that has no objection.

Two others criteria, environment and economy have similar result which economy approach has stronger support from respondent with more than 80 % and support for environmental sustainability has 63 % supporter.

From SWOT analysis, we consider the strengths of CBT Lawang as below:

a. Lawang has rich resources like perfect scenic, specific agriculture product (sugar cane)
b. Good cooperation and coordination among stakeholder in the village especially among legislative, yudicative and executive body
c. Enjoy economic growth through tourism activities
d. Start for English Village development
e. Have islamic boarding school for memorizing Al Qur'an
f. Has an experience to impelement CBT concept through homestay, plow paddy field tourist attraction and learn to cook rendang cusine as one of the best food in the world version CNN.

Beside many strengths that CBT Lawang has, it also has several weaknesses like:

a. Highly dependent on local champion person as driving force for CBT Lawang and probably this local champion has intended to create fake participation
b. Lack local government support to CBT Lawang
c. Synergy and cooperation of CBT Lawang was in difficult situation after releasing from local champion
d. CBT Lawang position in lawang tourism is not well defined
e. Low standard for food and culinary places
f. Lack of Human resource and management development of CBT Lawang

CBT Lawang has opportunities like:

a. The increasing number of CBT in West Sumatera
b. The province government support CBT as the main objective of tourism regional development in West Sumatera
c. The development of ICT as a media communication, promotion and learning.

The Threat that CBT Lawang has faced:

a. Annual international and regional event that promote more to private owned tourist object: Lawang Park and Lawang Top
b. Tourism related activity or small business activity based tourism seems do not need the existence of CBT while they can run business without CBT support
c. Authority conflict between province government and district government to maintain tourism road to Lawang Park and Lawang Top
d. A competition with other village to get grant mechinery from local governement in drying and refining cane sugar product

4 Conclusion and recommendation

4.1 Conclusion

Lesson that we can learn from CBT Lawang as a pioneer of tourism based on community in West Sumatera:

a. CBT Lawang as starter for CBT in West Sumatera has unique condition because unconsciously compete with 2 private owned tourism object, 1 local government tourism object and several tourist object by community or village government.
b. CBT Lawang has no integration among tourism object, tourism activities and all toursim stakeholder.
c. Local participation can be concluded into level that community aspirates and communicates their willingness but relatively unexecutabale .

4.2 Recommendation

Finally we have something to learn about CBT development of CBT Lawang and in the future need major correction if we want this local initiative sustain in the long period.

a. Local government should make rules to regulate private tourism object in order to evade competition and to strengthen cooperation with CBT
b. Genuine local participation is a key to obtain sustainability
c. Degree of control is participation indicator that substantiate the sustainability.
d. Fake participation must be eluded in any way: to get government funding and elit interest [10].

References

1. Guzman, T. Lopez, Canizares, S. Sanchez, P. Victor, J. of Tourism, **6,** 1 (2011)
2. Dangi, B. Tek, J. Tazim, An Integrated Approach to "Sustainable Community-Based Tourism". (2016)
3. R. Sugi, Pengembangan Pariwisata Berbasis Masyarakat (Community Based Tourism) di Kabupaten Kulon Progo Daerah Istimewa

https://doi.org/10.1051/e3sconf/202233906007

Yogyakarta, (Laporan Penelitian Hibah Bersaing. Universitas Negeri Yogyakarta, 2015)

4. Tolkach, Denis, K. Brian, Strengthening Community-Based Tourism in New Resource-Based Island Nation: Why and How?, Tourism Management, (2014)
5. Nurhidayati, S. Endah, F. Chafid, Jejaring Adminsitrasi Publik, **1**(2012)
6. Creswell, W. Jhon, Research Methods: Quliatative, Quantitave and Mix Method Approach, (4nd edition. SAGE Publication, 2014)
7. Forum Agrowisata Nagari matur@wordpress.com, accessed on November 3, 2020, at 15.15
8. R. Phillips, R. Sherma, Tourism Planning and Community Development, (1st edition. Routledge., 2013)
9. O. Etsako, Journal of Sustainable Tourism, **16**, 5 (2008)
10. K. Kalson, Asia Pacific Journal of Tourism Research, **7**, 2 (2002)

 https://doi.org/10.1051/e3sconf/202233906008

An analysis of lighting design in a football stadium

Nyak Amir[1], *Saifuddin*[2], *Muhammad* Muhammad[3*], *Riyan* Maulana[4]

[1]Teacher Training and Education Faculty, Sport Department, Universitas Syiah Kuala, Banda Aceh, Indonesia
[2]Teacher Training and Education Faculty, Sport Department, Universitas Syiah Kuala, Banda Aceh, Indonesia
[3] Department of Marine Sciences, Faculty of Marine and Fisheries, Universitas Syiah Kuala, Banda Aceh, Indonesia
[4] STMIK Indonesai Banda Aceh, Information System, Banda Aceh, Indonesia

Abstract. Lighting is an essential element for human life. Based on the decree of the Minister of Health No.1405/Menkes/SK/XI/2002 Date: 19 November 2002 concerning the requirements and procedures for the implementation of healthy working environment. Indonesian National Standard 16-7062-2004 states that the quality of the inadequate lighting gives bad effects for visual function, the environment around the workplace, and psychological aspects that can be felt as fatigue, feeling less comfortable, less alertness until the effects of the most serious such as accident. The purpose of the study was to determine the intensity of the lighting, the spread of light, lighting design, type of lamp, light bulb conditions, efficiency of energy usage, and maintenance of the football stadium lights. This study employed a qualitative descriptive approach using the type of evaluation research. The subject of this research is Harapan Bangsa Football Stadium in Banda Aceh. Data analysis techniques are: data reduction, data display, and drawing conclusion. The findings of the research showed that the lighting intensity of Harapan Bangsa Football Stadium was 341 lux. Meaning that it was deficit of 458.8 lux of the standard set by PT Liga Indonesia (800 lux), deficit of 959 lux of the standard set by AFC (1.200 Lux), and deficit of 1659 of the standard set by FIFA 2000 Lux. Lighting in the Harapan Bangsa Football Stadium spread unevenly. The lighting intensity design of the stadium was 594 lux. The kind of light bulb used was mercury. The electric power needed to support all of lamps was 377. 200 Watt. The type of light bulb used was Philips trademark. Most of the light bulbs were broken and expired. The use of mercury lamps has not been efficient in energy use. The maintenance of the lamps has not maximized yet.

1 Introduction

Stadium is a training center as well as a football game center. The establishment of a stadium building has to meet several criteria, one of the supporting criteria of football stadium is to have lighting. Says that lighting and visual clarity is a prerequisite for a person in the visual system [1]. Says that lighting plays a very important role in human life, even other living beings also have a heavy reliance on light. Human beings need light to obtain visual information about conditions around them [2]. Based on the decree of the Minister of Health No. 1405 / Menkes / SK / XI / 2002 Dated November 19, 2002 regarding the requirements and procedures for the implementation of office working environment health [3].

In order to make the lighting meet the health requirements, it is necessary to do the followings: 1) natural lighting or artificial lighting does not cause glare and has the intensity as intended. 2) placement of the light bulbs can produce optimum lighting and the bulb cleaned frequently. 3) The light bulbs which are not working properly must be immediately replaced [4]. Any kind of activity has a minimal standard value of light intensity. Lighting is a lighting tool that helps clarify our vision in the workplace while working, without an illumination device work activity could be disrupted, but the lighting must also meet the appropriate standard of light intensity [5, 6]. So is the light intensity at the football stadium. Eligibility standards of stadium lighting according to Regulation of the Indonesia Super League is 800 lux for the match in the evening [6]. While the AFC criteria says that the whole stadium must be illuminated with recommended minimal force is from 1,200 to 1,400 lux or more. Stadium should also provide a generator in case of a blackout; the power of the generator is at least 900 lux for the entire stadium [7]. The lighting standard set by FIFA is 200 lux for the practice field or the daily games, 500 lux is recommended by FIFA for amateur competitive matches. FIFA also recommends 750 lux for national scale matches that are not televised, while for the televised national game must have an illumination minimum of 2000 Lux [8, 9].

Banda Aceh has one largest football stadium called Harapan Bangsa Football Stadium lies in Lhong Raya, Banda Aceh. It has a capacity of 45.000 seats. It is a community pride of the rencong land (Acehnese people) that was one of the most magnificent stadiums in Indonesia in 1997. For the match in the evening, Harapan Bangsa Football Stadium becomes the best choice. It should have been equipped with the set standard light bulbs. Indonesian National Standard 16-7062-2004 stated that the bad quality of the lighting

* Corresponding author: muhammad@unsyiah.ac.id

gave bad effects to visual function, the environment around the workplace, as well as psychological aspects that can be felt as fatigue, feeling less comfortable, less alertness that leads to the toughest effect such as accident.

2 Research Metodology

This research is a kind of evaluation. That the evaluation research is a scientific procedure which systematically measures the results of a program or a project whether it is in accordance with the plan or not by collecting data, analyzing data and reviewing the implementation of the program fairly [10].

The subject of research was Harapan Bangsa Football Stadium Banda Aceh. The reason as to why the researchers took the stadium as the subject of the research was that it is frequently used for the national-level soccer match in the evening. The Instrument used to measure light intensity of the football stadium was a light meter. The instrument is as follows.

Fig 1. Light Meter instument

Observation format was used to obtain data on the condition and state of the headlights and the lights on stands, while interview was used as a support instrument in collecting data. This study was conducted on May 21, 2015 in Harapan Bangsa Football Stadium, Lhoong Raya Banda Aceh.

3 Results and Discussion

3.1 The Results of Lighting Intensity Measurement

After measuring 77 points of the lighting intensity in Harapan Bangsa Football Stadium Banda Aceh, it was found that light intensity on the average was 341Lux.

3.2 Lighting intensity scattering scheme

Scheme for distributing the lighting intensity of Harapan Bangsa Football Stadium Banda Aceh can be seen in the image below.

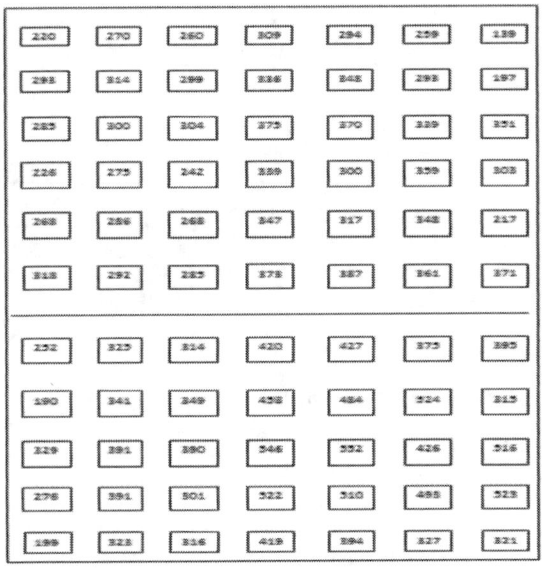

Fig 2. The scheme of scattering of light at Harapan Bangsa Stadium

3.3 Interview Data

Tabel 1. Interview Data

No	Questions	Answers
1	What types of light bulbs used for lighting the Harapan Bangsa Football Stadium ?	The main light on the pole using a mercury spotlight model. The light on the stands using ordinary mercury light bulbs.
2	What was the trademark the light bulbs used for lighting in the Harapan Bangsa Football Stadium ?	Both Light bulb brand and panel such as ballasts, philips, ignitor, condenser and NCB that used were trademarks or registered
3	How long have these light bulbs used ?	Since the Harapan Bangsa stadium was built in 1997
4	How many watts of main light bulb was on a pole ?	2000 watts for a single lamp
5	How many watts of light bulbs were on the stands	200 watts for a single lamp
6	What causes all of the light bulbs did not work well ?	Damage of the light panel due to the usage time was expiredDamage of the light panel due to seepage of rain as the stadium roof was leaking and caused the lamp panels charred
7	How many light bulbs were there in the spectator stands ?	The western stand were 59 light bulbs Eastern stand were 59 light bulbs Northern stand were 44 light bulbs Southern stand were 44 light bulbs However, all the light bulbs in the stands did not work anymore due to damaged panels

 https://doi.org/10.1051/e3sconf/202233906008

8	Usually, how many years / months are the light bulbs replaced ?	Erratic, depending on the needs and usage. Usually, they were replaced when there were events.
9	What is the height of main light poles ?	30 meters
10	As a manager, what steps should be taken to overcome the bad condition of the lighting ?	Making a proposal to the supervisor namely the Department of Sports. Replacement of the light bulbs highly depends budgeting

3.4 Observasion Data

Tabel 2. Observation Data

No	Things Being Observed	Observation Results
1	The conditions of the light bulbs	Most of the main lights did not not work as well as those lights on the stands
2	Number of the main lights	The main lights that were on poles totaling 148 lamps. The Harapan Bangsa Football Stadium has four poles and each of which has 42 light bulbs.
3	Number of the stand lights	There were 206 lamps on the stands. All the lights scattered on four stands with details of 59 lights on the western stand, 59 lights on the eastern stand, 44 lights on the northern stand, 44 lights on the southern stand
4	The working main lights	The condition of the lights on the main poles, namely : 1) the condition of the north lights in the east side were only 30 of 42 lights worked well. 2) The condition of the north lights in the west side were 34 of 42 lamps worked well. 3) The conditions the south lights in the west side were only 35 of 42 lamps worked well. 4) The conditions of the south lights in the west were only 21 of 42 lamps worked well
5	The working stand lights	The results of observation showed that there was none of the stand lights worked. He lights on the spectator stands did nor work.

4 Discussion

4.1 Standard Lighting Intensity of the Harapan Bangsa Stadium

Referring to the regulations set by PT Liga Indonesia, the minimum intensity of a football stadium was 800 lux or 42.65% [10]. Thus, the light intensity of the Harapan Bangsa stadium Banda Aceh did not meet the criteria set by the PT Liga Indonesia because it deficit 458.8 lux or 57.35%. According to AFC, the minimum standard lighting of a football stadium is 1200 lux, it means that the Harapan Bangsa stadium deficit 959 lux or lack of 71.58%. Lighting standard regulations set by FIFA for international matches lighting specifications televised is 2000 lux, So, the Harapan Bangsa stadium deficit 1659 lux or 82.95%. The light intensity of the Harapan Bangsa stadium at this time can interfere visual clarity of soccer players and can create conditions which are not convenient for the players. If these conditions continue to occur in the long term period then the players will have problems in their vision systems

4.2 The Lighting Spread

The scheme data of the measurement results above show that the spread of the light of the Harapan Bangsa Football Stadium Banda Aceh was uneven. The chart of the spread of the lights of the stadium can be seen in the graph below

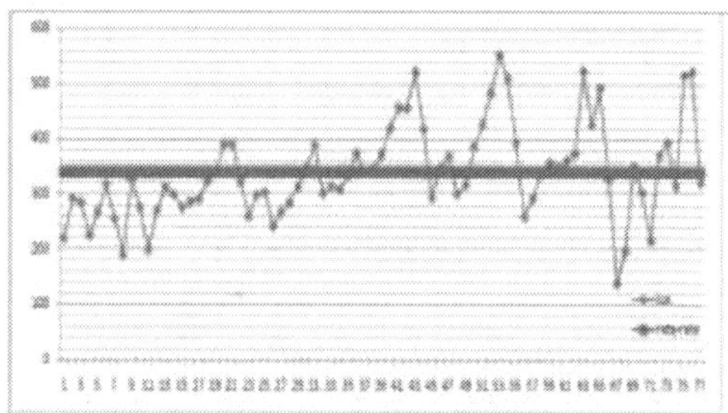

Fig 3. The spread graph of the light spred

4.3 Lighting Design of the Harapan Bangsa Stadium

The result of the design of the lighting intensity of the Harapan Bangsa stadium Banda Aceh is 594 lux. The intensity can be used for a non-televised league games. See the lighting specifications set by FIFA below.

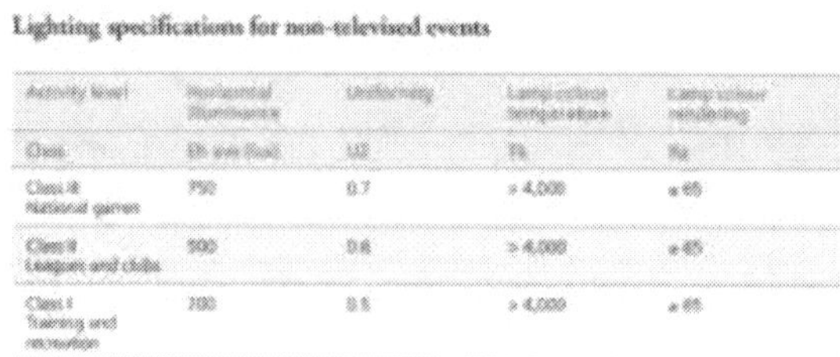

Lighting specifications for non-televised events

[illegible]	750	0.7	> 4,000	[illegible]
[illegible]	500	0.6	> 4,000	[illegible]
[illegible]	200	0.5	> 4,000	[illegible]

Fig 4. Lighting specifications for non-televised events. Source: (*Football stadiums technical recommendations and requirements* :89).

4.4 Lamp Type

The main types of lamps used in the lighting pole is the high light itencity discharge lamp. Mercury vapor

https://doi.org/10.1051/e3sconf/202233906008

lamps are the oldest models of HID lamps. Although they have long life and low initial cost, these lamps have a poor efficacy of 30 to 65 lumens per watt. These lamps have the lowest efficiacy of HID family.

4.5 Lamp Condition

Illumination of the stands from the observation showed that none of the lights could be used due to the damage to the circuit panel of lamps. The total number of the main lamps on all of poles was 148. The Harapan Bangsa Stadium Banda Aceh has four lighting poles with each of which has 42 lamps. The results of the observations showed that none of the lamps was working. Malfunctioning of the lamps was due to the damage to the lamp panels.

4.6 Use of Electricity

The total power of the lighting at Harapan Bangsa Stadium Banda Aceh was 377.200 Watts. PLN (the state electrical power supplier) as a major supplier of electrical power to the stadium must provide a minimum of 377.200 Watts.

4.7 Efficiency of Energy Use

Energy use of High Intensity Discharge lamps at the Harapan Bangsa Stadium Banda Aceh is 2000 watts for one lamp, the amount of the main lamps on the poles are 148. If they are calculated, 2000 watts times 148 lamps equals to 296.000 Watts. Emition Diode Light lamp is a solid state functioning rectify the electric current [11]. Comparison of power consumption between LED lights and mercury lamps can be seen in the figure below :

Some silent comparison Watt Calculator (electrical power consumption)

- 1W LED - LOW Incandescent, example LOW LED LIGHT replaces 100W
- Shining light
- 1W LED • 6W halogen, eg LOW LED floodlights replace halogen floodlights
- 60W
- 1W LED • SW mercury lamp. An example of a high bay sow led lamp using mercury
- Ugh 2SOW
- 1W LED - 2W high pressure mercury lamp. Example of 90W Ngh bay light get
- 180W. high pressure mercury lamp
- '1W LED • 1W metal halide lamp.
- 1W LED- 2.5W fluorescent lamp. Example of 14w led tube replace fluorescent 36W light light

Based on the comparison in the figure above, it can be concluded that the use of 148 pieces of mercury lamps consume power of 296.000 Watts. While the use of 148 LED light bulbs only consume power of 14.800 Watts.

4.8 Lamp Trademark

Based on interviews, the brand of lamps used in the Harapan Bangsa Stadium Banda Aceh is the Philips trademark. The use of lamps of philips brands in the sports arena is very prevalent as the use of LED vision spotlights on Chelsea FC soccer field that provides the convenience for the football players. Football club KAA Gent has a big stadium with a capacity of 20,000 peoples also use Arena Vision spotlights of Philip brand for high-quality sports field lighting in order to produce a clear visibility for players and spectators. Ensure flicker-free performance super slow motion camera can record footage perfectly [12].

4.9 Maintenance of Lamps

Interview results showed that the maintenance of the lamps under the management of Harapan Bangsa Stadium Banda Aceh was not regularly done. The process of maintenance was done only when it was needed and used for certain events. There are several key factors that led to the loss of light in the lighting system as described, that the lamp lumen depreciation, light soiling and armature, low voltage shutdown, inefficient armature and dirty depreciation rooms [13-18].

5 Conclusions

1. Lighting intensity of Harapan Bangsa Stadium Banda Aceh was 341 lux, It is very much less than the lighting intensity standard set by FIFA, 2000 lux.
2. Spreading light in the Harapan Bangsa Stadium Banda Aceh was not even due to unequal distribution of the amount of light that produced by the light bulbs on the main poles.
3. Design of the lighting intensity of Harapan Bangsa Stadium Banda Aceh reached an average of 594 lux that can be used for non-televised league games.
4. The type of lamp used at Harapan Bangsa Stadium Banda Aceh was mercury lamp which is the oldest types of lights from High Intensity Discharge, these lamps have a poor efficacy of 30 to 65 lumens per watt.
5. Condition of the most of lamps both those on the main poles and those on the spectator stands of the Harapan Bangsa Stadium lights Banda Aceh were damaged.
6. Power consumption of Harapan Bangsa Stadium Banda Aceh amounted to 377 200 Watt.
7. The efficiency of energy use of the mercury lamps was bigger than the current modern lamps. Such conditions lead to wasteful usage cost. To conserve power usage it is advised to use models of LED lights.
8. The Brand of the lamps used in the Harapan Bangsa Stadium Banda Aceh was Philips.
9. Maintenance of lamps of the Harapan Bangsa Stadium Banda Aceh was not good yet.

References

1. Suptandar. *Lighting Systems for Interior Design.* Jakarta: Universitas Trisakti 2006.
2. Parmonangan, M. *Architectural Lighting Design: Concepts of Artificial Lighting on the Exterior Space.* Yogyakarta: Andi Offset 2009.
3. Standardization Agency of Indonesia. *Lighting Intensity Measurement Standards at Work.* SNI 16-7062-2004.
4. Karlen, M dan Benya, J. 2004. *Lighting Design Basics*. (Kindle Edition). Jhon Wiley & Sons: New York.
5. Karam A and Stein. *Light and Sound (Science Foundations*). New York: Chelsea House Publisher 2011.
6. Mursal *et al* 2018 *J. Phys.: Conf. Ser.* **1116** 032020
7. Standardization Agency of Indonesia. *Planning Procedures for Stadium Building Techniques.* SNI T25-1991-93.
8. E Handoko *et al* 2018 *J. Phys.: Conf. Ser.* **1080** 012002
9. Ismail Ismail *et al* 2018 *J. Phys.: Conf. Ser.* **1120** 012014
10. Hill, Mc.G. *Human Factor In engineering and Desaign*, Inc New York 1992.
11. FIFA. *Football Stadiums Technical Recommendations and Requirements* 2007.
12. Arikunto, S. *Research Procedures: A Practical Approach.* Jakarta: Rineka Cipta 2010.
13. www.ligaindonesia.co.id. Regulation Indonesia Super League 2014.
14. Ching, FDK dan Binggeli, C. *Interior Design with Illustration*. (Second Edition). Jakarta: PT Indeks 2011.
15. Zulkarnain Jalil *et al* 2018 *IOP Conf. Ser.: Earth Environ. Sci.* **105** 012098
16. Malahayati *et al* 2018 *J. Phys.: Conf. Ser.* **1120** 012061
17. www.Lighting.philips.co.id. Accessed on December 10, 2015.
18. Norbert, L. 2007. *Heating Cooling, Lighting. Method Design for Architecture*. (Second Edition). Jakarta: Raja Grafindo Persada.

Analysis of main facilities utilization level in Kuala Langsa Fish Landing Base, East Aceh, Aceh

Ratna Mutia Aprilla[1,2*], *Alvi* Rahmah[1,2], *Qurratu* Aini B[1], *Muhammad* Irham[3], *Ichsan* Rusydi[4], *Titien* Sofiati[5]

[1]Department of Fisheries Resources Utilization, Faculty of Marine and Fisheries, Universitas Syiah Kuala, Banda Aceh, 23111, Indonesia
[2]Research Centre for Marine and Fisheries, Universitas Syiah Kuala, Banda Aceh, 23111 Indonesia
[3]Departement of Marine Sciences, Faculty of Marine and Fisheries, Universitas Syiah Kuala, Banda Aceh, 23111, Indonesia
[4]Departement of Aquaculture, Faculty of Marine and Fisheries, Universitas Syiah Kuala, Banda Aceh, 23111, Indonesia
[5]Departement of Fishery Technology, Faculty of Fisheries and Marine Sciene, Pasifik Morotai University, Maluku Utara, 97771, Indonesia

Abstract. Fishing ports are an important factor in supporting various capture fisheries activities in Aceh. One of the fishing ports in East Aceh is the Kuala Langsa Fish Landing Base *(PPI)*. The problems in managing *PPI* Kuala Langsa are inadequate facilities and supporting infrastructure, therefore good port facility management is needed to be able to provide facilities and infrastructure more effectively and efficiently. This study aims to find out the actual condition of the fishing port dock and port pond facilities, and determine the utilization level of the fishing port dock and port pond facilities according to spatial and time in *PPI* Kuala Langsa. The research was conducted in August 2020 using survey methods and interviews with 18 fishermen. The actual conditions were then analyzed descriptively, whereas a level of used facilities was analyzed by using a formula from the Director-General of Capture Fisheries (1981), namely by using a percentage of the level of utilization. The results showed that the condition of the fishing port dock and port pond in *PPI* Kuala Langsa is good. The level of spatial utilization on the fishing port dock has crossed the limit (105%) and based on time was not optimal (45%). While the level utilization of the port pond facilities based on space and time level are still within optimal (100%).

1 Introduction

The Fishing Port is a fundamental part of fishery industry activities, especially in the Capture fisheries aspect. Fishing ports are classified into 4 classes, one of which is class D fishing port or fish landing base. The fishing port is a place consisting of the land and waters around it with certain boundaries as a place for government activities and fishery business system activities used as a place for fishing boats to dock, anchor, and load and unloads fish, equipped with shipping safety facilities and fishery support activities [1]. A fishing port equipped with facilities functions as a center for fostering and increasing fishery economic activities. Noverianto et al. [2] stated that if stakeholders such as fishermen, government, consumers, managers, and traders and fishery entrepreneurs utilize optimally existing port facilities, it gives ease and benefits for fisheries businesses.

The research conducted by Putri et al. [3] states that to achieve its functions and objectives as a fishing port, proper facilities and infrastructures are urgently needed so that all activities in a fishing port can be implemented properly. Fishing port facilities have to be built according to current and future needs. These facilities include basic facilities, functional facilities, and support facilities [4]. Basic facilities are facilities that must be provided in a fishing port consisting of breakwaters, revetments, grayness, docks, jetty, port ponds, shipping lanes, complex roads, drainage, and land [5].

The fishing port in East Aceh is classified in class D, namely Kuala Langsa fish landing base, where the port is equipped with several basic facilities, including a port pond, shipping lanes, docks, complex roads, drainage, and land. Other facilities include a fish auction place, cold storage, a gas station, an ice factory and, others. The availability of basic facilities in Kuala Langsa fish landing base is expected to provide benefits for fishermen in carrying out fishing activities. However, the condition of the basic facilities at Kuala Langsa fish landing base is still improper, such as a narrow port and neglected land. The available facilities in Kuala Langsa fish landing base are currently not being used optimally by fishermen, there are many neglected buildings. This condition is much unconducive when the fishing season arrives, the loading and unloading activities will become heavier, with the result that requiring a wider and more spacious landing area. Problems in fishing port management in Kuala Langsa fish landing base are limited facilities, weaknesses in human resources as managers, and lack of supporting infrastructure. The effectiveness and efficiency of port fishing facilities can

* Corresponding author: ratnamutia@unsyiah.ac.id

be obtained through good facility management. This study aims to find out the actual condition of the fishing port dock and port pond facilities in Kuala Langsa fish landing base, and determine the utilization level of the fishing port dock and port pond facilities according to spatial and time in Kuala Langsa fish landing base.

Research on the level of utilization of basic facilities at fishing ports has been conducted in different places. Research conducted by Mufadzal et al. [6] shows that the level of utilization of facilities at the Calang fishing port for the anchoring pond is 100% and the dock is 80%. Research by Sundari et al. [7] at the Wonokerto fishing port shows that the level of utilization of the port dock facilities is 76.4%, the port pond is 83% and the port land is 29.1%.

2 Materials and Methods

2.1 Location and time research

This research was conducted in August 2020. The research location is located in the Kuala Langsa Fish Landing Base, East Aceh, Aceh.

2.2 Data collection methods

The data collection method is a very important stage in the research. The correct data collection technique will produce accurate data and vice versa. The data used are primary and secondary. Primary data collection techniques are carried out by surveys, interviews, and documentation. The survey method was carried out by direct observation at the research location. Interviews were conducted using direct communication techniques based on questionnaires to obtain the required data. Documentation is done by taking the necessary pictures at the research location. Secondary data collection techniques are carried out with literature studies that aim to study theories that support research. The literature study in question can be obtained from related agencies, including the port administration office, books, and journals related to research.

The sampling method using the method of simple random sampling is sampling utilizing each population are given a number then the sample is drawn randomly by using a lottery. According to Usman and Akbar [8], the number of data samples of the research is taken about the following formula:

$$n = \frac{NZ^2P(1-P)}{Nd^2+Z^2P(1-P)} \quad (1)$$

Where:

n : Number of samples to study
N : Number of population
d : The maximum Error that can be accepted (0,1)
Z : Anormal Variable the standard (1,64)
P : The Percentage of the variance set (0.05)

The number of fishermen in Kuala Langsa fish landing base are 537 people. The 18 respondent were collected from the total number of fishermen. Fishermen who become the sample in this research is boat crew.

The number of samples is obtained from the following calculation :

$$n = \frac{537 . (1{,}64)^2 . 0{,}05\ (1\text{-}0{,}05)}{537 . (0{,}1)^2+(1{,}64)^2 . 0{,}05 . (1\text{-}0{,}05)}$$

$$n = \frac{97{,}98}{5{,}55}$$

$$n = 17{,}65 \rightarrow 18 \text{ respondent}$$

2.3 Data analysis

This research is only limited to an assessment of the utilization level of the main facilities, namely the fishing port dock and port pond.

2.3.1 Analysis of the actual condition of fishing port dock and port pond facilities in Kuala Langsa fish landing base

The actual conditions of the fishing port dock and port pond facilities in the Kuala Langsa fish landing base were analyzed using descriptive analysis. Descriptive analysis is an analysis that describes the object of research based on visible facts on observations. The data and information are described so that the actual condition of the main facilities can be clearly illustrated.

2.3.2 Analysis of facility utilization level

The analysis of the utilization level facilities using a formulation from the Directorate General of Fisheries [9], namely by using a percentage of the utilization level, it can be seen or assessed the utilization level of facilities in a fishing port. The utilization level of the facility is measured based on spatial and time function.

Facility utilization based on spatial function is on the ratio of the area of space used to the available space capacity of the facility. The facility utilization based on time function is based comparing on the number of vessel using the spacial (within a certain period) with the time capacity available in a facility.

The utilization rate of the fishing port dock and port pond based on the spatial function uses is as follows:

$$\text{The utilization level} = \frac{\text{Area of space used}}{\text{Available space capacity}}\ 100\% \quad (2)$$

The formula for the utilization level of the fishing port dock and port pond based on the time function is as follows:

$$\text{The utilization level} = \frac{\Sigma \text{ Vessel using spacial}}{\text{Time capacity available}}\ x\ 100\% \quad (3)$$

According to the Directorate General of Fisheries [9], the limits for knowing the use of facilities are as follows.

- If utilization > 100%, the utilization rate of port facilities exceeds the optimal limit.

- If the percentage of utilization = 100% then the utilization of port facilities is still within the optimal limit
- If the utilization percentage is <100%, the utilization rate of port facilities is not optimal.

3 Results and Discussion

3.1 The actual condition of fishing port dock and port pond facilities in Kuala Langsa fish landing base

Based on the research results, the Kuala Langsa fish landing base currently has 17 port facilities, namely basic facilities such as port land, port pond, fishing port dock, complex roads, and drainage. Functional facilities such as administrative offices, fish auction building, clean water facilities/water tanks, SPD installation (Solar Packed Dealer for Fishermen), electrical installations, ice factory, cold storage and parking lots as well as supporting facilities such as guard posts, meeting hall, public toilets buildings, and fishermen's shops. However, this research is only limited to an assessment the utilization level of the main facilities, namely the fishing port dock and port pond facilities. The conditions of the fishing port dock and port pond facilities in Kuala Langsa fish landing base are as follows:

3.1.1 Fishing port dock

The fishing port dock in Kuala Langsa fish landing base is used for catch lending activities with a fishing port dock length of 110 m and a width of 8 m, it is good conditions, currently, the fishing port dock is used by fishermen for loading and unloading activities, and mooring, and anchoring activities. The fishing port dock is made of reinforced concrete construction and piles that are equipped with additional tools, namely cranes that are used to transport fishermen's catches. The type of crane available in the Kuala Langsa fish landing base is the Harbor Mobile Crane. Harbor Mobile Cranes are usually used to transport light objects under 100 tones. The condition of the crane in the Kuala Langsa fish landing base is damaged, because of the lack of maintenance, so fishermen use local workers to transport their catch. The fishermen's catch is put into fiber and then pulled using a rope to a car carrying the catch by several workers at the fish landing base. The catch is then taken to the vessel owner's warehouse or brought to the market directly.

The fishing port dock in the Kuala Langsa fish landing base is also equipped with bollards and fenders. The bollard on this fishing port dock is made of cast iron at the end of the fishing port dock surface which functions to tie the mooring rope at the dock. This is following Mahyudin's [10] statement that a bollard is a form of a structure at the end of the port dock surface (quay edge) used to tie vessel mooring ropes. Bollard conditions are in good condition so it can make it easier for fishermen to moor their vessel.

Fenders are bearings that are placed in front of the port dock which is made of rubber. The main function of the fender is to prevent damage to the structure of the port dock and to protect the damage to the vessel's hull paint due to friction between the vessel and the port dock caused by motion due to waves, currents, and wind [11]. However, fishermen still use used car tires that are placed on the sides of the hull to protect the boat from hard collisions with the fishing port dock walls which can cause the hull of the vessel to be damaged. Some of the fishing port dock building conditions were damaged so that there was a pool of water in the fishing port dock building. Apart from being used for loading and unloading landings, the jetty in Kuala Langsa fish landing base is also used as a ferry dock. Other activities carried out in the Kuala Langsa fish landing base port dock include refuelling for boats, drinking water, and clean water.

3.1.2 Port pond

The port pond in Kuala Langsa fish landing base is currently a natural channel pond, where the port pond waters are formed naturally. Based on the results of interviews with the port administration office and fishermen, it was stated that the dredging ponds in the Kuala Langsa fish landing base had never been done. This is the same as the condition in the Ujung Seurangga fish landing base, which has a port pond, which is a natural pond, however, the dredging of the port pond at the fish landing base has not been used [12].

The port pond in Kuala Langsa fish landing base is quiet because it is located in the bay. According to Theo et al. [13], the condition of port waters is relatively protected naturally by the bay and surrounding islands so that the wave generation distance due to the wind is relatively short and the resulting waves are not too large. The area of the port pond is 1 ha with a depth of 8 m with good conditions so that it makes the vessel move without depending on the tides. Following the opinion of Armanda et al. [14], which states that the safety of vessel at anchor is highly dependent on the depth of the water pond at a fishing port, especially at the lowest water level.

The port pond in Kuala Langsa fish landing base is currently used as a shipping channel, a vessel swivel pool, loading and unloading activities, and mooring activities for vessel measuring 5-60 GT. The Kuala Langsa fish landing base port pond is also used as a shipping channel by outboard motorboats intending to carry out fishing activities. Outboard motorboats that are used as shipping lines measuring 0-10 GT catch fish using gill net, fishing rod, or traps. Apart from fishing, the Kuala Langsa fish landing base port pond is also used as a port dock for the crossing of Telaga Tujoh villagers, East Langsa District, Langsa.

3.2 The utilization level of fishing port dock and port pond facilities in Kuala Langsa fish landing base

3.2.1 The utilization level of fishing port dock facilities

The results of field observations during the study showed that the range number of vessel utilizing the dock was the vessel that >10-60 GT. The distance between vessel that land on the dock is 1-2 m. Based on the results of the calculation, it is found that the average number of vessel that frequently in and out of the port dock was 4 units per day. The length of fishing trips for fishing boats in PPI Kuala Langsa ranged from 12-24 hours, 2-5 days and some reaches 8-10 days. The PPI Kuala Langsa fishing port dock capacity is 110 m long and 8 m wide. The length of the fishing port dock required is currently 116.6 m. Research conducted by Aprilla et al. [15] at Sawang Ba'u fish landing base showed the length of the fishing port dock is 164 m, and the depth of the port dock is lower than 3 m, which the value has fulfilled the standard criteria based on regulation of the Minister of Marine Affairs and Fisheries No. 08 / MEN 2012.

Port activities in the Kuala Langsa fish landing base port dock are crowded in the morning because loading and unloading the catch activities were happening in the morning. The loading and unloading activities of the catch are carried out by workers at the port, by pulling each fiber from the vessel onto the transport car, the fiber is pulled by a rope by the workers with a total of 5-10 people. Vessel loading and unloading their catch is carried out at 07.00-18.00 WIB.

The Kuala Langsa fish landing base has one loading dock unit so that the vessel carry out loading activities in the same place. There are catch landing activities and preparation before going to sea in Kuala Langsa fish landing base, this shows that Kuala Langsa fish landing base has performed its function as a fishing port properly. This is under the regulation of the Minister of Marine Affairs and Fisheries No. 08 / MEN 2012 which states that fishing port exploitation functions include mooring and anchoring services for fishing boats, fish loading and unloading services, fishery product processing services, marketing and distribution of fish, utilization of facilities and land in the fishing port and others. The existence of fishing port facilities in Kuala Langsa fish landing base has been optimally utilized based on the spacial function where the results obtained are 105%. This is the same as the results of research by Ardandi et al. [4], which shows that the level of utilization of dock facilities in Tanjungsari fish landing base reaches 90% because the fishermen at Tanjungsari fish landing base have used the available fishing port dock facilities as basic facilities for unloading cargo, loading supplies, and berthing fishing boats. Research conducted by Darma et al. [16] shows that the utilization rate of the fishing port dock in Birea fish landing base is 90,5%, which means that the level of the facility utilization rate of Birea fish landing base is classified as high. Very high fishing port dock utilization rates also occur at Tanjung Sari Palembang fish landing base [4] and Lonrae Bone [17]. According to the decree of the Minister of Marine Affairs and Fisheries No.08/MEN 2012 that type D port must have a fishing port dock length of at least 50 m, while the length of the fishing port dock available in Kuala Langsa fish landing base is 110 m with 35 existing vessel and 4 vessel, the largest is 60 GT, so it takes an additional length of the fishing port of 5.4 m.

The use of fishing port dock facilities is used by fishermen for 11 hours, while the fishing port dock facilities are available for 24 hours. The utilization level of the fishing port dock based on the time function is 45%. This is because the vessel unloads around 07.00 - 18.00 WIB, so it can be concluded that the level of utilization of the fishing port dock based on time function is not optimal. The research result of Pujiastuti et al. [18] the less than optimal utilization of this post dock is due to a habit of the fishermen who tether their boats along the Pasilian River.

3.2.2 The utilization level of fishing port pond facilities

The port pond in Kuala Langsa fish landing base is used by fishermen for loading and unloading activities, supplies, mooring, and turning boats. Port ponds are also very important in the activities management, with a good pond condition, it should be very useful for fishermen. The data obtained on the total area of the port pond area that already exists in the Kuala Langsa fish landing base is 10.000 m^2, while the need for the Kuala Langsa fish landing base port pond needs reaches 14.606,85 m^2. The research of Primsa et al. [19] showed that the Pagurawan fish landing base requires a port pond area of 2.585,19 m^2 and a port pond depth of 2 m.

The number of vessel anchored and loading and unloading in the Kuala Langsa fish landing base is 35 purse seine. The vessel storage capacity in Kuala Langsa fish landing base is not by the decree of the Minister of Marine Affairs and Fisheries No. 08 / MEN 2012, regarding the classification of fishing ports for type D ports. Based on the decree of the Minister of Marine Affairs and Fisheries No. 08 / MEN 2012 that type D ports can accommodate there are at least 15 vessel with a size of $\geq$ 5 GT, while the number of vessel in Kuala Langsa fish landing base is 35 units and has 4 of the largest vessel measuring 60 GT which impacted towards the entrance and exit the port area. This is supported by the statement of Armanda et al. [14], where the size of the vessel and the number of vessel anchored in the port pond greatly affect the area of the port pond in a fishing port so that vessel leaving and entering the port area were not blocked.

The utilization rate of port ponds based on spatial and time function has reached 100% or the optimal limit. So that it can be done to develop the area of the port pond because the number of vessel in Kuala Langsa fish landing base is increasing. The development of the port pond area aims to support Kuala Langsa fish landing base activities. This is following the statement of Narindra et al. [20], the construction of fishing port infrastructure is urgently needed to help provide convenience and services so it is necessary to plan a fish landing base equipped with supporting facilities. The actions that need to be necessary for facilities whose level of utilization exceeds optimal conditions are increasing the area of facility capacity, while for facilities that have not yet reached optimal, it is necessary to increase the utilization of these facilities [21]. Research conducted by Faisal [22] shows that the utilization rate of the port pond at the Sawang Ba'u fishing port has also reached 100%.

4 Conclusions

1. The conclusions of the research results are:
2. The condition of the fishing port dock in Kuala Langsa fish landing base is good which has a fishing port dock length of 110 m and a width of 8 m, while the condition of the port pond is also good with an area of 4 ha and a depth of 8 m.
3. The level of spatial utilization on the fishing port dock was 105% yet based on time was 45%. This showed that the use of the port dock based on spatial utilization has crossed the limit and based on time was not optimal. While the level utilization of the port pond facilities utilization based on space level is 100% and based on time is 100%. It was indicated that the use of the facility is still within optimal limits.

Acknowledgments. The authors would like to express their utmost gratitude towards the Head of Kuala Langsa fish landing base and his staff for its full support in this research. Further researchers are welcome to use this published data and information.

References

1. N. Puspitasari, A. Soemarni, H. Juliani, Diponegoro Law Journal **5**, 4 (2016)
2. A. S. Noverianto, Ismail, A. N. Bambang, Journal of Fisheries Resources Utilization Managemet and Technology **5**, 1 (2016)
3. R. J. Putri, I. Y. Alit, Isnaniah, Jurnal online Fakultas Perikanan dan Kelautan Universitas Riau **1**, 1 (2018)
4. S. N. Ardandi, H. Boesono, A. Rosyid, Journal of Fisheries Resources Utilization Managemet and Technology **2**, 1 (2013)
5. E. Nurdyana, R. Abdul, B. Herry, Journal of fisheries resource utilization management and technology **2**, 2 (2013)
6. Mufadzal, R. Rizwan, I. Rusydi, Jurnal Ilmiah Mahasiswa Kelautan dan Perikanan Unsyiah **3**, 4 (2018)
7. R. Sundari, A. Rosyid, D. A. N N. Dewi, Journal of Fisheries Resources Utilization Managemet and Technology **4**, 4 (2015)
8. H. Usman, P. S. Akbar, Pengantar statistika (PT. Bumi Aksara, 2006)
9. Directorate General of Fisheries, Fishing Port Construction (Department of Agriculture, 1981)
10. Mahyudin, Tesis (Universitas Terbuka, 2016)
11. Fauzan, Jurnal Ensiklopediaku **1**, 1 (2018)
12. D. M. Wahyuni, Rizwan, O. Kandi, Jurnal Ilmiah Mahasiswa Kelautan dan Perikanan Unsyiah **4**, 2 (2019)
13. K. S. Theo, J. Longdong, M. R. P. Manoppo, Jurnal Sipil Statik **2**, 1(2014)
14. W. Armanda, Rizwan, A. Rahmah, Jurnal Ilmiah Mahasiswa Kelautan dan Perikanan Unsyiah **4**, 2 (2017)
15. R. M Aprilla, T. Rizwan, E. Hendrami, M. A. Chaliluddin, I. Rusdy, A. W Perdana, T. H. Iqbal, IOP Conf. Series: Earth and Environmental Science, 348 (2019)
16. Darma, Safruddin , Achmar Mallawa, Torani: JFMarSci **4**, 1 (2020)
17. A. E. P. Merdekawati, A. Mallawa, F. Amir, Jurnal IPTEKS PSP **6**, 12 (2019)
18. D. Pujiastuti, R. Iranawati, A. Rahmawati, Jurnal Perikanan dan Kelautan **8**, 1 (2018)
19. P. D. Primsa, J. Zain, Ronal M, Jurnal Online Fakultas Perikanan dan Kelautan Universitas Riau **1**, 1 (2014)
20. D. Narindra, Y. Muliati, F. Madrapriya, Jurnal Online Institut Teknlogi Nasional **4**, 4 (2017)
21. G. W. Sinag, A. Rosyid, B. A. Wibowo, Journal of Fisheries Resources Utilization Management and Technology **2**, 1 (2013)
22. A. Faisal, Z. Jalil, R. Juanda, Jurnal Ilmiah Mahasiswa Kelautan dan Perikanan **4**, 1 (2019)

Fishing vessel queue model in Kutaraja Fishing Port: Case study of Wharf Pias II

Rosi Rahayu[1], *Muhammad* Muhammad[2*], *Kurnia* Kurnia[3], and *Zakyatul* Muna[4]

[1] Department of Fisheries Resource Utillization, Faculty of Marine and fisheries, Universitas Syiah Kuala, 23111, Banda Aceh, Indonesia
[2]Department of Marine Science, Faculty of Marine and Fisheries, Universitas Syiah Kuala, 23111, Banda Aceh, Indonesia
[3]Department of Fisheries Resource Utillization, Faculty of Marine and fisheries, Abulyatama University, 23372, Banda Aceh, Indonesia
[4]Post Graduate Student of Marine and Fisheries Technology, Faculty of Fisheries and Marine. Bogor Agriculture University, 16680, Bogor, Indonesia.

Abstract. Kutaraja international Fishing Port is one of the biggest port with highest activity in Aceh Province. Nevertheless, when unloading process, fishing vessel often occur long queue of fishing vessel. This study aims to identify the vessel queue landed in Kutaraja Fishing Port, identify the degree of usefulness of the system utility of unloading fasilities and identify the vessel waiting time in the queue at Kutaraja Fisng port. The methods used was case study and standard formula of multi channel single phase. The results showed that the model queue that occurred in Kutaraja Fishing Port was (M/M/4): (FCFS / I / I). Based on calculations by the standard formula queuing model, it was obtained that the rate of arrival of the vessel unloading was 22 vessels/day and the rate of the service time was 16 vessels/day. Value of waiting time in the queue was 0 hour and the number of vessels that queue do not exist, while the waiting time in the system was 3,4 hours and the number of vessel arriving in the system was 1,42 units.The result of the calculation also showed that utility value of the system almost reached the optimum value (1,37) and the probably no vessel in a system was 0,08.The efforts that can be made to accelerate the performance of fish demolition in Kutaraja Fishing Port, namely using cranes as tools when unloading fish on fishing vessel.

1 Introduction

The increasing status of Kutaraja International Fishing Port from type C to type A brings many changes to the fishing port activities. in addition, after the fishing port relocated to a new place, Kutaraja International Fishing Port also has changes in terms, one of them is facilities [1] [2]. One of the facilities that has undergone changes is main facilities. The main facility is one of the most prioritized facilities in the Port because it serves as a protector for fishing vessels that land their catch from waves or other damage [3].

One type of basic facility that has undergone changes in Kutaraja International Fishing Port is the wharf. At the previous location, the length of the wharf was only 80 m and in the currently location became 100 m [4]. There is an increase in infrastructure that occurs, becoming an attraction for new vessel to land their catch in Kutaraja International Fishing Port, which initially only vessel 10-50 GT to >100GT, meaning that with the increase in infrastructure there is also an increase in activity. However, there is an increasing in infrastructure, the capacity of the new wharf of Kutaraja fishing port has not been able to compensate for the changes in activity that occur, especially if there arevessels with a size of≥100 GT that land the catch, where the vessel has a different vessel height size far from a medium vessel 20-50 GT so that vessels with large and medium sizes cannot queue in the same server.According to [5],vessels will queue in the same server if the size of the vessel on the server tends to be the same. In addition to the different vessel sizes, the 100 GT also has much more cargo than smaller vessel, so it takes longer in unloading [5] [6] [7]. Thus, in the initial observation there were many incidentsof vessel coming out of the queue because the queue was long enough that the commodity carried by the fish would rot and there was a lot of disposal of catches in the location. Based on this fact, anappropriate calculation is needed so that port service and performance can run optimally. Thus, the analysis of the queue of fishing vessels that unload catches at Kutaraja International Fishing Port needs to be done in an effort to improve service and port performance. The study aims to determine: the model of the queue of fishing vessels that land catches at Kutaraja International Fishing Port; the utility level of the system or the utilization rate of catch loading and unloading service facilities; and the waiting time of vessel in the queue at Kutaraja International Fishing Port.

2 Methods

* Corresponding author: muhammad@unsyiah.ac.id

https://doi.org/10.1051/e3sconf/202233907002

2.1 Location and time

Research in the field was conducted in March-May 2016, but the discussion was renewed through the study until 2021. Research location at Kutaraja Ocean Fishery Port, Banda Aceh, Aceh Province. Details of the research location can be seen through Figure 1.

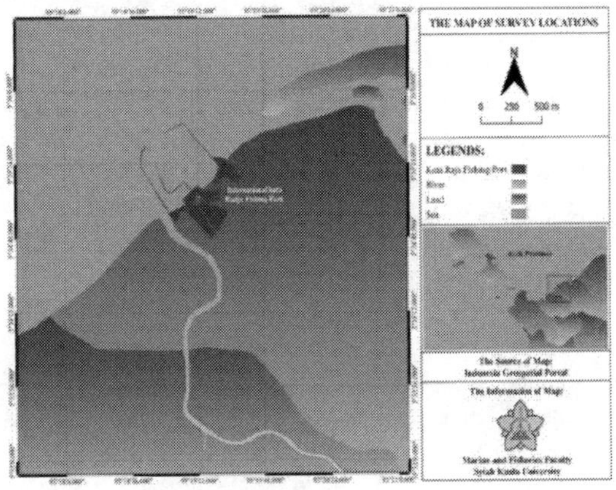

Fig. 1. Map of survey location.

2.2 Data Collection

The research method used is the observation method. Observation methods are methods used to obtain information in real terms by utilizing the five senses. The steps taken to obtain the data are as follows: primary data collection in the form of number of vessel arrivals, vessel arrival time, time of service of vessel union time [8]. Data collection is carried out during the port rush hour period that are in the morning on 06.00 am till afternoon on 04.00 pm, while secondary data in the form of data on the size of facilities and activities of Kutaraja International Fishing Port.

2.3 Data Analysis

To get the appropriate model, it must be done the selection of model functions that are in accordance with the circumstances in the field.Here are the steps used to determine the right model, namely grouping data based on arrival and service rates [9], looking for frequencies, finding raw deviations, testing data distributions, establishing parameters of the selected distribution. To establish the queue model, then there are several notations used, one of the notations used is kendall notation with the [10] following format:

2.3.1. Queue Models Identifications in Kutaraja International Fishing Port

Identification of the vessel queue model was analyzed using the Kendall DG model. The notation system is as follows:
(a/b/c):(d/e/f)

Where:a is the distribution of arrivals, b is the distribution of service time, c is the number of service facilities, d is the discipline of queue, e is the maximum number of customers in the system, and f is the population size of the customer

Standard notation that replaces a and b is: M is the distribution of arrivals and the rate of service following the distribution of poisson, D is service time, G is distribution of public service time, GI is distribution of arrivals

Notation that replaces e and f is: I is The maximum number of customers (Infinite), F is The maximum number of customers (finite). The discipline used is First Come First Service (FCFS)

2.3.2. Vessel arrival data distribution testing

Distribution of vessel arrival data can be analyzed using kolmogorov smirnov with is hypothesis.
H0: arrival rate of poisson-distributed vessel
H1: the arrival rate of the vessel is not distributed poisson

Taraf significant α 0.05
Statistical test D=| SoupS(x)–F0(x)|
where: S(x) is Cumulative distribution of vessel arrival data, F0(x)is Cumulative distribution of hypothesized distributions

Test criteria
H0is accepted at a significant level α 0.05, if the D value> thevalue of D*(a/2)
where: D*(a/2) = Is the critical value obtained from The Kolmogorov Smirnov table.

2.3.3. Service-level distribution testing

Distribution of vessel arrival data can be analyzed using kolmogorov smirnov with the hypothesis:
H0: service level of poisson-distributed vessel
H1: service level of vessel not distributed poisson
Taraf significant α 0.05
Statistical test
D =Sup| S(x)–F0(x)|
where: S (x) is cummulative distribution from vessel servive data, F0(x) is cummulative distribution from distributuion that hipothesized.
Test criteria
H0is accepted at a significant level α 0.05, if the D value> thevalue of D*(a/2)
Information:
D*(a/2)
Where: D Is the critical value obtained from the table Kolmogorov Smirnov.

2.3.4. *Queue standard formula calculation*

In solving the queue problem, then previously there are assumptions that must be met. There are the following assumptions [11]:

Tabel 1. Type of assumption.

No	Type of Assumption
1	Arrival distribution must be distributed poisson
2	Service distribution mut be exponential
3	Queue dicipline follows FCFS
4	The number of consumers is unlimited
5	Service level > arrival rate

With the fulfillment of assumptions, the queue problem can be solved by the following equations:

$$p = \frac{\lambda}{\mu}$$

$$Po = \frac{1}{N}\sum_{N=0}^{c-1} \quad \frac{(\frac{\lambda}{\mu})}{Ni} + \frac{(\frac{\lambda}{\mu})}{c!(1\frac{\lambda}{c\mu})}$$

$$Lq = \frac{(\frac{\lambda}{\mu})^c \lambda\mu}{(c-1)!(c\mu-\lambda)^2} Po$$

$$Ls = Lq + \frac{\lambda}{\mu}$$

$$Wq = \frac{Lq}{\lambda}$$

$$Ws = Wq + \frac{1}{\mu}$$

Where: λ is number of arrivals of union vessel of time. μ is number of vessel served by union time, Q is system utility factor, Po is chance of 0 vessel in the system (no customers), Lq is average number of vessel in the queue, Wq is the average wait time for vessel in the queue, Ws is Average wait time for vessel in the system, and C is number of facilities in service.

3 Result and discussion

3.1 General conditions of Kutaraja International Fishing Port

Kutaraja International Fishing Port is equipped with various facilities to support fisheries business activities [12]. Such facilities include the main and supporting facilities of the three facilities, which play the most role in maintaining the safety of the vessel docked and ensuring the activity of the vessel at the time of entry to the port is the main facility [13]. According to [14],one of the main facilities that play the most role in maintaining the smooth operation of the vessel at the time of dismantling the catch is the wharf [15]. The wharf at Kutaraja International Fishing Ports is known to be divided into 4 pias. that each pias has a different size. Pias II is the Wharf that most often used to unload catches because of its location right in front of the fish auctions building. The size of this wharf is 100 m × 4 m with an open type.

During fishing seasons such as March-April, activity at Kutaraja International Fishing Port is more dense than usual, this tends to cause queues. The queues that occur can be seen with the vessel that are layered. Because the dominant activity is carried out on the pier pias II. In pias II, there are vessel of various sizes that land catches, this causes uneven vessel queues on each server, especially if there are vessel with a size of ≥100 GT that land, then the queue on the servers of other vessel smaller to be longer than 4 layers to 8 layers. This happens because large and small vessel cannot perform service on the same server due to the size gap of the vessel. In addition, large vessel also have a fairly long unloading time, this is due to the working hours of unloading Kutaraja International Fishing Port which is only 8 hours / day so that not all the contents of large vessel can be dismantled in a day. As a result, vessel that are too long in the queue can suffer losses caused by fish that decline in quality.

3.2 Identify Queue Models at Kutaraja International Fishing Port

The queue model on Kutaraja International Fishing Port follows the pattern of double lane queues. Idelanya Kutaraja International Fishing Port has 4 service servers for loading and unloading activities (Figure 3). The discipline applied is FCFS (First come first service) or called the first come first served [16]. At pias II, the dock, the server can be 4 if the vessel that landed is a medium-size vessel ≥50 GT, but if a large vessel lands, then the available for medium-sized vessel is only 2-3 servers, this results in increased queues.

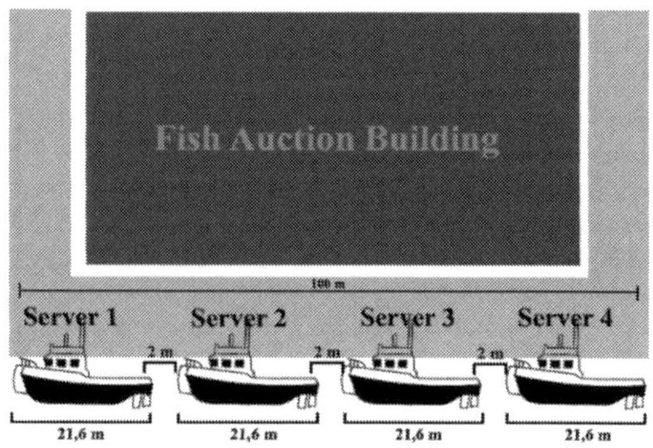

Fig. 2. Simulation Of Fishing Vessel Queue In Kutaraja International Fishing Port

Vessels visiting Kutaraja International Fishing Port generally do various activities, such as uncovering catches. The demolition of the catch was carried out at pias II pier. Based on the pictures, vessel visits in March-April tend to fluctuate.

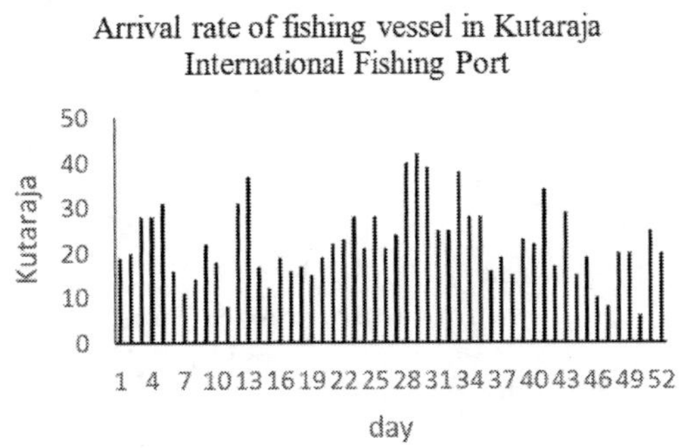

Fig. 3. Arrival rate of foshing vesssel in Kutaraja International Fishing Port

3.3 Queue System Utility Calculation at Kutaraja International Fishing Port

Seen in tabel 2, from the results of calculations, it is known that the arrival rate of fishing vessels is 22 vessel / day with an average service time of 16 vessels / day or 2 hours / vessel with 4 service servers. The results of the analysis showed that the arrival rate was much higher than the rate of service, indicating that there was a problem in the queue. This can be seen from the queuing intensity value that indicates the value of ≥1. The intensity value will be declared stable if it is worth 0.

Table 2. the result of utillity system queue

Queue model variables	Result
the average arrival rate into the system (λ)	22 units/day
M service level	16 Units/day
The number of servers (s)	4 units
Queue intensity ρ	1,37
Chances are there are no Vessel in the queue ρo	0,08
Average number of Vessel in Lq queue	0,05 units
Average number of Vessel in the Ls system	1,42 units
Average time of vessel in Wq queue	0,96 hours
Average time of Vessel in the Ws system	3,84 hours

The chance of no vessel in the system at Kutaraja International Fishing Port is 0,08. The average number of vessel queuing in the system is 0,05 vessel. Thus the vessel that come to Kutaraja International Fishing Port occur aof 0,96 hours or almost 1 hour. The average number of vessel in the system is 1,42 units with an average waiting time of 3,84 hours. The number of vessel in the system must be greater than the number of vessel in the queue because the number of vessel in the system is the number of vessel in the queue plus the vessel servedi. In addition, the average amount of time in the system should also be greater than the average time of queue [17].

Based on the description above, it can be concluded that what causes high queues is the entry of large vessel. As for the efforts that can be made to accelerate the demolition of fish in Kutaraja International Fishing Port, among others, with the addition of aids such as cranes so as to maximize the working time of fish demolition on high capacity vessels [18].

4 Conclusion

The queue model at Kutaraja International Fishing Port is (M/M/4):(FCFS/I/I) with a dual-lane queue system and one stage of service (MultiChannel SinglePhase). The characteristics of queues at Kutaraja International Fishing Port are unlimited population (infinite), queue length is also unlimited (infinite), the pattern of arrival of poisson-distributed vessel, exponential distribution vessel service and queue discipline is First Come First Served (FCFS). The average waiting act of a vessel in a queue (Wq) is 0 hours (no one is in line), while the average waiting time of the vessel in the system (Ws) is3,4hours.

Refferences

1. T. Rizwan, Z Jalil, A Akhyar, H Husaini. Oceanographic Factors As The Indicators for Shipyard Industry Development in Kutaraja Fishing Port: Preliminary Study. J. ecol. Eng, vol. 9, no. 22, pp. 237-245 (2021).
2. T. Rizwan, O. Kandi, Z. Jalil, I. Setiawan, R. Maulana, D. M. Ranti, I. Zulfahmi, Y. Muchlis, The Analysis of Clean Water Need For Fishing Activities in Kutaraja Fishing Port, Aceh Indonesia. Australian Journal of Maritime & Ocean Affairs., pp. 1-12 (2020).
3. P. E. Alayya, H. Boesono, D. Wijayanto. The strategies of Pekalongan fishing port development. IOP Conf. Series: Earth and Environmental Science, vol. 530, pp. 1-10 (2020).
4. R. A. Najah, E. Lubis, R. Muninggar. Existence of Facilities by Activities at Lampulo Beach Fishing Port, Banda Aceh. Marine Fisheries. vol. 3, no. 1, pp. 55-70 (2012).
5. D. Zhang, M. Sikveland, O. Hermansen, Fishing fleet capacity and profitability. Marine policy, vol. 88, pp. 116-121 (2018).
6. S. Hisamune, N. Kimura, K. Amagai, "Ergonomic Evaluationina in a Case Study on The Behaviour of Purse Seine Fishing Vessel workers. Fisheries Enginering. vol. 42, 1, 9-17 (2005).

7. R. Rahayu. The Utillization of Lampulo Fishing Port Main Facilities (Universitas Syiah Kuala, Aceh, 2016).
8. M. Rahardjo, The Methods of Data Collection of Qualitative Research (Malang: UIN Malang, 2011).
9. S. Syarif, N. Alimina, A. Mustafa, "The Models of Queue of Kendari International Fishing Port," Jurnal Manajemen Sumber Daya Perairan, vol. 4, no. 1, 231-242 (2019).
10. K. a. V. Vladimir. Mathematical Methods in Queuing Theory (Moscow: Kluwer Academic, 1994).
11. B. Murdiyanto. Fishing Port (IPB Bogor, 2004).
12. A. Rahmah, M. Rizayani, C.Chaliluddin. The influence of the service quality of the International Lampulo Fishing Port on the smooth operation of fishing operations. Depik, vol. 7, no. 1 (2018).
13. E. Lubis and Pujiono. The Needs for Main Facilities and Availability of the Capture Fisheries. JIPK, vol. 12, no. 2, pp. 224-235, (2020).
14. E. Lubis and N. Mardiana. Role of Fish Landing Place (PPI) Facilities on Fish Landing. JTPK, vol. 2, no. 1, pp. 1-10 (2011).
15. M. Wang, W Chen, A Epherimdes. Real-Time Reconstruction of a Counting Process Through First-Come-First-Serve Queue Systems. IEEE Explore, vol. 66, no. 7, pp. 4547 - 4562, (2020).
16. R. Khabibah, H. T. Sutanto, Y. P. Astuti. Queue System of Vessel Unloading Services at The Diamond Terminal, Tanjung Perak Port Surabaya. J. Ilmiah Matematika, vol. 2, no. 2, (2013).
17. A. A. Rahayu, Sugito, Sudarno. Model Analysis of Time Inter-Arrival and Service Time at The Cash Payment Section in Installation Installation Dr Kariadi Hospital, Semarang .Statistics Nasional Proceeding , pp. 249-261, (2013).
18. Johansen, N. S., Vänninen, I., Pinto, D. M., Nissinen, A. I., & Shipp, L, "In the light of new greenhouse technologies: 2. Direct effects of artificial lighting on arthropods and integrated pest management in greenhouse crops. Annals of Applied Biology, Vols. 159(1), 1–27., (2011).

AUTHOR INDEX